TUNNELS AND UNDERGROUND CITIES: ENGINEERING AND
INNOVATION MEET ARCHAEOLOGY, ARCHITECTURE AND ART

T0199378

PROCEEDINGS OF THE WTC2019 ITA-AITES WORLD TUNNEL CONGRESS, NAPLES, ITALY, 3-9 MAY, 2019

Tunnels and Underground Cities: Engineering and Innovation meet Archaeology, Architecture and Art

Volume 11: Urban Tunnels - Part 1

Editors

Daniele Peila
Politecnico di Torino, Italy

Giulia Viggiani
University of Cambridge, UK
Università di Roma "Tor Vergata", Italy

Tarcisio Celestino
University of Sao Paulo, Brasil

CRC Press
Taylor & Francis Group
Boca Raton London New York

CRC Press is an imprint of the
Taylor & Francis Group, an **informa** business

A BALKEMA BOOK

Cover illustration:

View of Naples gulf

CRC Press/Balkema is an imprint of the Taylor & Francis Group, an informa business

© 2020 Taylor & Francis Group, London, UK

Typeset by Integra Software Services Pvt. Ltd., Pondicherry, India

All rights reserved. No part of this publication or the information contained herein may be reproduced, stored in a retrieval system, or transmitted in any form or by any means, electronic, mechanical, by photocopying, recording or otherwise, without written prior permission from the publishers.

Although all care is taken to ensure integrity and the quality of this publication and the information herein, no responsibility is assumed by the publishers nor the author for any damage to the property or persons as a result of operation or use of this publication and/or the information contained herein.

Published by: CRC Press/Balkema
        Schipholweg 107C, 2316XC Leiden, The Netherlands
        e-mail: Pub.NL@taylorandfrancis.com
        www.crcpress.com – www.taylorandfrancis.com

ISBN: 978-0-367-46899-6 (Hbk)
ISBN: 978-1-003-03185-7 (eBook)

Tunnels and Underground Cities: Engineering and Innovation meet Archaeology,
Architecture and Art, Volume 11: Urban
Tunnels - Part 1 – Peila, Viggiani & Celestino (Eds)
© 2020 Taylor & Francis Group, London, ISBN 978-0-367-46899-6

Table of contents

Tunnels and Underground Cities: Engineering and Innovation meet Archaeology,
Architecture and Art, Volume 11: Urban
Tunnels - Part 1 – Peila, Viggiani & Celestino (Eds)
© 2020 Taylor & Francis Group, London, ISBN 978-0-367-46899-6

Preface

The World Tunnel Congress 2019 and the 45th General Assembly of the International Tunnelling and Underground Space Association (ITA), will be held in Naples, Italy next May.

The Italian Tunnelling Society is honored and proud to host this outstanding event of the international tunnelling community.

Hopefully hundreds of experts, engineers, architects, geologists, consultants, contractors, designers, clients, suppliers, manufacturers will come and meet together in Naples to share knowledge, experience and business, enjoying the atmosphere of culture, technology and good living of this historic city, full of marvelous natural, artistic and historical treasures together with new innovative and high standard underground infrastructures.

The city of Naples was the inspirational venue of this conference, starting from the title Tunnels and Underground cities: engineering and innovation meet Archaeology, Architecture and Art.

Naples is a cradle of underground works with an extended network of Greek and Roman tunnels and underground cavities dated to the fourth century BC, but also a vibrant and innovative city boasting a modern and efficient underground transit system, whose stations represent one of the most interesting Italian experiments on the permanent insertion of contemporary artwork in the urban context.

All this has inspired and deeply enriched the scientific contributions received from authors coming from over 50 different countries.

We have entrusted the WTC2019 proceedings to an editorial board of 3 professors skilled in the field of tunneling, engineering, geotechnics and geomechanics of soil and rocks, well known at international level. They have relied on a Scientific Committee made up of 11 Topic Coordinators and more than 100 national and international experts: they have reviewed more than 1.000 abstracts and 750 papers, to end up with the publication of about 670 papers, inserted in this WTC2019 proceedings.

According to the Scientific Board statement we believe these proceedings can be a valuable text in the development of the art and science of engineering and construction of underground works even with reference to the subject matters "Archaeology, Architecture and Art" proposed by the innovative title of the congress, which have "contaminated" and enriched many proceedings' papers.

Andrea Pigorini
SIG President

Renato Casale
Chairman of the Organizing Committee WTC2019

Acknowledgements

REVIEWERS

The Editors wish to express their gratitude to the eleven Topic Coordinators: Lorenzo Brino, Giovanna Cassani, Alessandra De Cesaris, Pietro Jarre, Donato Ludovici, Vittorio Manassero, Matthias Neuenschwander, Moreno Pescara, Enrico Maria Pizzarotti, Tatiana Rotonda, Alessandra Sciotti and all the Scientific Committee members for their effort and valuable time.

SPONSORS

The WTC2019 Organizing Committee and the Editors wish to express their gratitude to the congress sponsors for their help and support.

Tunnels and Underground Cities: Engineering and Innovation meet Archaeology,
Architecture and Art, Volume 11: Urban
Tunnels - Part 1 – Peila, Viggiani & Celestino (Eds)
© 2020 Taylor & Francis Group, London, ISBN 978-0-367-46899-6

WTC 2019 Congress Organization

HONORARY ADVISORY PANEL

Pietro Lunardi, President WTC2001 Milan
Sebastiano Pelizza, ITA Past President 1996-1998
Bruno Pigorini, President WTC1986 Florence

INTERNATIONAL STEERING COMMITTEE

Giuseppe Lunardi, Italy (Coordinator)
Tarcisio Celestino, Brazil (ITA President)
Soren Eskesen, Denmark (ITA Past President)
Alexandre Gomes, Chile (ITA Vice President)
Ruth Haug, Norway (ITA Vice President)
Eric Leca, France (ITA Vice President)
Jenny Yan, China (ITA Vice President)
Felix Amberg, Switzerland
Lars Barbendererder, Germany
Arnold Dix, Australia
Randall Essex, USA
Pekka Nieminen, Finland
Dr Ooi Teik Aun, Malaysia
Chung-Sik Yoo, Korea
Davorin Kolic, Croatia
Olivier Vion, France
Miguel Fernandez-Bollo, Spain (AETOS)
Yann Leblais, France (AFTES)
Johan Mignon, Belgium (ABTUS)
Xavier Roulet, Switzerland (STS)
Joao Bilé Serra, Portugal (CPT)
Martin Bosshard, Switzerland
Luzi R. Gruber, Switzerland

EXECUTIVE COMMITTEE

Renato Casale (Organizing Committee President)
Andrea Pigorini, (SIG President)
Olivier Vion (ITA Executive Director)
Francesco Bellone
Anna Bortolussi
Massimiliano Bringiotti
Ignazio Carbone
Antonello De Risi
Anna Forciniti
Giuseppe M. Gaspari

Giuseppe Lunardi
Daniele Martinelli
Giuseppe Molisso
Daniele Peila
Enrico Maria Pizzarotti
Marco Ranieri

ORGANIZING COMMITTEE

Enrico Luigi Arini
Joseph Attias
Margherita Bellone
Claude Berenguier
Filippo Bonasso
Massimo Concilia
Matteo d'Aloja
Enrico Dal Negro
Gianluca Dati
Giovanni Giacomin
Aniello A. Giamundo
Mario Giovanni Lampiano
Pompeo Levanto
Mario Lodigiani
Maurizio Marchionni
Davide Mardegan
Paolo Mazzalai
Gian Luca Menchini
Alessandro Micheli
Cesare Salvadori
Stelvio Santarelli
Andrea Sciotti
Alberto Selleri
Patrizio Torta
Daniele Vanni

SCIENTIFIC COMMITTEE

Daniele Peila, Italy (Chair)
Giulia Viggiani, Italy (Chair)
Tarcisio Celestino, Brazil (Chair)
Lorenzo Brino, Italy
Giovanna Cassani, Italy
Alessandra De Cesaris, Italy
Pietro Jarre, Italy
Donato Ludovici, Italy
Vittorio Manassero, Italy
Matthias Neuenschwander, Switzerland
Moreno Pescara, Italy
Enrico Maria Pizzarotti, Italy
Tatiana Rotonda, Italy
Alessandra Sciotti, Italy
Han Admiraal, The Netherlands
Luisa Alfieri, Italy

Georgios Anagnostou, Switzerland
Andre Assis, Brazil
Stefano Aversa, Italy
Jonathan Baber, USA
Monica Barbero, Italy
Carlo Bardani, Italy
Mikhail Belenkiy, Russia
Paolo Berry, Italy
Adam Bezuijen, Belgium
Nhu Bilgin, Turkey
Emilio Bilotta, Italy
Nikolai Bobylev, United Kingdom
Romano Borchiellini, Italy
Martin Bosshard, Switzerland
Francesca Bozzano, Italy
Wout Broere, The Netherlands

Domenico Calcaterra, Italy
Carlo Callari, Italy
Luigi Callisto, Italy
Elena Chiriotti, France
Massimo Coli, Italy
Franco Cucchi, Italy
Paolo Cucino, Italy
Stefano De Caro, Italy
Bart De Pauw, Belgium
Michel Deffayet, France
Nicola Della Valle, Spain
Riccardo Dell'Osso, Italy
Claudio Di Prisco, Italy
Arnold Dix, Australia
Amanda Elioff, USA
Carolina Ercolani, Italy
Adriano Fava, Italy
Sebastiano Foti, Italy
Piergiuseppe Froldi, Italy
Brian Fulcher, USA
Stefano Fuoco, Italy
Robert Galler, Austria
Piergiorgio Grasso, Italy
Alessandro Graziani, Italy
Lamberto Griffini, Italy
Eivind Grov, Norway
Zhu Hehua, China
Georgios Kalamaras, Italy
Jurij Karlovsek, Australia
Donald Lamont, United Kingdom
Albino Lembo Fazio, Italy
Roland Leucker, Germany
Stefano Lo Russo, Italy
Sindre Log, USA
Robert Mair, United Kingdom
Alessandro Mandolini, Italy
Francesco Marchese, Italy
Paul Marinos, Greece
Daniele Martinelli, Italy
Antonello Martino, Italy

Alberto Meda, Italy
Davide Merlini, Switzerland
Alessandro Micheli, Italy
Salvatore Miliziano, Italy
Mike Mooney, USA
Alberto Morino, Italy
Martin Muncke, Austria
Nasri Munfah, USA
Bjørn Nilsen, Norway
Fabio Oliva, Italy
Anna Osello, Italy
Alessandro Pagliaroli, Italy
Mario Patrucco, Italy
Francesco Peduto, Italy
Giorgio Piaggio, Chile
Giovanni Plizzari, Italy
Sebastiano Rampello, Italy
Jan Rohed, Norway
Jamal Rostami, USA
Henry Russell, USA
Giampiero Russo, Italy
Gabriele Scarascia Mugnozza, Italy
Claudio Scavia, Italy
Ken Schotte, Belgium
Gerard Seingre, Switzerland
Alberto Selleri, Italy
Anna Siemińska Lewandowska, Poland
Achille Sorlini, Italy
Ray Sterling, USA
Markus Thewes, Germany
Jean-François Thimus, Belgium
Paolo Tommasi, Italy
Daniele Vanni, Italy
Francesco Venza, Italy
Luca Verrucci, Italy
Mario Virano, Italy
Harald Wagner, Thailand
Bai Yun, China
Jian Zhao, Australia
Raffaele Zurlo, Italy

Urban Tunnels - Part 1

Tunnels and Underground Cities: Engineering and Innovation meet Archaeology,
Architecture and Art, Volume 11: Urban
Tunnels - Part 1 – Peila, Viggiani & Celestino (Eds)
© 2020 Taylor & Francis Group, London, ISBN 978-0-367-46899-6

Geomechanical behavior evaluation of a bio-constructed limestone in an urban tunnel

S.W. Abreu
PCE Projetos e Consultorias de Engenharia Ltda, Rio de Janeiro, Brazil

ABSTRACT: This work presents an evaluation of the geomechanical behavior of a bio-constructed limestone layer formed by coral sedimentation near the surface on a Caribbean coastal plain. These rocks are denominated Caliza and Caliche. Caliza overlaps Caliche. Since Caliza is a less deformable rock than Caliche, the displacements observed on the surface were negligible when compared to those occurring in in-depth excavations. The study was based on convergence and settlement reference pin measurements arranged in cross sections along a 1km shallow tunnel in a very dense urban area of Santo Domingo – Dominican Republic. Tunnel overburden varies from 4 to 20m. The measurements were evaluated considering several aspects of a partial excavation, the excavation front positioning and the GSI values of each excavation front.

1 INTRODUCTION

Many underground excavations in urban areas are shallow, either for the purpose of urban mobility, such as road tunnels or subway systems, or for infrastructure, such as water supply or depletion, for instance. Hence, the occurrence of deformations as a result of excavation is one of the main project concerns. Monitoring and prevention are necessary and underground excavations in urban areas often have an instrumentation system to control settlements on the surface and in structures as well as at depth near excavations.

From a geological point of view, Calizas and Caliches cover a large part of Santo Domingo (Dominican Republic) and its surroundings. These limestone materials formed by the sedimentation of corals have long occurred on the Coastal Plains of the Caribbean Sea.

Little is known regarding the geomechanical behavior when applied to underground excavations considering their specific conditions, in areas where Calizas (stronger and less deformable rocks) appear over Caliches.

This study aims to evaluate the geomechanical behavior of carbonate rock layers when submitted to the underground excavations of a shallow 1km road tunnel in Santo Domingo city.

For this reason, monitoring was planned to verify the tunnel excavation influence on the surface and surrounding the excavation to avoid settlements on the streets and damage to constructions and buildings above the excavation.

The monitoring data were evaluated according to the proposed geomechanical classification. Other aspects related to the construction method were also considered in this work.

2 GENERAL CHARACTERISTICS, EXCAVATED MATERIALS AND GEOMECHANICAL CLASSIFICATION

The road tunnel analyzed is a 71 m² excavated cross section, 940m in length and has an overburden varying between 4 and 20m. Inaugurated in 2012, it is situated downtown starting in Ortega e Gasset Avenue at the North portal and arriving at Santo Tomás de Aquino Avenue

Table 1. Summary of the geological characteristics of Calizas and Caliches.

Lithology	GSI	Weathering degree	Coherence degree	Fracturing degree	Cavity occurrence	Additional materials
C-I	45–55	A1–A2	C1–C2	F1–F2	Porous with sporadic larger cavities	-
C-II	35–45	A2–A3	C2–C3	F2–F3	Porous with high frequency of larger cavities	-
C-IIA	30–35	A3	C2–C3	F3–F4	Cavities of all sizes	Very hard crystals and carbonates
CR-1	25–30	A4	C3–C4	–	Cavities of all sizes	Silts, clays, sands and crystals
CR-2	20–25	A5	Brittle	–	Cavities of all sizes	Silts, clays, sands and crystals

by the South portal. The UASD tunnel is recognized as the first urban road tunnel in the Dominican Republic.

Running along the whole tunnel length, the Calizas and Caliches are limestone materials made up of coral sedimentation. Its matrix is naturally porous due to its formation process, with the presence of various sized cavities. The cavities observed were generally associated with the presence of fractures in the rock mass. The rock mass weathering and cavity formation process is a result of water percolation in the fractures. Table 1 presents a summary of the geological features of the region the tunnel crosses through. The Calizas, more resistant materials, were separated into three groups, namely Caliza I, Caliza II and Caliza II-A. The Caliches were divided into two groups: Caliche 1, self-sustaining and Caliche 2, not self-sustaining.

The geomechanical classification system that served as basis for the work performed by the field team, acting as a guideline for the mapping of excavation fronts is known as GSI - Geological Strength Index of MARINOS and HOEK (2000).

Since the GSI system was especially improved for use on soft rocks, it was understood that it could be adapted to the situation at hand.

Thus, the system "Geomechanical Classification of the UASD Tunnel" was created, which aimed to formulate conventions and criteria for better adaptation to the original GSI system. These adaptations refer to: 1 - percentage of each lithology per excavation front, composition and structure of the materials, 2 - degree of weathering and decomposition and the condition of the adjacent mass. Thus a classification value is obtained resulting from the weighting of the parameters related to the abovementioned adaptations, which then determines the mass quality at the excavation front.

Considering the coherence of each type of rock, as well as the weathering, porosity and presence of cavities in all lithologies, a direct correspondence can be drawn with Caliza and Caliche to the original GSI system presented in the MARINOS and HOEK table (2000). A case of this type of adaptation can be seen in MARINOS and HOEK (2001) for a heterogeneous and sedimentary origin material called Flysch.

Figure 1 presents the proposed geomechanical classification for this material (ABREU, 2013).

Figure 2 presents the geomechanical profile of the tunnel. It can be noted that fractures appear concentrated where there is less resistant material. It can also be seen that Caliza I, the more resistant material, is located near the surface, evidence of an unusual condition in which more resistant rocks appear above less resistant rocks.

The true behavior of the materials was identified when beginning the excavations. The Caliche, a major concern during the design of the project, presented better strength and self-support capability than expected.

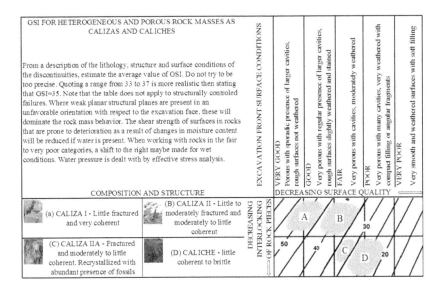

Figure 1. GSI Geomechanical Classification proposed by ABREU (2013).

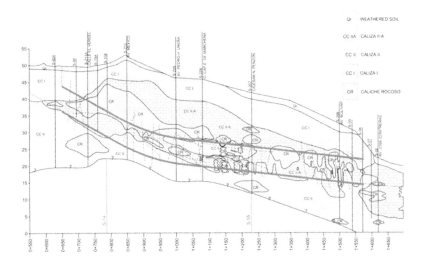

Figure 2. Geomechanical Profile.

Caliche and Caliza II-A resistance contributed to a change in the tunnel cross-section, eliminating the temporary and definite inverts. This is the reason for the unusual shape of the cross-section performed.

This tunnel was instrumented with convergence sections at every 10m, settlement markers and settlement gauges (where possible, since it was built in a very urbanized region).

The surface instrumentation presented very small measurements, always close to 0mm. This fact, associated with the occurrence of the "Caliza Slab" on the surface, shows that tunnel excavation even close to the portals where overburdens were 4 and 6m - the smallest along the entire tunnel - were not able to mobilize deformations on surface and affect nearby structures. For this reason, measurements of settlement markers and settlement gauges will not be

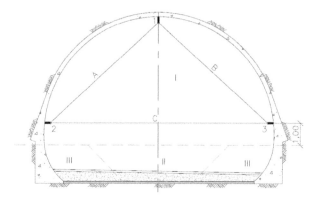

Figure 3.   Cross section.

considered in this work and emphasis will be given to the measurements obtained from the pins installed in convergence and settlement cross sections within the tunnel.

Figure 3 presents the position of three settlement/convergence pins (1, 2 and 3), convergence distances (A, B and C) and excavation stages (I –top heading excavation, II - Central bench excavation and III – right and left bench excavation).

3   DATA PROCESSING

Surface and deep settlements were measured with instruments installed in the public areas and streets and internally, the convergences and vertical movement were measured through the pins' distance and settlements.

The equipment used by the monitoring team was a Topcon model 433 station with angular accuracy of 3".

The reading results analyzed in this work are internal, as previously mentioned. These were read considering the possibility of systematic reading errors, as well as instrument damage which occurred during the construction.

This treatment was applied to relative unit values between the readings. For instance, individual negative convergence or settlement greater than 1mm was limited to 1mm. This and other criteria were used to give a better graphic to view movement tendency.

The following criteria were used for data processing (units in millimeters):

$r^{\pm}$: Positive or negative individual settlement measurement
$c^{\pm}$: Positive or negative individual convergence measurement
$R^{\pm}$: Positive or negative individual treated settlement
$C^{\pm}$: Positive or negative individual treated convergence
and:
i: 0, 1, 2 . . .n-1, n, n+1
Rules:

$$R^-_{\max} \; or \; C^-_{\max} = -1$$
$$if \; r^+_i > 0.2 \; R^+_i = 0.2$$
$$if \; c^+_i > 0.5 \; C^+_i = 0.5$$

$$if \; r^+_i > 0.2 \, and \; \left|r^-_{i-1}\right| = r^+_i \pm 0.1$$
$$or \; r^+_i > 0.2 \, and \; \left|r^-_{i+1}\right| = r^+_i \pm 0.1$$
$$or \; r^+_i > 0.2 \, and \; \left|r^-_{i-1}\right| + \left|r^-_{i+1}\right| = r^+_i$$
$$then \; \left|R^-_{i-1}\right| + \left|R^-_{i+1}\right| = 0.2$$

$$\text{if } c_i^+ > 0.5 \text{ and } |c_{i-1}^-| = c_i^+ \pm 0.1$$
$$\text{or } c_i^+ > 0.5 \text{ and } |c_{i+1}^-| = c_i^+ \pm 0.1$$
$$\text{or } c_i^+ > 0.5 \text{ and } |c_{i-1}^-| + |c_{i+1}^-| = c_i^+$$
$$\text{then } |C_{i-1}^-| + |C_{i+1}^- = 0.5|$$

4 GEOMECHANICAL BEHAVIOR

The expected behavior of measurements coming from settlement markers and settlement gauges was confirmed by the readings and did not present significant values. This insignificant superficial settlement was attributed to the existence of a layer of Caliza I near the surface and above the tunnel heading. It is understood that this layer has acted as a slab in not allowing displacements above the tunnel heading, both in surface and in depth.

The convergence and settlement readings for all the cross sections were treated as mentioned in item 3.

After that, two sections with different behaviors were selected to be studied in depth. The selected sections were section 14 and section 55.

Region 14 had the largest displacements measured while the region of section 55 showed reduced displacements.

Table 2 presents the rock mass characteristics of the regions S14 and S55.

4.1 Displacement evaluation according to rock mass quality

For each section, the settlements were analyzed considering the mapped GSIs. Figure 4 presents the measurements of settlements for the whole tunnel and the assumed GSI. Thus, it can be observed that the cross section S14 represents a large segment of poor rock mass quality with GSI equal to 25 while S55 is in a segment with better rock mass quality with GSI equal to 35.

From the S14 region evaluation, the predominance of Caliche leads the monitored section to present bigger deformations when compared to regions where Calizas predominate. This can be attributed to the reduced resistance of Caliche when compared to Calizas.

The presence of Calizas in this section is basically at the bottom and for this reason it is possible that all three pins have been fixed in Caliche. This would explain the high deformability values. In addition, as there was pin settlement, it is reasonable to consider that the tunnel in this section exhibited rigid body behavior.

In the Section 55 region, the settlement values were less than 2mm. The convergence reached 2.7mm. The rock mass in this region is very homogeneous, being made up basically of Caliza CC II and Caliza CC II-A. The displacements less than 5 mm can be correlated to high rock mass quality, and therefore correspond to the expected behavior.

Table 3 presents a summary of settlements and convergence values obtained from the regions close to where section S14 and S55 were monitored.

Table 2. Instrumented region characteristics.

Section	Composition	GSI	Density of cavities	Cavity filling	Weathering
14	80% Caliche (top cross section portion) 20% C-IIA	25 to 30	Walls in C-IIA - Low of small cavities. heading in Caliche - High of small and medium cavities	Silt, crystals and Caliche on the walls and Silt and crystals on the heading	High
55	70% C-II 30% C-IIA	35 to 40	High of micro and small cavities	Silt and crystals	Slightly weathered walls

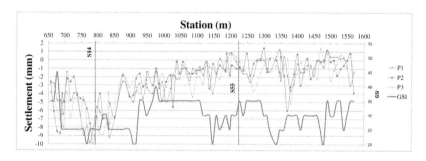

Figure 4. All Maximum settlement measures and the GSI.

Table 3. Settlements and convergences in the regions of Sections S14 and S55.

Sections	Progressive	Modify GSI		Maximum settlement			Maximum Convergence		
12	0+767	20	25	-8	-5.8	-7.7	-0.2	-2.6	-1.5
13	0+777	25	30	-9.8	-7.6	-8.4	-2.8	-2.6	-1.4
14	0+787	25	30	-8.2	-10.2	-9	-1.3	-0.3	-3.9
55	1+228	35	40	-1	-0.8	-0.2	-0.1	-0.8	-1.2
56	1+238	30	35	-1.6	-1.2	-0.8	1.2	-1.4	0.4
57	1+248	35	40	-1.6	0.8	-1.6	-2.7	-0.6	-1.9

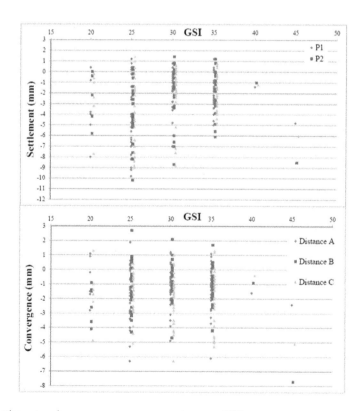

Figure 5. Settlement and convergence measurements versus GSI.

Regarding the convergence measurements, for all mapped GSI classes, the values were restricted to 5mm. Regarding the settlements, the majority of sections with GSI greater than 30 present values limited to 5mm. In the sections of 20 to 25 GSI, there is no maximum settlement value tendency, since the measures are dispersed. However, displacements are also considered to be dependent on the overburden and the construction method (advance and excavation stages), but only geological mapping (GSI) is considered in this evaluation. Figure 5 presents the settlement and convergence measurements compared with GSI.

The cross section S47 is 1147m where it can be seen that there is a GSI equal to 20 but the adjacent segments is GSI equal to 30 or higher. For this section the deformation readings did not present significant values.

4.2 Displacement evaluation as a function of excavation advance

The influence of excavation advance steps on the settlement and convergence values can be evaluated by the construction method analysis for each section. Table 4 presents average advances for a segment of one hundred meters S14 and S55adjacent regions.

The major advances (between 7 and 9 meters approximately) were executed between sections S09 and S12. Thus, considering the segment of the adjacent regions, section S14 had the highest average advance (considering a 100m segment) executed during the entire excavation. Considering the low GSI, this high advance may have influenced the settlement measurements in this section, since this section presented the greatest measurements throughout the tunnel.

Still on the subject of the advance influence on instrumentation measurements, when comparing other segments with the same geological characteristics but with smaller advances than section S14, much smaller measurements are observed. Thus it can be understood that an advance of 2 meters for rock mass with GSI below 30 is reasonable to maintain low deformations in this type of material.

For S55, where the GSI average is about 32, the average advance of 4 meters does not seem to have influenced the displacements, since these were low.

4.3 Displacement evaluation due to excavation fronts

For an individual analysis of bench execution influence on the displacement values, the left side of Figure 6 presents the settlement and convergence measurements of S14 pins when executing the benches.

The measurements of both settlements (10mm) and convergences (4mm), if evaluated considering the central bench execution, increased progressively after they had already presented a tendency to stabilize prior to the execution of the bench.

For section S55, the right side of Figure 6 shows that the bench execution does not influence the excavation displacements, keeping them at very low levels.

Table 4. Average avances.

Station	Section	Average advance (m)	Average GSI
0+737 to 0+837	S14	4.80	25
1+177 to 1+277	S55	3.90	32

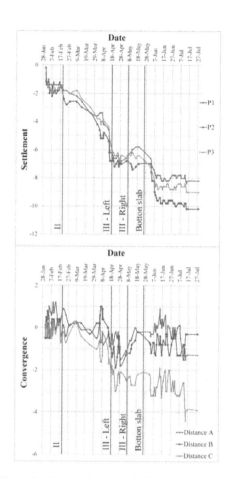
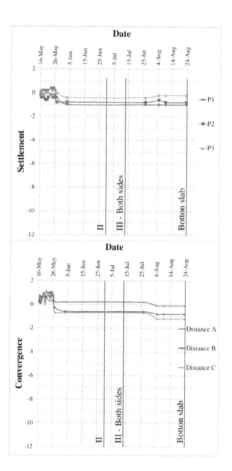

Figure 6. Excavation stage influence on the evolution of the displacements in Section S14 (left) and in Section S55 (right).

5 CONCLUSION

Analysis of convergence and settlement pin instrumentation data verified that the most significant deformation measurement component was the pin settlement.

It can be seen that the deformations measured along the entire tunnel are directly influenced by three factors: Dimension of the advances, Quality of the rock mass and recess execution.

The significant deformations occur in materials with GSI below 30, which were mapped in long stretches as was the case in the region of section S14.

However, the excessive deformations assessed through section S14 can also be understood as a response to the executive process at the site where there was excavation with advances reaching 9.0 meters. This fact associated with the long segment of low rock mass quality took the deformations to 10 millimeters.

Table 5. Suggested advance values in function of GSI.

GSI	Advance excavation step (m)
20	1.0
25	2.0
30–35	3.0
Above 40	5.0

Most displacements greater than 5 mm occurred in regions of GSI between 20 and 30. Therefore, the higher the excavation front GSI, the lower the displacements.

A general displacement analysis considering the advance values related to geological characteristics presented a greater advance tolerance value with acceptable displacement levels greater than the GSI.

Bench execution may have influenced the stabilized displacements growth resumption.

After detailed displacement analysis, it can be suggested that the advances presented in Table 5 are reasonable in order to maintain the displacements around 5.0 mm.

REFERENCES

Abreu, S. W. 2013. *Análise de comportamento geomecânico de túnel escavado em rochas carbonáticas*. Rio de janeiro:UFRJ/COPPE.

Hernaiz Huerta, P. P.; Díaz de Neira, J. A. 2004b. *Mapa geológico de la hoja a e. 1:50.000 n° 6272-iii (monte plata) y memoria correspondiente. Proyecto l-zona so de cartografía - geotemática de la República Dominicana. Programa sysmin.*Santo Domingo: Dirección general de minería.

Hoek, E.; Brown, E. T. 1997. Practical estimates of rock mass strength. *International journal of rock mechanics and minning science* 34 (8): 1165–1186.

Hoek, E.; Carranza-Torres, C.; Corkum, B. 2002. Hoek-Brown criterion. *Proc. Narms-tac conference 2002*: 267–273.

Hoek, E. Marinos, P. 2000a. Predicting tunnel squeezing Problems in Weak Heterogeneous Rock Masses – part 1: estimating rock mass strength. *Tunnels and tunnelling international*:45–51.

Hoek, E.; Marinos, P.2000b. Predicting tunnel squeezing Problems in Weak Heterogeneous Rock Masses – part 2: estimating tunnel squeezing problems. *Tunnels and tunnelling international*: 34–6.

Marinos, P.; Hoek, E. 2000. Gsi:ageologically friendly tool for rock mass strength estimation. *Proc. Geoeng2000 conference.*

Marinos, P.; Hoek, E. 2001. Estimating the geological properties of heterogeneous rock masses such as flysch. *Bulletin of the engineering geology & the environment (iaeg)* 60: 85–92.

PCEProjetos e Consultorias de Engenharia, 2011. *Túnel Ortega &Gasset - uasd - informe geológico - geotécnico nov/2010 a set/2011.* Santo Domingo: PCE Projetos e Consultorias de Engenharia.

Serra-kiel, J.; Ferràndez-Cañadell, C.; García-Senz, J.; Hernaiz Huerta, P. P. 2007. Cainozoic larger foraminifers from Dominican Republic. *Boletín geológico y minero* 118(2): 359–384.

Tunnels and Underground Cities: Engineering and Innovation meet Archaeology, Architecture and Art, Volume 11: Urban Tunnels - Part 1 – Peila, Viggiani & Celestino (Eds)
© 2020 Taylor & Francis Group, London, ISBN 978-0-367-46899-6

Vibration-oriented tunnel design

M. Acquati
MM Spa, Milan, Italy

W. Stahl
Chair and Institute for Road, Railway and Airfield Construction, Technical University Munich TUM, Germany

ABSTRACT: The paper makes a survey on the general criteria for tunnel design with regard to vibration attenuation, with contributions from various experiences. The criteria used in this paper are a collection of knowledge arriving from the Technical University Munich TUM– since decades a strong point of reference for track design and validation – and MM SpA – the leading Italian company for metro design and construction. The most effective method to reduce the mitigation of noise and vibration at the source is the installation of mass-spring-systems. A resilient supported mass is decoupling the track from the tunnel construction. The perfect operation of a mass-spring system essentially depends on the correct design with regard to the natural frequencies due to bending of the track slab. However, the trends in design are opposed regarding divided planning procedures of tunnel structures with minimized cross sections for standard track. Constricted cross sections are limiting the installable mass, connected with increased bending of the track slabs and increased strain of the resilient bearings. Bigger tunnels bring about another advantage, which is the higher mass of the tunnel itself that acts as a dynamic decoupler of vibrations from tunnel to ground. Vibration mitigation has to be involved during the first planning phases of a tunnel, to optimize technical and economical features of tunnel and track.

1 INTRODUCTION

Rail traffic is one of many different sources of vibrations and certainly not the only source that causes the highest vibration levels.

In buildings near underground railways, the immissions are caused by structure-borne transmission. The effects caused by structure-borne transmission can be divided into ground-borne vibrations (below the hearing threshold) and audible structure-borne noise (secondary airborne sound). The ground-borne vibrations, which occur within a frequency range of around 5 Hz to 20 Hz, are particularly relevant in "soft buildings" with wooden joist floors and ceilings or "lightweight steel girder floors and ceilings". If structures of a building, walls, ceilings or other parts were excited to vibrate, so they could be heard. The structure-borne noise (activated by the structure-borne vibration) can be heard as a muffled rolling noise, mainly with frequencies between 40 Hz and 80 Hz.

Metro networks are increasingly colonizing urban areas, passing below buildings for human work or residence. Vibrations are the main reason of complaint after service initiation in tunnels. Technology can help to mitigate vibrations. However, many people are confident that mitigation measures are items to be added in the transmission path, hence ignoring that the tunnel itself can be a mitigation measure.

The main aim of this paper is to identify and explore the positive effects of tunnel design in mitigating vibrations. This paper draws the path to a positive holistic approach to tunnel design that may take proper account of the dynamics of the tunnel.

2 A NEGLECTED RISK

There may be many uncertainties during tunnel design phases: quality of geotechnical data, design cost, time and quality. Similarly, many other risks may happen during construction: tunnel collapse or flooding, finding of unpredicted existing structures or extensive deformations.

Notwithstanding these risks, there is a risk that is usually neglected during design and construction of tunnels – in particular during design. This risk is the effect of tunnel shape, dimension and depth in governing vibration type and levels.

In fact, it is general convincement that vibration attenuation is a side effect of tunnel design and construction, in particular in urban areas. The back face of this convincement is that vibrations can be mitigated with works that can be added on, something like a workaround.

This fact is in consequence of a bad conception of structural design. Tunnels are topics for structural engineers, whereas vibrations is the field of conquest for structural dynamics. However, it is rather intuitive that once vibrations are generated inside tunnels, the tunnel itself is part of the vibration transmission path. Therefore, vibration levels outside tunnels do depend on the way the tunnel is built.

This paper puts together experience of an engineering company (MM Spa) and a technical university (TUM), whose experts have faced in years the consequences of this misconception in third party's projects: tunnels too close to buildings to impede any possible way of vibration mitigation, tunnels too small to allow any possible mitigation measure to put in, etc.

The consequences of neglecting this risk can be economical too. Recognizing specific issues early in their development is critical to the ability of taking preventive actions and decoupling the contingencies from the work process before they impact any project performance factors. The difference is like the one between fire prevention and fire fighting.

This paper is not aimed to add another complex factor to tunnel design, which is already complex on its own. Considerations about vibration generation mechanism is rather a simple passage within the phases of tunnel design and construction. The paper will move along a paved pathway, whose stones are the various contributions to the dynamic phenomena. Most of them will appear obvious, even to non-experts. The contradiction between the obviousness of these factors and the fact that in tunnel design these factors sometimes are neglected is the tangible demonstration that assuming these factors as a part of tunnel design is not a heavy job. However, this paper is not thought to be the history of negligence found in others' project, but a technical and, as much as possible, simple explanations of the technical factors to be assumed.

3 VIBRATION GENERATION MECHANISM

Vibrations originate from the unevenness of either one (or both) of two surfaces in rolling contact with each other (wheel and rail). This unevenness can be inherent to the surface (corrugation on a railhead) or due to variation in the support stiffness (e.g. hanging sleepers or soft spots in the subsoil of the track). Due to this unevenness, a dynamic force is applied to the two bodies which then respond with movement (they vibrate). They will be more responsive to this excitation at their natural frequencies where the impedance (resistance to the force) is relatively low.

Both the train and the track represent a complex structure, responding to dynamic forces as resonating bodies. This means that some excitation frequencies "fit" to the bogie, so that the bogie, when excited, will vibrate strongly and almost without damping. Other frequencies "fit" to the track so that either the sleepers or the rails will respond strongly and almost without damping. The vehicle track system may withstand excitation of yet other frequencies, because they do not fit to the response of the system.

All in all, the source of vibrations is a complicated interactive system, which makes it very difficult to accurately predict the generation of vibrations, mainly because it is difficult to know all the relevant parameters with sufficient certainty.

In rail traffic, the track is excited by the rolling wheel on the rail and vibrations are induced in the track. Especially if there are irregularities in the rail running surface, out-of-roundness of the wheels or flat areas on the running surfaces. Furthermore, so-called imperfections such as rail corrugation or gaps in the frogs (common crossings) of switches, also cause strong excitation of the track.

If the tunnel structure of underground railways is excited to vibrate, energy is transferred in the form of elastic waves. With an increasing distance of overlying buildings from the emission point, the wave amplitudes reduce due to material attenuation and due to propagation of the interference. However, the elastic waves can induce vibrations in a building foundation. The relevant frequency range for this based on available experience can be up to 200 Hz.

In some case vibrations originating from sources outside the house may be felt inside dwellings. This applies for example to heavy road traffic, trams and railway lines, both on surface lines and in tunnels. Whether or not the vibration can be perceived depends on many factors, including distance to the source, speed and type of the traffic, quality of the road or track, type and build-up of the ground, the way the building is supported by its foundation and the construction of the building itself. Among all the factors, this paper brings focus on the tunnel structure, geometry and dimensions.

4   VIBRATION ATTENUATION MECHANISM

The vibrations generated by rail traffic are transmitted through the track bed into the soil. There the vibrations are propagating in the form of waves travelling through the ground. The tunnel structure is as well part of this travelling path of vibrations. Some of these waves run on the surface of the soil, more or less like water waves. Other wave forms travel through the deep ground. The propagation of these waves, in particular the speed, is influenced by the ground properties like e.g. density and stiffness. The ground is typically not a homogeneous medium; there are big differences between layers which may include sand, clay, rock and ground water.

Vibration amplitudes decrease with increasing distance from the source. This attenuation is frequency dependent and is more significant for the higher frequencies. At larger distances, the low frequencies dominate.

In case of an underground railway (or metro), the tunnel itself can radiate vibrations in any direction and the tunnel structure can modify the characteristics of vibrations. The preferable route of radiation may depend on the cross-sectional shape of the tunnel. Vibrations radiated upward can easily excite foundations of buildings and from there run up to the building floors. Higher vibrations can reach buildings when the tunnel alignment is directly below the buildings, compared to alignments running under street axes. Boundaries between soil layers (also water table) can partially reflect vibration waves. This boundary can be above the railway and hence vibrations can travel very far from the railway and reach receivers that may not even suspect the source of incoming vibrations is that. Railway tunnels are usually built in urban areas where the transmission medium (soil) usually hosts many underground services (e.g. sewage, water, electricity, etc.) which may alter the normal radiation of vibration waves. Often, it may happen that a part of the tunnel touches a part of a building foundation. This constitutes a rigid bridge for vibration transmission and hence it shall be avoided absolutely in tunnel design and construction. The same can happen when soil concreting or stiffening techniques are used.

In general terms, for areas with stiff soils the maximum vibration levels at buildings near a railway tend to be in the 35–70 Hz range and the usual result is audible noise. For medium stiffness soils, the maximum vibration transmission tends to be in the 20–40 Hz range and the

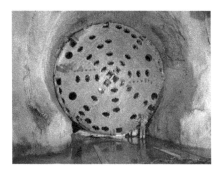

Figure 1.   The front shield of a TBM: circular tunnels can properly attenuate vibrations with 3 dB by doubling distance [Source: Wikipedia – Gotthard Base Tunnel].

vibration may be either feel able or audible, sometimes both. In soft soils, the frequency of maximum vibration tends to be in the range of 15–25 Hz with the result being perceptible vibration, usually with no audible noise even at very short distances from the track structure.

Attenuation of vibrations is in consequence of two physical phenomena. The first is the geometrical spreading. If the tunnel has circular cross-section, the vibration energy is spread around the tunnel circumference and the further the vibration energy goes from the source, the lower the level of energy. It is like a stone in a pond.

If the source of vibration is assumed to be a linear source, the attenuation is 3 dB per doubled distance, i.e. if the level is 72 dB @ 1 m from the source, it is 69 dB @ 2 m, 66 dB @ 4 m, 63 dB @ 8 m, etc.

If the tunnel has not circular cross-section and e.g. rectangular, the energy spreading is not on a circumference basis, therefore the law is not properly applicable.

The second attenuation mechanism is provided by internal dissipation of the ground material. The vibration energy is dissipated by friction between the particles in the soil.

In general, the equation governing this dissipation is as follows:

$$L = L_0 - \alpha \cdot (R - R_0) \cdot \frac{f}{V} \tag{1}$$

Where: L is the level at distance R; L_0 is the level at distance R_0; α is the loss factor, depending on soil; V is the wave speed, depending on wave type and soil type; f is the frequency.

The equation is clearly depicting some dissipation features:

- Unlike geometrical spreading, soil dissipation depends on frequency. The higher the frequency, the higher the dissipation. Such dependency is strong or low in consequence of soil mechanical characteristics.
- The wave speed is the main factor governing such dependency. Slow wave speeds contribute to steeper increase of attenuation by increasing frequency. Waves moving into the soil with high speeds are minimally attenuated at high frequencies.
- Wave speed is directly depending on the soil stiffness. If the soil is hard, the speed is high. If the soil is soft, the speed is low.

The latest point can better help to understand why in stiff soils the transmitted frequencies are in a higher range. Typically for well-consolidated clay, dissipation is very low, and the geometric spreading governs most of the behaviour.

5 VIBRATION MITIGATION MEASURES

The purpose of mitigation measure is to protect people, structures and mechanical systems from shock and vibration by taking action between the source and the receiver. The purpose may include ensuring:

- the static safety condition of the building close to the railway systems;
- the comfort of people in temporary or permanent structures subject to vibration excitation;
- the presence of sensitive equipment in these structures;
- the correct operation of any existing isolated equipment;
- the fulfilling of legal requirements.

If a mitigation measure is applied at the source then the entire neighbourhood is protected. Such measures can be applied:

- when designing and constructing new railway system near buildings or structures;
- when traffic conditions (e.g. increase of traffic per day, speed) are modified;
- when railway structure is modified;
- when the propagation path between railway structure and the environment is modified;
- when receiving complaints from people working or living in the area of vibration sources;
- when there are limiting values for vibration in legislation which are exceeded;
- when isolation of receiver is difficult, expensive or impossible.

In the case of impossibility or non-satisfaction of mitigation measure at source, receiver isolation is applied. Sometimes it is an economical compromise. It may concern:

- the new building or elements of the building in the neighbourhood of a railway or tunnel;
- the sensitive building (music halls, laboratories, or sensitive installations);
- the support of sensitive equipment (laser tables, computer discs, electronic microscopes, etc.).

Mitigation effects are soil dependent and distance dependent. This is due to a number of effects. The underlying ground influences the behaviour of the train-track system, of which the vibration mitigation measure is a part. It is for this reason that measures for stiff ground do not give the same results as for soft ground. Additionally, the propagation in the soil introduces frequency and wave type dependent filtering, which influences the mitigation effect spectra. Distance dependency also comes from "near field" and coherence effects due to the change of influence of the excitation mechanisms.

The most important mechanism in the generation of ground borne vibration from rail vehicles is the interaction between the wheel and the rail. The vibration is generated by the rail interacting with the unsprung mass of a wheelset. To obtain a low level of emitted vibration, it is necessary to reduce the amount of vibration generated at the wheel rail interface.

The generation of environmental vibration at receiver locations away from the railway is usually dominated by the dynamic excitation. This is generated by the combined roughness of the wheels and rails interacting with the unsprung mass.

The roughness of the track is a combination of a number of different factors that affect the vibration generation at the wheel rail interface. The factors that are of primary importance are:

- rail roughness
- track geometry
- rail joints
- switches and crossings

All these factors have an influence on vibration levels, and since all have a frequency dependent nature, the speed of the train will have an effect.

Yet, all these factors are non-related to tunnel geometry and depth, therefore they will not be considered in the exploration of the effects of tunnel design on vibration performance. All they will result the same, no matter the type of tunnel is there.

The most used mitigation measures are implemented at track level. By application of the correct choice of stiffness and of masses, very good performance can be achieved.

The mitigation effects can be calculated (principal and specific design rules) for various track systems. With the principal design rule a calculation method is found for all elastomer-based mitigation measures.

In the frequency range < 100 Hz every resiliently supported vibrating system can be modelled as a single-degree-of-freedom (SDOF) system with rigid masses and springs where the exciting force acts on the mass. Above a natural frequency f_0 the spring reduces the forces acting on the ground but on the other hand leads to their increase at the resonance frequency.

Crucial parameters for the mitigation effect of a vibration isolation are:

- the effective dynamic mass of the resiliently supported system above the spring;
- the stiffness and damping of the resilient elements;
- the stiffness and damping of the ground.

From the equation of motion of a SDOF system the ratio of force acting on the subgrade, F, to the exciting force, F_0, becomes

$$\frac{F}{F_0} = \frac{k_k}{k_k - \omega^2 . m} \tag{2}$$

with resonance at $\omega_0^2 = \frac{k_k}{m}$

where m is the effective dynamic mass (wheelset unsprung mass, rail and fastening components, sleepers, localized track mass, concrete slab); k_k is the complex dynamic stiffness of the elastic elements involved which comprises damping:

$$k_k = k_{re} + i \cdot k_{im} = k_{re}(1 + i\eta) = k_{re} + i\omega d. \tag{3}$$

The damping can be modelled either by a viscoelastic damping coefficient, d, or by a loss factor, η, (structural damping).

The complex dynamic stiffness, $\underline{k}_k$, is the geometrical sum of the stiffnesses of all elastic elements (in series) in the transmission path of the force:

$$\frac{1}{k_k} = \frac{1}{k_{kPAD}} + \frac{1}{k_{kMAT}} + \frac{1}{k_{kSG}} + \cdots \tag{4}$$

Where $\underline{k}_{kPAD}$ is the rail pad stiffness; $\underline{k}_{kMAT}$ is the stiffness of the isolation mat; $\underline{k}_{kSG}$ is the subgrade modelled as a spring.

Therefore, the effectiveness of the track form as a mitigation measure is controlled by:

- dynamic mass, that shall be high enough;
- dynamic stiffness, that shall be low enough.

6 EFFECT OF TUNNEL GEOMETRY

According to Capponi (1998), the tunnel shape can influence the structure response. Rectangular-shaped tunnels can show very high response at 60 Hz, be they with or without central half-span central wall. Although the high response at 60 Hz is typical for the structure, the effect of it on the overall level is depending on the input spectra by the running trains. On the assumption that the input spectrum is more energetic at frequencies higher than 60 Hz, the tunnel response can have a beneficial effect, because it cuts down the major contributions to the transmitted vibrations.

Youshoka (1996) carried out important studies that reported some empirical correlations between vibration levels and various parameters, tunnel shape included. Difference between vibration levels above shield (circular) tunnels and box-shaped tunnels is constantly around 2.5 dB.

TUM studies concluded that vibration velocity in box-shaped tunnels – measurement point at mid-span of tunnel soffit – are significantly higher than vibration velocity at tunnel soffit of circular tunnels. Conclusions are bringing to judge the box-shaped tunnels more prone to absorb part of the vibrational energy, compared to circular tunnels. However, the low frequency bending of tunnel soffit in box-shaped tunnels can generate waves moving vertically and hence critical for receivers located directly above the box-shaped tunnel. Yet, excavation technology for box-shaped tunnels do not allow buildings above the tunnel axis.

7   EFFECT OF TUNNEL MASS

The higher the tunnel mass, the lower the level of vibration transmitted to the surface. This point is fully agreed about by many research works. Since 1982 (Saurenman et al.) concluded that the equation governing tunnel mass and vibration level is the following:

$$L = L_0 - 56 \, log\left(\frac{m}{m_0}\right) \tag{5}$$

Where: L is the level of vibration in decibels, with L_0 reference level; m and m_0 are the masses the tunnel masses per length units, with m_0 associated with L_0

The equation is valid for circular tunnels. For box-shaped tunnels, the equation is different. Therefore, for doubling mass, the vibration reduction is 15 dB.

Sitou (1997) carried out studies on box-shaped tunnel – double line. He developed a quite complex formula that illustrates the effects of many factors:

$$\ddot{y} = \frac{k_1^{0.3} . k_2^{0.5} . k_3^{0.04}}{I^{0.2} . m_2^{0.3} . m_3^{1.2}} \tag{6}$$

Where:

- k_i are the various stiffnesses of elastic components in track
- I is the bending rigidity of the rail
- m_2 is the sleeper mass
- m_3 is the tunnel mass.

Keeping constant the other factors and varying only the tunnel mass, Sitou's equation brings to conclude that for doubling tunnel weight, the vibration reduction is 7 dB.

Therefore, for box-shaped tunnels the effect of tunnel weight is less.

Increasing tunnel mass has a minimal effect on tunnel construction cost, but a great effect on vibration reduction.

8   EFFECT OF TUNNEL DEPTH

As said at point 4, vibrations generated at trackbed go decreasing in amplitude once they move away from the tunnel. The mechanism is double-faced: geometric spreading and internal dissipation of ground material. The valuable consequence of this is that the further is the tunnel from receivers, the lower is the level of vibrations that may affect receivers.

This effect is predominant when the tunnel is very deep into ground. Generally speaking, we may say that the first mitigation measure to nullify vibrations is by deepening the tunnel as much as possible. The performance is more effective if the tunnel is in soft soil, not rock. This is because – as said – soft soils can dissipate vibration energy better.

We do know that tunnels cannot be excavated very deeply into ground, because there can be other issues arising.

However, the beneficial effect is not only limited to the direct one, because joint advantages can arise.

9 JOINT EFFECTS

Tunnel deepening's outcome is not only to put tunnels further from any receiver. It realizes also cascade effects, all beneficial.

Our examples are predominantly inherent to circular tunnels, which are frequently used in recent works.

Deeper circular tunnels need thicker linings and, on their turn, thicker linings increase the total mass of the tunnel. As said above, a heavier tunnel is more capable to reduce vibrations.

Therefore, by doubling depth, vibrations are reduced at least by 3 dB, assuming only geometrical spreading. Beyond that, doubling depth can at least double tunnel mass, which has the beneficial effect of 15 dB of additional reduction. To summarize, bringing the tunnel down into soil by a double length has the impressive outcome of more than 15 dB reduction. This is a value that can be hardly achieved with a very expensive mass-spring system – see below.

Another scenario to be explored is tunnel enlargement, by increasing the diameter. Tunnel engineers are quite reluctant in increasing tunnel diameter, because it has direct effect on excavation costs. Nevertheless, for long excavation sections, wider diameter can be worth the initial higher investment.

Widening the tunnel diameter can provide a joint effect, whose benefits are beyond the direct effect. The primary advantage caused by diameter widening is that the tunnel circumference is bigger and the tunnel mass is bigger. However, there is not only one geometrical equation that increases tunnel mass. Bigger diameter requires thicker structures. Moreover, typically a wider tunnel implies a more massive trackbed. The final result is that the increase of diameter does not simply yield to increase of tunnel mass with the power of two, but with the power of three.

10 MITIGATIONS MEASURES ONCE TUNNEL EFFECTS ARE NOT ENOUGH

Mass-spring systems: This measure (also called floating slab track) is generally the most effective, but in most cases it is the technically most complicated and therefore the most expensive method. Implementation examples exist as:

- Ballastless tracks
- Track panel supported on ballast
- Single precast units
- In-situ concrete slabs.

In mass-spring systems (MSS), the track is supported elastically via special spring elements. If a MSS is correctly adjusted, noise and vibration nuisance can be avoided even if there is a direct structural connection between the tunnel structure and the property to be protected. The perfect operation of a mass-spring system essentially depends on the correct design with regard to the natural frequencies due to bending of the track slabs and precise construction and assembly work.

Deflection z under dead-load of the mass (good rule of thumb):

$$f_0 = \frac{1}{2\pi}\sqrt{\frac{c}{m}} \; simplification \rightarrow f_0 = \frac{5}{\sqrt{z[cm]}} \rightarrow z[cm] = \left(\frac{5}{f_0}\right)^2 \tag{7}$$

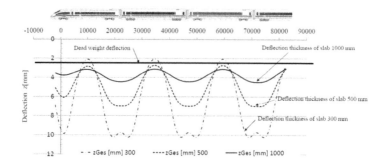

Figure 2.   Deflection diagram for MSS with three slab thicknesses.

Hence, lower natural frequencies imply increased deflection. Moreover, they need increased mass and increased cost.

The natural frequency f_{eig} should be square root 2 below the natural frequencies of the surrounded buildings or structures. To avoid overloading of the elastic elements, the maximum deflection under traffic load should not exceed 1.5 to 2 times the value of the static deflection (dead weight deflection) z. Therefore, the mass and the springing of the system have to be in balance. For instance, a 10 Hz-System shows a dead weight deflection z of 2.5 mm, the additional deflection under running trains should not exceed 5.0 mm. The wheelset force dynamic and wheel force transfer in a curve also have to be taken into account when defining the loading range of the MSS and the bearings. In general, the wheelset dynamic can be allowed for by applying a factor of 1.5 to 1.6 - and the wheel force transfer in the curve by a factor of 1.1 to 1.2.

The following figure shows the deflections of concrete slabs with different thickness of a 10 Hz MSS under the load of an ICE 3. The very thin slab (zGes= 300 mm) is showing high deflections. This is connected with a strong bending of the slab itself and strain of the elastic elements which are installed under the slab. With a slab height of 1000 mm, the dead weight deflection is in good balance with the traffic load deflection and will lead to a sustainable layout of the system.

Depending on the required natural frequency of the system and wheelset force (80 to 210 kN), a clear height between the top of the rail and tunnel invert of between 80 and 150 cm is required to lay a mass of 4 to 9 t/m (~half of the axle load of the running trains).

10.1   *Geometry constraints of mitigation measures on tunnel geometry*

In order to allow the installation of a mass-spring system, the tunnel shall be big enough. In fact, as written above, the mass-spring system to be effective needs enough space between the top of the rail and tunnel invert.

Sometimes, it happens that tunnels are built very small, with tight diameter, so as to limit costs of excavation. The direct consequence is that such tunnel cannot host any kind of MSS. This example is another support to the demonstration that vibration impact is not taken into proper account when choices on tunnel layout are made.

Making tunnels bigger is advantageous first because on its own can create conditions to limit vibrations and – what's more – if the conditions are not enough, it can host a MSS, which make the final brick to vibration definitive elimination.

In case the tunnel is not sufficient to be place for a MSS, other track solutions are available but not as efficient as MSS. Very resilient fastening systems can work to absorb a significant part of vibration energy. However, they are generally very expensive, both for installation and for maintenance.

11 ECONOMICS

Some last words are left for the economical point of view.

Undoubtedly, any vibration mitigation measure has a cost, no matter whether it is by adaptation of the tunnel geometry or by adding a MSS inside the tunnel. Thus, any decisional process, even in this field, should be supported by cost-benefit analyses.

Widening of tunnel diameter, deepening of tunnel alignment, etc.: all of them are expensive layouts. However, a mass-spring system is expensive too. Yet, any last-minute workaround is definitely more cost-impacting than planned response to the problem.

Moreover, no choice has effect directly on to the factor for which the choice was taken. Deepening tunnels will make less accessible stations (long escalators or elevators to reach the station platform). As well as mass-spring systems shall be studied also from the point of view of vehicle-track interface, because the dynamics of the track can interfere with a correct runability of the trains. Therefore, any choice shall be duly weighted, considering all the interactions and interfaces, because there may be a cost-impacting effect.

12 CONCLUSIONS

Early recognition of undesirable effects is a critical, precondition for managing projects. In addition, project leaders must not only recognize potential factors in general, but also should know when they will most likely affect the project. Recognizing specific issues and risks before they occur or early in their development is critical to the ability of taking preventive actions and decoupling the risks from the work process before they impact any project performance factors.

As said vibration impact is a neglected risk in tunnel design. Early recognition is fundamental and this paper strongly wishes to make tunnel designers aware of it. Efforts required to manage properly dynamics of tunnels are minimal but benefits are worthy.

Authors hope that this paper can make tunnel designers aware that a possible approach to vibration-oriented design is feasible and the relevant expert support, for whoever may need, can be found in the experience and knowledge whose dissemination has been done here in the care of the authors.

REFERENCES

Capponi, G & Murray, M. 1998. Theoretical modelling of vibration reduction in urban rail transport. *Gallerie e grandi opera sotterranee*, July, pp.55–63.

Saurenman, H.J., Nelson, J.T., Wilson, G.P. 1982. Handbook of urban rail noise and vibration control, *U.S. Department of Transportation*, Washington D.C.

Sitou, K. 1997. Applicable prediction methods for environmental vibrations (subway induced vibrations), *INCE/JAPAN Technical Committee*, Tech Report no. 20, October.

Unterberger, W. & Hochgatterer, B. & Poisel, R. 1997 Numerical predictions of vibrations caused by trains in tunnels. *Tunnels & Tunnelling International*, December, pp. 45–47.

Yoshoka, O. 1996. Prediction analysis of train-induced ground vibrations using equivalent excitation force, *Quarterly Report of Railway Technical Research Institute*, Japan, Vol 37, No. 4, December, pp. 216–224

Wilson, G.P. 1987. Developments in control of ground-borne noise and vibration from rail transit systems. In INTER-NOISE and NOISE-CON Congress and Conference Proceedings, *InterNoise87*, Beijing CHINA, pages 431–834, pp. 619–622(4)

*Tunnels and Underground Cities: Engineering and Innovation meet Archaeology,
Architecture and Art, Volume 11: Urban
Tunnels - Part 1 – Peila, Viggiani & Celestino (Eds)*
© 2020 Taylor & Francis Group, London, ISBN 978-0-367-46899-6

Long-term behavior of the tunnel under the loads of consolidation and tunnel temperature

A. Afshani, W. Li & H. Akagi
Waseda University, Tokyo, Japan

O. Shigeaki
TEPCO Holdings Corporation, Japan

ABSTRACT: Tunnel excavated in low-permeability soil may experience large deformation and increase in lining load if water drains into the tunnel. Long-term dissipation of tunneling-induced excess pore pressure into the tunnel consolidates soil around tunnel with low permeability and imposes extra load on the lining. Meanwhile, seasonal temperature can also expand and shrink joint opening between rings and segments and alters the rate of water leakage into the tunnel. The lining deformation can increase under the incremental increase of consolidation load, but it might also show fluctuating pattern because of the seasonal temperature. In this study, combined effect of long-term consolidation load and seasonal temperature load on the joint opening are discussed. An empirical relationship between water discharge into the tunnel and the amount of joint opening is proposed using the field measurements data of an old tunnel. Then, the joint opening under the influence of both consolidation and temperature load are investigated using a soil-water coupled consolidation model and a thermal stress model. The numerical results show that by employment of these numerical models, the behavior of joints between rings can be simulates very close to those of measured ones.

1 INTRODUCTION

The load on tunnel lining is an important factor for tunnel design. The initial lining load after installation could alter by time. Among the influencing factors, long-term consolidation of low-permeability soil around the tunnel and temperature variation inside of the tunnel is quite important. The dissipation of tunneling-induced excess pore water pressure and change in the water head elevation around the tunnel by leaking of water into the tunnel could consolidate the soil around the tunnel and increase the liner load. The leaking of water into the tunnel happens because the tunnel has introduced new drainages routes to the cohesive soil round tunnel (Mair, 2008, Laver et al, 2016). The leaking starts to flow into the tunnel mainly through the joint opening, deteriorated bolt and grouting holes (Shin et al., 2012). Consolidation load affects the opening amount and crack propagation by time in the tunnel. On the other hand, the daily temperature inside the tunnel is affected by the outside temperatures especially near the entrance and exit ramps and ventilation shafts. The variation of the temperature inside of tunnel affect the joint opening between rings and segments in the lined tunnel and therefore water leakage into tunnel alters which finally leads to the change of liner load. In fact, the liner behavior in these circumstances is under the influence of combined soil-water coupling interaction problem and tunnel temperature. The decision on the amount of water leakage that can infiltrate into the lining, i.e. liner permeability, is not known readily and needs accurate long-term field monitoring of water leakage.

The analytical solution for stresses and displacements in the lined and unlined tunnel due to thermal load in elastic soil were already discussed by several researchers such as Tao et al.

(2016) and Elsworth (2001). Considerably less study has been focused on the long-term response of the joint opening under a load of internal tunnel temperature and consolidation load. In this study, time dependents field measurement data of an old tunnel are used to demonstrate the interaction between the tunnel lining, the soil around the tunnel, and the temperature inside of the tunnel. The data include temperature changes inside of the tunnel, the magnitude of joint opening-closing between rings and segments, and the volume of water leakage into the tunnel. An empirical relationship between water discharge into the tunnel and the amount of joint opening is proposed using the field measurements data. Then, the joint behavior under the influence of both consolidation and temperature load is investigated using a coupled soil-water model and a thermal stress model. Finally, numerical results of the opening-closing of joints between rings are compared with that of measurement ones.

2 MEASUREMENT DATA

2.1 Introduction of a tunnel and types of measurement data

The tunnel used in this study was excavated in 1981 by using an earth pressure type shield machine near to Tokyo metropolitan. The tunnel has an outer diameter of 4 m and length of 2217 m and its path is perpendicular to an above ground railway line situated on top of an embankment. Along the tunnel, five ventilation shafts (VS) were located to facilitate access to the tunnel. In recent years, a large volume of water accumulated in tunnel invert, numerous

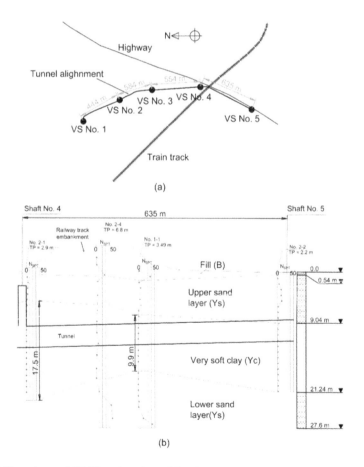

Figure 1.   (a) Plan view and (b) Elevation view of the site.

5255

signs of cracks and wet patches were observed along the tunnel at the internal surface of lining mainly between VS No. 4 and 5. These events understood to be the direct consequence of the increase in consolidation load on the tunnel. This study focuses on the part of the tunnel between VS No. 4 and 5. Figure 1(a) shows the plan view of the tunnel, railway line and locations of the five shafts, and Figure 1(b) displays longitudinal section of tunnel between VS No. 4 and 5.

In Figure 1(b), name, depth from ground level and SPT values of the soil layers are also shown. The tunnel was built in a very soft alluvial clay known as Yurakucho clay (Yc) with a range of thickness from 10 to 17.5 m and is sandwiched between two sandy layers (Ys). The tunnel is supported by a lining made of reinforced concrete with the length of 0.9 m, the thickness of 0.25 m, and outer and inner diameters of 4 m and 3.5 m.

In the tunnel under this study and between the VS No. 4 and 5, the temperature changes inside of the tunnel, the magnitude of the ring to ring and segment to segment joint openings, and monthly water leakage during one year were measured. The details of the data are explained in the following parts.

2.2 Temperature changes inside of the tunnel

Figure 2 shows the temperature variation in 2002 at three different locations inside of the tunnel. It is seen that from May to September, temperature increases and then it decreases from September to January. From January to the end of April, it is almost constant and temperature in these months do not contribute to the changes of the joint opening-closing. The maximum change in temperature in one year is less than 15 °C.

2.3 Joint opening-closing

Thermal changes inside of tunnel alter segment to segment and ring to ring joint opening. Figure 3 shows measured ring to ring joint opening of two gauges (DC2-6 and DC2-7) located between rings 317 and 318 for one year. Based on the figure, maximum changes in the opening of gauges are 114 and 237 μm. Figure 4 shows similar data of segment to segment joint opening named as DC2-1. The measured data show that for the first 5 months of the year, the opening of the joints does not vary by time, however, by rising and falling of the temperature, the joints between neighbor rings close and open respectively.

2.4 Water leakage

The joints between rings and segments of tunnel lining and the seal materials are possible seepage routes for the water leakage and most of the strain generally develop at the joints (Shin

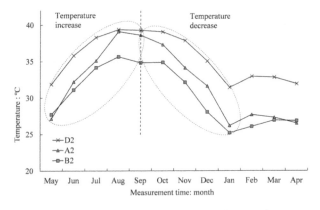

Figure 2. The pattern of annual temperature variation inside of the tunnel.

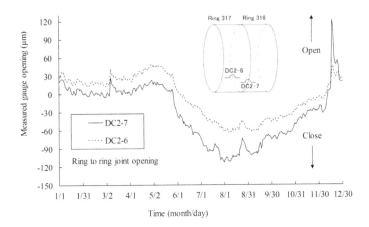

Figure 3. The measured ring to ring joint opening in one year.

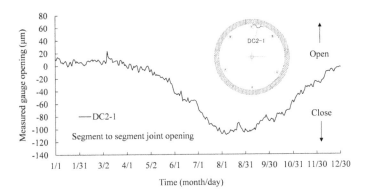

Figure 4. The measured segment to segment joint opening in one year.

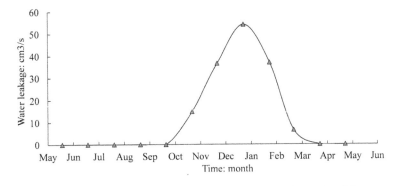

Figure 5. Monthly water leakage values collected between vertical shaft No. 4 and 5.

et al., 2012). By monitoring of the measured data, it is confirmed that the leakage mainly happens through joints, and bolt holes. If the increase in lining load was only because of consolidation of low-permeable soil, depending on the magnitude of the opening and soil type, after several years, drainage into tunnel reaches to a steady state condition. However, temperature indirectly influences the behavior of soil consolidation and tunnel liner load. Figure 5, shows

5257

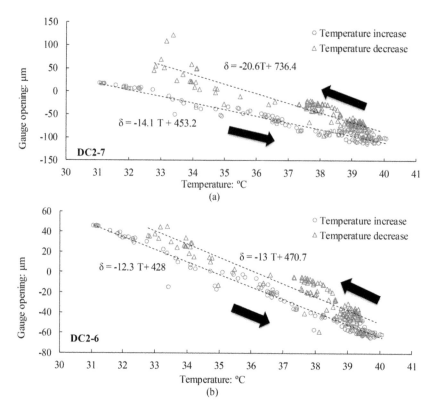

Figure 6. Relationship of gauge opening by temperature in measurement device (a) DC2-7, and (b) DC2-6.

the total monthly collected water in the tunnel between the VS No. 4 and 5. The amount of the maximum water leakage happens at the mid of winter between December and February.

3   CORRELATIONS BETWEEN FIELD MEASUREMENTS

The amount of joint opening under the influence of both consolidation and tunnel temperature loads can be an indication for water leakage amount. The measurement data shows that almost a linear relationship can be drawn between temperature and joint opening for the months with temperature increase and decrease pattern. By investigating the field measurement data of the tunnel under this study, Oka et al., (2017) demonstrated a relationship between joint gauge opening by temperature as shown in Figure 6. This relationship considers the months of the year with the noticeable change in temperature. The months with constant temperature do not contribute to the opening of the joint and therefore excluded in this equation. The monthly discharged water volumes shown in Figure 5 represents the total collected water for the distance between VS No. 4 and 5. It is practically very difficult to measure the water leakage from a specific opening by time. Figure 7 shows a relationship between water discharge, tunnel temperature, and joint opening. The data of joint opening by temperature is recorded from DC2-6 and DC2-7 devices. Equations (1) and (2) show a relationship between water discharge (Q) and tunnel temperature (T) for the months of the year with the increase and decrease temperature patterns.

$$Q = 0 \qquad \text{For the month with temperature increase} \qquad (1)$$

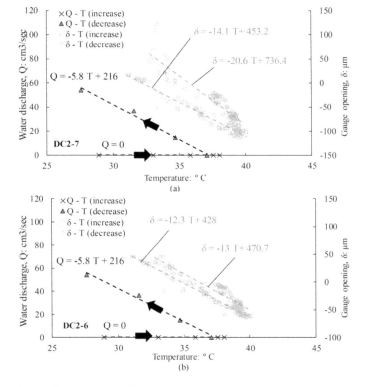

Figure 7. Relationship between water discharge, tunnel temperature and joint opening.

$$Q = -5.8T + 216 \qquad \text{For the month with temperature decrease} \qquad (2)$$

Using these equations and already presented joint opening-temperature relationship shown in Figure 6 for both of DC2-6 and DC2-7 gauges, Equations (3) and (4) are proposed for water discharge (Q) and joint openings (δ) for the section of the tunnel case under this study.

$$Q = 0 \qquad \text{For the month with temperature increase} \qquad (3)$$

$$Q = 0.363\delta + 7.32 \qquad \text{For the month with temperature decrease} \qquad (4)$$

The empirical relationship between water discharge into the tunnel and the joint opening was obtained for the long-term behavior of the tunnel in low-permeability soil under the influence of both consolidation-induced load increase on the liner and the internal temperature of the tunnel.

4 NUMERICAL MODELING OF CONSOLIDATION AND TUNNEL TEMPERATURE

4.1 Introduction of numerical models

The tunnel installation and aftermath tunneling induced long-term soil consolidation along with thermal analyses are modeled numerically. In this part, the details of the numerical calculation are explained. The consolidation analyses were performed using a 3D coupled soil-water and thermal calculations were done using thermal stress finite element analyses. All the

Table 1. Type and steps of numerical modeling based on the construction sequences.

No.	Construction sequences (year)	Type of analyses
1	Tunnel construction (1981)	Soil-water coupling analyses
2	Tunneling induced consolidation (1981–2001)	Soil-water coupling analyses
3	Thermal analyses (2001–2002)	Thermal stress analyses

calculations were carried out using Midas-GTS program (MIDAS IT Japan Co., Ltd., 2018). Numerical modeling was carried out in chronological order of actual construction as listed in Table 1. The coupled soil-water model does not simulate step by step excavation of the tunnel and focuses on the long-term behavior of the soil and tunnel. The tunnel rings were modeled in one step by deactivating soil elements and activating solid type lining elements. The longitudinal soil profile is shown in Figure 1(b) is used in the coupled soil-water model. The modified Cam-clay model was adopted to model Yurakucho clay layer and elastoplastic model with Mohr-coulomb failure criteria are used to model the rest of soil layers; the model parameters of all soil layers are listed in Table 2. Parameters of soil models were obtained using laboratory tests.

The length and thickness of the lining are taken to be 0.9 m and 0.25 m respectively. The lining was modeled as a linear elastic and its properties are also given in Table 2. Figure 8 (a) shows the numerical mesh for coupling soil-water analyses, and Figure 8 (b) shows the model for thermal stress analyses. The thermal and coupled soil-water analyses are performed separately according to the steps shown in Table 1.

Table 2. Parameters of soil layers and tunnel lining used in numerical analyses.

Soil parameters	Symbol	Units	B Backfill	Ys Sand	Yc Soft clay	Ys Sand
Thickness		m	1.9	5.5	12.3	4.6
Unit weight	γ	kN/m^3	17.5	17.5	16.4	17.5
Young Modulus	E	kPa	9.3	4	-	4
Cohesion	c$'$	MPa	-	-	-	-
Internal friction angle	φ'	Degree	-	-	-	-
Poisson ratio	ν	-	0.33	0.33	0.317	0.33
Void ratio	e	-	1.105	1.015	1.353	1.105
Permeability coefficient	k	m/sec	4.11×10^{-9}	4.24×10^{-5}	4.11×10^{-9}	5.9×10^{-6}
Slope of the isotropic normal compression line	λ	-	-	-	0.233	-
Slope of the unloading reloading line	κ	-	-	-	0.018	-
Slope of the critical stateline in p: q plane	M	-	-	-	1.41	-

Lining parameters	Symbol	Units	Value Concrete lining	Steel bolt	Joint seal
Unit weight	γ	kN/m^3	28	78.5	11.4
Young Modulus	E	kPa	$3.2 \times 10^{+7}$	$2.1\times 10^{+8}$	$7.0 \times 10^{+3}$
Poisson ratio	ν	-	0.17	0.3	0.35
Thickness	t	m	0.25	-	0.0064
Length	d	m	0.9	-	-
Outer/Inner diameters	D_0/D_i	m/m	4.0/3.5	0.02	-
Thermal coefficient	α		1.5×10^{-5}	-	-
Initial temperature	$'$	Degree	27.5	-	-

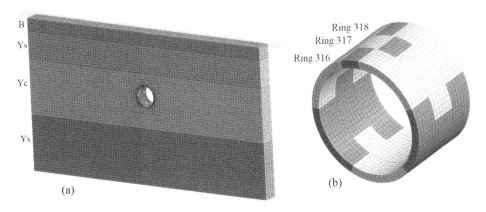

Figure 8.   (a) Soil-water coupling model, (b) thermal stress model.

4.2   Coupled soil-water numerical model

Wongsaroj, et al. (2006, 2013) discussed that the stress and pore pressure profiles in the region near the monitoring plane are approximately constant in the longitudinal direction and simulations of the long-term consolidation can be adequately modeled in two dimensions. According to this hypothesis, the model dimension in the longitudinal direction was limited only to a few rings with measurement data. As shown in Figure 8 (a), 3D consolidation model in a longitudinal direction includes the section related to the ring numbers 316, 317, and 318 between vertical shaft No. 4 and 5. The long-term consolidation modeling was started since tunnel construction in 1981. The tunnel lining was set to be a drainage boundary and was modeled by the 3D solid element having permeable characteristics. The drainage condition of the tunnel boundary is controlled by the relative magnitude of the tunnel lining permeability and the surrounding soil permeability. This permeability model originally was proposed by Wongsaroj, et al. (2006) and it is employed in this study too. By changing the permeability coefficient of lining, impermeable lining with no water leakage, fully permeable lining with a large amount of leakage, or partially permeable lining with limited leakage can be simulated. The pore water pressure inside of the line is set to zero at atmospheric pressure was considered inside the tunnel. The pore water pressure outside of lining is unknown and is decided by consolidation calculation. The difference in pore water pressure at both side of the lining triggers the soil consolidation outside of the liner. The initial permeability of the Yurakucho soft clay layer is set to be constant as listed in Table 2. The permeability of tunnel lining is an important factor in this calculations and cannot be known readily. The liner initial permeability is assumed to be $k_{lining} = 1 \times 10^{-9}$ m/sec in 1981 and then was changed gradually up to the end of consolidation analyses. The initial value and changes of liner permeability were decided by matching the tunnel invert displacement at the end of consolidation analyses with the field measurement results. Based on this comparison results, the final value of lining permeability is found to be 3 times of the initial ones (i.e. $k_{lining} = 3 \times 10^{-9}$ m/sec in 2001).

4.3   Thermal stress numerical model

After performing coupled soil-water analyses from 1981 to 2001, the imposed loads on the tunnel liner is read and transferred to the thermal stress model as shown in Figure 8 (b). The thermal stress analysis is run for one year started from January 2002, and it is assumed that during one year, the change in loads caused by consolidation is negligible and steady-state water drainage has been formed. This is a true assumption as most of the changes in liner load induced by the consolidation of low-permeable soil happen during several first years in which water drainage into the tunnel is in a transient condition. The initial gap between the rings is set to be 2 mm (JSCE, 2010). Rubber joint and steel bolt between three rings were also

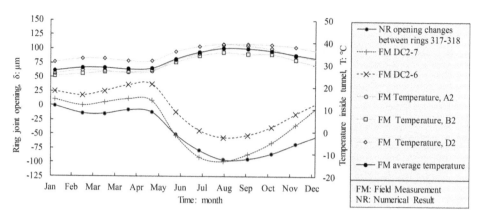

Figure 9. Joint opening modeling results and comparison with field measurements.

modeled and their properties are listed in Table 2. The configuration and thickness of rubber joint and diameter of steel bolt were decided based on the field measurements and their properties are set using proposed values in previous similar studies (Jin et al., 2004, Zhang et al., 2018). The initial temperature of the thermal stress model is set to be same as the average of field measured temperature value inside of a tunnel in January 2002 as shown in Figure 2 and Table 2. Then, the thermal-induced joint opening-closing between ring number 316 and 317 and also ring numbers 317 and 318 was modeled by introducing the average of the measured temperature changes shown in Figure 2.

4.4 Modeling results

Figure 9 shows the comparison of the joint opening between numerical results and field measurement. The temperature data of Figure 2 and their average values, measured values of the joint opening in gauges DC2-6, and DC2-7 are also shown in Figure 9. The average of temperature along with imposed consolidation load from coupled soil-water model is introduced into the thermal stress model, and amount of joint opening between rings 317 and 318 are read and compared with measurement values. The results show that by the employment of the two separate models, the ring to ring joint opening-closing under the influence of both consolidations induced increase in liner load and temperature load can be simulated very similar to that of the measured ones. Field data shows that the opening-closing pattern of the joints is repeated every year as shown in Figure 9. It is understood that the consolidation load has a larger influence on liner behavior from the time of tunnel installation until the moment that water drainage around tunnel reaches to a steady-state condition, but after that, the temperature influence on the liner becomes more important. The numerical analyses performed here is planned to be continued to study the liner load under the influence of combined consolidation and tunnel temperature.

5 CONCLUSION

The combined effect of long-term consolidation load and seasonal temperature load on the opening of the tunnel lining joint were discussed in details by showing the field measurement data and performing of the numerical analyses. An empirical relationship between water discharge into the tunnel and the amount of joint opening was proposed using the field measurements data of an old tunnel. The joint opening under the influence of both consolidation and temperature load were investigated using a coupled soil-water model and a thermal stress model. The findings of this study are as follow:

- During the months of the years with increasing temperature pattern, the joints stay closed and lining becomes almost impermeable.
- The proposed empirical equation between water discharge into the tunnel and the joint opening showed that by increasing of the joint opening in the cold months of the year, the amount of the water discharge into the tunnel increases linearly.
- The behavior of liner joint under the combined influence of the consolidation and temperature load can be investigated by coupled soil-water and thermal stress numerical analyses.
- The consolidation load has a larger influence on liner behavior from the time of tunnel installation until the moment that water drainage around tunnel reaches to a steady-state condition, but after that, the influence of temperature on the liner becomes more important.

REFERENCES

Elsworth, D., 2001. Mechanical response of lined and unlined heated drifts. Rock mechanics and rock engineering, 34(3),pp.201–215.

JSCE, 2010. Design of Segment for Shield Tunnel (Revised version). JSCE.

Laver, R.G., Li, Z. and Soga, K., 2016. Method to evaluate the long-term surface movements by tunneling in London clay. Journal of Geotechnical and Geoenvironmental Engineering, 143(3), p.06016023.

Mair, R.J., 2008. Tunnelling and geotechnics: new horizons. Géotechnique, 58(9),pp.695–736.

MIDAS IT Japan Co., Ltd., 2018. GTS NX User's manual V1.1.

Oka, S., Ito, Y., Yokota, A., Saito, J., Kaneko, S. and Akagi, H., 2017. Long Term Prediction of Vertical Earth Pressure Loading on a Shield Tunnel in Soft Clay Ground. Journal of Japan Society of Civil Engineers, Ser. F1 (Tunnel Engineering), 73(3).

Shin, J.H., Kim, S.H. and Shin, Y.S., 2012. Long-term mechanical and hydraulic interaction and leakage evaluation of segmented tunnels. Soils and Foundations, 52(1),pp.38–48.

Tao, F. and Bobet, A., 2016. Effect of temperature on deep lined circular tunnels in transversely anisotropic elastic rock. Underground Space, 1(2),pp.79–93.

Wongsaroj, J., 2006. Three-dimensional finite element analysis of short and long-term ground response to open-face tunnelling in stiff clay (Doctoral dissertation, University of Cambridge).

Wongsaroj, J., Soga, K. and Mair, R.J., 2013. Tunnelling-induced consolidation settlements in London Clay. Géotechnique, 63(13), p.1103.

Jin, X.L., Guo, Y.Z. and Ding, J.H., 2004. Three dimensional numerical simulation of immersed tunnel seismic response based on elastic-plastic FEM. In Key Engineering Materials (Vol. 274, pp. 661–666). Trans Tech Publications.

Zhang, D.M. and Liu, J., 2018, May. Shearing Behavior of Segmental Joints of Large-Diameter Shield Tunnel. In GeoShanghai International Conference (pp. 351–360). Springer, Singapore.

Tunnels and Underground Cities: Engineering and Innovation meet Archaeology,
Architecture and Art, Volume 11: Urban
Tunnels - Part 1 – Peila, Viggiani & Celestino (Eds)
© 2020 Taylor & Francis Group, London, ISBN 978-0-367-46899-6

Identifying the geological hazards during mechanized tunneling in urban areas – the case of Tehran alluvium conditions

A. Alebouyeh & A.N. Dehghan
Islamic Azad University (Science and Research Branch), Tehran, Iran

K. Goshtasbi
Tarbiat Modares University, Tehran, Iran

ABSTRACT: The current study addresses the various types of challenges and geological hazards during tunnel excavation by mechanized tunneling method (TBM) in Tehran Alluvium conditions. In addition to introduction of tunnel construction projects in the city of Tehran, challenges existing in their implementation along with strategy and solutions to deal with them, which are considered as valuable experiences of the projects, are described in this study. The results of this study showed that among the various hazards that occur during the excavation of a tunnel, eleven hazards are more important than other hazards, such as abrasion of the machine cutting tools, clogging around the machine cutting tools, soil mass displacement in the tunnel face, liquefaction, sand lenses, soil-sand tunnel face, severe water column fluctuations, collapse of the tunnel ceiling and instability in its walls, high soil penetrability, lack of fine-grained materials, and presence of boulders and cobbles. Also, in this study, Tehran's zoning maps for two standard parameters, standard penetration test (SPT) N-value and percentage of fine-grained particles are provided. Some of the most important features of the sediments in Tehran which can be problematic for tunnel excavation are presented by the help of these maps.

1 INTRODUCTION

In recent years, various researchers have conducted a variety of studies on the hazards of excavation in terrestrial lands, but most of these studies are qualitative and all the factors involved in the occurrence of hazards have not been addressed in their studies. Usually high costs and the need for initial investment in mechanized excavation are significantly higher than other excavation methods. Any geological hazard that reduces machine productivity may economically result in the use of TBM. Therefore, it is always necessary to provide a guide model to accurately identify the hazards of excavation. The dissociation and identification of ground along the tunnel route and the prediction of geological, geotechnical and hydrogeological conditions associated with tunneling (traditional or mechanized) are the most important work to be done before designing and constructing a tunnel (Dehghan et al. 2012; Sadeghi et al., 2016). Any unplanned problem from hydrological and geological conditions can increase construction time and cost and create risks for employees. In addition to environmental damage and land constraints, if the condition of the earth, geotechnical hazards and problematic soils are not identified or predicted, its output will often be associated with costs, delays and differences of opinion during the construction of the tunnel. In mechanized tunneling, appropriate geological predictions and identification of geotechnical risks are necessary for the design and use of TBM (Thewes, 2007; Ball et al. 2009). A comprehensive assessment of the ground conditions is required in order to correctly identify the complex risks, which in addition to the underlying public data; the assessment should include all information gathered in connection with happened rare risks, such as the problematic conditions of underlying soils of the area. The prediction of geotechnical conditions and geological hazards along the tunnel is critical for selecting the excavation machine. Some recent research work focused on evaluating problematic soils and predicting geological-geotechnical risks on EPB-based tunneling including cohesive and clogging risk (Thewes and Burger, 2004; Sass and Burbaum, 2009) and soil abrasion (Nilsen et al. 2007; Tarigh Azali and Moammeri, 2012). Also, in traditional tunneling, accurate prediction of geological conditions can prevent problems that stop excavation. For example, dealing with boulders in soft grounds is one of the most important problems in traditional excavation, if it is not identified, can cause great problems.

Therefore, considering the ongoing urban tunnel projects in Iran, especially metro tunnels, the study of soil hazards, both at the stage of study and at the stage of implementation of the tunnel, is very important for designing and ordering the TBM and for calculating the contract fees of the contractors. The studied projects in this research are related to some of the excavated tunnels in Tehran. These projects, designed and constructed in recent years with the latest technologies in Iran, include lines 3 (Northeast-Southwest), 6 (Northwest-Southeast), and 7 (North-South and East-West) from Tehran metro lines. The total length of these projects is over 100 kilometers, and many of them are investigated for understanding the hazards of excavation.

2 CASE STUDY

Tehran is located on Quaternary alluvium called Tehran Alluvium. Rieben (1966) divided this alluvium into four formations based on geological characterizations, identified as A, B, C and D formations, where A is the oldest and D the youngest (Figure 1). In this study, the geological and geotechnical characteristics of the three projects implemented in Tehran have been investigated to identify the ground hazards during mechanized tunneling in urban areas (Figure 2). The following three projects will be presented briefly.

2.1 Tehran subway line 7

The subway line 7 is divided into two parts: North-South and East-West. The East-West part is to be excavated and segmented with an EPB machine. The North-South section of Tehran metro line 7 is about 12 km in length and is to be excavated and segmented by an EPB machine. To study subsurface conditions along the course of this tunnel, more than 70 observation wells and boreholes are excavated with different depths. The minimum depth of these observation wells and boreholes is 10 meters and the maximum is 50 meters. According to the results obtained from the field and laboratory studies and also considering the scale of the map, the soil layers in the tunnel route are divided into six units of engineering geology (ET-1

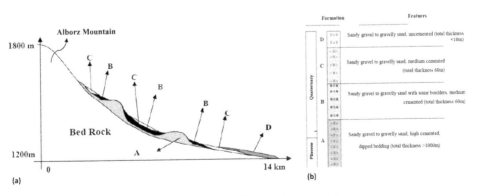

Figure 1.   Sequence of Tehran alluvia formation and its stratigraphic features.

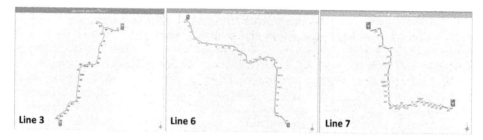

Figure 2.   General plan of the tunnels route of Tehran subway- lines 3, 6 and 7.

to ET-6), which are presented in Table 1. The physical and mechanical properties of different units of engineering geology are estimated by statistical analysis obtained from laboratory tests and field studies (boreholes log and field experiments). A summary of the statistical parameters for each unit is presented in Table 2.

2.2   Tehran subway line 6

Tehran subway line 6 is divided into three South, Middle and North parts. The southern part of Tehran's line six is more than 16 kilometers. The length of the two parts of the middle and north parts of the line 6 of Tehran are about than 20 kilometers. The northern part of this line has been excavated traditionally. To study subsurface conditions along the course of this tunnel, more than 80 boreholes and observation wells with different depths were excavated. The minimum depth of these boreholes and observation wells is 10 meters and a maximum of 45 meters. According to the results obtained from the field and laboratory studies and also considering the scale of the map, the soil layers in the tunnel route are divided into four units of engineering geology (ET-1, ET-2, ET-3, ET-4), which are presented in Table 3. The physical and mechanical properties of the units of engineering geology are summarized in Table 4.

2.3   Tehran subway line 3

Tehran subway line 3 the longest subway line in the Middle East is 37 kilometers long. In order to study subsurface conditions along the course of this tunnel, more than 90 boreholes and observation wells with different depths have been dug. The minimum depth of

Table 1.   Specifications of separated engineering geology units in the tunnel route of line 7.

Engineering geology unit	ET-1	ET-2	ET-3	ET-4	ET-5	ET-6
Soil description	Sandy gravel & gravely sand	Very gravely sand with silt & clay	Very silty clayey sand with gravel, very sandy clay (or silt) with gravel	Clayey silty sand with gravel	Clayey silt & silty clay with sand, very sandy clay (or silt)	Very soft clayey sandy silt
Percentage passing through sieve No. 200 sieve No. 200	%50<	%60<	34-22%	60-30%	30-12%	12-3%
Soil type (USCS)	ML	CL, ML & CL-ML (rarely CH)	SC, SM	SC, SM & CL	SC, SC-SM & GC	GW, GW-GM, GP-GC, SW & SP

Table 2. Average values of geotechnical parameters for Tehran subway line 7.

Engineering geology unit	C' (kg/cm^2)	Φ' (degree)	C_{cu} (kg/cm^2)	$Φ_{cu}$ (degree)	E (kg/cm^2)	υ	unit weight Dry (g/cm^3)
ET-1	0.14	34	0.16	29	800	0.30	1.86
ET-2	0.15	33	0.18	29	850	0.30	1.84
ET-3	0.30	33	0.40	27	500	0.32	1.90
ET-4	0.22	32	0.28	26	500	0.30	1.82
ET-5	0.34	28	0.43	19	350	0.35	1.70
ET-6	0.0	27	0.0	25	100	0.35	1.70

Table 3. Specifications of separated engineering geology units in the tunnel route of line 6.

Engineering geology unit	ET-1	ET-2	ET-3	ET-4
Soil description	Sandy gravel & gravely sand	Very gravely sand with silt & clay	Very silty clayey sand with gravel, very sandy clay (or silt) with gravel	Clayey silt & silty clay with sand, very sandy clay (or silt)
Percentage passing through sieve No. 200 sieve No. 200	3-12 %	12-30 %	30-60 %	>60 %
Soil type (USCS)	GW, GW-GM, GP-GC, SW & SP	SC, SC-SM & GC	SC, SM & CL	CL, ML & CL-ML (rarely CH)

Table 4. Average values of geotechnical parameters for Tehran subway line 6.

Engineering Geology Unit	C' (kg/cm^2)	Φ' (degree)	C_{cu} (kg/cm^2)	$Φ_{cu}$ (degree)	E (kg/cm^2)	υ	unit weight Dry (g/cm^3)
ET-1	0.14	34	0.16	29	800	0.30	1.86
ET-2	0.15	33	0.18	29	750	0.30	1.84
ET-3	0.30	33	0.40	23	500	0.32	1.90
ET-4	0.31	28	0.43	19	350	0.35	1.70

Table 5. Average values of geotechnical parameters for Tehran subway line 3.

Engineering geology unit	C' (kg/cm^2)	Φ' (degree)	C_{cu} (kg/cm^2)	$Φ_{cu}$ (degree)	E (kg/cm^2)	υ	unit weight Dry (g/cm^3)
ET1-coarse- grained soils (gravel)	0.14-0.20	30-36	0.13-0.15	27-32	700-950	0.30-0.32	1.80-1.90
ET2 - coarse- grained soils (coarse sand)	0.15-0.25	32-35	0.18-0.21	28-33	700-900	0.30-0.33	18.10-1.85
ET3 -medium- grained soils (medium sand and silt)	0.30-0.35	30-35	0.38-0.45	20-26	450-600	0.32-0.35	1.85-1.95
ET4 -fine- grained soils (clay)	0.31-0.38	25-30	0.40-0.45	17-23	300-450	0.35-0.38	1.65-1.75

these boreholes and observation wells is 15 meters and the maximum is 55 meters. According to the results obtained from the field and laboratory studies, and also considering the scale of the scale, the soil layers in the tunnel route are classified into four units of engineering geology (Table 5). A summary of the physical and mechanical properties of the units of engineering geology that are estimated by statistical analysis is presented in Table 5.

3 GEOLOGICAL HAZARTS DURING EXCAVATION

3.1 Evaluation of geological conditions

Ground classification based on the geological properties is among the most important geological activities for detecting excavation hazards in a tunneling project. Various methods have been proposed for classification of the ground engaged in mechanized excavation. In this regard, Hashem Nezhad and Hassan Pour (2017) proposed a method, which is based on the geotechnical data obtained from borehole logs, in situ tests, and other laboratory tests such as grain size distribution analysis and Atterberg limits. The raw data needed for this purpose are analyzed using the common statistical methods and the results are presented as some charts and profiles for each tunnel.

In the present study, soil particle size distribution was analyzed in some boreholes and wells through the sieve and hydrometry analyses. The sieve analysis was carried out in accordance with the unified soil classification system (USCS). The soil samples collected for this purpose are representative of the actual soil conditions along the tunnel route. Figure 3 presents the grain size distribution analysis of the soil particles along the tunnel routes.

Another parameter important in the classification of soft grounds is the plasticity index (PI), which may have different effects on soils with similar grading. The PI values of the soil samples collected from different projects studied in this research are graphically presented in Figure 4. As can be seen from Figure 4, the majority of these soil samples have a medium PI of 7 to 17%.

Figure 5 presents the distribution of the consistency index (CI or I_C) for the soils collected from the studied tunnel routes. As can be seen, the soils in these tunnels have the "non-plastic" to "hard to very hard" conditions.

3.2 Hazard detection and prediction

Mechanized tunneling hazard is referred to any mutual interaction of the ground and tunnel boring machine (TBM) that can potential decline excavation efficiency in both economic and time aspects. This mutual interaction is indeed the ground reaction or behavior against removing a part of ground materials that is dependent on engineering geology properties of the

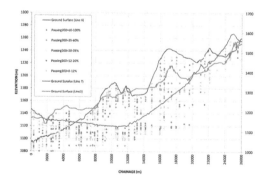

Figure 3.   Grain size distribution along the tunnel route.

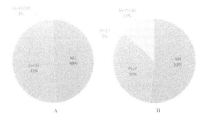

Figure 4. Distribution of liquid limit (LL) and plasticity index (PI) in the soils collected from the tunnels' routes.

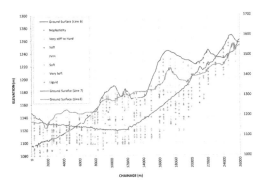

Figure 5. Distribution of consistency index values along the studied tunnel routes.

tunnel route materials. In the present study, based on the available results and some common empirical relations, we investigate the occurrence probability of various hazards and poor geological conditions along the route of these tunnel projects (lines 3, 6, and 7 of Tehran Metro). The most important hazards detected in the alluvial grounds include stickiness of soil and TBM clogging, soil mass displacement in the tunnel face, liquefaction, sand lenses, soil-sand tunnel face, severe water column fluctuations, collapse of the tunnel ceiling and instability in its walls, high soil penetrability, soil abrasion, lack of fine-grained materials, and presence of boulders and cobbles.

Tables 6 and 7 summarize the excavation hazards in soil grounds around the City of Tehran. As shown in Table 6, the detected geological hazards are dependent on various geotechnical parameters, with soil grain size distribution being among the most effective ones. Also, Table 7 presents the geological hazards detected in three tunneling projects in Tehran Metro (lines 3, 6, and 7) in various alluvial formations of Tehran. Figure 6 shows some photos of the hazards observed along the mechanized excavation route of the Metro tunnels in Tehran alluviums.

In soil environments, excavation hazards can be divided into three categories: geomechanical, geo-hydraulical, and hydro-mechanical. Soil gradation is an important parameter in the occurrence of these hazards such that it almost plays a pivotal role in this regard. For instance, clogging and friction behaviors are observed only in fine-grained sediments (deposits of Formations C and D). The results show that Regions 10 (Tehran Bazar area) to 17 (Javadieh and Yaft Abad), which have both fine-grained sediments and a high groundwater level, provide favoring conditions for the occurrence of this phenomenon. Hazard such as abrasion, lack of fine-grained materials, the presence of coarse-grained particles, collapses, wall and ceiling instabilities, permeability, severe groundwater fluctuations, and liquefaction are typically seen in coarse-grained sediments (deposits of Formations and B and, to a less extent, Formation C). The accumulation of

Table 6. Geotechnical and engineering geology parameters effective on excavation hazards in soil environments.

Geological hazards	The effective geotechnical or engineering geology parameters					
	Grains petrography	Soil particle size distribution	Percentage and type of clay minerals	Deformability properties	Strength properties	Moisture content
Clogging and stuck of excavation tool	✓	✓	✓			✓
Rock mass movements in the tunneling face		✓		✓	✓	✓
Liquefaction		✓				
Sand lenses		✓				✓
Mixed clay/sand face		✓				
Severe water column fluctuations		✓				
The collapse of the tunnel ceiling and instability of its walls		✓		✓	✓	
High soil permeability		✓				
Soil abrasion	✓	✓	✓			✓
Lack of fine-grained materials		✓				
Presence of boulders and rock blocks		✓				

coarse-grained sediments, high groundwater level, and other favoring conditions for the occurrence of this phenomenon in Region 1 (Tajrish Square area) to 6 (Enqelab) make them liable to geohazards. Furthermore, some hazards are induced due to the simultaneous presence of fine-grained and coarse-grained particles, providing that other conditions also exist.

To predict the hazards detected in Tehran alluvium, zoning maps of the city (Figures 7 to 10) were prepared using two information layers for depths 0-10 m and 10-20 m: 1) standard penetration test (SPT) number (N value) and 2) percentage of fine-grained particles (those passing through sieve #200).

Zoning map of Tehran based on the fine-grained conditions for depths 0-10 m and 10-20 m are presented in Figures 7 and 8. As can be seen from these figures, the coarse- and fine-grained soils are dominant in the northern and southern parts of Tehran. Figures 9 and 10 illustrate SPT-based zoning map of Tehran for depths 0-10 m and 10-20 m, respectively. The SPT numbers (N-values) of the coarse-grained soils from the northern part of the city are much greater than those of the fine-grained soil in the southern part of the city.

As shown in Figures 7 and 8 (zonation based on fine-grained content) and Figures 9 and 10 (zonation based on N-values), a weak separation between coarse- and fine-grained soils is seen at depths 0-10 m in the northern and southern parts of the city. One of the main reasons for such an inaccurate boundary may be the presence of fill soils recently produced by various civil engineering projects.

Accordingly, the obtained zonation maps show that some hazards during the excavation specific to coarse-grained soils (e.g., abrasion, encountering boulders and large particles, severe water level fluctuations, and absence of fine-grained soils) are more probable in the northern parts of the city (Regions 1 to 9) and Regions 21 and 22 and some others (e.g., clogging) have a higher probability in the southern parts of the city.

Table 7. Geological hazards and soils involved in the projects related to the Tehran Metro.

Project	Machine type	Geological hazards and involved soils										
		Clogging and stuck of the excavation tool	Soil mass movements in the tunneling face	Liquefaction	Sand lenses	Mixed clay-sand tunneling face	Severe water column fluctuations	The collapse of the tunnel ceiling and instability of its walls	High soil permeability	Soil abrasion	Lack of fine-grained materials	Presence of boulders and rock blocks
Metro Line 7	EPB	D	A,B	A,C	B,C	B,C	A	A	A	A,B	A	A
Metro Line 6	EPB	D,C	A,C	A,B	A,C	B,C	A,B	A	A,B	A,B	A,B	A,B
Metro Line 3	EPB	D,C	A,B	C	A,B	B,C	A	A,B	A	A,B,C	A	A,C

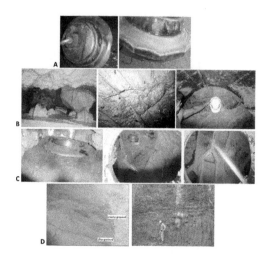

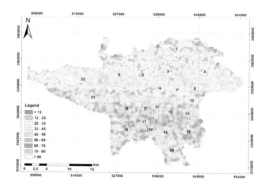

Figure 6. Images taken from some hazards involved in tunnel excavation in the Tehran alluvium: (a) abrasion of the machine cutting tools, (b) boulders and rock pieces in the tunnel face, (c) clogging around the cutting tools, and (d) mixed face along the tunnel excavation route.

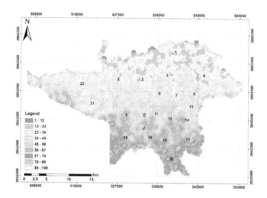

Figure 7. Zoning based on fine-grained content (passing through sieve #200) of the soils for depths 0 to 10 m.

Figure 8. Zoning based on fine-grained content (passing through sieve #200) of the soils for depths 10 to 20 m.

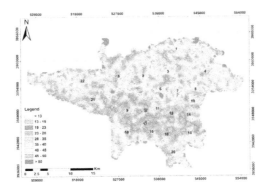

Figure 9.   Zoning based on SPT number (N-value) of the soils for depths 0 to 10 m.

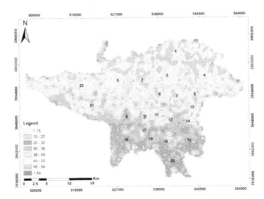

Figure 10.   Zoning based on SPT number (N-value) of the soils for depths 10 to 20 m.

4   CONCLUSIONS

Hazard during tunnel excavation are among the major issues emerged when constructing an urban tunnel. These hazards can highly affect the excavation efficiency and thus the total costs during a tunneling project. The main results of the present study can be outlined as follows:

1. Mechanized excavation hazards in soft soils are classified as two main groups. The first group consists of those hazards that are induced due to the intrinsic properties of a given soil. These hazards can be investigated in terms of particle size and hardness level of the soils. Hazards such as abrasion and clogging are among the typical hazards of this category. The second group consists of those occurring due to the simultaneous presence of different types of soils and the existence of other favoring conditions; e.g., mixed tunneling face and sand lenses.

2. Presence of cohesive soils and clogging, mass displacement in the tunneling face, liquefaction, sand lenses, mixed clay-sand tunneling face, severe water column fluctuations, collapse and instability of tunnel ceiling, high soil permeability, soil abrasion, lack of fine-grained materials (passing through sieve #200), and presence of bounder and rock blocks are among the major hazards encountered during the tunnel excavation in the Tehran alluvium.

3. The most important facto effective on grains' abrasion is the Mohs hardness of the grains. Generally, the soils with a higher Mohs hardness show high abrasion.

4. The most important parameters affecting the clogging phenomenon are grain size and hardness of the soils. Typically, soils with higher hardness are more liable to clogging.
5. Encountering mixed zones, the presence of sand lenses, and presence of boulders and large rock particles in the tunneling face are more probable in areas with a high particle size distribution index.
6. High soil permeability and groundwater level fluctuations are only seen in coarse-grained soil units.
7. Based on the prepared zoning maps of the Tehran alluvium, the highest clogging and tool abrasion hazards are seen in the southern and northern parts of the city of Tehran.

REFERENCES

Ball, R.P., Young, D.J., Isaacson, J., Champa, J. & Gause, C., 2009. Research in soil conditioning for EPB tunneling through difficult soils. In Rapid Excavation and Tunneling Conference (RETC), Las Vegas, USA (pp. 320–333).

Dehghan, A.N., Shafiee, S.M. & Rezaei, F., 2012. 3-D stability analysis and design of the primary support of Karaj metro tunnel: Based on convergence data and back analysis algorithm. Engineering geology, 141, pp.141–149.

Hashemnejad, A., & Hassanpour, J., 2017. Proposed soil classification based on the experiences of soft-ground tunneling in Iran. Bulletin of Engineering Geology and the Environment, 76(2):731–750.

Nilsen, B., Dahl, F., Holzhäuser, J. & Raleigh, P., 2007, June. New test methodology for estimating the abrasiveness of soils for TBM tunneling. In Proceedings of the rapid excavation and tunneling conference (RETC) (pp. 104–116).

Rieben, E.H., 1966. Geological Observation on Alluvium Deposits in Northern Iran. Geology Survey of Iran, Report No. 9.

Sass, I. & Burbaum, U., 2009. A method for assessing adhesion of clays to tunneling machines. Bulletin of engineering geology and the environment, 68(1), pp. 27–34.

Sadeghi, M., Pourhashemi, S.M., Dehghan, A.N., & Ahangari, K. 2016. The Effect of Excavation Progress on the Behavior of Hakim Highway Tunnel Using Geotechnical Instrumentation. ITA-AITES World Tunnel Congress 2016 (WTC 2016), 22–28 April 2016, San Francisco, California, USA.

Tarigh Azali, S. & Moammeri, H., 2012. EPB-TBM tunneling in abrasive ground, Esfahan Metro Line 1. In ITA-AITES world tunnel congress (WTC), Bangkok, Thailand.

Thewes, M. & Burger, W., 2004. Clogging risks for TBM drives in clay. Tunnels & Tunnelling International, 36(6).

Thewes, M., 2007. Tbm tunnelling challenges-redefining the state-of-the-art-keynote lecture at the 2007 world tunnel congress in prague. Tunnel, 16, pp.13–21.

Tunnels and Underground Cities: Engineering and Innovation meet Archaeology,
Architecture and Art, Volume 11: Urban
Tunnels - Part 1 – Peila, Viggiani & Celestino (Eds)
© 2020 Taylor & Francis Group, London, ISBN 978-0-367-46899-6

Studying the effects of gypsum karst cavities in tunneling

M.R. Al Kaabi & R.L. Sousa
Khalifa University of Science and Technology, Abu Dhabi, United Arab Emirates

ABSTRACT: Abu Dhabi is expecting a huge growth over the next 20 years and is planning a massive integrated transportation system, including several modes of transportation to ensure that Abu Dhabi becomes a sustainable city on a global scale. Most of these structures are expected to encounter difficult ground conditions, which are a concern for construction both at the surface and underground. This study focuses on a ground hazard common in Abu Dhabi, gypsum dissolution and karst formation, and its impacts on subsurface infrastructure development. Numerical simulations to study the effect of gypsum cavities on tunneling are conducted using the finite element analyses with the PLAXIS 2D software and the geologic and construction data from the construction of the Strategic Tunneling Enhancement Programme (STEP) tunnels, in Abu Dhabi, UAE. Results of the analyses are presented graphically in charts that show the tunnel lining bending moment as functions of the tunnel geometry, cavity characteristics, construction parameters and properties of the ground. These graphs allow one to identify the most critical scenarios for tunneling through ground where cavities are present.

1 INTRODUCTION

Abu Dhabi is planning a massive integrated transportation system, which includes several modes of transportation such as bus, rail and metro, to ensure that it can accommodate a growing population. Most of these structures are expected to encounter difficult ground conditions, which are a concern for construction both at the surface and underground. In this paper we focus on a geologic hazard, gypsum dissolution and karst formation, and its impacts to infrastructure development. We study the effect of gypsum cavities on tunneling, using geologic and construction data from the construction of the Strategic tunneling Enhancement programme (STEP) tunnels, in Abu Dhabi, which is the largest pipeline infrastructure project in the entire GCC region, with around 41 kilometers of deep sewer tunnels. To investigate this problem, finite element analyses using the software PLAXIS 2D are conducted, where the size of the cavity, of the tunnel, the depth and distance between the two openings is systematically varied.

2 GYPSUM KARST HAZARD IN ABU DHABI

Karsts are topographies formed through the dissolution of the bedrock caused by surface or groundwater. Sinkholes, caves, large springs, dry valleys and sinking streams are typical features of karst landscapes. These features cannot be easily detected at the surface, but they generally occur within certain bedrock-geologic formations such as Carbonate rocks (limestone, marble, dolomite) and evaporate rocks (gypsum and rock salts) (Veni, 2001; Williams, 2008). Gypsum is one of the most soluble common rocks, for instance when gypsum is subjected to water flow, it dissolves 100 times faster than limestone. Even though the karst features that form within gypsum are typically the same as the ones that form within limestone or dolomite,

Figure 1. Cavities in Abu Dhabi, UAE. On the left: cavity/dissolution features in carbonate rock encountered during STEP project, Abu Dhabi, and on the right: cavity in weathered gypsiferous layer during site excavation, Masdar city, Abu Dhabi (adapted from MacDonald, 2010 and private communications ADSSC authorities).

Figure 2. Homes in Khalifa City where the ground has swelled under the pavement and cracked the foundation (adapted from Al Khan, 2008; The National, 2009).

the main difference is that those features form more rapidly within gypsum, i.e. within a matter of weeks or years. This higher solubility of gypsum, as well as, the fact that voids within gypsum layers could occur at any depth, poses a threat to infrastructure development. Voids within gypsum formation are openings where groundwater can be stored, and if connected provide pathways for groundwater flow. In Abu Dhabi, the existence of the presence of gypsum that occurs within the tertiary bedrock, as persistent quasi-horizontal bands, at different depths, poses a threat to construction. In fact, the dissolution of gypsum is also a cause for cavities that can be found within this formation in greater Abu Dhabi (Figure 1 and 2).

3 STEP PROJECT

STEP project was developed by the Abu Dhabi Sewerage Services Company (ADSSC) to improve the sewerage systems in Abu Dhabi, United Arab Emirates. The project and its major components are described in the next sections. Focus will be on the geotechnical issues face during construction, particularly occurrence of karst voids. Finally, the STEP project is used, in subsequent sections, as a case study to investigate the effect of karst cavities on tunneling.

3.1 Project Description

The Abu Dhabi Sewerage Services Company (ADSSC) has developed a comprehensive Capital Investment Plan (CIP). The key aspects of the plan are two major investment programs that cover important strategic and conceptual aspects, which support the Urban Planning Council's (UPC) goal to manage Abu Dhabi's growth in a sustainable manner and

Figure 3. Layout of STEP. Adapted from (communications with Eng. Al Nuaimi from ADSSC, 2017).

accommodate the projected sewage flow in the long-term future. The first is the tactical investment plan (TIP), a program that focuses on infrastructure rehabilitation and construction projects. The second major capital investment is the Strategic Tunnel Enhancement Programme (STEP), which has been launched by ADSSC to address the aging sewerage infrastructure and rapid population growth (CH2MHILL, 2014)

The layout of STEP as shown in Figure 3, consists of: (ZUBLIN, 2013)

- Deep sewer tunnel that extends from Abu Dhabi Island (ADI) to the Al Wathba Independent Sewage Treatment Plant (ISTP).
- One large pumping station adjacent to Al Wathba ISTP at the end of the deep tunnel sewer.
- Linking the existing sewerage system with the new deep sewer tunnels through a series of link sewers. This will allow for the removing of several of the existing pumping stations.

The deep sewer tunnel is 41 km long running from the northern part of Abu Dhabi Island to the Mainland, drilled by eight Tunneling Boring Machine (TBM). The entire deep tunnel is divided into 3 contracts: T-01, T-02 and T-03, as shown in Figure 3.

Link sewers, totaling 50 km long with varying diameters (max. of 3.1 m), were constructed under two contracts, one covers the Abu Dhabi Island (LS-01) and the other (LS-02) covers Abu Dhabi mainland (Residential and Industrial Area) as shown in Figure 4. These link sewers are intended to stop the flow into the existing pumping stations and divert it to the deep tunnel using gravity. Contract LS-01, covers about 35.7 km of the whole link sewer system with diameters ranging from 200 mm to 2800 mm. while, Contract LS-02 will consists of 15.4 km sewerage network with diameters varying from 200 mm to 3100 mm. Furthermore, LS-01 and LS-02 will include 247 and 95 shafts, respectively, for manholes with a depth between 8 m to 26 m. Micro-tunneling and pipe jacking methods were adopted for the deep link sewers while open cut excavation was used for the shallow ones (ZUBLIN, 2015 and ZUBLIN, 2017).

3.2 STEP project geology and cavity issues

The geology of Abu Dhabi comprises of superficial deposits of marine sands and silts, and Sabkha deposits which overlay alternating beds of claystone and mudstone, calcarenite, sandstone and gypsum.

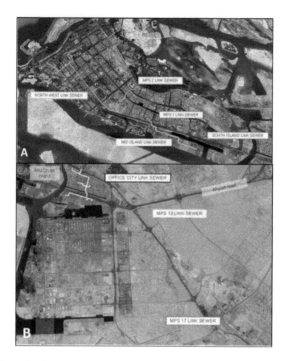

Figure 4. Layout Plan of Link Sewers A) Contract LS-01 and B) Contract LS-02. Adapted from (ZUBLIN, 2017).

Cavities that can be frequently found within the subsurface Abu Dhabi are a significant hazard for tunnel construction. These cavities are associated with the presence of evaporate minerals (salt, gypsum and anhydrite) within the subsurface of Abu Dhabi. The high solubility of these minerals allows cavities to form within days to years. In addition, the presence of carbonate minerals (limestone and dolomite) within the bedrock also leads to the formation of cavities; however, the timeframe for dissolution of these types of rocks is normally in the order of centuries. The formation of subsurface cavities is normally triggered by groundwater flow. Figure 5 shows groundwater flow from some of these cavities within the rock mass in Abu Dhabi (A.C.E.S., 2009).

Geotechnical surveys performed prior to the construction of STEP also identified, along several stretches of the project, typical indicators for karstic features, such as voids, loss of drilling fluid. Figure 6 shows the location of a gypsum cavity encountered during the construction of one of the link sewers.

4 CAVITY AND TUNNELING INTERACTION

Geophysical surveys prior and during the construction of the STEP project showed the cavities found during tunneling to be within or below the pipe/TBM of the sewer links and could reach dimensions of 3 m in diameter. Driving a tunnel close to existing cavities may lead to important interaction effects and affect stress-redistribution and ground deformation around tunnel. This led us to study the interaction between cavity and tunnel through numerical simulations of a shield (TBM) tunnel construction, using the STEP tunnels as a case study. The analyses are done using Finite Element Method (FEM) in Plaxis 2D software.

4.1 Numerical simulations

There are different approaches to simulate tunnel installation with a 2D model that allow one to capture the missing behavior in the third dimension. Some of these approaches include the

Figure 5. Groundwater flow from cavities. Adapted from (A.C.E.S., 2009).

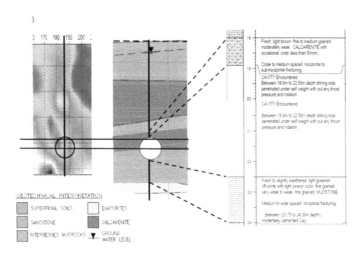

Figure 6. Invert of TBM is within a gypsum cavity. Adapted from (ZUBLIN, 2014).

contraction method, stress reduction method and modified grout pressure method (Moller, 2006). In this study, the contraction approach developed by (Vermeer and Brinkgreve, 1993) is used to allow for the 3D tunneling effect to be taken into consideration in the 2D numerical models.

We have modelled different scenarios of cavity-tunnel interaction by varying the ground materials, cavity location and dimensions. In this paper we will present the results obtained when considering that the ground is composed of one single material, mudstone, which is a predominant layer within the underground of Abu Dhabi. Mudstone is normally associated with the existence of an adjacent gypsum layer. At the interface of these two layers it is common for voids to exist. The Hoek-brown model was used to model the behavior of mudstone. The material properties were obtained from STEP (ZUBLIN, 2015) as well as literature review (Alejano and Alonso, 2005; El Khan, 2008) and are shown in Table 1.

Different scenarios were generated to simulate the cavity-tunnel interaction at different stages. The basic staged construction steps followed in this study are presented in Figure 7 and described below:

1. Initial phase: at this phase no deformations are calculated. This step is only used to set the initial stresses of the model as a starting point for numerical calculation.
2. Tunnel excavation: the ground is removed from inside the tunnel to simulate the excavation and at the same time tunnel lining is placed (the contraction method is used)
3. Lining contraction: the tunnel lining lift to contract until it reaches the pre-assigned contraction value (0.5%) (the contraction method is used)

Table 1. Material properties of the soil layer (Mudstone) used in the numerical simulations.

Material Model	Hoek- Brown
Drainage type	Drained
γ_{unsat} (k N/m^3)	17
γ_{sat} (k N/m^3)	20
E (kN/m^2)	163200
ν	0.3
σ_{ci}*	1600
m_i**	9
GSI***	50
K_0	1

Young's modulus
* Uniaxial compressive strength
** Material constant for the intact rock
*** Geological strength index

4. Grouting: used to account for the tail void injection effect, which is modeled by a pressure applied to the gap between soil and tunnel lining. Therefore, in this stage tunnel lining and interface are deactivated to simulate that.
5. Final lining: the tunnel size in the beginning of excavation is different due to some factors as contraction, etc. therefore applying final lining through activating lining and interface will present the final tunnel shape and size.

The goal of the numerical simulations was to determine the effect of the existence and proximity of cavities in the tunneling construction (e.g. stability of the tunnel, lining moments and deformations). To achieve this, 270 different cases (varying cavity size, orientation, distance) were performed. The paper presents a portion of these simulations. The schematic of model containing the parameters that were varied during simulations is shown in Figure 8. Finally, the cavity effect on tunneling is analyzed at two different construction stages: the first scenario case is the Before Grouting (BG), i.e. before stage 4 in Figure 7, which simulates the cavity effect on the tunnel before applying the grouting which provides the tunnel support against soil. The second case is the After Lining (AL), i.e. cavity simulated after stage 5, when the tunnel is fully supported. Practically this means that the cavity was activated before stage 4 in the BG scenario and after stage 5 in the AL scenario.

The numerical 2D FE model was defined as plane strain, and consists of an 8m diameter circular tunnel that was excavated within Mudstone at a depth of 22 m. The model dimensions and boundary conditions are shown in Figure 9.

A parametric study was conducted to study the effect of cavity characteristics and construction parameters on tunnel construction. In this paper, it was assumed that the voids (cavity) are happening in the Mudstone medium as a result of gypsum dissolution. To capture the

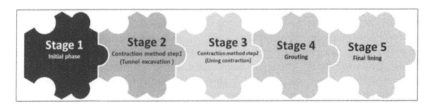

Figure 7. The basic staged construction steps followed in this study.

Figure 8.   Schematic of the tunnel interaction simulations (d: distance between tunnel and cavity; D: cavity diameter; α: angle of the cavity).

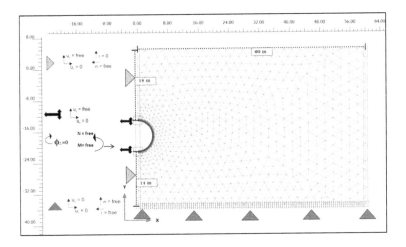

Figure 9.   2D FE meshing dimensions and boundary conditions.

Figure 10.   The labeling system of the analysis.

effect of cavity at different stages of tunneling, cavity diameter (0.5,1 and 2m), distance and orientation (0°,90°and -90°) from the tunnel were systematically varied, as shown in Figure 8. The case labelling used in this work is shown in Figure 10.

4.2   Numerical simulations results

Figure 11 to 13 show the results of the effect of the cavity on the maximum bending moment on the tunnel lining for different simulation cases. When comparing the results of the cavity simulations (AL and BG scenarios) with the base case (BC), i.e. tunnel construction without cavity, one can observe that the highest maximum bending moment on the tunnel lining occurs for a cavity with diameter 2m (AL scenario, i.e. after the placement of the tunnel lining). It is also clear that, as expected, the maximum bending moment observed on the tunnel lining decreases as the distance between cavity and tunnel increases, however the effect of the cavity on the tunnel is generally negligible when a cavity occurs a distance greater than

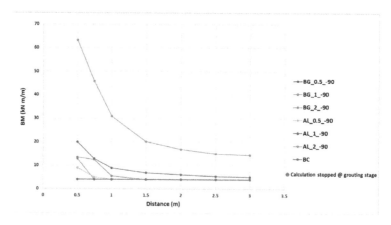

Figure 11. Maximum bending moment in the tunnel lining for cases where the cavity diameter takes on values of 0.5m, 1m, and 2m at two different calculation stages: (BG) and (AL), and the cavity is below the tunnel invert (–90°).

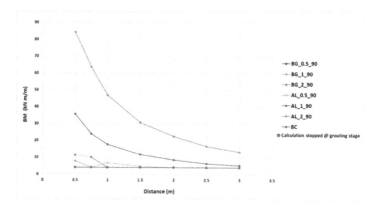

Figure 12. Maximum bending moment in the tunnel lining for cases where the cavity diameter takes on values of 0.5m, 1m, and 2m at two different calculation stages: (BG) and (AL), and the cavity is above the crown of the tunnel (90°).

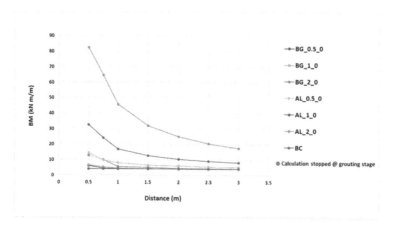

Figure 13. Maximum bending moment in the tunnel lining for cases where the cavity diameter takes on values of 0.5m, 1m, and 2m at two different calculation stages: (BG) and (AL), and the cavity is cavity is at the tunnel side (0°).

2m from the tunnel. Also, *soil body collapse* (i.e. FEM calculation stops) regularly happens during the BG stage with cavity size of 2m at distance 0.5 m and 0.75 m. This is probably related to the excessive deformations occurring as the stress is very high between the two opening at close distance. Finally, the orientation of the cavity seems to also influence the bending moments of the tunnel lining. This effect is maximum when the cavity is located either above the crown (90°) or on the sidewall of the tunnel (0°)

5 CONCLUSIONS

Karst terrain poses serious challeges to tunneling construction. In Abu Dhabi, the existence of gypsum rocks, as persistent quasi-horizontal bands, at different depths, has caused problems during the construction of the STEP tunnels. An extensive numerical study conducted by (Al Kaabi, 2017) uses data from the construction of the STEP project to study the interaction between cavity and tunneling. In this paper we present a portion of the results obtained during AlKaabi's study. In particular we focus on the effect of the existence of voids and their characteristics on the maximum bending moment of the tunnel lining. The major observations of the results presented in this paper can be summarized as follows:

- As the cavity size increases, the maximum bending moment on the tunnel lining increases, and this effect is quite high if the cavity occurs at the final lining (AL) stage and for large diameters of 2 m.
- As the distance between the cavity and the tunnel increases, the maximum bending moment on the tunnel lining increase.
- The effect of the cavity on the bending moment on the tunnel lining is almost negligible when the distance between the tunnel wall and the cavity is greater than 2m, even for large diameter cavities.
- Cavity orientation also influence the bending moments on the tunnel lining. The impact of the cavity on the bending moment is the highest at angles 0° and 90°, i.e. when the cavity is located at the side wall and when the cavity is located at the crown of the tunnel, respectively.

ACKNOWLEDGEMENTS

This research was funded by the MIT & Masdar Institute Cooperative Program and the ADEK Award for Research Excellence Award, AARE 2016. The authors we like to acknowledge and thank the ADSSC authorities for generously allowing us to use data from the construction of the STEP program, without which this work would not have been possible.

REFERENCES

A.C.E.S. 2009. Geotechnical Investigation for 15.3 Km of Mainland Link Sewers between Officer City and Al Wathba Project, Abu Dhabi, UAE, in S08000364 – Link Sewer Contract June.

Alejano, L. and Alonso, E. 2005. Considerations of the dilatancy angle in rocks and rock masses. International Journal of Rock Mechanics and Mining Sciences, 2005. 42(4): p. 481–507.

Al Kaabi, M. 2017. *Geologic Hazard Investigation for Abu Dhabi*. Masdar institute of Technology. PhD thesis, December 2017

Al Khan, M. N. 2008. Living on the edge: Holed up in a crack [cited 2015 20-April]; Available from: http://gulfnews.com/news/uae/general/living-on-the-edge-holed-up-in-a-crack-1.448917

CH2MHILL, 2014. 3D VisualizationStudio (2014), STEP – The Sequel 2014.

El Ganainy, H., et al. 2016 Stability of Solution Cavities in Urban Developments: A Case Study towards Enhancing Geohazard Risk Assessment. Geotechnical and Geological Engineering, 2016. 34(1): p. 125–141.

MacDonald, M. 2010, MASDAR Site-wide Infrastructure Design Geotechnical Interpretive Report

Möller, S.C. 2006. Tunnel induced settlements and structural forces in linings. Univ. Stuttgart, Inst. f. Geotechnik.

Private communications with Eng. Nasser Khalfan Ali Al Nuaimi (ADSSC), D.P.M., April, 2017.

The National. 2009. It looks like the street was hit by an earthquake [cited 2017 20-jun]; Available from: http://www.thenational.ae/news/uae-news/it-looks-like-the-street-was-hit-by-an-earthquake.

Veni, G. 2001. *Living with karst*: American Geological Institute.

Vermeer, P. and Brinkgreve, R. 1993 Plaxis version 5 manual. AA Balkema, Rotterdam.

Williams, P. 2008. World heritage caves and karst. IUCN, Gland, Switzerland, 2008: p. 57

ZUBLIN 2014a Presentation: Cavities NWLS Line L (private communications)

ZUBLIN 2013. Geotechnical Design Basis Report. Strategic tunnel enhancement programme (STEP) – Link sewer contract LS-01.

ZUBLIN 2014b Final draft report on surface geophysical survey for lines G to L. Strategic tunnel enhancement programme (STEP) – Link sewer contract LS-01.

ZUBLIN 2015. Additional cavity borehole grouting report at NWLSL between shafts L01 to L02. Strategic tunnel enhancement programme (STEP) – Link sewer contract LS-01.

ZUBLIN 2017. Step Link Sewer Contract 01 and 02. Report.

Tunnels and Underground Cities: Engineering and Innovation meet Archaeology, Architecture and Art, Volume 11: Urban Tunnels - Part 1 – Peila, Viggiani & Celestino (Eds)
© 2020 Taylor & Francis Group, London, ISBN 978-0-367-46899-6

The design of a segmentally lined tunnel for a large sewer outfall. "Lote 3 – Emisario Planta Riachuelo, Argentina"

R.D. Aradas & D. Tsingas
Jacobs, Buenos Aires, Argentina

M. Martini
J.V Salini Impregilo SpA – SA Healy Company – Jose J. Chediack SA, Buenos Aires, Argentina

ABSTRACT: The Riachuelo Sanitation Scheme in Buenos Aires (Argentina) foresees the execution of 12km of outfall tunnel (hydraulic capacity of 27 m³/s) into the La Plata River, one of the largest estuaries in the world. The tunnel lining design had to allow for very demanding structural requirements since during the project operation the internal pressures inside the tunnel may exceed the hydrostatic external pressures, up to 1.5 bar, throughout a service life of 100 years.

An extensive analysis of possible design alternatives was carried out, the solution of a single-pass 4.3m ID segmentally lined tunnel being finally implemented. Such solution eliminated the need for a secondary cast-in-situ lining which is commonly adopted for similar cases.

This paper describes the innovative tunnel lining design conceived to guarantee high performance levels of structural strength and durability. This considers a load-sharing scheme between radial bolts and circumferential dowels, backed up with extensive laboratory and full scale testing of the mechanical connections, detailed assessment of tensile and shear loads (from bolts and dowels), transmission into the entire reinforcement, crack width, joint opening and construction tolerances, to meet the foreseen structural mechanisms. Geotechnical monitoring and back analysis program was implemented to confirm critical design assumptions.

The removal of the internal lining led to additional conveyance capacity, lower pumping heads and a simplification in the construction process reducing significantly health and safety risks.

1 INTRODUCTION

The Matanza-Riachuelo River flows for a length of 64 kilometers draining the riparian territories of the City and Province of Buenos Aires (Argentina), before it discharges into the La Plata River, one of the biggest estuaries in the world. The Riachuelo river has historically suffered the discharge of several untreated industrial and waste water dry weather flows, which has transformed it into one of most polluted water courses of the World, often compared to the Thames Estuary case.

The poor conditions of the Matanza-Riachuelo Basin triggered the implementation of a comprehensive sanitation program to improve sewerage services to over seven million inhabitants. This program includes the construction of a sewerage interceptor tunnel over the left bank of the river, a treatment plant and final disposal through a long deep outfall 12 km into the La Plata River. Figure 1 presents the location of the project.

This project follows the trend of construction of long sewerage interceptors and outfalls, a piece of infrastructure which has become increasingly common for modern urban cities facing a large water front and which are under a significant environmental burden imposed by decades of sustained growth.

This paper describes the innovative tunnel lining design conceived to provide a structural solution to the conveyance and diffusion sections of the tunnel. Such design approach

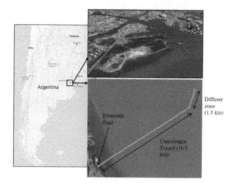

Figure 1. Location of Riachuelo Outfall (Buenos Aires – Argentina).

permitted optimization of the original solution based on a double lining system, hence reducing construction time and costs. The implemented solution delivered high performance levels in terms of structural strength and durability. It was based on a load-sharing scheme between radial bolts and circumferential dowels, backed up with extensive laboratory and full-scale testing of the mechanical connections. Geotechnical monitoring and back analysis program was implemented to confirm critical design assumptions.

2   CONCEPTUAL ASPECTS OF THE DESIGN

The Rio de la Plata Outfall Project comprises the final disposal of the conveyance and collection sewerage system of Buenos Aires. The tunnel presents a 4.3 m internal diameter, 12 km long, being the last 1.5 km a diffuser zone comprised of 34, 25 m long risers. The system works by gravity with an in shore deep load chamber serving as a launching shaft during construction and thereafter controlling during operation the upstream hydraulic head for the tunnel and also attenuate the down surge wave due to the upstream pump stoppage. Figure 2 illustrates the conceptual design.

As a result, the tunnel will be subject to a total internal pressure of approximately 5.5 bar and an external water pressure of 4.3 bar (depending on the phreatic groundwater level as well as the river level), resulting in a net internal pressure in excess of 1 bar. The hydraulic design considered a peak flow rate of 27 m³/s. Detailed hydraulic studies were carried out, by using amongst others CFD simulations, to optimize the hydraulic functioning of the system in order to minimize pumping head requirements and the resulting internal pressure and, ultimately tensile loads.

The presence of a net internal pressure is common to the design of all outfalls, but it becomes critical for the design of large and deep systems. In such cases the hydraulic requirements impose the adoption of large hydraulic sections and thereby the implementation of

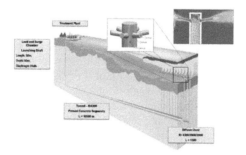

Figure 2. Riachuelo Outfall – Conceptual design (caption shows CD results to optimize diffusers).

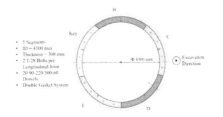

Figure 3. Segmental lining adopted – Overall segment arrangement.

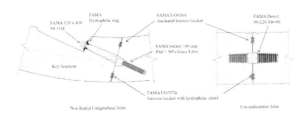

Figure 4. Segmental lining adopted – Segment and ring connections.

tunneled concrete solutions as opposed to traditional plastic solutions, which are still conditioned to diameters no bigger than 2500 mm.

Historically, many outfalls that ought to fulfill similar operational conditions are designed to withstand internal forces with a "decoupled" secondary lining, adopting a double pass lining, often considered to be an optimum structural solution as the primary lining responds adequately in compression whilst the continuously reinforced secondary lining resists the tensile axial forces, performing satisfactorily to durability restrictions. However, this concept cannot be generalized, as the design needs to account for the relative coupled stiffness of both linings, the geotechnical considerations around the tunnel, and the relative movement of the primary lining imposing point load concentrations in rather thin secondary lining. This in turn can lead to underestimation of flexural loads endangering the desired crack control and durability (Aradas et al, 2016)

This project was no exception and in fact was originally based on this same traditional concept, with a primary segmental lining complemented by an in-situ secondary concrete lining for approximately 10 km before discharging into a transition offshore shaft from which a 1.5 km of piled diffuser pipes in trench emerged into the river bed.

However, a comprehensive risk analysis was carried out in order to assess different construction approaches. Such analysis showed that the original tender would jeopardize the implementation of the overall sanitation scheme since it would present several construction risks and would compromise environmentally the operation of the nearby water intakes of Buenos Aires due to the intensive off shore dredging activities. As a consequence of what it was described above, a one-pass lining solution was selected as the final solution to address the aforementioned constraints, but at the same time led to significant design challenges. To date, there has been a very limited implementation of one-pass segmental linings for these type of projects with reliance on complex joint or segment arrangements to provide an adequate tensile strength; such the one adopted for the South Bay Ocean Outfall in San Diego (Kaneshiro et al, 1996) as well as other proposals such as Bay Delta in Sacramento (Bednarski & Lum, 2013) and Brightwater (Drake et al, 2006)

A common feature of these very limited case history was the recommendation of completely neglecting the external ground confinement contribution as a safety redundancy. Neglecting the effective contribution of ground, despite conservative, is unrealistic but also can overlook the relative vertical-horizontal distortion of the segmental lining and therefore the resulting

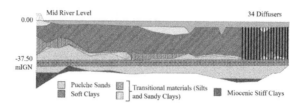

Figure 5. Geological profile.

bending moments. However, the safe adoption of an effective external ground pressure is also very challenging as it not only relies on soil mechanic principles as it also largely depends on the interaction of the ground with the TBM shield and its operation. The estimate of effective ground pressure was a critical feature of the design and it is addressed in subsequent sections of this paper.

The main lining design criteria were focused into two aspects: constructability and robustness in terms of structural strength and durability. The adopted solution consisted of a universal segmental lining with a 4.3m internal diameter tunnel, 30 cm thickness, two traditional inclined T28 AISI 316L stainless steel bolts with 160 mm long plastic sockets (Fama, 2017) and four high strength (220 kN) shear dowels (with a steel bar core) per segment, to provide short and long term interaction of both elements in order to take tensile axial forces. A sealed-double gasket system was adopted to guarantee high water tightness performance; the outer gasket was coupled with a continuous hydrophilic gasket; an hydrophilic strip was also specified in the circumferential joint to prevent water flow inside both gaskets during construction stages. The following figures illustrate the segmental lining adopted. Figures 3 and 4 illustrate the design.

3   GEOTECHNICAL CONDITIONS

The regional geology of the area is characterized by a sequence of glacial and inter-glacial sedimentary deposits comprising primarily sand and clays. The two main geological layers encountered in the alignment are: the Puelche Formation, characterized by dense sand deposits of high permeability, overlaid by the Postpampeano Formation, composed of soft to very soft, normally consolidated, silts and clays (Figure 5).

An initial SPT geotechnical offshore campaign was carried out, supplemented by further offshore CPT testing. The geotechnical studies were instrumental to define a longitudinal profile to suit the operational needs and to determine the required computational parameters, in particular the effective ground pressure acting on the segmental lining.

4   DESIGN APPROACH

4.1   General workflow design procedure

The design of the one pass segmental lining had to address several challenges as described in more detail in the following subheadings. One of the critical aspects was the 100-year service life without practical possibilities for access and clear maintenance activities once the outfall is in operation. Further, in order to deliver the required robustness and quality of the final product, the contractor gave particular attention to the various constructability aspects, intended as simplicity of ring build and installation of the connection between segments and rings. The following Figure 6 illustrates the conceptual approach adopted:

Material assessment: detailed laboratory tests were carried out to assess the behavior of PA6 polyamide plastic sockets under unique operating conditions: full saturation and permanent tensile loads which result in rheologic effects.

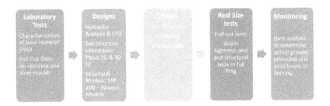

Figure 6. Summary of design approach.

Structural design and checks: assessment of the resulting net internal force leading to tensile stresses in the lining; this implied the evaluation of load share between radial and circumferential connections which also varied under short and long terms conditions. Results showed that radial connections (bolts) prevailed under short term loads with a progressive transfer of load to the dowels acting on shear as the plastic bolt sockets relax through time. Conceptual design was followed by compliance with all normative requirements, in particular crack control and verification of stress due to the introduction of punctual forces from the mechanical connections. Several load schemes had to be considered, including, construction stages, operational stage with internal pressure, operating internal pressure with river bed dredging.

Real size testing: prototype evaluation of water tightness as well as the integrated functioning of radial and circumferential connections to confirm load share predictions.

Monitoring and back analysis: finally, given the uniqueness of the design conditions and uncertainty related to the effective ground collaboration, the ultimate client approved the design and construction ad referendum of an assessment milestone after completion of the first 1000 m of excavation. This chainage coincided with the most critical sections due to the larger internal pressures and therefore the need to validate, through monitoring and back analysis of internal forces, the parameters adopted in the design. Should the evaluation resulted unsatisfactory the client had the choice to enforce a secondary lining for at least the critical reach with net tensile forces on the lining.

4.2 *Assessment of effective pressure acting on the lining: relaxation aspects.*

One of the critical parameters in the determination of effective pressure acting on the lining is the ground relaxation due to the excavation process, TBM shield conicity, ring build process and annular grouting; this in an inherent soil structure interaction problem. Traditionally, conservative assumptions are adopted when assessing soil-structure interaction and full overburden conditions are often adopted to verify the segmental lining; however, when assessing a lining under net positive internal forces, an opposite criteria needs to be adopted and the estimation of a minimum contributing ground pressure is required for safety purposes. Analytical models (Muir Wood, 1975; Duddeck&Erdmann, 1985), as well as Finite Element models (Plaxis 2D and 3D) were adopted. For FE Plaxis modelling, Hardening Soil Model was adopted, and several relaxation parameters have been evaluated in order to obtain an envelope of internal forces, throughout different chainages along the tunnel.

For this project, ground relaxation was determined with a 3D approach simulating all construction stages predicting ground displacement above crown and the associated effective pressure. Complementary, ground relaxation curves using a 2D approach were also developed to infer specific relaxation conditions to each relevant geotechnical set up along the alignment of the tunnel. The following figure summarizes the main outcome of the soil-structure interaction models. Effective pressure, after initial relaxation due to the ring build process followed by an additional relaxation due to deformation of the lining, accounted for some 65 kN/m^2 (full overburden is 270 kN/m^2); a sensitivity analysis was carried out which showed that despite variations of this estimate, the resulting compression axial forces in the lining remained fairly invariant. Maximum resulting axial forces reached 148 kN/m, which implied 228 kN of ultimate force required for the bolts and sockets (the latter including force to maintain the gasket seals compressed).

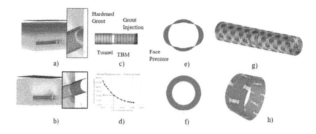

Figure 7. a) and b) effective pressure from Plaxis 3D, d) Ground relaxation curve from Plaxis 2D, e) and f) Bending moments and axial loads from Plaxis 2D. g) and h) Detailed 3D modelling of the structural behavior of the ring (SAP 3D).

4.3 *Assessment of materials – tests on bolt sockets*

The adoption of inclined bolts with plastic sockets obeyed flexibility and constructability reasons as opposed to other solutions like curved steel bolts or spear bolts but with steel sockets. These alternative solutions would have led to a more predictable force-displacement constituent law. However, curved bolts always cause difficulties during their installation in the ring build process and require larger holes to avoid blockages, but the bigger the gap, the larger the deformation required for the bolt to start taking loads, hence bigger the joint openings. On the other hand, steel sockets can sometimes limit flexibility during the installation of the bolts.

Nonetheless, the final adoption of the plastic sockets posed significant challenges as there was no track record of their use under full saturation conditions and long-term sustained tension loads. Several tests were conducted to determine the actual capacity of the threaded socket under sustained load in submerged conditions and allowed to inform the design and production of the base material mix. A blend of PA6 reinforced with 50% fiber glass was the final selection after the tests (University of Trento, 2017).

The following tests were implemented to characterize the behavior of both, the base material and the actual threaded socket (University of Trento, 2017; University of Roma, 2017):

a) Pull out tests of actual sockets in dry or conditioned sockets (absorbing ambient humidity) embedded in steel blocks;
b) As above for fully saturated sockets;
c) Pull out tests of actual sockets embedded in concrete blocks but in ambient conditions (not in full saturation);
d) Tensile tests under long-term sustained loads and under full saturation conditions over raw material specimens (not in actual sockets) for various tensile stresses (60Mpa was the main reference testing stress)

The main conclusions from the tests were:

- Base material exhibited a nominal strength of 185 MPa at a 5% elongation rupture; under full saturation the strength reduced to 100 MPa but the deformation was preserved;
- Similarly, pull out tests of dry sockets on steel blocks reached over 700 kN of resistance; under full saturation the strength reduced to almost half its initial value, also preserving the same deformation of the dry conditions (Figure 8);
- An analysis of the deformation-load behavior of the sockets in steel blocks showed a deformation limit of around 1.6 mm at the ultimate strength under full saturation; a similar pattern was observed when testing the sockets embedded in concrete, despite the material could not be fully saturated. However, there was a clear distinction in terms of the failure mode: sockets in steel failed along the internal thread while, when in concrete, the failure obeyed to shear and crushing of the concrete (Figure 9).

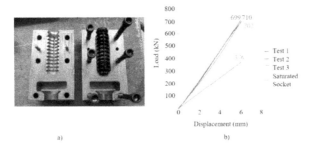

Figure 8.   a) Typical steel molds used for pull-out tests. b) Load displacement diagram for both conditioned and saturated sockets on steel molds (measured on the testing apparatus, not on the bolt).

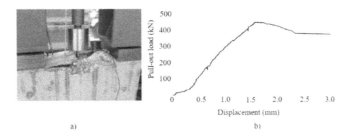

Figure 9.   a) Typical concrete molds used for pull-out tests. b) Load displacement diagram conditioned sockets on concrete molds.

- Creep tests on dry and saturated specimens showed that, at a sustained stress of 60 MPa and under saturation conditions, deformations double after 100 years of service. This correlates with the Elasticity modulus which was 16,000 MPa under dry conditions, 8000 MPa under short term saturated conditions and around 4000 MPa after 100 years of sustained load (Figure 10)
- The final adopted long-term sustained load resulted in the range of 190 kN to 220 kN, satisfying design requirements (Figure 11)

4.4   *Real size testing*

Besides computer models, extensive real size tests, have been carried out to validate the design; these included: pull-out test of bolts but placed in concrete blocks reproducing actual segments (Figure 12), and in situ full ring tension test, to test both structural behavior, load share between connection and water tightness, tests were executed with up to two bar pressure.

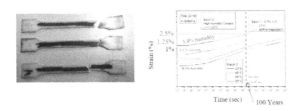

Figure 10.   a) Typical base material specimens. b) Tests under tension in saturated condition requirements.

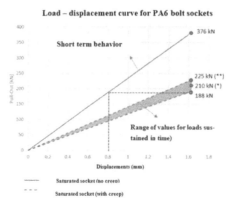

Figure 11. Summary of force-displacement design chart.

Figure 12. Left: tension tests on segments; Right: full scale ring tests.

The first set of tests aimed to analyze the transfer of point tension loads from bolts into the segment and the full rebar cage and to develop a tension-displacement curve at joints. Several extensometers were placed in the internal rebar cage, as well as in the bolts so as to track the full tension path. As predicted, some segment twisting was observed along longitudinal joints due to the combined effect of the inclined bolts, leading to secondary torsional loads taken by the dowels with radial shear. As a result, the rebar cage was improved incorporating local additional reinforcements around bolt pocket and anchorage cones. Measurement of bolt forces due to 48 hours sustained loads were obtained and used as part of the long-term performance evaluation of the system. The full ring tests, under 2 bar internal pressure, proved that gaskets performed well under the most stringent conditions for a sustained period of 24 hours (Figure 12). The tests also permitted to assess the performance of the bolts proving, for example, that friction and compression between rings, helped to diminish the loads in the bolts.

5 MONITORING AND BACK ANALYSIS

Monitoring and back analysis was a component of the design approach adopted as there was a contractual milestone to assess structural behavior after the first 1000 m of excavation. The main objective was to assess the actual effective ground contribution onto the tunnel lining in comparison with the values used for the design, as obtained from numerical modelling.

A set of 4 strain gauges were installed in each segment of the instrumented rings, combined with a pressure cell located at the exterior face of the tunnel segment. Results were taken periodically every fifteen days during the construction of the tunnel.

Graphs below show the behavior of the instruments at a relevant design section. First graph below corresponds to the measurement of soil pressure in the tunnel exterior face; as it was reported in several case studies (Schotte et al, 2011; Ninic & Meschke, 2017), pressure cells are not entirely reliable and difficult to interpret but are still an important means to estimate direct effective pressures acting on the lining. For this project, given the importance to arrive at a reliable estimate of pressures and the difficulties to install other instrumentation off shore like deep borehole gauges, made the pressure cells almost inevitable.

Despite apparently random measures were observed through the various cells, including misfunctioning elements, a fairly clear pattern was observed throughout. Overall, it was observed that pressure is initially zero while the ring is still inside the TBM, then normally peaking up to a value around hydrostatic pressure followed by a clearly marked decrease of load explained by complex processes related to the interaction of the cells (embedded in the lining) with the layer of annular grouting; rheology plays a key role as the grouting hardens. As reading time goes beyond approximately a period of 30 days, pressure increases seems to start detecting soil pressures (Figures 13 and 14)

The readings from the strain gauges exhibited a far more consistent and reliable pattern of information used to back calculate the axial forces and bending moments in the lining (Figure 15). It was interesting to observe that all rings instrumented showed a steady increase of deformations, reflecting a rise in the effective pressures acting on the lining, what further reassures the conservative assumptions made in the design. Measurements were also influenced and correlated to the grouting pressures and number and location of injection points. Figure 15

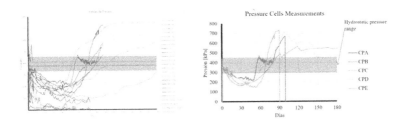

Figure 13.   Pressure cells. Left: all results; Right: pressure cell in ring #38.

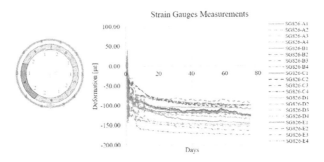

Figure 14.   Readings from strain gauges at ring #38.

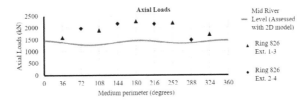

Figure 15.   Back analysis – comparison of compression axial loads from monitoring against the values used in design.

shows how the actual monitoring predicts axial compression loads larger than those used in the design to estimate, once in operation, the effective net tensile loads acting on the lining; this permitted to give the satisfactory final go ahead of the construction with the one pass solution.

6 CONCLUSIONS

The adoption of segmental lining solutions as a one-pass solution implemented to date have been limited to a very few applications with relied on complex joint or segment arrangements to provide an adequate strength in tension. The Riachuelo outfall design had to address some critical challenges: 100 years useful life, extremely difficulties to access and repairs once in operation, complexity of installation of any kind of second pass lining due to the long length of the system, high durability and water tightness requirements.

This paper presented the main challenges, conceptual approach and results of the solution adopted which consisted in the design of a segmental lining that satisfied very stringent requirements but at the same time adopting a standard solution in terms of its constructability.

This innovative design led to significant reductions both in terms of costs and time for the ultimate client. It also led to a bigger tunnel what implies additional conveyance capacity or, for the same design flow, a lower initial driving head and less pumping head requirements and operational costs. The construction schedule will be reduced by two years; with a direct impact on the overall sanitation. The elimination of the secondary lining also caters for evident savings during construction as it reduces the quantities of construction materials with an ultimate impact to the ultimate client.

Very detailed and thorough laboratory and prototype testing were carried out to assess the mechanical elements of radial and longitudinal connectors under permanent exposure to tensile loads and permanent contact with sewage effluent. These elements had been never considered before for a design situation like this.

Finally, monitoring, followed by subsequent back analysis, is always very important but in this case it was also an integral element of the design approach and contractual arrangement with the ultimate client, yielding satisfactory results and full reassurance of the premises adopted in the design.

REFERENCES

Aradas, R. D., Fernandez, J. M., Harding A. & Tsingas, D. 2016. Challenges in the design of segmentally lined tunnel for combined sewer outfalls. *World Tunnel Congress 2016 (WTC 2016)*. *San Francisco, U.S.*
Bednarski, J. & Lum, H. 2013. Update on the "Pipeline/Tunnel Option" for the Delta Habitat Conservation and Conveyance Program. In: *Rapid Excavation and Tunnelling Conference Proceedings 2013*: 300–312
Drake, R.D., Dugan, D. & Pooley, A.J. 2006. Brightwater Conveyance Project – Soft Ground Tunnels and Shafts. In: *Tunnel Association of Canada Conference Proceedings 2006*.
FAMA S.p.A. 2017. Design of high strength and durability socket T28x160 Close Fit Threaded Socket.
Ninić, J. & Meschke, G. 2017. Simulation based evaluation of time-variant loadings acting on tunnel linings during mechanized tunnel construction, *Engineering Structures 135:* 21–40
Kaneshiro, J.Y., Navin, S.J. & Korbin, G.E. 1996; Unique Precast Segmented Liner for the South Bay Ocean Outfall. *North American Tunnelling '96* 1:267–276. 1996. Rotterdam: Balkema.
Duddeck, H. & Erdmann, J. 1985. On Structural Design Models for Tunnels in Soft Soil. *Underground Space 9*: 246–259.
Muir Wood, A. M. 1975. The circular tunnel in elastic ground. *Géotechnique 25* (1): 115–127. March 1975.
Schotte K, De Backer, H, Nuttens, T, De Wulf, A. & Van Bogaert, P. 2011. Monitoring strains in the Liefkenshoek railway tunnel. In: *World Tunnel Congress 2011 (WTC 2011)*: 254-255. *Helsinki, Finland.*
Universita degli Studi di Roma Tor Vergata. Civil Enginnering and Computer Science Department, TERC. Tunneling and Engineering Research Centre 2017. Relaxation test under pullout load on connection systems for precast segments for tunnels.
Universita degli Studi di Trento. Dipartamento di Ingegneria Industriale 2017. Study of durability and creep resistance of Domamid 6G50H1 (Polyamide PA6 reinforced 50% glass fibre)

Tunnels and Underground Cities: Engineering and Innovation meet Archaeology, Architecture and Art, Volume 11: Urban Tunnels - Part 1 – Peila, Viggiani & Celestino (Eds)
© *2020 Taylor & Francis Group, London, ISBN 978-0-367-46899-6*

The relations between surface settlements and TBM excavation parameters

U. Ates
Istanbul Technical University, Istanbul, Turkey

I.S. Binen, F. Kara, M. Akca, B. Turker, T. Cinar & T. Acilioglu
Gulermak-Nurol-Makyol Metro Cons. J.V., Istanbul, Turkey

ABSTRACT: TBMs are widely utilized in urban areas, however, keeping ground movements at minimal rates in the areas with complex geological conditions require comprehensive preliminary studies and calculations with very sensitive excavation procedures. For avoiding ground or volume loss, which is the main cause of surface settlements, TBM chamber pressure must be kept under control, however, it is not the only important parameter and many other excavation parameters are related with settlements such as excavated muck volume. This paper shows relations between TBM excavation parameters, geological conditions and surface settlements. To investigate the relations, the data obtained from the excavation of Mecidiyekoy-Mahmutbey Metro Line was used. Excavation parameters of 3 TBMs were compared with settlement data obtained from the surface measurement stations. For each geological zone, corresponding settlement curve was calculated. Measured settlements from the settlement stations of that impact zone were then compared with TBM excavation parameters for each ring.

1 INTRODUCTION

Istanbul is the one of the most crowded cities in the world and its large population concentrated in a relatively small area which affects transportation and causes serious traffic issues. Increasing demand on infrastructures leads using soft ground tunneling methods in heavily urbanized areas. Mecidiyekoy-Mahmutbey Metro Line, having 18.5 km length, is one of the seven lines in Istanbul Metro which is about to be completed (Figure 1).

The area is densely populated and the buildings around the line are generally old, moreover, on some parts, problematic sandy sections exist. As the geology is complex, it is very common to use TBMs for metro projects in Istanbul, due to their various advantages over other tunneling methods. Gulermak-Kolin-Kalyon Metro Construction J.V. chose to construct the line with a combination of NATM and TBM methods. The line is divided into two sections from Veysel Karani station in accordance with the feasibility studies and geological considerations. It has been decided to adopt the use of NATM for Mecidiyeköy – Veysel Karani section, while the utilization of TBMs has been selected for Veysel Karani – Mahmutbey section where the complex geological conditions were expected.

Under suitable ground conditions for the selected TBM, it may excavate two or four times faster than drilling and blasting (Barton 1999, Guglielmetti et al. 2008, Bilgin et al. 2014). However, mixed-face conditions could cause excessive surface settlements, low TBM performance and increased costs (Steingrimsson et al. 2002, Blindheim et al. 2002, Zhu & Ju, 2005, Bilgin et al. 2014, 2016, Ates et al. 2017). There are mainly three settlement prediction approaches for mechanized tunnel excavations which are numerical analysis such as finite element method, analytical method, and semi-theoretical method. With the careful calculation, planning and excavation, settlements could be kept at minimum levels. However, it may

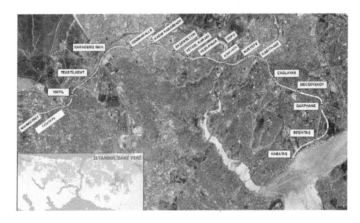

Figure 1. Location of the metro line.

not always be possible to cover the tunnel route completely during the investigation phase, thus some local geological changes may go unnoticed. Since excavation parameters, especially face pressure, directly affects the settlement rates, TBM crew must follow the geological conditions and react to sudden changes in short time.

The present paper focuses on the settlements recorded during the excavation of Mecidiyekoy-Mahmutbey Metro Line and their relationship between geological changes and TBM excavation parameters.

2 GEOLOGY OF THE AREA

The tunnel route lies on east-west direction on European side of Istanbul which has complex geological features due to its geological history and development. These complex features mainly consist of fault lines, saturated sand lenses, highly weathered and blocky zones, frequent formation transitions, and partial face conditions where two different formations are present. Numerous active and abandoned water wells are also present on tunnel route. Some of them were located right above the lines, even some reaching the tunnel elevation, which caused additional problems (Ates et al. 2016, 2017). In general, two dominant formations have been observed during tunneling activities which are Cekmece formation and its Cukurcesme and Gungoren members and the other is Trakya formation.

Dominant formation of the line is Trakya formation, which is basically sandstone and shale alternation. Sandstones are observed as rich in quartz, mid strong - strong and sometimes fractured (Polat 2014, 2015). Shales in lower sections contains clay. According to tests conducted on samples taken from the belt conveyor or directly from the face during cutter tool replacements, Trakya formations uniaxial compressive strength has been observed up to 150 MPa on some chainages. It should be mentioned that the formation also has very low strength and fractured sections.

Cekmece-Gungoren-Cukurcesme formation is mostly composed of sand and silt with green and gray colored clay dominated zones (Polat 2014, 2015). Saturated, loose sandy-silty zones are common and significant for TBM tunneling considering face pressure and surface settlements, while clay containing zones possess the threat of clogging and other adhesion related problems during the excavations. However, it should be noted that characteristics of this formation was different between Yenimahalle-Mahmutbey and Yenimahalle-Veysel Karani sections. On Yenimahalle-Mahmutbey section sand was more cohesioneless comparing with the other section, which was caused bigger problems.

In this study, lithological definitions such as sandstone, weathered/blocky sandstone, clay-stone/siltstone, clay, sand, etc. have been used instead of geological formations since

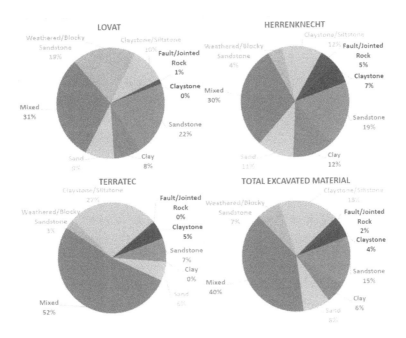

Figure 2.   Geology of the metro line.

the above-mentioned formations contain frequent interbedding of different rock and soil formations. Percentage variations of these lithological zones for each TBM are given in Figure 2. Geological data for the categorization has been gathered by combining preliminary geotechnical survey with samples taken during tunneling, and face observations made during cutterhead inspections and maintenances.

3   TUNNEL ALIGNMENT AND TBMS USED IN THE PROJECT

As the most of the route passes Trakya Formation which formed from sandstone, claystone and siltstone, EPB TBMs were found suitable, thus selected for the project. Two of the three TBMs, TBM 1 and TBM 2 were refurbished and TBM 3 was new (Figure 3). Table 1 summarizes TBMs properties.

Figure 4 summarizes TBM alignment. TBM 2 excavated Line 1 from Yenimahalle Station to Mahmutbey Station and TBM 1 excavated Line 2 at the same direction until 200 meters after 100 Yil Station. At the beginning it was planned to excavate all section with TBM 1 similar to TBM 2, however, due to problems (Ates et al. 2017), it was decided to use TBM 3 for the remaining part.

Table 1.   TBM specifications.

TBMs	TBM 1	TBM 2	TBM 3
Diameter	6,57 m	6,55 m	6,56 m
Thrust	55000 kN	32000 kN	40000 kN
Torque	4500 kNm	4400 kNm @ 1.8 RPM	5440 kNm
		2100 kNm @ 3.8 RPM	
Cutters	38 x 15,5" 5 xv16", 3 x 16,5"	41 x 17"	43 x 17"
CH motors	8 x hydraulic	8 x hydraulic	6 x electric (VFD)
CH Speed	0-4 RPM	0-3,8 RPM	0-3 RPM
CH Opening Ratio	29%	20%	37%

Figure 3.   TBMs used during the project, TBM 1 (left), TBM 2 (middle) and TBM 3 (right).

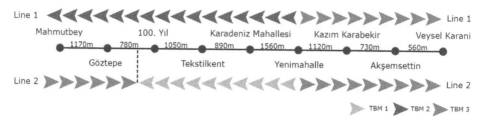

Figure 4.   TBM alignment.

TBM 3 was at the opposite direction with the others from Yenimahalle to Veysel Karani Stations. The TBM excavated Line 1 and disassembled then transported and reassembled at Yenimahalle Station for the Line 2. After completing the second section, TBM, again, disassembled and transported to Mahmutbey Station then reassembled for the excavation of the remaining part from TBM 1 which will be referred as return drive in this paper for the TBM 3.

4   SURFACE SETTLEMENTS AND METHODOLOGY

4.1   Surface settlements

Tunneling under densely populated areas is always a delicate job due to the risk of ground deformations which then affects both surface and subsurface structures. Therefore, a methodology for constant monitoring of deformations should be adopted and the deformations should not be allowed to exceed the pre-determined, acceptable levels. While short terms settlements generally occur in a few days after excavation, long term settlements may take few months to few years and are mostly due to creep, stress redistribution, and consolidation of soil after drainage of the underground water and elimination of pore water pressure inside the soil (Attewell et al. 1986).

Main parameters that affect ground deformations can be listed as ground conditions, tunnel depth and geometry, tunnel diameter, single or double tube lines, neighboring structures, excavation method, face support pressure, advance rate, stiffness of support system, excavation sequence and ground treatment (O'Reilly & New 1982, Arioglu 1992, Karakus & Fowell 2003, Tan & Ranjit 2003, Minguez et al. 2005, Suwansawat & Einstein 2006, Bilgin et al. 2014).

The primary reason of surface settlements is convergence of the ground into the tunnel after excavation. This changes in situ stress states of the ground and results in a stress relief. Convergence of the ground is also known as ground loss or volume loss. The volume of the settled ground on the surface is usually assumed to be equal to the volume loss inside the tunnel (O'Reilly & New 1982).

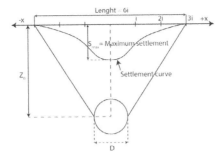

Figure 5. Surface settlement parameters.

The second tube of the tunnel line is thought to be more prone to settlements due to the disturbance of the ground from the excavation of the first tube (Bilgin et al. 2018). Increased risk of settlement in twin tube tunneling should also be considered while investigating the deformation behaviors of the second tube excavations. It should also be noted that, experience gained from the excavations of the first tube can be useful and guiding and might help mitigating the problems for the second line.

Ground loss can be classified as radial loss around the tunnel periphery and axial, face, loss at the excavation face (Attewell et al. 1986, Schmidt 1974). It is possible to minimize the face loss in full-face mechanized excavations by means of applying face pressure, as in this metro project where EPB TBMs were utilized. The ground loss is usually more in granular soils than in cohesive soils for similar construction conditions. The width of the settlement trough on both sides of the tunnel axis is wider in the case of cohesive soils, which means a lower maximum settlement for the same amount of ground loss (Bilgin et al. 2014).

Surface settlement parameters are presented in Figure 5. As seen, the shape of the settlement curve is very similar to Gaussian probability curve and the maximum surface settlement occurs over the centerline of tunnel.

4.2 Calculation Methodology

Before the excavations, during the preliminary works, all surface structures in the possible tunnel settlement zone area are equipped with prisms and reference measurements were taken by using total stations. After excavation begin, deformation measurements started to be taken and difference with reference was calculated a couple times in the day for informing tunnel crew about the settlement rates at the excavation area. Especially on risky zones, with bad quality buildings and unfavorable geological conditions, measurements were done in every 2 hours. After the TBM passes, measurements were continued to be taken with a reduced frequency. For the calculations in this paper, maximum settlement values were used for each prism, which can also be identified as long term settlement values, since generally the data includes 2 or 3 months after excavation. In addition to deformation data, all TBMs used in the project were equipped with data recording systems, which records excavation parameters as well as ring coordinates, thus making this study possible.

To investigate the relationship between the excavation parameters and the surface settlements, geological zones on the tunnel alignment were separated using in-situ geological data. Then, TBM data were divided according the geological zones and average depth of the tunnel, Z_0, was calculated for each zone. Then inflexion point, i, of the settlement curves were calculated for each geological zone by using equation suggested by Arioglu (1992):

$$\frac{i}{R} = 0.9\left(\frac{Z_0}{2R}\right)^{0.88} \tag{1}$$

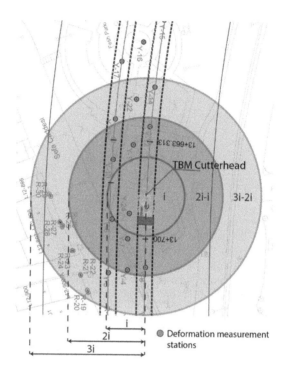

Figure 6. Settlement calculation area.

where R = radius of the excavation diameter; i = inflexion point of the settlement curve; and Z_0 = depth of the tunnel axis.

By using the ring coordinates and inflexion points for each zone, effected areas were determined for each excavation. The measurement prisms which are in the corresponding circular region with the diameter of i, 2i, and 3i were identified for each excavation. Eventually for each excavation, 3 different circular settlement zones were obtained, which are unique to that particular excavation. That way, it was possible to identify and associate data obtained from the deformation measurement stations with the excavations and related excavation data.

To carry out this process automatically and go through the huge amount of data in short time, a custom MATLAB script was written. With the help of the script, deformation measurements obtained from approximately 3000 different deformation measurement stations were separated, classified, and associated with the related excavation and excavation data. Figure 6 shows a simple schematic of circular deformation regions.

Since the project was composed of twin tubes, influence of the first tube should be reduced from the second tube. To be able to fulfil this reduction, a reference deformation measurement was required for each deformation station that resides on the second tube. Therefore, deformation measurements which were taken only 2 days before the arrival of TBM to that point have been accepted as a reference deformation value.

5 CALCULATIONS AND COMPARISONS WITH TBM EXCAVATION PARAMETERS

The calculated deformation values were divided in accordance with the same zones that the excavation data were divided. The graphs in this section shows average deformations and maximum deformations for each zone for i, 2i-i and 3i-2i regions separately. Since face pressure is one of the most important factors affecting the settlements, it was also included into graphs. Settlement values were also compared with other TBM excavation data and mentioned if any anomalies were present at the highly settled zones.

TBM 1 excavated the shortest length among all TBMs. The average deformations were recorded up to 14mm on clayey zone, however, most of these deformations occurred when entering sandy zone and exiting sandy zone (Figure 7). Due to the settlement problems occurred on Cukurcesme Formation, which was composed of mostly cohesionless sand, extensive grouting application was conducted in this area (Ates et al. 2016, 2017, Bilgin et al. 2016). Thus, average settlements on this area were kept at minimum. The maximum deformation recorded for TBM 1 were close to 90 mm which were occurred in clayey sandy area in a local schoolyard. By inspecting TBM data it was revealed that this settlement was related with an

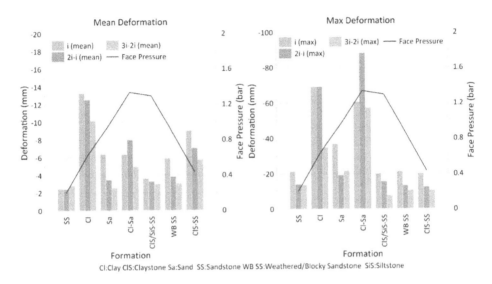

Figure 7. Average and maximum settlement and face pressure data for TBM 1.

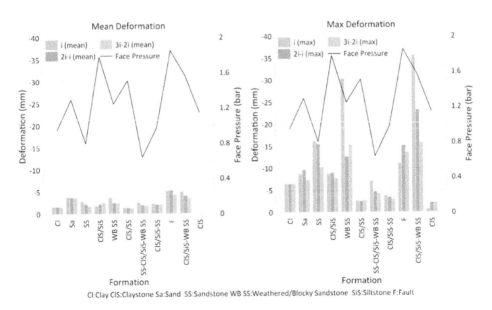

Figure 8. Average and maximum settlement and face pressure data for TBM 2.

over excavation. It should also be noted that TBM 1 was excavating the second tube after TBM 2, thus it was operating in a disturbed zone.

TBM 2 was the best TBM regarding minimum average deformations, which was equal or less than 5mm among all line (Figure 8). It was excavated side by side with TBM 1 until 200 meters from Yuzyil Station and excavated with TBM 3 in opposite directions at the remaining part after the TBM 1's excavation. However it was the first TBM to excavate the most of the tunnel route except for the last 500 meters, thus it was generally excavating in an undisturbed zone.

One of the maximum deformations which was approximately 30 mm has occurred in the transition zone between blocky sandstone and claystone-sandstone. Due to the transition zone, ground became unstable. In addition to that, keeping EPB pressure steady became difficult and pressure dropped to 1 bar from 1.5 bar. Figure 9 shows TBM data as well as settlement data for the affected zone.

The other high settlement rates recorded at the beginning of the tunnel from Goztepe station to Mahmutbey station. Immediately after starting excavation from the station, TBM entered a fault zone followed by claystone-siltstone-blocky sandstone interbedding. At this area settlement values reached to 35 mm.

Similar to TBM 2 maximum deformations of the TBM 3's return drive were occurred close to Goztepe station, which were up to 70 mm. The ground was already disturbed by TBM 2's excavation and station construction, thus TBM 3's settlement rates on this were higher than TBM 2 as expected (Figure 10). The TBM's data recording system did not record the excavation data for this area due to a malfunction, however since deformations occurred on line 1, TBM crew were aware of the unstable zone and kept face pressure high until the entrance of the station. Other than this faulty zone, surface deformations were minimal for TBM 3's excavation for this line.

Figures 11 and 12 shows settlement values for TBM 3's first and second tube excavations. Even though the horizontal distance between the two lines is not more than 30 meters, there were some slight differences between the geological conditions. The average deformations were occurred on sandy part of the route as expected, since EPB TBMs' were not suitable for these kind of conditions, however it is possible to say that TBM crew successfully completed

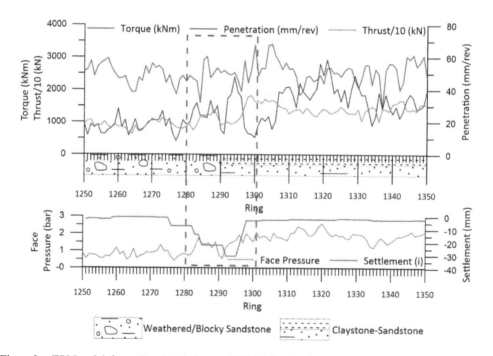

Figure 9.   TBM and deformation data belonging TBM 2 showing between rings 1250 and 1350.

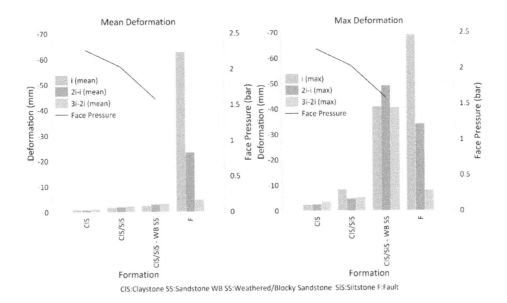

Figure 10.   Average and maximum settlement and face pressure data for TBM 3's return excavation.

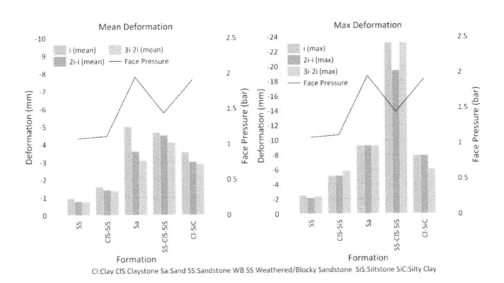

Figure 11.   Average and maximum settlement and face pressure data for TBM 3 Line 1.

excavation of these areas since the average deformations are below 5mm. The maximum deformations on this area were recorded at the beginning of the formation change from clays-tone-siltstone to sand and entrance of the station, which can be described as disturbed zone.

On the other hand, maximum deformations were slightly higher than average deformations and reached approximately 24mm on line 1 and 10mm on line 2. These deformations were occurred on different locations, for the first line formation was composed of sandstone-claystone-siltstone alteration and effected a wider area since even in 3i-2i distance deformations were close to i distance. This settlement were occurred in a park, related with local ground conditions since face pressure was similar to previous excavations. On the second line, maximum deformations

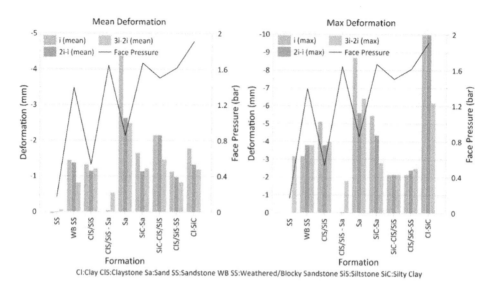

Figure 12. Average and maximum settlement and face pressure data for TBM 3 Line 2.

occurred very close to station entrance of the tunnel, which can be described as disturbed zone due to station construction.

In addition to these calculations artificial neural networks were also used to find any relationship between excavation parameters, geological conditions and settlement values. However, correlation coefficient between predicted and measured values was around 0.60, which indicates more detailed studies should be done to obtain better results. For keeping this paper more simple these calculations were not added to the paper.

6 CONCLUSIONS

In urban tunneling, one of the main considerations is the effect on the neighboring surface and underground structures in terms of deformations. Especially in twin tube projects, controlling deformations at an acceptable level may require extra efforts. Understanding the relationship between the excavation parameters and the deformations in accordance with the geology on a large-scale data set rather than event-based data could be difficult due to challenges of associating the deformation data with the related excavation.

In this study, an approach in the form of a back analysis was intended to understand the relations between the excavation parameters and deformations. A custom MATLAB script was developed to cope with the extensive amount of data obtained from the deformation stations. Unique deformation regions which were identified separately for each excavation and geological zone were found to be helpful for associating the deformation measurements with the related excavations.

It was seen that the maximum deformations were occurred in either unsuitable ground conditions for EPB TBMs such as highly saturated sandy zones or in transition zones where maintaining face pressure at a steady level becomes difficult. It was also seen that the previously excavated/disturbed areas caused by the twin tube tunneling turned out to be more problematic for TBMs. However, as it was in the case of TBM 3 and TBM 2, twin tube method was also helpful to identify some geological features such as fault lines which otherwise are very difficult to spot during the preliminary investigations. In overall, most of the deformations occurred in the project were related with unfavorable ground conditions or sudden changes of the geological conditions rather than TBM operational parameters. There were only a couple of instances related with excavation parameters, and considering the length of the line and number of occurrences, it is possible to say that tunneling experience of the crew was suitable for the project, which should be taken into consideration while selecting workers and engineers.

REFERENCES

Arioglu, E. 1992. *Surface movements due to tunneling activities in urban areas and minimization of building damages.* Short course notes, Istanbul Technical University, Mining Eng. Dept.

Ates, U., Binen, I.S., Acun, S., Murteza, M. & Celik, Y. 2016. EPB Tunneling Challenges in Sandy Ground. *2ⁿᵈ International Conference on Tunnel Boring Machines in Difficult Grounds, Istanbul, 16–18 November 2016.*

Ates, U., Binen, I.S., Acun, S. & Celik, Y. 2017. TBM Performance and Challenges Faced During Excavation of Mecidiyekoy - Mahmutbey Metro Line. *International Tunneling Symposium in Turkey: Challenges of Tunneling (Tunnel Turkey, Istanbul), Istanbul, 2–3 December 2017.*

Attewell, P.B., Yeates, J. & Selby, A.R., 1986. *Soil Movement Induced by Tunneling and Their Effects on Pipelines and Structures.* New York: Chapman & Hall.

Barton, N. 1999. TBM performance estimation in rock using QTBM. *Tunnels and Tunnelling International* 31: 30–34.

Bilgin, N., Copur, H. & Balci, C. 2014. *Mechanical excavation in mining and civil industries.* CRC Press.

Bilgin, N., Copur, H. & Balci, C. 2016. *TBM Excavation in Difficult Ground Conditions: Case Studies from Turkey.* John Wiley & Sons.

Bilgin, N., Acun, S., Ates, U., Murteza, M. & Celik, Y. 2017. The factors affecting the performance of three different TBMs in a complex geology in Istanbul. *Proceedings of the World Tunnel Congress 2017 – Surface challenges – Underground solutions. Bergen, Norway, 9 – 15 June 2017.*

Bilgin, N., Acun, S., Korkmaz, O., Murteza, M. & Celik, Y. 2018. The Effects of TBM Operational Parameters on Excavation Performances in Complex Geology when Driving Twin Tunnels in Istanbul. *Proceedings of the World Tunnel Congress 2018. Dubai, UAE, 20 – 26 April 2018.*

Blindheim, O.T., Grøv, E. & Nilsen, B. 2002. The effect of mixed face conditions (MFC) on hard rock TBM performance, *Proceedings of the 28ʳᵈ ITA-AITES world tunnel congress, Sydney.*

Guglielmetti, V., Grasso, P., Mahtab, A. & Xu, S, 2008. *Mechanized tunnelling in urban areas: design methodology and construction control.* CRC Press.

Karakus, M. & Fowell, R.J., 2003. Effects of different tunnel face advance excavation on the settlement by FEM. *Tunnelling and Underground Space Technology* 18:513–523.

Minguez, F., Gregory, A. & Guglielmetti, V. 2005. Best practice in EPB management. *Tunnels and Tunnelling International, November: 21–25.*

Suwansawat, S. & Einstein, H.H. 2006. Artificial neural networks for predicting the maximum surface settlement caused by EPB shield tunnelling. *Tunnelling and Underground Space Technology* 21:133–150.

O'Reilly, M.P. & New, B.M., 1982. Settlement above tunnels in the United Kingdom—Their magnitude and prediction. *Proc. the Tunneling 82 Conference, Brighton.*

Polat, F. 2014. *Mecidiyeköy – Mahmutbey Metro Line, Depot, Maintenance Area and Depot Connection Lines Construction Works Geological – Geotechnical Works Report for the Section Between Yenimahale - Mahmutbey (Km 15+535 – 22+763), Istanbul.* (Unpublished).

Polat, F. 2015. *Mecidiyeköy – Mahmutbey Metro Line, Depot, Maintenance Area and Depot Connection Lines Construction Works Geological – Geotechnical Works Report for the Section Between Yenimahale – Alibeykoy River (Km 13+780 – 10+540), Istanbul.* (Unpublished).

Schmidt, B., 1974. Prediction of settlements due to tunneling in soil: Three case histories. Proc. Rapid Excavation and Tunneling Conference, Vol. 2, pp. 1179–1199.

Steingrimsson, J.H., Grøv, E. & Nilsen, B. 2002. The significance of mixed-face conditions for TBM performance, Mixed Face TBM Performance. *Sydney: World Tunnelling, pp. 435–441.*

Tan, W.L. & Ranjit, P.G. 2003. Parameters and considerations in soft ground tunneling. *The Electronic Journal of Geotechnical Engineering* 8(D):344.

Zhu, W.B. & Ju, S.J. 2005. Mix ground and shield tunnelling technology in Guangzhou. *Proceedings of 2005 Shanghai International Forum on Tunnelling, Shanghai.*

Tunnels and Underground Cities: Engineering and Innovation meet Archaeology,
Architecture and Art, Volume 11: Urban
Tunnels - Part 1 – Peila, Viggiani & Celestino (Eds)
© 2020 Taylor & Francis Group, London, ISBN 978-0-367-46899-6

Effects of construction and demolition of a TBM excavated tunnel inside existing diaphragm walls

S. Autuori, M.V. Nicotera & G. Russo
University of Napoli Federico II, Naples, Italy

A. Di Luccio & G. Molisso
Ansaldo S.T.S., Naples, Italy

ABSTRACT: The Line 6 of Naples Underground is part of the underground network system. The paper reports aspects of the design and monitoring data during the tunnel construction and demolition inside the San Pasquale Station shaft. The station shaft has a rectangular shape in plan of 85 × 24 m; the maximum excavation depth is about 27 m, and the water table is about 1 m below the ground level. The excavation is supported by large T-section diaphragm walls. The diaphragm walls were installed before the tunnel excavation using an EPB TBM crossed the station area. Two rectangular chambers located on both short sides of the shaft were realized by treating with jet grouting the soil volume confined between specially constructed plane diaphragm panels. During the tunnel construction and the lining demolition, effects on the peripheral diaphragm panels and further on the close buildings were observed and are summarized in the paper.

1 INTRODUCTION

With an urban population of about 1.2 million people more than 5 million in the metropolitan area Napoli is the third largest city of Italy. The density of population is among the highest in Europe and accounts for nearly 2,000 inhabitants per square kilometre. In 1997 the Municipality of Napoli approved a new City Transportation Plan, that has led to a significant pressure for the construction of new underground train lines, stations and car parks. With nearly 64 km of track currently operating and more to open as part of the expansion plan, Napoli Underground had a daily ridership of 470,000 in 2005, risen to over 700,000 in 2013 (Russo et al., 2012). The line 6 of the network can be broken down into 3 stretches:

- stretch between Mostra and Mergellina stations, partially connecting the borough of Fuorigrotta to the city centre, is already operating;
- stretch between Mergellina and Municipio stations with three new intermediate stations, Arco Mirelli, San Pasquale and Chiaia is nearly finished;
- stretch between Mostra and Porta del Parco stations is fully designed and waiting for financial support.

In the following, attention will be focused on some details of the design and the construction problem of San Pasquale station of Line 6. A more general and complete description of the station is available elsewhere (L'Amante et al., 2012; Russo et al., 2012; Autuori, 2016; Russo et al. 2016). The focus in the paper is on the effects of the construction with the EPB and the following demolition of the tunnel inside the station area previously surrounded by huge peripheral diaphragm T walls.

2   SAN PASQUALE STATION: SITE AND DESIGN ISSUES.

The main body of San Pasquale station has a rectangular shape in plan of 85.50 m × 24.10 m and the maximum excavation depth is approximately 27 m (i.e. 26 m under the groundwater table) (Figure 1). The long side of the station is parallel to the longitudinal tunnel axis and the closest buildings are located on the north side keeping approximately a unique alignment which is again parallel to the long side of the station. The main and rather large shaft contains the passenger platforms and eliminates the necessity of excavating underground platform tunnels.

The excavation is supported from T-section diaphragm walls made by reinforced concrete and built using hydromill equipped by a 90° rotating drilling head. Each panel of the diaphragm walls was built by intersecting two separate rectangular excavations. The depth of the panels is about 50 m allowing a large embedment in the Neapolitan Yellow Tuff formation (NYT). The upper portion of each panel was excavated under the protection of Cutter Soil Mixing (L'Amante et al., 2012) to avoid hole instability and to reduce induced displacement in the surrounding area.

At the design stage the decision to build the station according to a top-down procedure was taken. This choice allowed to avoid the troubles connected to drilling ground anchors at large depth below the groundwater table. The r.c. diaphragms have been executed first, leaving soft eyes with fiberglass reinforcement bars to be drilled by TBM. The passage of the TBM was the second main step while the excavation of the station was executed as the final step and required the demolition of the tunnel lining within the station box. At the design stage geotechnical investigations were carried out in the area occupied by the station (i.e. 2000 m²). The site is inserted in an urban area bounded even by historical buildings on the north side and by the sea with the interposition of a public garden on the South side. The area is relatively flat

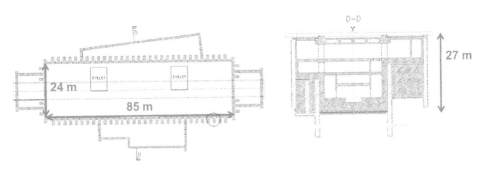

Figure 1.   Plan view and transversal section of the station.

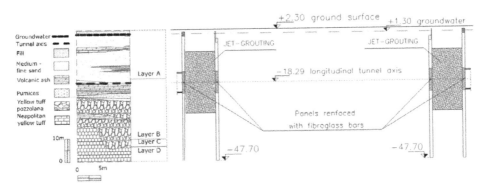

Figure 2.   Longitudinal section with layers and geological info.

with the ground level located between +2 m and +2.30 m a.s.l. while the groundwater table is at +1.30 m a.s.l.

Both in situ and laboratory geotechnical tests were carried out and more details may be found elsewhere (Russo et al. 2012; Fabozzi et al. 2017).

Figure 2 shows the geological soil profile corresponding to the transverse central cross-section. The upper main layer consists of marine sand underlying a few meters of hand-made ground and extending down to a depth of 17 m from the ground surface (layer A). It is followed by a layer of pyroclastic products consisting of silty sands, or ashes and pumices from the depth of 17 m down to a depth of 41 m (layer B). Going deeper a thin layer of uncemented yellow tuff pozzolana (layer C) separates the layer B from the lithified facies of the NYT formation (Layer D).

3 TBM AND TUNNEL CONSTRUCTION STEPS

At the design stage according to the prevailing ground conditions along the stretch an Earth Pressure Balance Tunnel Boring Machine (EPB-TBM) was selected for the construction of the stretch of Line 6 between Mergellina and Municipio Station. This machine turns the excavated material into a soil paste that is used as pliable, plastic support medium once it is mixed also with a conditioning agent. This medium is used to balance the pressure conditions at the tunnel face, avoids uncontrolled inflow of soil into the machine and creates the conditions for a fast advancement while keeping low settlement. The EPB-TBM used is produced by WIRTH, model TB816H/GS T with a maximum diameter D = 8.15 m. More details on the tunneling machine and the construction process of the lining maybe found elsewhere (Bitetti, 2010, Bilotta & Russo, 2013). The TBM entered the station shaft boring the tunnel through a first diaphragm wall an intermediate grouted soil portion and finally the T shaped main diaphragm wall reinforced by fiberglass rebars. In Figure 2 this sort of entry plug is graphically described and sketched and is similar to the exit plug on the other side of the station. The function of the plugs is to reduce or preferably to avoid perfectly water inflow at the back of the tunnel face and around the tunnel shield before the watertight lining is installed. Check on the uplift of the empty tunnel was carried out at the design stage, comparing the stabilizing actions of soil weight above tunnel lining and upwards action of buoyancy. The procedure to build the inside station box was such that a progressively reduction of the thickness of the soil covering was expected with the ongoing excavation process; at the same time buoyancy action was expected to reduce due to the groundwater lowering operations. However in order to guarantee an adequate safety factor against the uplift in any step of the construction procedure it was decided that the section of the tunnel had to be almost entirely filled with a poor cement mortar. The filling of the section allowed also to proceed to safe demolition of the tunnel lining approaching the opening of the ring without inducing any structural collapse which could result in a serious risk of injuries for the workers. In Figure 3 a picture of the ongoing excavation process while approaching the upper portion of the tunnel lining is reported. Just for the sake of completeness the following list reports the main construction stages (at least by a geotechnical point of view) of the whole station:

a. construction of all the diaphragm walls;
b. bottom plugs of jet-grouting below the escalator shaft and on the short sides of the station (entry and exit plugs for the TBM);
c. main tunnel excavation via TBM;
d. drains installation;
e. excavation with preliminary groundwater lowering, following top down technique; in this stage is included the demolition of the tunnel lining;
f. installation of waterproof membrane at the bottom of the main shaft;
g. concreting of the bottom slab;
h. stop of groundwater lowering;
i. partial cutting of intermediate slabs and finishes.

Figure 3.   Picture of the tunnel lining demolition during the excavation process of the station.

The whole construction lasted approximately 4 years while the tunnel construction lasted about 1 month and the demolition step of the tunnel lining lasted about 6 months.

4   MONITORING LAYOUT

The monitoring plan was based on a rather huge number of instruments and on a rather tight schedule for the readings which were in most cases automated via wireless data-loggers. The schedule was obviously adjusted several times during the whole construction process according to the velocity of the construction sequence. In Figure 4 the layout of the some instrumentation adopted is sketched on the plan view of the station with a partial representation of the surrounding buildings, more details about monitoring layout and used instruments are available (Russo et al., 2012; Autuori, 2016; Russo et al. 2016, Russo et al., 2018). A total of sixteen inclinometers

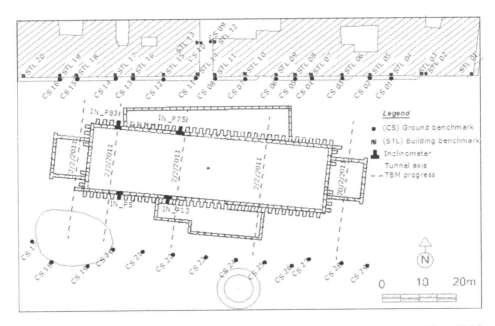

Figure 4.   Plan view of the station with the monitoring layout and location of the tunnel face (TBM) at several dates along the station box.

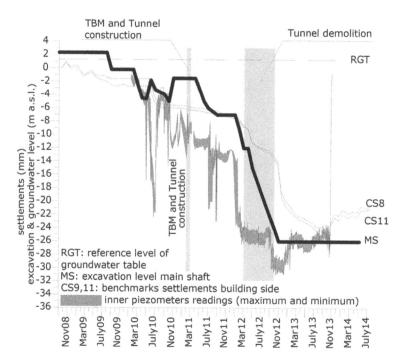

Figure 5. Average settlement on the building side during the whole construction process.

were installed inside the r.c. panel of the diaphragm walls with twelve out of them embedded in the huge T section panels of the main station and only four in the rectangular panels of the escalator shaft. Inclinometers were installed also in the ground close to the diaphragm walls. The numbering of the inclinometers pipe showed herein is reported in Figure 4. Both vibrating wire piezometric cells and standard open pipe piezometers were installed to follow the groundwater movements and to measure the change of the piezometric head induced by natural fluctuations and mainly by the heavy dewatering used to prevent the water inflow during the excavation of the main shaft. The use of piezometric vibrating wire cells was limited to the station inside to avoid the additional difficulty of handling and preserving the integrity of plastic pipes during the top down excavation in a relatively small working area. Tens of optical survey points consisting of marks on sidewalks and streets were monitored to measure the ground movements on both the sides of the main station shaft. On the side opposed to the sea, where several buildings are located, benchmarks directly installed on structural members at about 3 m above the ground surface were also monitored. The location and the numbering of the marks are represented in Figure 4.

In Figure 5 just for completeness a plot of the average settlement on two central marks is reported as measured during the whole construction process whose main steps are listed in the above section from a) to i). More details are reported elsewhere (Russo et al., 2016; Autuori 2016, Russo et al., 2018). As shown by the plot the maximum value of the settlement recorded during the whole construction process was as high as 24.75 mm recorded rather uniformly in at least two benchmarks. In the present paper the focus is limited to the effects induced by the tunneling construction and the subsequent demolition steps.

5 DISPLACEMENT INDUCED BY TBM PASSAGE INSIDE THE DIAPHRAGM WALLS

As mentioned before the TBM passed inside the station box when the peripheral diaphragm walls were already built. The monitoring actions partially automated and partially executed

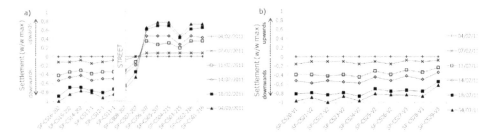

Figure 6. Settlement induced by TBM passage a) on the building side b) on the seaside.

by technical operators on a weekly schedule allowed to identify the effects of the passage of the TMB on the existing structure including the r.c. diaphragm walls and the outside buildings. On the seaside the settlements were measured on benchmarks installed at the ground level in the public gardens of *Villa Comunale* facing the sea.

The effects of the TBM were previously monitored with several greenfield sections located in the stretch between Arco Mirelli and San Pasquale station. The well conducted TBM was capable of producing very small volume losses and sometimes even small uplift movements were recorded during and after the passage of the machine. These readings have been already discussed elsewhere (Bitetti, 2016) and some data are also reported in another paper submitted to this conference (Bilotta et al. 2018)

In the plan scheme of Figure 4 the location of the TBM face at different dates is represented.

As shown by Figure 6 a) the pattern of the settlement induced in different locations along the building side seems not strictly depending on the position of the TBM face (i.e. different dates).

The vertical displacements increase with time (and consequently with the development of the excavation process) but they are directed downwards in a first part of the overall station side while they reverse the sign and are directed upwards in the last part of the stretch. The same trend is not observed on the seaside (see Figure 6 b)) where all the displacements are directed downwards. A possible cause of the trend shown on the building side could be just the stiffening presence of the buildings (Aversa et al, 2015; Bilotta et al., 2017). Of course even variations on the pressure in the face chamber of the TBM could be responsible of such a behavior. Detailed data on this variable are unfortunately not available. As discussed elsewhere (Bilotta & Russo 2011) even the shield effect due to the presence of the diaphragm walls could have influence on the recorded results. The data plotted are normalized dividing the settlement for the maximum value which is in any case very small and in the order of a few tenths of a millimeter (i.e. 0.3/0.4 mm).

In Figure 7 few inclinometers profiles are plotted. They were selected among those derived from the readings performed during the passage of the TBM. The profiles are referred to two monitoring sections along the station box, the inclinometers 83 and 75 being on the building side and the inclinometer 5 and 13 being on the opposite side of the shaft (sea side).

The maximum horizontal displacement was rather small and not particularly significant by an engineering point of view, but, however, much larger than the vertical ones measured at the ground surface, the values ranging between 2 mm and 4 mm.

The shape of the profiles on the opposite sides was nearly symmetrical confirming the good quality of the measurements setup.

In all the four cases a maximum inwards movement was registered approximately at the depth of the tunnel crown while the reverse occurred at the top of the profile close to the ground surface, where generally an outward horizontal movement was observed.

It is important to keep in mind that the displacements were measured inside the structural diaphragm walls which at the time of the passage of the TBM were already restrained in the top part by the thick slab closing the station box. This could at least partially explain the

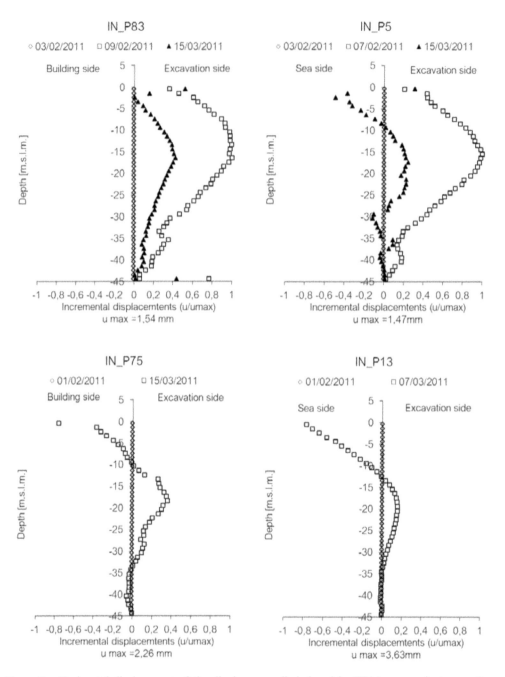

Figure 7. Horizontal displacement of the diaphragm walls induced by TBM passage in two section along the station box.

bulging shape showed by the thick diaphragm walls which in turn acted as a mitigating measure for the displacements of surrounding buildings whose vertical movements were indeed even smaller than the small horizontal displacement revealed by the inclinometers investigations (Bilotta & Russo, 2011).

6 DEMOLITION OF THE TUNNEL LINING DURING THE EXCAVATION STAGE

As shown by plot in Figure 5 the demolition of the tunnel lining occurred almost one year after its construction. The main steps of the safe process of the demolition were described in the section 3 of the paper.

One crucial aspect of the whole procedure was the filling of the tunnel section with lightweight concrete in order to prevent uplift of the section during the excavation of the station box.

No significant movements however, neither outside nor inside the diaphragm walls, were recorded during these steps. After the excavation of the tunnel cover the workers ended on the crown of the lining and since that moment the demolition of the tunnel coincides with the excavation of approximately 8 m of soil inside the station box. The tunnel section was located at depth between –14 m asl at the crown and –22 m asl at the invert.

Some concern was arisen at the design stage about the demolition phase and particularly about the sub-step corresponding to the ring opening following the removal of the key segment of the prefabricated lining. However, the measurements reported in Figure 5 demonstrate that the steps corresponding to the tunnel removal caused an increase of the settlement in the surrounding buildings of about 6 mm (from 8 mm to 14 mm).

The period involved in the tunnel demolition spans approximately over 6 months from March 2012 to September 2012. In the same period the results of the investigations of the couple of opposite inclinometers pipes IN_P83 and IN_P5 already adopted for showing the horizontal displacement caused by TBM passage are plotted in Figure 8. A maximum horizontal displacement directed inward the excavation in the range from 6 mm to 7 mm was registered at the depth of tunnel axis approximately.

These values are in good agreement with the settlement increase caused by the tunnel demolition. Furthermore the shape of the normalized profiles are similar to those already shown

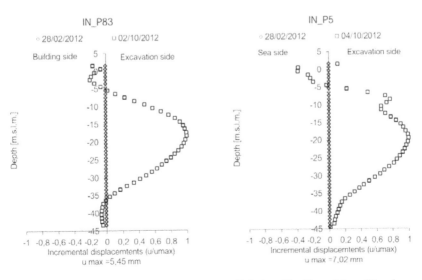

Figure 8.   Horizontal displacement of the diaphragm walls induced by Tunnel demolition in one section along the station box.

for the steps of the TBM passage during tunnel construction. In both cases the prevailing action is an excavation process inside the station peripheral diaphragm walls.

7 CONCLUSION

In the paper a significant phase of the construction of the whole San Pasquale station of the new Line 6 of Napoli Underground is described in detail. The overall construction with all the finishes and the technical plants was successfully completed in the 2017. The focus of the paper was first on the passage of the TBM, a full section EPB with a diameter of approximately 8 m inside already existing r.c diaphragm walls where soft eyes with fiberglass bar as reinforcement were predisposed.

This option was chosen because at the design stage it was not possible to establish reliably whether the station box or the tunnel would be completed first. When it was clear that the TBM had to pass after the diaphragm walls construction but before the shaft excavation some concerns about the effects of the TBM passage and subsequent demolition of the tunnel lining were present among the technician involved in the construction process. Some additional measures for precautions as an increased groundwater lowering and the partial infill of the tunnel section with light poor cement mortar were adopted. As testified by the accurate monitoring results reported in the paper, also thanks to these further precautions, the operations were safely conducted and in the end, significant or more precisely unforeseen effects were not registered.

ACNOWLEDGEMENTS

The Authors wish to thank to Metropolitana di Napoli Spa and to Ansaldo S.T.S. for providing access to monitoring data and to the site during the whole construction process of the San Pasquale Station.

REFERENCES

Autuori S. 2016 La stazione San Pasquale della metropolitana di Napoli. Costruzione, monitoraggio ed analisi dei comportamenti osservati (ph.D. Thesis) - *Università degli Studi di Napoli Federico II, Dottorato in Ingegneria delle costruzioni, marzo 2016*; http://www.fedoa.unina.it/10823/

Aversa, S., Bilotta, E., Russo, G., Di Luccio, A. 2015. Ground movements induced by TBM excavation under an historic church in Napoli. In: *Proceedings of the XVI ECSMGEGeotechnical Engineering for Infrastructure and Development. ICE Publishing*, pp. 425–430. ISBN 978-0-7277-6067-8.

Bilotta, E., Paolillo, A., Russo, G., & Aversa, S. 2017. Displacements induced by tunnelling under a historical building. *Tunnelling and Underground Space Technology*, 61, 221–232.

Bilotta, E., Russo, G. 2011. Use of a line of piles to prevent damages induced by tunnel excavations *Journal of Geotechnical and Geoenvironmental Engineering* ISSN 1090-0241 137(3) pp. 254–262 doi:10.1061/(ASCE)GT.1943-5606.0000426

Bilotta, E., Russo, G. 2013. Internal forces arising in the segmental lining of an earth pressure balance-bored tunnel. *Journal of Geotechnical and Geoenvironmental Engineering*, 139 (10), pp. 1765–1780 DOI: 10.1061/(ASCE)GT.1943-5606.0000906

Bitetti B. 2010. Effects of tunneling in urban areas. (ph.D. Thesis) - Università degli Studi di Napoli Federico II, Dottorato in Ingegneria delle costruzioni, 2010; http://www.fedoa.unina.it/id/eprint/8445

L'Amante D., Flora A., Russo G., Viggiani C. 2012. Displacements induced by the installation of diaphragm panels. *Acta Geotechnica* 7(3): 203–218; doi: 10.1007/s11440-012-0164-9

Fabozzi S., Licata V., Autuori S., Bilotta E., Russo G., Silvestri F. (2017) Prediction of the seismic behavior of an underground railway station and a tunnel in Napoli (Italy) *Underground Space*, Volume 2, Issue 2, June 2017, Pages 88–105 open access

Russo, G., Autuori, S., Corbo, A., & Cavuoto, F., 2015. Artificial Ground Freezing to excavate a tunnel in sandy soil. Measurements and back analysis. *Tunn. Undergr. Space Technol.* (50), 226–238.

Russo G, Autuori, S, Nicotera M.V. 2018. Case Study: Three-dimensional performance of a deep excavation in sand. Accepted for publication on *J Geotech Geoenv Eng* ASCE

Russo, G., Nicotera M.V., Autuori S., 2016. San Pasquale Station of Line 6 in Naples: Measurements and Numerical Analyses, *Procedia Engineering*, Volume 143, 2016, Pages 1503–1510, *open access*, https://doi.org/10.1016/j.proeng.2016.06.177-Advances in Transportation Geotechnics 3. The 3rdInternational Conference on Transportation Geotechnics (ICTG 2016)

Russo, G., Viggiani, C., Viggiani, G.M.B. 2012. Geotechnical design and construction issues for lines 1 and 6 of the Naples underground. *Geomechanik Tunnelbau* 5 (3), 300–311.

Santangelo V., Di Martire D., Bilotta E., Ramondini M., Russo G., di Luccio A., Molisso G. 2019 Use of DInSAR technique for the integrated monitoring of displacement induced by urban tunneling. *Proc. WTC 2019*, Napoli.

Tunnels and Underground Cities: Engineering and Innovation meet Archaeology, Architecture and Art, Volume 11: Urban Tunnels - Part 1 – Peila, Viggiani & Celestino (Eds)
© 2020 Taylor & Francis Group, London, ISBN 978-0-367-46899-6

Ground movement problems in EPB tunneling through confined station boxes

T. Babendererde, L. Babendererde & R. Hasanpour
BabEng GmbH, Lübeck, Germany

ABSTRACT: In the recent years, the shielded TBMs are implemented for tunneling in the shaft boxes of the subway stations to provide a direct connection between the tunnel routes throughout the shaft. This ensures no machine stoppage within boring process that decreases the operational costs and then a short distance between surface and the underground platforms is created. However, this adaption can result in some adverse effects such as break-ups of the overburden or blow-outs of the support medium. In this paper, challenges regarding a shallow tunneling using a shield machine through a rectangular shaft box with water bearing soil are described. A case study was considered for evaluation of the instabilities during tunnel excavation, and the relevant calculations regarding the face support pressure are given. Finally, two proposed approaches are presented in order to apply in the similar situations to avoid surface instabilities and some discussions on the recommended solutions are given.

1 INTRODUCTION

Tunneling using shielded TBMs is widely applied in metropolitans area for construction of underground infrastructures such as metro, wastewater tunnels and etc. This is due to the fact that using shield tunneling mitigates considerably the risk of ground settlements within excavation and consequently damage to the existing buildings. When the tunnels are excavated near enough to the ground surface, the tunnel construction costs and the compulsory depth of the underground stations can be decreased significantly. Furthermore, the other merits of tunnel borings at shallow depth can be addressed by the lower operational costs and short distance between surface and the underground platforms (Vu et al., 2015).

In the recent years, shielded TBMs have been utilized in many cases for tunneling through subway stations. This provides a direct connection between tunnel routs via the shaft box and without any stoppage at the boring operation. The positive effect of implementing this excavation method is that the remaining construction works can be started immediately after completing the tunnel route without wasting long time for exit procedures into the shaft. Although, the process was only implemented in dry soil or in high cohesive soils with high bearing capacity, now it can be adapted to the ground with water bearing soils. However, this adaption can result in some adverse effects such as break-ups of the overburden or blow-outs of the support medium, recalling that some precautions measures should be considered and some lessons should be learned.

The difference between the dry/cohesive approach compared to the water bearing soils can be addressed by the applied face pressure and its consequences. While in dry or highly cohesive soils the face pressure is more robust, it should be maintained at all times in the water bearing soft grounds throughout the machine operation. Additionally, the maintaining of a face pressure becomes more difficult in the approaching structures like shafts or even more in tunneling through shafts. Therefore, this challenging process is not an issue at all regarding tunneling on the stretch away from all construction shafts, it is only related to the inventiveness of the tunneling team.

Assessing the impact of underground construction on surface issues that can be observed as break-ups of the overburden or blow-outs of the support medium in the urban areas should be considered at the design stage (DAUB, 2016). Studies in this regard have mostly focused on the ground movements around shield tunneling and the settlement on the surface, but research focusing on the ground displacements and its mechanism in mining through a shaft box that can affect nearby buildings or cause to surface instabilities have not been reported yet. In this paper, the challenges regarding a shallow tunneling using a shield machine through the water bearing soils in a shaft box are explained. Furthermore, evaluations of the instabilities during tunnel excavation and discussions on the mechanism of the observed failures are given for a subway project. Finally, the proposed solving approaches in order to employ in such situations and discussions about the offered methods are presented.

2 CHALLENGES OF SHIELD TUNNELING THROUGH A RECTAGULAR SHAFT BOX

The normal construction sequence of a shaft box relies on closing the sides with the shaft walls (piles) and installing slabs at bottom and top before approaching TBM to the shaft. The water pressure in the shaft remain similar to its pressure outside the shaft box along the tunnel route, providing that no water penetration from the surface into the shaft box is allowed (see Figure 1). Once the TBM breaks through the shaft wall, the machine should maintain the pressure from the outside until the tunnel is fully sealed around in the shaft walls (see

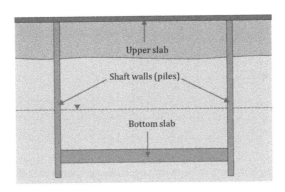

Figure 1.   Construction of a rectangular shaft box for a subway station.

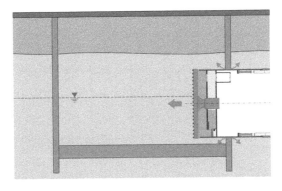

Figure 2.   Entering cutterhead into the shaft while the shield is still outside of the shaft wall.

Figure 2). While the water pressure outside the shaft can regulate itself with the surrounding groundwater, the water level inside of the shaft box, which is corresponding directly to the TBM operation, is being trapped into the rectangular box. The water pressure inside of the shaft box is relatively equal with the pressure level outside of the shaft walls until fully sealing of whole TBM into the shaft (see Figure 3).

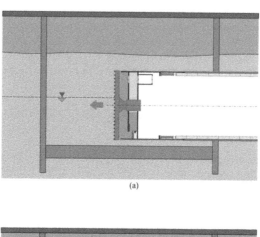

(a)

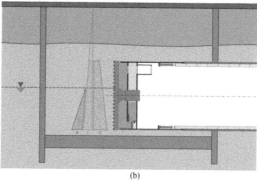

(b)

Figure 3.   a) TBM driving into the shaft in which backfilling is placed near the shaft wall, b) face pressure is in balance with earth and water pressures (groundwater level into the shaft box is equal with the pressure level at the tunnel route).

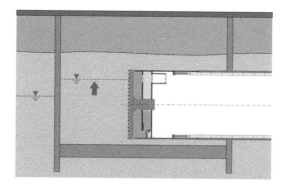

Figure 4.   Construction of a rectangular shaft box of a subway station.

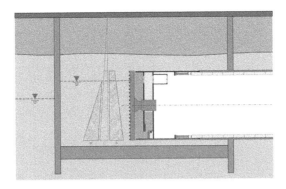

Figure 5. Face support pressure required to be increased to keep a balance with earth and water pressures as groundwater level increases by machine advancing through the shaft box.

TBM advancing through the shaft box is resulted in removing the certain volume of the soil inside the box. As the face support pressure applied by machine is always higher than the surrounding earth and water pressure, then the water inside the shaft would not be moved away. Therefore, it remains relatively unchanging, despite the fact the TBM removes the soil inside the box. The removed soil is replaced with the TBM volume. With TBM advancing through the box, the higher volume of the TBM and back-up system are placing into the box. This causes to gradually raising at groundwater level, consequently demand for higher support pressure to stabilize the tunnel face (see Figures 4 & 5). The increase at water level inside of the shaft can cause to flooding around the shafts as the water was confined above the soil into stairways or onto the surface.

More critical situation is when the TBM is driven close to the surface. If the water level hits the roof, no further relieve is possible. This situation equals a hydraulic cylinder. The whole surrounding ground is fully filled with a liquid. Whenever the TBM tries to move on, the pressure around the TBM raises and causes to the face collapse or water blow-out (See Figures 6 & 7). The similar mechanism have been observed in several TBM projects over the world.

3 DESCRIPTION OF THE MECHANISM

Figure 6a shows an example of shield tunneling through a shaft box that has been resulted in an increase in water level to a certain value. Figure 6b also shows the raising and dropping of the water level for several sequences of the excavation phases and lining installations. The incrementation of the water pressure during tunneling and equalization of it during installing of the segmental rings can be seen clearly in this Figure. Once the machine is fully inside of the shaft walls and digs through the ground, the water pressure increases until completion of one excavation phase and starting installation of the segmental lining. During lining installation, the water pressure is dropping to the natural level until the excavation phase is commenced once again.

In tunneling through water bearing soils inside of a closed shaft box, the water pressure can be raised to a level and lead to surface collapses. Figure 7 shows an example of such failure, it can be seen that the water pressure exceeded to the anticipated level and by this it resulted in a sinkhole. The same events have been observed in several tunneling projects around the world.

Figure 8 shows the blow-out mechanism because of increasing groundwater level and its trapping beneath the sealed soil layers with TBM advancing. The mechanism is always the same and it can be demonstrated not only for water in the confined spaces but also with the compressed air under sealed soil layers. Once the face support medium or parts of it is either

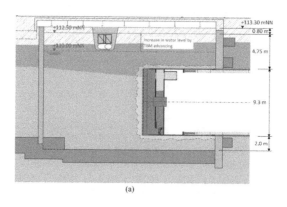

(a)

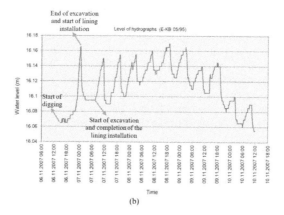

(b)

Figure 6.   a) Increase at water pressure due to TBM excavation b) Corresponding changing in water level during tunneling operation and lining installation.

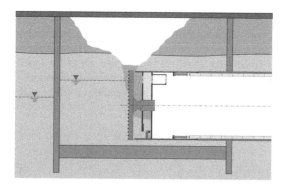

Figure 7.   Face collapse and break-ups of the overburden in the shaft box.

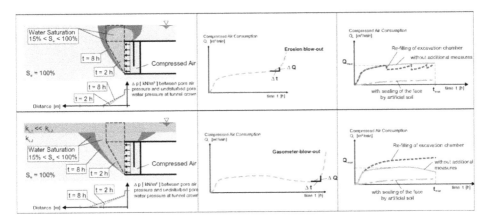

Figure 8. The blow-out mechanism in excavating through permeable ground inside of the shaft box.

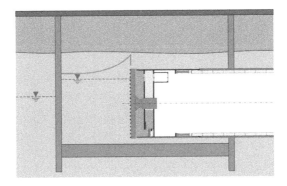

Figure 9. Pressure increase at the tunnel face with gradient.

captured in a confined area (shaft or layer of sealing material) the necessary gradient of the face support cannot establish itself. The pressures balance themselves at the face and due to the higher weight the soil ravels into the excavation chamber (see Figure 9).

4 DISCUSSION AND SOLUTION METHODS

To avoid such a problematic situation, the mechanism described above should be evaluated in detail at the design stage and also during the preparation phase. The relevant calculations should consider the volume of groundwater that will be drained off within excavation in the closed box. Only the additional water should be allowed to be injected by slurry, and if applicable, through injecting the backfill grouting. This additional injected water and the reduced space for the remaining water is changing the water level noticeably.

To overcome this situation, a water balance should be carried out within excavation process. As long as the TBM is driving into the shaft box, the water level should be monitored and lowered to an acceptable level using water pumps in case of any increase in the water level. This should be prepared and calculated before displacing the volume that is highly depending on the tunnel diameter and the excavation speed.

In case of performing compressed air under sealing layer, the confined air should be evacuated and drained by means of implementing some special drainage tools. Figure 10 shows a

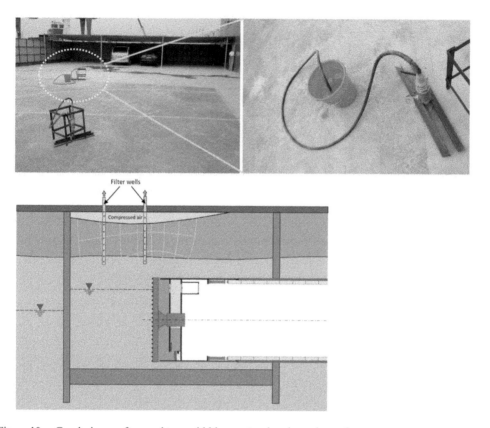

Figure 10. Gas drainage of ground to avoid blow-out or break-up the surface.

case of evacuating process of the compressed air for a shield tunneling project in china. In this method, some usual water wells are drilled under the sealing layer. The "filter area" collects the trapped compressed air and allow reducing the pressure beneath the sealing layer. By applying this method, the pressure gradient from the tunnel face into the ground can establish itself and supports the face.

5 CONCLUSION

In this paper, the mechanism of surface instabilities due to shield tunneling crossing though a rectangular shaft box was studied. Special attention was paid to the controlling parameters of TBM and ground that influence on the tunnel-shaft interaction during the tunnel excavations and lead to blow-out or break-up at ground surface.

It is found that the groundwater level is increasing with advancing machine through the ground inside of the shaft box. It confined between the slabs and shaft walls and can be resulted in instabilities like break-ups of the overburden or blow-outs of the support medium. To overcome these problematic situations, a water balance should be performed within TBM operation. As long as the TBM is advancing into the shaft, the water level should be monitored and lowered by pumps to an acceptable level. In case of being compressed air under the sealing layer, evacuating and draining the compressed air would be very beneficial. By applying one of the abovementioned methods, a safe shallow tunneling in the shaft box or under sealing layers will be provided.

REFERENCES

ITA-AITES. 2016. Recommendations for Face Support Pressure Calculations for Shield Tunnelling in Soft Ground. *Deutscher Ausschuss für unterirdisches Bauen (DAUB)*.

Vu, M.N., Broere, W., & Bosch, J. 2015. Effects of cover depth on ground movements induced by shallow tunnelling. *Tunnelling and Underground Space Technology 50*:499–506.

Tunnels and Underground Cities: Engineering and Innovation meet Archaeology,
Architecture and Art, Volume 11: Urban
Tunnels - Part 1 – Peila, Viggiani & Celestino (Eds)
© 2020 Taylor & Francis Group, London, ISBN 978-0-367-46899-6

Evaluation of long-term tunnel behaviour in preparation for construction over existing tunnels, in The Netherlands

K.J. Bakker
Delft University of Technology, The Netherlands & WAD43 bv

K.J. Reinders
Delft University of Technology, The Netherlands

C.J.A. van der Wilt
IV-Infra, The Netherlands

J. Jonker
Movares, The Netherlands

A.F. Pruijssers
ExAequo The Netherlands

J.J. Bogaards
ProRail, The Netherlands

ABSTRACT: In 2006 the Betuwe Route including three bored tunnels; was opened. Since then nearly 10 years of operation had passed when in 2015, new construction works in the vicinity of the tunnels was undertaken. During constructions the interaction needed to be evaluated. For that, systematic surveys by laser scanning in the tunnels where performed and additional ring joint rotations were evaluated. It appeared that the soft soil conditions at the Botlek tunnel leads to larger deformations, whereas the design of the rubber sealings at Pannerden is more critical for additional deformations. For the Botlek tunnel the unfavourable situations was Ramps being built in close vicinity of the tunnels, whereas at Pannerden at one cross section clay deposits from an adjacent Stone forging factory need to be relocated and at another location a cross passage needs to be built over the tunnel approach. Characteristic data from both tunnels is shown.

1 INTRODUCTION

In 2015, the A15 motorway was widened and a new Botlek bridge was constructed, close to the existing Botlek rail tunnel. Shortly after that, an extension of A15 motorway connecting it to the A12 was planned near the tunnels at Pannerden. Both these tunnels are part of the Betuwe route, a cargo railroad connecting the Rotterdam Harbour with Germany that was opened in 2006.

The interaction between the new infrastructures and the bored tunnels needed to be evaluated. In order to evaluate to what extent, the tunnels would be influenced by the loads of the new infrastructure in order to prevent leakage to the present water bearing rubber sealings.

1.1 *The Betuwe Route for Rail cargo*

The developments for the Betuwe Route for rail cargo started in the middle 1990's. The initiative was triggered by the forecast that a further growth of road transport to Germany without

Figure 1. The Betuwe Route as part of the European Network for Cargo Rail.

Figure 2. Betuwe route for Rail Cargo, Botlek tunnel under the "Oude Maas" (1), The Sophia Tunnel under the "The North" (2) and the "Pannerdensch Canal tunnel", under the Pannerdensch Canal (3).

any intervention would congest the road system. Politics decided that the best possibilities for extension of the transport capacity would be a new Rail Road for cargo from the harbour of Rotterdam to the Hinterland, see Figure 1. The so called the Betuwe route.

In the same period there was a major discussion in the Netherlands that new infrastructure above ground was unfavourable as it would destroy the open landscape. Therefore, at 3 locations in the Betuweroute a bored tunnel was planned and additional research into this construction method was arranged, see Bakker (2008)

1.2 The tunnels in the Betuwe route

The three bored tunnels in de Betuwe Route are: the Botlek rail tunnel, the Sophia Tunnel and the tunnel under the Pannerdensch canal. See figure 2. All three are twin tunnels with an inner diameter of 8.65 m and a lining thickness of 0,40 m. The lining consists of seven segments and a keystone. The maximum inclination for the approaches of the tunnels is 2,5 %, which is equal to the inclination of others tunnels in the Netherlands. The first tunnel built, was the Botlek tunnel under the Oude Maas river, adjacent to the old Botlek Rail bridge. It was built with an Earth Pressure Balance Machine.

The decision to use EPB was influenced by the desire to monitor differences between the EPB and the Slurry shield method, that was used at the 2nd Heinenoord tunnel, a bored road tunnel, not far from the location where the Botlek tunnel.

The Botlek tunnel is 1835 m long and apart from the middle section underneath the river the tunnel is mainly located in soft layers of silt and clay. The east approach crosses layers of peat, which where improved by injection of grout to facilitate boring. See figure 3.

The Sophia Rail tunnel is located about twenty kilometres to the East of the Botlek rail tunnel and passes underneath the River "The north" and what is called "The Reed lands". The tunnel has a total length of 8115 m, including the open U-shaped approaches. The bored part of the tunnel is 4240 m long. Unlike the Botlek tunnel, the Sophia tunnel is mainly located in the stiffer Pleistocene Sand layers, which are located much shallower, at a depth of around NAP -13 m, compared to the location at the Oude Maas where the Pleistocene sand is located at around NAP -20 m. (NAP is Amsterdam Ordnance datum at about average sea-level).

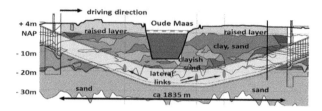

Figure 3. The underground at the Botlek rail tunnel.

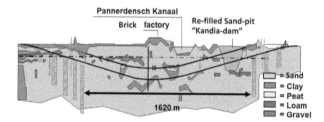

Figure 4. Geologic Length profile at the Tunnel Pannerdensch Canal.

Finally, the Tunnel at Pannerden, see Figure 2 and Figure 4. has a length of about 1620 m and underpasses a canal which has a river like appearance, connecting the river Rhine with the Lower Rhine, the river Lek and the river IJssel. Although the "river" here is canalized, there is a Summer and a Winter bed, that may be flooded when there is a high river flow.

The tunnel underpasses a brick factory that has an elevated greenfield level and is located within the Winter bed. At the East Bank the tunnel was bored through an old sand pit, called Kandia Dam, that was back-filled and densified prior to construction to let the bore machine pass. The slopes of the Kandia dam may consist of sand with a low relative density.

2 10 YEARS AFTER COMPLETION

After about 10 years of train service along the Betuwe route, new infrastructures near the tunnels were planned and developed. These new developments may influence the structural behaviour of the tunnels and therefore an evaluation was needed, considering the present state of the tunnels.

2.1 *The Botlek rail tunnel*

In Rotterdam at the Oude Maas, a new Botlek bridge was built, within the framework of the reconstruction of the A15 Motorway. The old bridge that was built in 1955 was replaced by a new larger one next to the old one. The new bridge was opened in 2015. The bridge has different lanes for road traffic, a railroad, and a side lane for bicycles. These different functions require different approach structures, such as embankments and piled foundations adjacent to the tunnel and embankments on top of the tunnel.

2.2 *The Sophia Rail tunnel*

At the Sophia rail tunnel no new structures are constructed next to the tunnel and therefore this tunnel is not discussed in this paper. For more information, see Jonker 1999 and Stive 1999. The main concern is the East Portal, where lateral deflections where measured. The U-shaped trench is founded on vertical piles only, that give little resistance to lateral load.

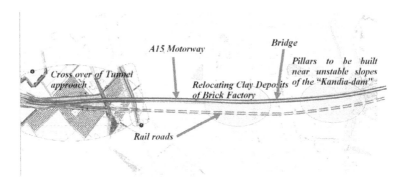

Figure 5. Reconstruction works related to the construction of a bridge for the A15 Motorway and a crossover of the Betuwe Route rail connection, see also Figure 4.

2.3 *Crossover of Motorway A15 and moving clay deposits at tunnel Pannerdensch Canal*

The developments near the tunnels at Pannerden are more complex. At the West Entrance a crossover of the motorway will be constructed near the tunnel approach. At the centre of the tunnel, clay deposits from an adjacent Stone forging factory are present. In order to build the main piers of the bridge for the A15, these clay deposits need to be relocated. At the East Entrance, bridge pillars will be built in what now still is the Kandia lake, close to where the tunnel underpasses the Kandia dam, see Figure 5. It needs to be kept in mind that the dam itself is an artificial fill in what before had been a Winter bed that had been 'mined' for sand.

3 MONITORING DATA AND EVALUATION OF LIMIT STATES

Any new structure needs to be evaluated both for Ultimate and for Serviceability Limit states. This evaluation is generally quite straight forward applying the Codes of Practice and guidelines that are indicated in a contract. However, bored tunnels are flexible structures and the loading is interactive with displacements in the underground. For these tunnels, the Ultimate Limit state is easier to comply with than Serviceability Limit states, see also Bakker et al 2009. Moreover, the evaluation of the Serviceability Limit states and especially on issues that relate to the tunnel's function; as support for a railroad and the feasibility of leakage, is less straight forward.

3.1 *Criteria for deformations and rotations*

3.1.1 *Criteria related to deformation of the rail*
If the rails are displaced horizontally with respect to the design track alignment, an additional sideway acceleration may be imposed to the train, which might lead to derailment. Without going into detail, the acceleration may be estimated assuming a sinusoidal shape of the deformation according to Figure 6. With this configuration a maximum amplitude in the sideways acceleration may be calculated according to equation 1.

$$\hat{\ddot{u}} = -2\pi^2 \left(\frac{v}{L}\right)^2 \delta a \tag{1}$$

Due to the acceleration, a force will be imposed on the rail bed. Based on various analyses ProRail, the Rail Infrastructure provider in the Netherlands, has inferred a critical value for the sideways deformation of 0.02 m and for normal track conditions a limit of 0.015 m calculated as a deviation from the design track over an evaluation wavelength of 10 m.

However, the situation in a tunnel is considered more critical as a train derailing in a tunnel might seriously harm the structure and therefore derailment guiding profiles are constructed as a standard measure, see also Figure 7. These guiding profiles consist of concrete steps at the side of the railbed and create a limitation that is stricter. Therefore, a maximum allowable deformation was set to 0.0015 mm, and a critical deformation of 0.005 m; i.e. 5 mm. Beyond this value the derailment guiding system would need to be reconstructed, meaning serious structural and operational costs.

3.1.2 Criteria related to deformation of the lining

Due to changes in the loading of the tunnel lining, either by additional loading, e.g. placing embankments on top or adjacent of the tunnel or due to piling operations near the tunnel, the tunnel might be displaced and may deform. This deformation may increase the bending in the lining segments and may lead to larger rotations in the joints, than originally taken was into account in the design. As mentioned in the previous section, small deformations, up to about 5 mm may be compensated by adjusting the rail on its fixation, larger deformations would require more serious reconstruction of the derailment guiding benches.

Leakage test were performed, with an artificial segment configuration that is composed of the T-shaped connection of a longitudinal joint that connects to a ring joint, see Figure 8. Based on the Stuva governed guidelines (Stuva is the German Society for Research to Tunnel and Road Engineering), testing diagrams can be established that give a relation between the feasibility of leakage for the profile for various water pressure, gap opening and offset distance.

Because of the soft soil conditions at the Botlek tunnel, that may lead to larger deformations compared to the stiffer soil conditions at Pannerden, different rubber sealing gaskets for both locations were used. At the Botlek tunnel a Datwyler P86-260 seal, which has characteristic profile height of 23.5 mm was applied, whereas for the tunnel at Pannerden, a smaller seal, called Datwyler M385 "Lesotho" seal was used, with a height of only 14.5 mm. Based on the Stuva tests, a maximum gap opening of 0.0076 m was determined for the Datwyler P86-260 seal and a maximum gap opening of 0.0036 m for the Datwyler M385 "Lesotho". For gap openings lesser than this value it is assumed that no leakage will occur.

3.2 Limit values for the ring rotations at the Botlek tunnel and Tunnel Pannerdensch canal

Then the critical rotation was calculated by dividing the critical gap opening by the arm with respect to the axis of rotation, see Figure 14, acc. to: $\theta_{crit} = \frac{Gap_{crit}}{h}$. Considering a critical gap opening of 0.0076 mm, the critical rotation at the Botlek rail tunnel is 0.0076/0.2 = 0.038. For the tunnel at Pannerden, with a critical gap opening of 0.0036 m, the critical rotation is 0.018 rad.

Figure 6. Sinusoidal shaped imposed horizontal deformation on the rail.

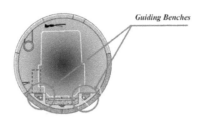

Figure 7. Characteristic cross section of tunnels in the Betuwe route, with a lining composed of seven segments and a keystone. Further the inlay at the bottom with the derailment guiding benches are indicated.

As the flexibility of the lining has a serious influence on the soil loading and the interaction of the tunnel with the ground, several methods may be applied to model the flexibility of the lining in a 2D FEM model. One way is to assume a substituted reduced stiffness for the joint, assuming Janssen's equations (1983). Alternatively, one might use the simplified equation by Muir Wood (1975), which in our project showed a good corroboration with measured data, who formulated a reduced stiffness according to equation 2:

$$I_e = I_j + (4/n)^2 I \qquad (2)$$

Where I_e is the effective lining stiffness, I is the second moment of inertia of the segments, I_j is the reduced stiffness due to the reduced contact in the joint and n is the number of segments that make a ring. For the situation that $I_j \ll I$, and 8 segments the effective lining stiffness reduces to

$$I_e \approx I/4 \qquad (3)$$

An analysis based on reduced stiffness' normally will not yield direct results expressed in joint rotations. However, it is assumed that the flexibility in the lining is mainly concentrated at the joints because the segments themselves are relatively stiff, For the situation of the Botlek Rail tunnel, flexibility was assumed only by four segments in the upper part of the tunnel because of a stiff inlay at the bottom of the tunnel, See figure 9.

It was inferred that the maximum rotation would be in the order of

$$\theta_{\max} \approx \frac{4\Delta u}{L} \qquad (4)$$

Where Δu is the maximum deflection and L is the segment length. The latter relation as a first assumption for the order of the rotations was confirmed with the data retrieved from the tunnels by monitoring at the Botlek Rail tunnel in 2016, further described in the next section.

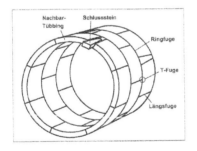 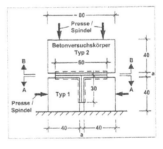

Figure 8. Evaluation of the feasibility of leakage; T-joint configuration and test configuration (Stuva 2005).

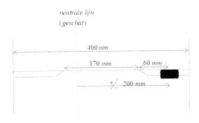

Figure 9. Rotation in a segmental joint.

Alternatively one may assume there is a neglectable rotation stiffness in the joint which would mean that a flexibility is assumed to be in the order of $I_e \approx I/4$, which in practice will not be far off of reality as we have seen from back analysis of the data. This approach was e.g. taken by the party that consulted for the project, for the evaluations that were done for the Passage at Pannerden, which simplified the further evaluation drastically as rotations could now be taken directly from the output data of the analysis

3.3 Monitoring of ring displacements

In 2014, at the Botlek tunnel, laser scanning was done after construction of the new Botlek bridge with its elevated approaches and before pile foundations next to the tunnel where introduced, the displacement of rings with respect to their ideal position was monitored with a moving 3D laser scanner on a trolley. This was done for cross sections at every 5 m along the whole tunnel length. See figure 10 for measurements of a critical cross section. Then, a complete 3D point cloud of the tunnel was made.

The point cloud was cleaned by removing the measurements of the cables and ducts and railings and supports in the tunnel. After good results were achieved at the Botlek tunnel, during the preparations for the construction works at Pannerden this same approach was introduced and performed. Unfortunately, after completion of the tunnels, no zero-measurement was performed and thus the ideal CenterPoint for the tunnel ring was unknown. Therefore, the CenterPoint of the tunnel was first back analysed, using the construction inner radius of the tunnel of (R=4,325m). Second, for each point of the point cloud, the distance between this point and the back calculated CenterPoint was calculated. Third, the inner radius of the tunnel was subtracted of this distance and a deflection curve was derived. see Figure 11.

3.4 Monitoring of tunnel deformation at Botlek rail tunnel and tunnel Pannerdensch Canal

Subsequently, a cross section with a relatively large deformation was selected and further analysed. It must be mentioned that it was only possible to get deformation data of the four segments in the tunnel crown, which is only 2/3 of the cross section of the tunnel, because at the bottom section there is a concrete inlay to support the derailment guiding benches and the ballast bed, see also Figure 7.

Evaluation of the graph in Figure 11 and also in Figure 12, and considering a segment length of $L = \pi D/n$, where the number of segment in a ring, n = 7, it was recognized that the sharp bends at x = 8 and x = 12 may be recognized as the joint. Back reasoning revealed that the more distributed bend at x = 3 must be due to the keystone. Thus, the four segments could be identified clearly. See figure 13. Then, recognizing the different inclinations of the four segments, linear regression for parts of the curve and subtraction revealed joint rotations, see also Figure 14

The data as shown in Figure 14 is seen as evidence that deformation, primarily concentrates in the joints. At the Botlek tunnel, the maximum joint rotation at different locations along the tunnel, was found to be in the order of 0.018 rad. At the Pannerdensch canal tunnel, the maximum joint rotation at different locations along the tunnel, was found to be in the order of 0.012 rad.

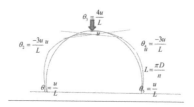

Figure 10. Inference of relation between maximum of deflection and maximum of joint rotation in the lining.

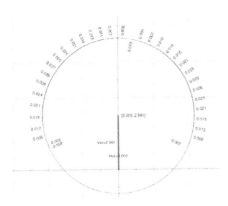

Figure 11. Monitoring data and graph characteristic for the most critical cross sections at the Botlek rail tunnel;

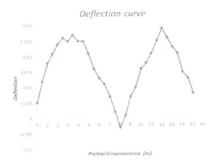

Figure 12. Monitoring data and graph characteristic for the most critical cross sections at the Botlek rail tunnel; the data of Figure 11 displayed as a deflection curve.

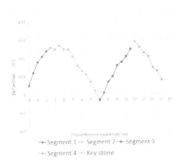

Figure 13. Identifying the segments in the deformation data.

3.4.1 Botlek tunnel

Based on monitoring data, the maximum joint rotation at different locations along the tunnel, was found to be in the order of 0.018 rad, see section 3.3. Then, Finite Element analysis (FEM) analyses with Plaxis were by performed, simulating the old situation with only the tunnel and the new situation with the approaches of the new bridge. The calculated deformations of Plaxis for the situation at present were compared to the observed measurements and the parameters for the analysis where adjusted to obtain a good fit with the measurements. Subsequently the additional rotation for the future was calculated for different locations and

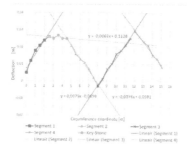

Figure 14.   linear regression of the deformation data.

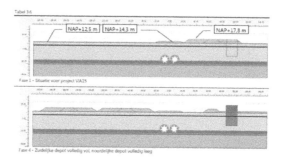

Figure 15.   Relocating the Clay deposits, and indicative the location of a Bridge pier, before at the top and after.

e.g. for ramp 200 and 300 appeared to be in the order of 0.002 rad. Adding up the expected additional rotation from the Plaxis analyses to the measurements, the total expected value is 0.02 rad, which is much smaller than the allowable critical rotation of 0.038. Thus, in general at ramp 700 on the West bank, no leakage is expected.

However, this result was not general, e.g. for Ramp 700 at the East bank the predicted rotation exceeded the critical values and therefore, the application of a light weight material to raise the embankment was introduced, using Yalibims i.e. a pumice stone lightweight grained material, to minimize the additional weight, was applied and at ramp 500 the application of an under piled structure, with the restriction that only vibration free, non-driven piles could be used. Cased piles were applied where concrete under pressure was pumped into a bored hole, while the casing was simultaneously pulled upwards. Still some deformations were measured at the tunnel presumably because the concrete was pumped into the cavity in a closed circuit under high pressure.

3.4.2 Tunnel Pannerdensch Canal

At the Pannerdensch Canal several different issues must be considered. To begin there is the Kandia dam. The slopes of the Kandia-dam are composed of sand with a low relative density and therefore may be susceptible to flow slides. Further, Geocontainers filled with sand are still present at the toe of the slope and a flow slide might be triggered, in case of foundations works such as piling. These issues need to be attended to by the contractor that will design and build the bridge piers near the slope. The analysis of the Kandia Dam was not part of this paper.

The second issue relates to the relocation of the clay deposits of the brick factory at the middle section of the tunnel. These clay deposits need to be relocated to build the main piers of the new bridge for the A15, See Figure 14. The new situation, including relocation of the clay deposits and considering the effect of one of the main piers of the bridge, was modelled with FEM analysis, using Plaxis. The result reveals an additional rotation in joints of the bore tunnel of about 0.004 rad, which is slightly more than was found at Botlek. This origins for the larger loads to be considered and asymmetry, removing a load at one side and filling up on the other side of the tunnel. With a monitored largest rotation of about 0.012 rad, this addition will lead to an expected rotation of about 0.016 rad. This expected rotation is still within the critical value but close to and therefore during construction the monitoring data will be closely followed.

Finally, there is the issue of the new road crossing with piled foundation over the tunnel entrance at the West side, see Figure 6, which is still under analysis at the time of writing this paper. The design of piled foundations near the tunnel approach gives concern in relation to driven piles which by favour need to be avoided due to the horizontal loads on the piles of the U-shaped concrete box, which might lead to overload and too much horizontal displacement of the rails.

4 CONCLUSIONS

In this paper, the effect of the loads of new infrastructures on two tunnels in different ground conditions is analysed. Characteristic monitoring data and empirical relations where presented.

Up to now, overall the tunnels show only very slight numbers of leakages that in general are thought to be related to the construction process at the start and not to deformation due to operational use; however, the new construction works give concern to maintain this favourable situation. For that an evaluation procedure in combination with monitoring was developed and explained. Based on these evaluations, the following conclusion where drawn:

- Modern new laser technology enabled the measurement of deformations of a tunnel along its total length.
- Based on those measurements, the location of the tunnel that relatively show the largest deformations could be selected and evaluated.
- Based on the deformation measurement, the joint rotations could be calculated relatively easy.
- During the lifetime of a tunnel, the construction of new infrastructure near this tunnel is a likely event.
- Compared to the Botlek tunnel the joint capacity at the Pannerdensch canal leaves less rotation before it becomes critical. For future tunnel construction it must be advised to consider the feasibility of future construction works to leave a more redundant structure.
- Careful investigation of the effects of new constructions work is recommended, on the one end to prevent too much deformation of the tracks in the tunnel, where deformations are limited due to derailment guiding benches, on the other hand too much deformation of the lining needs to be avoided to prevent leakages.

REFERENCES

Bakker, K.J. & Bezuijen, (2008), Ten years of Bored tunnels in the Netherlands, Proc. Geotechn. Aspects of Underground Construction in Soft Ground, CRC press, Shanghai, China

Bakker, K.J. and C.B.M. Blom, (2009), Ultimate Limit State Design for linings of Bored Tunnels, Tübbing Bemessung im Grenzzustand der Tragfähigkeit beim Schildvortrieb, Geomechanics and tunneling, 2009

Jonker, J.H. (1999), Bored tunnels on Holland's Betuweroute, Tunneling and Underground Space Technology, Vol 14, issue 2

Janssen, P, (1983), Tragverhalten von tunnelausbauten mit gelenk Tübbings, PhD Thesis Braunsweig

Muir wood, A.M. (1975), The circular Tunnel in elastic ground, Geotechnique 25, No 1, 115–127.

Stive, R.J.H. (1999), Design features of the Sophi Rail Tunnel in the Betuwe Route, Tunnelling and Underground Space Technology, Vol 14, issue 2.

Stuva, (2005), Stuva Recommendations for Testing and Application of sealing Gaskets in segmental Linings, Tunnel 8

Tunnels and Underground Cities: Engineering and Innovation meet Archaeology,
Architecture and Art, Volume 11: Urban
Tunnels - Part 1 – Peila, Viggiani & Celestino (Eds)
© 2020 Taylor & Francis Group, London, ISBN 978-0-367-46899-6

Surface settlement induced by shield tunneling–complete 3D numerical simulation on two case studies

M. Beghoul & R. Demagh
Department of Civil Engineering, Batna, Algeria

ABSTRACT: Surface subsidence can be minimized due to tunnelling with tunnel boring machines (TBMs). However, ground settlement cannot be avoided during shield TBM tunnelling. In this study, surface settlement induced by shield TBM excavation is simulated by using finite differences code Flac-3D. The proposed three dimensional simulation procedure is taking into account in an explicit and complete manner the main sources of movements in the soil mass. It is illustrated in two different cases of Lyon's subway for which well experimental data are available. The comparison of the numerical simulation results with the in-situ measurements shows that the 3D procedure of simulation proposed is relevant, in particular in the adopted representation for the different operations achieved by the tunnel boring machine.

1 INTRODUCTION

In urban areas, pressurized-face tunnel boring machines (TBMs), such as slurry shields, are increasingly employed to overcome the congestion at the ground surface. However, shield tunneling inevitably causes some ground movements that can have adverse effects on both buried and aboveground infrastructures and buildings. Therefore, it is a major concern in the underground works to estimate tunneling-induced ground movements due to shield tunneling. These ground movements are strongly influenced by many machine control parameters: Excavation, front support, taper of the TBM shield, over-cutting, injection procedure and rheology of the grout. This complexity makes an explicit numerical simulation very difficult. In plane deformation, we can cite (Pantet (1991), Moroto (1995), Kastner (1996), Benmebarek (2000)). For these authors, the deformations are controlled using a single parameter, the deconfinement rate. In addition, empirical prediction such as numerical approaches that directly simulate the final phase may be insufficient. Several 3D numerical simulation procedures of tunnel excavation using pressurized-face tunnel boring machines, in soft and saturated soils, have been proposed by various authors: Mroueh (1999), Dias (2000), Broere (2002), Kasper (2004), Bezuijen (2006), Demagh (2008-2009). The confrontation with field results shows that, in spite of advances in terms of computing resources, the phenomena induced by the passage of a tunnel-boring machine are still insufficiently known.

This paper presents a complete 3D numerical simulation of a tunnel boring process. This one is applied on two different sections of Lyon's subway, for which well experimental data are available.

2 GROUND CONDITIONS AND TBM

2.1 *Overview of the site*

The project is part of the extension of the Lyon metro network, through an extension of the D line between the Gorge de Loup district and the Vaise station (Fig. 1). Two tunnels approximately 6.27 m in diameter and 900 m long were excavated successively between June 1993 and February 1995. The geology of the site is constituted of a superficial layer of fill, a layer of silty clay alluvium, a layer of sand and gravel, a base of gneiss and some conglomerate lenses. The silt that is most often located at the crown of the tunnel has a natural water content very close to the liquid limit and a low consistency index, making it a very sensitive soil for reshuffling.

2.2 *Experimental plots*

It is both to provide the group of companies with elements for the control of their tunnel-boring machine and to build a database to qualify numerical simulation procedures. The experimental device implemented on two plots (Fig. 1) located in two different geological zones consisting of continuously measuring the displacements both on the surface and inside the massif in order to identify the origin of the movements as well as to follow their evolution.

a) Plot 1: which was intended to test the operating parameters of the tunnel-boring machine in a break-in area. It consists of two sections S1 and S2 of similar geologies. The presence of a retaining wall retaining an embankment six meters high make the geometry of these two sections dissymmetrical. The section S2 is selected to qualify the 3D simulation procedure (Fig. 2a).

b) Plot 2: (consists of a single section) is located at the end of the route about 800 meters further. Its geometry is symmetrical, but we note the rise in the lower part of the bedrock outcropping three meters under the tubes (Fig.2b) Demagh (2009).

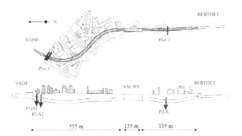

Figure 1.   Plan view and profile of the site.

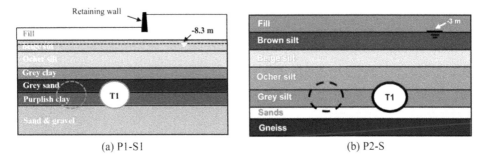

(a) P1-S1   (b) P2-S

Figure 2.   Geological profile of the three sections.

Table 1.   Mohr-Coulomb parameters of section P1-S2.

Layer	Depth m	G Mpa	K Mpa	γ kNm-3	c' kPa	φ' degrees	ψ' degrees
Fill	0-7	2.92	7.8	18	30	30	17
Beige silt	7-10	2.74	7.3	19.5	12	33	20
Ocher silt	10-13	2.74	7.3	21	15	25	14
Grey clay	13-15	1.57	4.2	16.5	35	27	15
Grey sand	15-17.5	10.5	28	21	5	35	21
Purplish clay	17.5-20	5.16	13.8	18.5	35	27	15
Sand and gravel	20-30	10.5	28	21	0	34	20

Table 2.   Mohr-Coulomb parameters of section P2-S.

Layer	Depth m	G Mpa	K Mpa	γ kNm-3	c' kPa	φ' degrees	ψ' degrees
Fill	0-3	2.92	7.8	19	30	30	17
Brown silt	3-5	2.74	7.3	21	10	25	15
Beige silt	5-8	2.74	7.3	21	15	32	20
Ocher silt	8-12	1.57	4.2	21	15	25	14
Grey silt	12-15	10.5	28	21	5	30	14
Sands	15-18.5	5.16	13.8	21	0	34	20
Gneiss	18.5-20	10.5	28	21	150	45	30

The geotechnical characteristics of the two sections are summarized in Tables 1 and 2.

2.3   Slurry shield TBM

The knowledge of the nature of the terrain, the route located largely under the built-up area (old masonry buildings) and the low coverage led the group of companies to propose the technique of the slurry shield TBM. This tunnel-boring machine consists of a shield (slightly conical) of about 7 meters long and 6.27 meters in diameter (Tab. 3), a 50-meter long back-up train and weighing nearly 300 tons. The machine is equipped with about fifty sensors recording the parameters of continuous operation. The tunnel is covered with prefabricated concrete segments (35 cm thick) placed under the shelter of the shield tail: when the tunnel boring machine moves forward and the ring escapes from the shield tail, there remains an annular void of 13.5 cm (thickness of the shield tail) between the soil and the lining (Fig. 3) . In order to minimize settling, this void is immediately injected with a filling grout made of bentonite, sand and water.

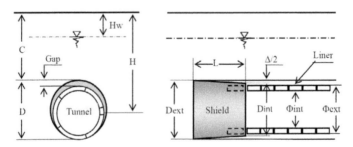

Figure 3.   Tunnel, shield and liner parameters.

Table 3. Tunnel parameters.

Section	H m	L m	C/D -	Hw m	Dext m	Dint m	D/2 mm	Φext m	Φint m	Liner width cm	Liner thick cm
P1-S2	18.9		2.50	8.3	6.27	6.24	15	6.0	5.3	100	35
P2-S	13.6		1.67	3							

3 COMPUTATIONAL MODELLING

3.1 *Finite difference model*

To assess the ground movements during slurry shield tunneling, a finite difference (FD) model is developed using Itasca Flac3D. The FD model is developed (Fig. 4) with zero transverse (x-axis) displacement at x= -120 m and x = 120 m, and zero longitudinal (y-axis) displacement at y = 0 and y = 120 m, respectively. The top of the model boundary (z = 0) is set to be free, whereas the vertical movement at the bottom boundary (Z = - 30 m) is fixed. For this model, average soil parameters determined from the geotechnical tests are used (Tab. 1 and Tab. 2). The soil is modelled using solid elements with 8-grid points (155000 grid points for the section P1-S2). A linear elastic, perfectly plastic Mohr Coulomb constitutive model is employed with no-associative flow rule. The retaining wall in cross section P1-S2 is modelled by an elastic rule. The extent of the model, in the longitudinal direction, is conditioned by the position of the stationary section (Fig. 4). The model element lengths equal the ring width of 1m in the longitudinal direction within the tunnel excavation zone.

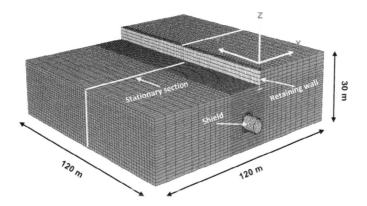

Figure 4. Flac3D finite difference model of slurry shield TBM tunneling.

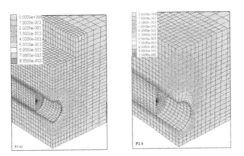

Figure 5. Contour of vertical displacements after complete installation of the shield.

5337

A shield with conical shape, perfectly rigid (the nodes are fixed according to the method called fixed center) (Benmebarek, 2000), modelled with thin volumetric elements, is installed in a virgin ground solid mass (Fig. 5) for which an initial state of geostatic stresses is imposed with K0 equal to 0.5 (normally consolidated soils), Benmebarek (2000), Demagh (2008, 2009).

The section P1-S2 made the three-dimensional simulations more difficult, due to the presence of a retaining wall on the surface. This singularity forced us to consider the tri-dimensional model in its entirety. On the other hand, the three-dimensional numerical models found in the literature preferred vertically symmetric models, which made the elaborate simulation procedures less constraining.

From the moment that the shield is completely installed (Fig. 5), the modelling sequence (Fig. 6a) could be applied.

3.2 *Modelling Sequence*

The excavation is simulated by the deactivation of disk element equal to the length of the lining segment (1 m). The stability of the front face is not the most important source of surface settlement Benmebarek (2000), so in our simulations, we considered that displacements at the front of the excavation are blocked by the shield. The shield passage, simulated by annulment of local tangential stresses, clears a volume loss that is immediately filled by soil convergence (very poor soils around the excavation). The interface which is on the shield is activated from the moment since a contact is established with the surrounding ground: the role of this interface is to block the radial ground convergence and also to allow the tangential convergence by arch effect (Fig. 6b).

The volume loss is partially compensated by the possible migration of the grout forwards the face of the shield (there is a great uncertainty on the post-closing shape of the ground around the shield). This migration is simulated by the correction of the shield conical shape, set so as to reproduce a vertical displacement recorded on construction site (back analysis on surface vertical displacement), Dias (2000) and Demagh (2008). The liner is simulated by shell elements. It is characterized by a weaker Young modulus in order to take account of the seals between the prefabricated rings of liner. The injection of the grout in the annular void is controlled in volume and pressure. The position of the injection pipes was decisive in the choice of the pressure diagram. The maximum value of the injection pressure is fixed on the measure of a maximum vertical displacement closest to the vertical axis of the tunnel (recorded on building site). It shows in particular that the pressure really transmitted to the ground remains lower than the average pressure measured at the exit of the injection pipes. This difference is due to the pressure loss by friction following the flow of the grout like to its impregnation of the surrounding ground (Demagh and al, 2008). Uncertainty on the rheology of the grout brings to consider two principal phases (liquid and solid phase) intercalated by a transitional one (Fig. 7).

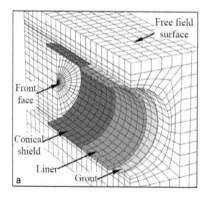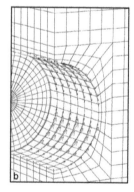

Figure 6.   Used mesh (a) View of different parts (b) Arch effect of the displacements at the soil/shield interface.

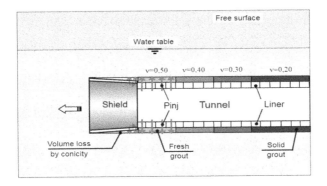

Figure 7. Complete phased simulation of TBM excavation process.

This procedure is repeated throughout the shield progression, until reaching a stationary section (Fig. 4) after a few tens of excavation steps (approximately 50 meters after the passage of the front face. The simulations are carried out in drained conditions (effective stresses with taking into account of the submerged volumetric weight corresponding to the long-term behavior). The simulation results are confronted with in-situ data collected on construction tunnel site.

4 RESULTS

Figure 8, compares the final transverse settlement trough against the corresponding simulation results. The settlement troughs resulting from the simulations are in agreement with the recorded measurements. .

On figure 8a, the final transverse settlement trough of the section P1-S2 is larger than the observed one: this difference in trough width can be partly explained by the use of the elastic linear Mohr-Coulomb model, which is not well adapted for this type of soils. Another singularity, the effect of the retaining wall, is also clearly represented by 3D simulations, what shows the relevance of the choices for the simulation, in particular the maximum value of the injection pressure, fixed on the vertical displacement of depth point on central extensometer recorded on building site. In addition, the choice of the injection pressure distribution is justified by the grout ports position at the back of the shield tail.

The second section, figure 8b, the simulation results are in agreement with the in-situ measurements, especially for the observed heave on the surface, and that has been well reproduced by three-dimensional simulations.

These settlements, for all the two sections, are very weak compared to the formed gap behind: this can be explained either by the migration of part of the grout injected under

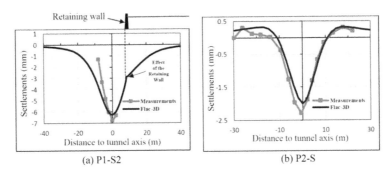

(a) P1-S2   (b) P2-S

Figure 8. Transverse settlement trough at final state.

pressure forwards by decreasing the ground decompression, or with the insufficiency of time so that the soil closes completely the annular void.

5 CONCLUSIONS

A 3D simulation procedure has been proposed to account for all the different operations achieved by a TBM. The procedure has been applied to reproduce by back-analysis the movements recorded on two different case studies.

The comparison of the results of simulations of the transverse troughs, with the recordings of two support sections, shows that the proposed 3D simulation procedure is pertinent, in particular in the representation of the different operations carried out by the tunnel boring machine.

Singularity such as the presence of the retaining wall or the heave on the surface were also well simulated.

However, uncertainties related to grout injection remain; if the migration of the grout seems to be well simulated by a corrected conicity, the different phases of the injection are still difficult to simulate.

REFERENCES

Benmebarek, S. & Kastner, R. 2000. Modélisation numérique des mouvements de terrain meuble induits par un tunnelier. *Revue Canadienne de Géotechnique*, 37: 1309–1324.
Bezuijen, A. et Talmon, A.M. 2006. *Proceedings of the Geotechnical Aspects of Underground Construction in soft Ground*, Bakker et al (eds) Taylor & Francis Group, London, 187–193.
Broere, W. & Brinkgreve, R.B.J. 2002. Phased simulation of a tunnel boring process in soft soil. *Numerical Methods in Geotechnical Engineering*, Mestat (ed.), Presses de l'ENPC/LCPC, Paris, 529–536.
Demagh, R., Emeriault F. & Kastner R. 2008. 3D modelling of tunnel excavation by TBM in overconsolidated soils. *JNGG'08* Nantes, 18-20 juin 2008, 305–312 (in french).
Demagh R., Emeriault E. & Kastner R. 2009. Shield tunnelling -Validation of a complete 3D numerical simulation on 3 different case studies. Euro: Tun 2009. *Proceedings of the 2nd International Conference on Computational Methods in Tunnelling*. Ruhr University Bochum, September 2009, 77–82.
Demagh R., Emeriault F. & Kastner R. 2009. Modélisation 3D du creusement de tunnel par tunnelier à front pressurisé – Validation sur 3 cas d'études. *Proceedings of the 17ème Conférence de Mécanique des Sols et de Géotechnique (17ème ICSMGE)*, 5-9 Octobre 2009, Alexandrie, Egypte, 77–82 (in french).
Dias, D., Kastner, R. & Maghazi M. 2000. 3D simulation of slurry shield tunnelling. *Proceedings of International Symposium on Geotechnical aspects of underground construction in soft ground*, Kusakabe, Balkema, Rotterdam, 351–356.
Kasper, T. & Meschke, G. 2004. A 3D finite element simulation model for TBM tunnelling in soft ground. *International Journal for Numerical and Analytical Methods in Geomechanics*, 28: 1441–1460.
Kastner, R., Ollier, C., et Guibert, G. 1996. In situ monitoring of the Lyons metro D line extension. Dans *Proceedings of the International Symposium on Geotechnical Aspects of Underground Construction in Soft Ground*, London, UK., 15-17 avril 1996.
Moroto, N., Ohno, M., et Fujimoto, A. 1995. Observational controlof shield tunnelling adjacent to bridge piers. Dans *Underground construction in soft ground*. A.A. Balkema, Rotterdam, pp. 241–244.
Mroueh, H. & Shahrour, I. (1999). Modélisation 3D du creusement de tunnels en site urbain. *Revue Française de Génie Civil*, 3: 7–23 (in french).
Pantet, A. 1991 Creusement de galeries à faible profondeur à l'aide d'un tunnelier à pression de boue. Mesures « in situ » et étude théorique du champ de déplacement. *Thèse de doctorat,*Institut National des Sciences Appliquées de Lyon, Lyon, France (in french).

*Tunnels and Underground Cities: Engineering and Innovation meet Archaeology,
Architecture and Art, Volume 11: Urban
Tunnels - Part 1 – Peila, Viggiani & Celestino (Eds)*
© 2020 Taylor & Francis Group, London, ISBN 978-0-367-46899-6

South extension of the metro Line 14 in Paris – focus on Limestone quarries grouting

A. Bergère, M. Coblard, J.P. Janin, F. Lanquette & P.L. Le Tolguenec
Setec, Paris, France

ABSTRACT: The Grand Paris Express is one of the biggest infrastructure projects of the century in Europe. Among the four new lines, which are planned to be completed before the Paris Olympic Games in 2024, the extension of line 14 will connect the capital and the Orly airport. Built entirely underground, the extension is composed of 14 km of underground tunnel, 6 stations and 12 shafts. One of the major technical complexities is the tunnel excavation under the limestone old exploitation quarries. To control ground settlements and to secure the excavation using an EPB TBM, it was decided to grout all the quarries above the tunnel. Finite element calculations have been performed in order to determine the extension of the ground treatment (size and type). The grouting work in a dense urban environment has been a major challenge requiring minimizing the disturbance on residents despite the coactivity of several grouting sites.

1 PRESENTATION OF THE PROJECT

1.1 *Project outline*

The south extension of the metro line 14 project is set to connect the existing terminus station « Olympiades » and its future substitute: a maintenance-station in Morangis. This underground project is 14km long, incorporating 7 new train stations and 12 ancillary shafts to the existing network.

Among these seven stations, three of them will be interchange hubs connecting the line 14 to other train lines, different urban transport systems and even aerial transport systems: the line 14 will connect the Paris-Orly Airport to the centre of Paris in less than 30 minutes.

The geometry of this double-track rail tunnel is circular with an inner diameter of 7.75m and an excavated diameter of 9.70m.

One of the major characteristics of this tunnel is that a 1.6 km portion of it is built under limestone quarries.

1.2 *A dense urban context*

The tunnel section involving the aforementioned presence of limestone quarries is located at Kremlin-Bicêtre, a council located in the south of Paris. This suburb can be divided in two types of urban areas:

– Groups of high building (5 to 12 storeys high) below which limestone quarries have been secured by injecting cement slurries or by building pillars,
– Small neighbourhoods, with local roads, which are often dead ends, and where the quarries remain without any reinforcement.

In this context, the location of drilling sites to inject from the surface was constrained by the urban context, not to mention the high density of buried services networks.

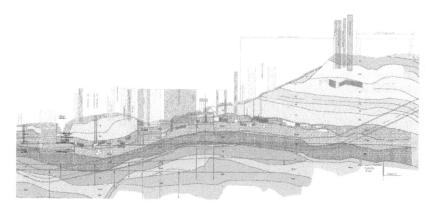

Figure 1. Vertical alignment of the tunnel under the 3 levels of quarries (in red/blue/purple) – scale 1H/ 10V.

2 FOCUS ON THE LIMESTONE QUARRIES

2.1 The Parisian Basin, South of Paris

The metro line 14 south extension project takes place in the geological context of the city of Paris, a succession of sedimentary formations that have been lightly disturbed thanks to a minimal tectonic activity. In this context, various soil types with different behaviours can be found. On the top of the stratigraphy there are:

– Gypsum marls: highly sensitive to dissolution,
– Lutetian soils: made of rock layers going from very stiff to very soft including the limestone layers, mined at the beginning of the century to construct Paris,
– Ypresian soils: including compact sands and over consolidated and swelling clays.

The limestone is made of three layers: the two highest have been mined while the lowest is more glauconitic and below the groundwater table. The average depth of the quarries' upper limit is 15 m and it reaches 40m deep below Villejuif's hillside.

2.2 The limestone quarries

The main mining method used nearby line 14 is an underground mining method named « hagues et bourrages », a specific French expression used in the quarries context. To optimise the exploitation, the miners excavated almost the entire stiff layer, built some stone pillars and elevated a dry wall named « hague » used to stock the quarries wastes called « bourrages ».
The quarries remained unchanged since they were abandoned:

– The cutting front and the corridors were left empty only with pillars measuring between 1.5 and 2.0 m, to support it;
– Behind the cutting front, around 10m away, the « hagues » can be found, with the « bourrage » behind it.

This exploitation method has led to a gradual subsidence of the quarries' upper limit on the « bourrages » and corridors sections. Therefore, some geological unknowns are to be considered:

– Subsidence above empty corridors that could have led to collapses or sinkholes,
– The « bourrage » is not a homogeneous layer. Sometimes it touches the upper limit and transmits the loads but sometimes voids remains between them. For our project, we have estimated a 50 cm average gap between the upper limit of the quarries and the "bourrage",

5342

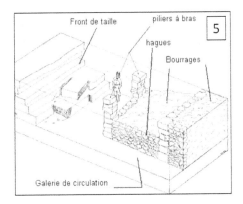

Figure 2. Schematic principles of mining by *"hague et bourrage"* – source: VACHAT, J.C. (1982). Les désordres survenant dans les carrières de la région parisienne. Etude théorique et pratique de l'évolution des fontis. *Thèse du CNAM*, 115–126.

– This height variation can lead to an intense flexion in the rock layer above the quarries which may cause an alteration of the geological characteristics inside the Marls.

The projected tunnelling works below the quarries level produce vibrations and the subsequent setling of their lower limit. These two combined factors may weaken the pillars that currently ensure the stability of the quarries' upper limit, leading to its subsidence, the creation of sinkholes or to the ultimate collapse of the quarries.

To secure the quarries and remain within the differential settlement tolerances, several methods can be used to strengthen the upper limit, such as the consolidation of the pillars or the injection of cement slurries to fill hollows and embed pillars.

2.3 *An indispensable tool: the maps of the "Inspection Générale des Carrières"*

The Inspection Générale des Carrières (IGC) is a French's state agency that was created in 1777 by Louis XVI after a chain of major collapses inside the capital. The office was tasked to find and map all the quarries below Paris, in order to be able to judge the feasibility of new constructions inside Paris.

Currently, its duty is to take care of the quarries risks below Paris and its suburbs. Thanks to years of experience, the IGC has produced several guidelines that define the rules to stabilise quarries with different methods like underground consolidations or injections from the surface.

The IGC maps are quite reliable and precise, giving a good insight of the subsoil state:

– Quarries areas are located with a 2–3 m accuracy,
– The number of level excavated (up to three levels for Limestone excavations and up to two levels for Gypsum excavations),
– The method used for the exploitation (« *hagues et bourrages* », corridors, access shafts, etc....) and the strengthening method used (pillars, injections...),
– State of the quarries, surface problems (cave-in, differential settlements).

These IGC maps can be read as 3D maps, with several layers superimposed. Each layer has a colour and represents an excavation layer and other information can be found, such as the stratigraphy or the existing sections.

However, even if these maps are precise, some of them are old and can be incomplete. Some zones remain unmapped and are called « white » zone because of how they look on maps. But this does not mean there is nothing there. Additionally, quarries may have been forgotten or excavated illegally. Therefore it is possible to find new quarries in that kind of area.

3 THE TUNNEL DESIGN TAKING INTO ACCOUNT THE QUARRIES

3.1 *Studied cases*

During the preliminary design, two scenarios were considered:

- The main scenario was to dig the shortest layout between all the vertical structures (stations and adjacent pits) leading the TBM below the quarries in the Seine Valley, then through them and finally above them in the Villejuif's hillside.
- The alternative scenario projected the tunnel going below the potential quarries for its entire alignment.

An intense geological assessment (1 drilling every 70 m) was performed to ensure that no quarries would be found below Villejuif's hillside, in the area where the TBM goes downward across the Limestone levels. This assumption was confirmed by the investigations.

For the main scenario, two methods were considered to bore through the quarries:

- Removing the « bourrages », filling them with concrete and securing the marls above with special injection before the boring with the TBM throught the quarries,
- Creating a 600 m meter conventional cavern across the quarries and the marls to slide the TBM inside it.

Moving the TBM above the quarries, as specified in the main scenario, also implied securing the quarries below it to prevent collapses or cave-in.

3.2 *Chosen solution*

The investigations, showing no quarries below Villejuif's hillside, have led to the choice of the alternative scenario. Indeed, going through quarries with the TBM inside a cavern was riskier, more expensive and longer to build.

The vertical alignment of the tunnel runs below the quarries along the aforementioned 1.6 km section. In some locations, the upper section alignment is at least 3 m away from the lower limit of the quarries: 2 m underneath it plus 1 m due to various security and uncertainty tolerances.

This lower limit level was already conservatively defined by making the hypothesis that three layers of quarries were excavated and by taking the deepest possible limit.

3.3 *Reconnaissance shaft*

A test shaft was performed during the design phase of the project. The main goal was to test the quality of the injection process. To do so, several types of tests were performed:

- Injection of cement slurries inside the « bourrage » and inside the Marls above the quarry. Then a corridor was dug from the end of the shaft into the injected area to check the result.
- An injection test: the dam slurry is very dense therefore, it was uncertain that it would flow inside a 45° pipe. Some 45° injection tests were made successfully. The inclination of the drillings was also measured, the deviation was about 1%

4 NUMERICAL MODELING

4.1 *The design criteria*

Quarries' stability may be found in critical condition; therefore, they must be reinforced with the solutions that we are capable of implementing. The goal of quarry injections is to avoid the collapse of the quarry roof.

Thus, we sought to fill the voids as aforementioned described. However, all quarries' levels are economically unfeasible to enhance. We undertook the numerical modelling required determining the width of the injections band, upon the following tolerance criteria:

- Absolute settlement of the quarry floor without injection below 10 mm
- Differential settlement of the quarry floor without injection below 0,7 ‰.

4.2 Description of the model

The tunnel geological profile provides with the following constrained interface layout:

- A full-face excavation in the limestone, a soft rock whose good characteristics make it possible to strongly limit the deformations of the soil around the excavation, but with a limited distance to quarries;
- A deeper excavation from the quarries, but necessarily in a geological condition more unfavourable to deformations, mixed-face between the limestones and the plastic clays, or full-face in the plastic clays.

A 3D finite element modelling of the tunnelling excavation was carried out to estimate the deformation of the quarry floor during excavation according to 3 types of geological conditions (full-face in the limestone with a minimum cover of 3 m from the quarries lower limit, mixed-face between the limestone and the plastic clays, full-face in the plastic clays).

It was sought to determine which configuration induces the least instabilities on the floor of the quarries:

- A higher cover but a face full or in part in deformable Plastic Clays;
- Reduced cover, therefore riskier, but with a full-face of excavation in good materials.

The 3D model also made it possible to understand the stresses in the pillars, main support of the top of quarries. For this, two types of modelling of pillars were implemented: the first with volume elements and the second with beams elements.

Both models showed similar results.

Additionally, 2D models were calibrated on these 3D models in order to extend the study to all the different interface configurations (geological, groundcover, frames) and to evaluate the confinement pressure values required to limit deformations on the floor of the quarry

4.3 Results

Based on the obtained results, we highlight the following points:

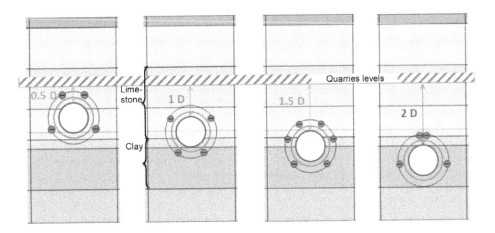

Figure 3.   Modelling Case Performed.

Beam or volume element

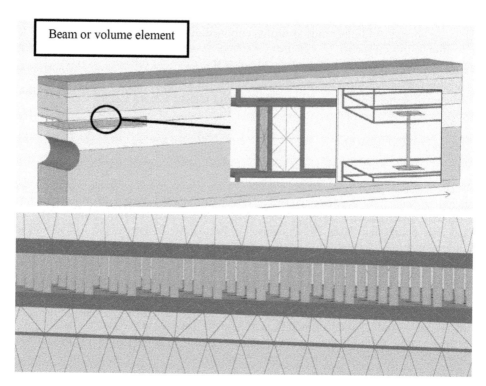

Figure 4.   Picture of the 3D model with beam element.

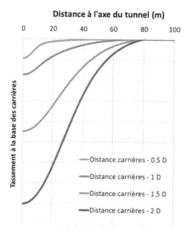

Figure 5.   Settlement of the quarry floor in the cases shown in Figure 4.

– When the TBM is in the clay layer the absolute settlements are higher even though the coverage is greater, thus resulting in the yielding of the plastic clays. On the other hand, the differential settlements are lower (the deflexion is wider);
– When the TBM is in a full face in the limestone, the settlements are lower in absolute value but higher in differentials;
– The reduction of tunnel cover reduces the width of the settlement curve and the differential settlements increase.

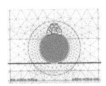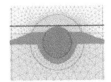

Figure 6.   Yielding criteria of the cases shown in Figure 4.

Plastic clays have relatively poor mechanical properties and the yielding around the tunnel becomes important when the face goes down in these clays, thus considerably increasing the deformations at the level of the quarries.

4.4   *Implementation of these results on the project*

Following the results, a risk analysis was conducted to complete this modelling, and the following principles were deduced:

– It is necessary to fill the quarry voids in order to confine the existing pillars and to avoid their instability when the TBM passes below;
– Filling the quarry voids makes it possible to limit the cave-in and deflexions;
– A minimum spacing of 20 m (i.e. 10 m on either side of the tunnel axis) of injection is implemented. This spacing is increased according to the percentage of the face in the plastic clays, to reach a maximum of 30 m.

According to the modelling, the prescribed injection works for the quarries are the following:

– Systematic filling at a minimum spacing of 20 m, focused on the tunnel;
– Filling where the modelled settlement is greater than 10 mm in absolute or 0.7 mm/m in differential at the level of the quarry floor. These values are the ones that have provided with the final design injection spacing (20 to 30 m).

5   METHODOLOGY OF TREATMENT ADOPTED: GRAVITARY FILLING INJECTION

The chosen solution to treat the quarries before the TBM excavation is filling by gravity injection from the surface.

The time available to build the tunnel and the access conditions do not allow filling the quarry from the bottom. Also, and for safety reasons because of not being able to inspect the entire length of quarries, this kind of treatment is not acceptable.

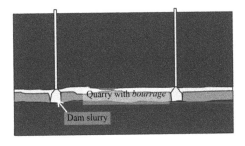

Figure 7.   Step 1 - seal the quarry.

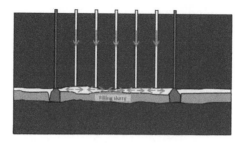

Figure 8.   Step 2 - injection of the filling slurry.

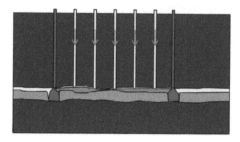

Figure 9.   Step 3 - pressurized slurry is injected.

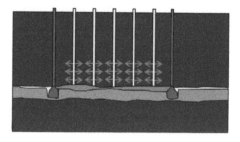

Figure 10.   Step 4 - pressurised slurry is injected to fill the marls.

The main method used to secure a quarry by injecting cement slurries from the surface is a 4 steps method as follows:

–   First, the quarry area that needs to be secured must be sealed. Otherwise, if it is not sealed, the slurry may flow outside the area to secure. To prevent that, dam slurry may be used. This slurry is very dense and permits to seal the area.
–   Then, filling slurry, more liquid than the aforementioned dam slurry, is injected inside the area. Once the filling slurry is injected inside the quarry, it requires one week for hardening.
–   Once the filling slurry has hardened, it shrinks leaving small hollows. These hollows need to be filled so pressurized slurry is injected to fill them.
–   Then, if necessary to stabilise the marls above the quarry, other pressurised slurry with a different composition is injected.

Some of the buildings that are located above the tunnel have been designed taking the presence of the quarries into the account in their design. According to the kind of foundation, the treatment to choose from is as follows:

- Foundations on masonry pillars: no treatment is envisaged, in order to avoid any destabilization of the pillars;
- Foundations on bored piles: these piles are undertaken in filled quarries, so no treatment is envisaged for the tunnel project;
- Shallow foundations: test drillings are envisaged in order to check the filling and a specific treatment is undertaken if no filling is found under the building.

To control the quality of the filling, one test drilling every ten injections drillings is performed, one month after the end of the injections at the soonest.

6 CONCLUSION: DESIGN OF TREATMENTS, AND FIRST WORKS

Once the treatment methodology was decided to be from the surface, a great work was performed to make it possible to fill the 1.6 km length of quarries in a dense urban context while maintaining pedestrian and vehicle circulations as much as possible and guaranteeing the resident accesses during the duration of the works. All the identified buried services along the quarries length were incorporated to the drawings in order to check the feasibility of high dense drillings.

To conclude, 14 injection locations have been determined, in order to plan the injection works, both spatially and in time. It was decided that no more than 3 locations can be treated simultaneously.

The contractor chosen to perform the works is a Vinci and Spie Batignolles joint venture. The quarries treatment has begun in June 2018 and are planned to be completed by the beginning of 2020.

REFERENCES

Tritsch, Jean-Jacques (2007). Guide Technique – Mise en sécurité des cavités souterraines d'origine anthropique, Surveillance, Traitement. *Ministère Francais de l'Ecologie, de l'Energie, du Developpement durable et de l'Aménagement du territoire.*

Notice technique de l'IGC (2004). Travaux de consolidation souterraines executes par piliers maçonnés dans les carriers de Calcaire Grossier situées en region parisienne. *Insitut générale des carrières.*

Notice technique de l'IGC (2003). Injection gravitaire, clavage et traitement des fontis préalables à la mise en oeuvre de fondations profondes, de type pieux ou micropieux de type supérieur ou égale à II, en zone sous minée par d'anciennes carrières souterraines ou à ciel ouvert. *Insitut générale des carrières.*

Notice technique de l'IGC (2003). Travaux de consolidation souterraines exécutés par injection pour les carrières de Calcaire Grossier, de gypse, de craie et les manières.

Vachat, J.C. (1982). Les désordres survenant dans les carrières de la region parisienne. Etude théorique et pratique de l'évolution des fontis. *Thèse du CNAM* 115–126.

Triclot, J. & Berraud, J-P. Réalisation du creusement de la tête et du tunnel sous la terrasse de Saint-Germain. *Travaux souterrains: des techniques et des hommes. Journées d'études internationals de Chambéry.*

Tunnels and Underground Cities: Engineering and Innovation meet Archaeology,
Architecture and Art, Volume 11: Urban
Tunnels - Part 1 – Peila, Viggiani & Celestino (Eds)
© 2020 Taylor & Francis Group, London, ISBN 978-0-367-46899-6

High density slurry for shallow bored tunneling in Singapore

H. Bernard, M.K. Lau, M. Senthilkumar, Y.S. Sze, M. Marotta & C.N. Ow
Land Transport Authority, Singapore

G.T. Senthilnath
Geoconsult Asia Singapore, Singapore

ABSTRACT: Since 1995, the Land Transport Authority of Singapore has been constructing and planning the expansion of the underground roads and rails network.

The Thomson-East Coast Line (TEL) is a fully underground Mass Rapid Transit in Singapore that is currently under construction and approximately 43km long with 31 stations connecting the north, to the south and the east coast of Singapore. The construction of the TEL involved the usage of 24 Slurry and 38 Earth Pressure Balance TBMs.

The use of slurry TBMs, selected for rock and mix ground conditions, is an engineering challenge in shallow cover and in soft ground, where there is a higher risk of slurry blowout or spill out to surface. This could result in loss in pressure with very serious consequences. The paper discusses how the usage of High Density has mitigated those risks and elaborates on the slurry mix design, rheological properties and its effectiveness.

1   INTRODUCTION

1.1   *General Information*

In conjunction with Singapore's vision to have sustainable, efficient and extensive Mass Rapid Transit (MRT) system across Singapore Island, more and more underground rail tunnels are being planned and constructed. With the target to expand current rail networks from 178km in length to about 360km by year 2030, eight in 10 households will live within 10 minutes from MRT station. A total of 92 TBMs, both Slurry type and Earth Pressure TBM were extensively used for tunneling under past projects in North East Line (NEL), Circle Line (CCL) and Downtown Line (DTL). As for the on-going projects in Thomson-East Coast Line (TEL), there are 51 TBMs being deployed.

Table 1 indicates the numbers of different types of TBMs used since 1995. The fact that Singapore is an island which contains a wide range of rapidly changing geology, a selection of the TBM becomes important for tunneling under various geological condition such as tropically weathered sedimentary, low grade metamorphic and igneous deposits incised by channels of very soft marine clay and fluvial deposit. (Yee et al. 2017).

At planning stage, it is possible to minimize the tunnel impact to the surface structure by designing the tunnel alignment at a depth with sufficient overburden and more competent geological condition. However, there are limitations to the vertical alignment of a tunnel, in terms of depth of station, tunnel maximum vertical gradient, interface with existing underground infrastructures, etc. Therefore, urban tunnels are often constructed under shallow overburden.

This paper presents two case studies which used high density slurry as a mitigation measure to maximize the use of slurry as a support medium and to prevent any possible blowout/ slurry-spill during the shallow TBM drive.

Table 1. Type of TBM used in Singapore MRT from 1995 to 2017.

MRT Line	EPB	Slurry	Open Face
Phase 1–2	1	1	
NEL	14 (2 dual mode)	0	2
CCL 1–5	19	8	0
DTL/CCL extension	3	0	0
DTL2	8	11	0
DTL3	29	0	0
TEL (ongoing)	27+1 Rectangular	24	1
TOTAL	**102**	**43**	**4**
GRAND TOTAL		**149**	

1.2 Theoretical background on the application of higher density slurry in terms of face pressure

In order to design the Target Face Pressure to be applied during tunneling operation, different conditions need to be analyzed, including upper and lower limits conditions.

A maximum allowable support pressure is determined to prevent blow out, which occurs when the support pressure at the tunneling face is too high and the soil column above is pushed upward and may lead to escape of support medium. This will cause support pressure to decrease and tunneling face can collapse. This phenomenon has been described by (Broere 2001 and Ngan et al 2015).

Before the classical blow-out (as described above) could actually happen, there are other limitations which could be governing for limiting the Target face pressure as the Slurry spill pressure.

Slurry-spill pressure can be defined as the pressure at which the support medium/slurry can raise up through the existing voids, for example, when the TBM advancement intercepts with an old borehole which was not properly sealed. The phenomenon may also occur when the slurry permeates through a very permeable geological formation or geological interface and spill/flow/bubble at the surface.

In shallow tunnels, there is a narrow range between the target pressure and the limiting maximum pressures, and the risks related to this conditions can be mitigated by the use of High Density Slurry.

The slurry-spill pressure is expressed by linear relationship as defined by the equation:

$$P_{slurry-spill} = \gamma_{Sl} \times C$$

Where: $P_{slurry-spill}$ = Slurry-spill pressure; γ_{sl} = density of slurry; C = depth of the tunnel crown

With reference to Figure 1, if the target pressure calculated at the tunnel crown is equal to the tunnel cover C, the use of HD slurry would increase the pressure required to send the slurry up to surface under open path condition and would also allow for higher margin which can be useful during operation (fluctuation).

However, the minimum pressure is generally governed by ULS or SLS. The Target Pressure is often much higher than the hydrostatic pressure, therefore also the use of an HDS is not effective in reducing the Target Face Pressure below the blowout pressure.

Face pressure in soil is generally designed between the hydrostatic and the maximum overburden pressure (heave), which is generally higher than the open path slurry-spill pressure.

Therefore, it is important to understand that the usage of HD slurry is not the mitigation measure to prevent open path slurry-spill. In any potential open path such as old borehole, a proper grouting/seal is required.

However the HD slurry can be effectively used under certain conditions, as in shallow tunnels in permeable soil in which the slurry might escape to surface in a manner similar (but not equal) to an open path.

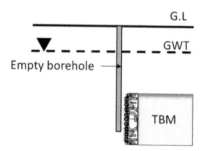

Figure 1.   Sketch of the Open Path Condition in Tunneling.

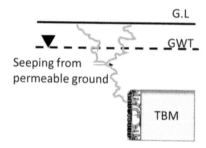

Figure 2.   Sketch of the slurry permeate out to the surface.

2   APPLICATION OF HIGH DENSITY SLURRY IN CONTRACTS T211 AND T219

2.1   *Background Information of T211 and T219*

Contract T211 includes the Design and Construction of Bright Hill Station and associated 3.9 km tunnels.

Tunnelling included two drives towards the east wall boundary of Mayflower Station undercrossed the Kallang River (ABC Canal) and a 4-storey residential building. Another two drives undercrossed a residential estate and a highly trafficked arterial Road. The total length of the tunneling is 3910 m. An overview of T211 tunnel alignment is shown in Figure 3.

3   GEOLOGICAL CONDITION

Singapore ground conditions are complex and varied due to its diverse depositional environments and subsequent geological events. There are five different types of formations in Singapore. The ground conditions during the construction of both Contract T211 and T219 are mainly from two formations, Kallang Formation and Bukit Timah Granite as shown in Figure 4.

3.1   *Kallang formation*

A Holocene and late Pleistocene deposit which can be encountered within buried river channels. The formation is governed mainly by Marine Clay. It also contains loose fluvial sands, moderately stiff fluvial clay and soft organic clay or peaty clay.

3.2   *Bukit Timah Granite*

Comprising various igneous rocks, the oldest formation in Singapore was formed since an early to middle Triassic. The igneous strata also include acidic rocks which predominantly

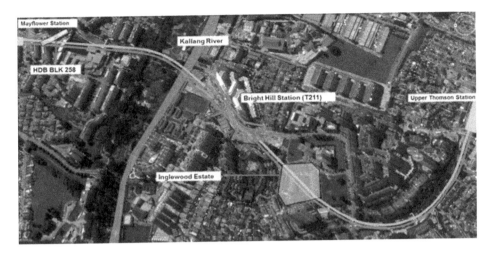

Figure 3.   Overview of Contract T211 Bright Hill Station.

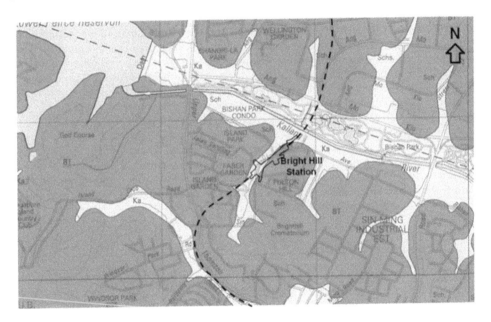

Figure 4.   Geological condition at Contract T211.

granite and granodiorite. It has a wide range of strength and high abrasiveness which pose difficulty in the tunneling work as cutter tool wear is high.

3.3   Old Alluvium

An early Pleistocene deposit which consists of alluvial sands and clays. It generally behaves or constitute strength of a weak or very weak rock and is subjected to tropical weathering which leads to its soil properties in term of strength and permeability.

3.4   Jurong formation

The Jurong formation, which is located mainly at the western and south western of Singapore, consisting of late Jurassic and early Triassic sedimentary strata. It also includes mudstone,

Table 2. Geological condition for the section between Bright Hill Station to Mayflower Station.

Section	Soil (GV, GVI)	Rock (GI – GIV)	Mix Face	Total Length
T211 to MFL (Changi Bound)	530m	70m	420m	1020m
T211 to MFL (Woodlands Bound)	400m	100m	410m	910m

siltstone, sandstone, conglomerate and limestone. The rock exhibits varying degrees of metamorphism and has been subjected to intense tropical weathering. Where shale and quartzite were also encountered at this location, the mudstone and shales fragment as they weather, while quartzites are relatively resistant to weathering. This may result in a highly variable conditions for tunneling (Shirlaw et al. 2003).

3.5 Fort canning boulder bed

It is a Pleistocene colluvial deposit which comprises strong to very strong sandstone and quartzite boulders. It is derived from the Jurong Formation and limited to are within the Central Business District in Singapore. With the boulders having a diameter of up to 7m, it poses significant challenges in term of machine advancement and wear (Osborne 2008).

The type of ground condition encountered during the tunneling towards May Flower Station (MFL), varies such as 7–11% of Bukit Timah Granite Rock (GI-GIV), 44–52% of residual soil (GV, GVI) and 41–45% of Mixed Face Geology as shown in Table 2.

Both the TBM drives undercrossing the river passed through a shallow cover with the minimum of 8.5m from the crown level. As shown in Figure 5 that the subsurface strata consists of fluvial sand which is apparent as a highly permeable sand, the consideration of using the HDS has become more pronounced. The surface layout comprises many reserved and protected trees, compelling the team to plan for all contingency work from within the tunnel. Figure 6 shows the condition of the river and the flora surrounding it which explains that grouting from the surface is not a feasible option.

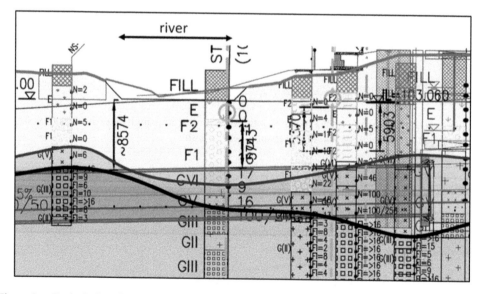

Figure 5. Geological profile for river undercrossing tunneling stretch.

Figure 6. Surface conditions of the river during dry weather and heavy rainstorms.

4 DESIGN MIX AND SLURRY PROPERTIES OF HIGH DENSITY SLURRY

In T211, the reason why contractor decided to use HDS for the canal undercrossing is that the canal is located close to the launching shaft (15 meters). The STS pump were not originally de-sign for HD slurry, however at initial stage (short distance from pumps to TBM during river undercrossing) the capacity of the pumps was deemed as sufficient.

Slurry quality is strictly controlled during tunneling, as the application of face pressure in permeable soil imply the formation of a filter cake. This can be also defined as a Membrane Model, as shown in Figure 7, indicating the formation of a thin impermeable filter cake through which the support pressure can be applied.

For the above reasons, slurry rheological properties, including fluid loss, and filter cake, yield point, are regularly monitored during tunneling. Contamination of the slurry during the tunneling work is relatively frequent and will affect the slurry properties. It is essential to recycle the slurry with fresh slurry or adding polymer to maintain the KPI throughout the entire mining works. Hence, the process of the KPI test was conducted in three times for every ring advancement. It was carried out before, during, and after one ring of excavation to regularly monitor the changes in the slurry properties and to provide ample time to improve the slurry.

The key aspect in HD slurry is to maintain high density while ensuring its other parameters with-in the design slurry KPI. Another consideration was the process of transition to change-over between the fresh slurry and the HD slurry. A trial and error in the mixing process were carried out to examine the properties of the slurry before applying it into the ground. Calcium carbonate was used as the key element to increase the slurry density. The mix was also used with a polymer to maintain its viscosity to the favorable KPI. The mixing process took approximately one work shift depending on the volume of the HD slurry.

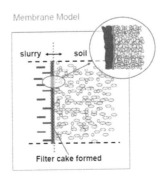

Figure 7. Membrane Model, Maidl et al. (2012).

During the mixing process, it was apparent that the viscosity of the slurry changed as the fluidity of the slurry might be altered after the reaction with the calcium carbonate.

A proper agitation in the tank was also crucial to prevent coagulation beneath the tank which may impact to the overall tank capacity.

Figure 8 below explains the process of the mixing in contract T211 which involved two separate batching of the slurry in order to achieve the HD slurry.

The fresh slurry batched was stored in the fresh bentonite tank. After sending it to the active/regulating tank, a polymer was added in to maintain the viscosity to the desired level.

On the other section, 2 holding tanks of 8m³ each were used before transferring the HD slurry into the regulating tank.

It functioned not only as a temporary storage tank but also to prevent direct contamination to the existing fresh slurry before the transition into shallow overburden area.

With a certain proportion of the mix required, the Calcium Carbonate will then be pumped into the regulating tank and agitated.

A slurry KPI will be tested afterwards to ascertain the characteristic of the HD slurry before the mining work.

The design mix of the fresh normal density slurry and the HD slurry is shown in Table 3.

The High Density Slurry is not about density but also about the rheological properties of the slurry which are classified in accordance to Key Performance Indicators (KPI) based on the function of the slurry (Marotta 2010). As one of the fundamental element in the slurry operation, it was established and monitored throughout the process of the mining.

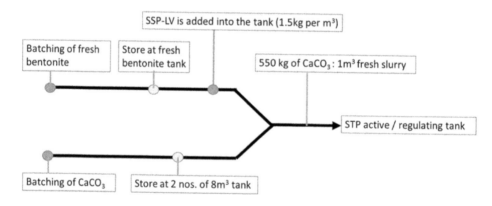

Figure 8.   Sequential method of HD slurry mixing.

Table 3.   Design mix for normal and HD slurry.

Fresh Normal Slurry

Component	Weight (in kg)	Volume (in l)	Density (in T/m³)
Water	1000	1000	1
Bentonite	50	18.5	2.7
Target Density			1.03

Fresh HD Slurry

Component	Weight (in kg)	Volume (in l)	Density (in T/m³)
Water	1000	1000	1
Bentonite	40	14.8	2.7
Calcium Carbonate	550	200	2.75
Target Density			1.30

The density of the fresh HD slurry is maintained to be higher as compared to the normal slurry. It serves as a benchmark to where the density of the non-contaminate HD slurry should have despite that it would increase due to fines contamination which was not entirely treated by the plant.

The Marsh Funnel Viscosity is critical in the HD slurry where the speed of the fluid is needed for transportation of the solid particle. Polymer addition is necessary to maintain the figure of the flow within the range of 40 – 65 s/l.

To reduce the risk of fluid permeating into the ground, the fluid/water loss was reduced to a value lower than that of the normal slurry. Hence, the essential of this parameters are emphasized under the stretch of tunneling where highly permeable sand such as fluvial sand (F1) is present. The arrangement of the particles would also determine the thickness of the filter cake which is thin and flexible to improve its impermeable surface.

The yield point is one of the key parameters which affect the slurry penetration and to allow the capability of restarting the flow of the slurry. The pressure to break the slurry within the pipe after it has stiffen will increase proportional to the yield point. The parameter is aligned to the plastic viscosity where more solid particles are moving in the fluid and will determine the rate of pumping. Table 4 summarize the KPI under difference condition used for normal slurry and HD slurry.

KPI are tested in accordance to API standard. (1 lbs/100ft^2 = 0.47 Pa)

As discussed in the previous section, the face pressure is designed assuming a membrane model where filter cake is formed at the tunnel face. Slurry KPI are generally set based on this assumption. Despite the target to create filter cake by controlling slurry rheological properties by determining KPI as the holding point, a filter cake may not be always be formed in coarse-grained ground. This will create the possibility of bentonite slurry penetration into the face.

The penetration distance of slurry depends on the soil particle distribution, face pressure and inversely proportional to the yield point of slurry. Adopting HD slurry with relatively higher yield point value during the undercrossing of sensitive location will act as an additional mitigation measures to the risk of slurry infiltration. Hence, by achieving higher yield point, it will reduce the slurry infiltration.

Therefore the usage of the HD bentonite based slurry is important not only for achieving high density (which might increase by excavated earth), but also for achieving higher rheological properties and in particular high Yield Point value, as shown in Figure 9 on the average recorded during T211 undercrossing of the canal.

Table 4. Key Performance Index (KPI) adopted.

Tunneling Operation				CHI
Slurry Properties	Bad Ground	Very Bad Ground	High Density slurry	
Density (SG)	≥ 1.03	≥ 1.03	1.25 (fresh) -1.55	≥ 1.03
Marsh Funnel Viscosity	≥ 40	≥ 45	40–65	≥ 55
PH	7–10	7–10	7–10	7–10
Fluid/water loss (cc/30 mins)	≤ 35	≤ 30	≤ 20	≤ 25
Filter cake (mm)	≤ 5	≤ 4	≤ 3	≤ 3
Yield Point (lbs/100ft^2)	≥ 7	≥ 10	8–35	≥ 10
Sand Content (%)	≤ 3.5	≤ 4	≤ 5	≤ 4
Gel Strength (lbs/100 ft^2) – 10 sec	≥ 7	≥ 8	≥ 5	≥ 8
Gel Strength (lbs/100 ft^2) – 10 min	≥ 16	≥ 7	≥ 10	≥ 17
Plastic Viscosity	≥ 6	≥ 7	≥ 10	≥ 7

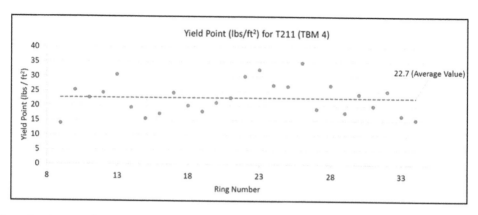

Figure 9.   Average yield point in Contract T211.

Three tests were carried out during each ring advancement to monitor the changes in KPI before, during and after each ring excavation. The plotted graph indicated the average value for each ring. In order to maintain the important rheological properties, a HD slurry mix design shall be developed.

5   CONCLUSION

Tunneling in shallow overburden is particularly challenging in poor geological conditions and when there is concern related to the serviceability limit of above ground structures.

One of the Key Parameter for tunneling under such condition is the applied Target Face Pressure. Due to the shallow overburden, any deviation from the Target pressure is more critical than a similar deviation with deeper tunnel.

Among the various risk mitigation measures, the use of High Density (HD) slurry has been successfully applied in Contract T211 and T219 of the Singapore Thomson East Coast Line.

In general, where slurry is needed as a support medium, the Target Pressure is higher than the slurry-spill pressure, where the slurry spill pressure is only marginally higher than the hydrostatic pressure to surface. Hence, using the HD slurry cannot fully mitigate the risk of slurry coming into surface through open paths conditions, as unsealed existing borehole intercepting the tunnel alignment. Surface inspection to identify and to seal off the potential open path such as old boreholes shall still be carried out prior to tunneling.

However, where tunneling is done under shallow cover with the presence of high permeability of soil, the usage of HD slurry could allow some additional margin to the Target Pressure. It could minimize the penetration of the slurry into the ground and could even prevent the slurry spill to surface.

Other rheological properties in the slurry shall be maintained in accordance to the designed Key Performance Indicators (KPI) to align with the assumption of the membrane model and assure the formation of an impermeable filter cake. The yield point plays a major role in determining the slurry penetration distance which may result in slurry infiltration into the coarse-grained particles. Therefore, the usage of the HD slurry does not emphasize only on the density, but also its rheological properties in performing into the permeable ground.

REFERENCES

Atkinson, J., Potts, D., (1977). Stability of a shallow circular tunnel in cohesionless soil. *Geotech-nique* *27 (2)*, 203–215.

Bernard, H., Lau, M.K., Senthilkumar, M., Sze, Y. S., Marotta, M., Senthilnath, G. T. (2018) Use of High Density Slurry for Shallow Tunneling in Thomson East Coast Line in (Ed,) *Proceeding of Underground Singapore 2018*, Singapore

Broere, W. (2001). Tunnel Face Stability & New CPT Applications. TU Delft.

Broms, B.B., Bennermark, H., 1967. Stability of clay at vertical openings. *J. Soil Mech. Found. Div.*

Chambon, P., Cort, J.-F., 1994. Shallow tunnels in cohesionless soil: stability of tunnel face. *J. Geotech. Eng.*

Chuan, J. Y. T., Marotta, M., Ow, C. N. (2017) Excavation Management System for Slurry TBM in Singapore. *ITA – AITES world Tunnel Congress WTC 2018*, Dubai

Maidl, B., Herrenknecht, M., Maidl, U. & Wehrmeyer, G. (2012) Mechanised Shield Tunnelling. Wilhelm Ernst & Sohn, Verlag fur Architektur und Technische Wissenchaffen GmbH & Company KG.

Marotta, M. (2010). Singapore Bukit Timah granite: Slurry quality control for TBM tunneling. *Proceedings of the World Urban Transit Conference* (WUTC 2010).

Ngan, M., Broere, W., & Bosch, J. (2015). The impact of shallow cover on stability when tunnelling in soft soils, 50, 507–515.

Osborne, N. H., Knight Hassell, C., Tan, L. C., Wong, R. (2008). A review of the performance of the tunneling for Singapore's circle line project. *World Tunnel Congress – Underground Facilities for Better environment and Safety*. India

Senthilnath, G. T. (2015). Probabilistic estimation of operating pressure for TBM Tunnelling. In D. Kolic (Ed.), *Proceedings of the ITA-AITES World Tunnel Congress 2015 (pp. 778–785)*, Du-brovnik: HUBITG, ISBN: 978-953-55728-5-5.

Shirlaw, J. N., Ong, J. C. W., Rosser, H. B., Tan, C. G., Osborne, N. H., Heslop, P. E. (2003). Local settlements and sinkholes due to EPB tunneling. *Proceedings of the Institution of Civil Engineers*, Geotechnical Engineering 156 October 2003 Issue GE4 p.193–211.

Tunnels and Underground Cities: Engineering and Innovation meet Archaeology,
Architecture and Art, Volume 11: Urban
Tunnels - Part 1 – Peila, Viggiani & Celestino (Eds)
© 2020 Taylor & Francis Group, London, ISBN 978-0-367-46899-6

Monitoring excavation-related ground deformation in London, UK using SqueeSAR™

C.A. Bischoff, P.J. Mason & R.C. Ghail
Imperial College London, London, UK

C. Giannico & A. Ferretti
TRE ALTAMIRA, Milan, Italy

ABSTRACT: InSAR technologies are becoming increasingly important in all phases of civil engineering projects. The high-resolution TerraSAR-X (TSX) data presented here demonstrate the unique advantages of using SqueeSARTM to monitor excavation-related ground deformation. The data reveal settlement troughs caused by Crossrail tunnels along their entire length. The settlement patterns are shown to be heterogeneous, both in extent and time. Ground-based monitoring data allow us to validate the remotely sensed measurements at Bond Street Station and the Limmo Peninsula. We also detect active dewatering, which has caused ground deformation over an area of several km². The TSX data show both ground settlement and the subsequent ground rebound caused by the different stages of dewatering. SqueeSARTM is used for historical ground deformation analysis and for ongoing monitoring. In general, InSAR is complementary to conventional ground-based measurement systems, and is the only way to create regional ground surface deformation maps with millimetre-scale accuracy.

1 INTRODUCTION

The remote sensing technique, InSAR (Interferometric Synthetic Aperture Radar), is increasingly used to complement conventional ground level monitoring in engineering projects, especially in urban areas (Giannico et al. 2013; Garcia Robles et al. 2015; Barla et al. 2016; Bozzano et al. 2018). High resolution SAR instruments such as TerraSAR-X provide measurements with millimetre precision, from SAR images acquired every 11 days. The large footprint of SAR images means that entire cities can be monitored with very high measurement point density (MP) of more than 1400 MP/km^2. Despite multiple, previous, successful validation studies (for example, Ferretti et al. 2007; Crosetto et al. 2008; Capes 2009; Crosetto & Monserrat 2009; Prats-iraola et al. 2016), additional validation studies under a variety of conditions are still important to prove the technique's reliability.

Any InSAR technique measuring ground deformation relies on detecting a phase shift between two or more SAR image acquisitions taken at different times, as shown in Figure 1. The most basic result of this technique is an interferogram, which is a map of the phase difference between two SAR images.

Ground movement causes a shift in phase by changing the travel path length of the radar pulse between subsequent image acquisitions. As shown in Figure 1, a ground displacement of magnitude Δr causes the travel path of the radar pulse to increase to R_2 for the

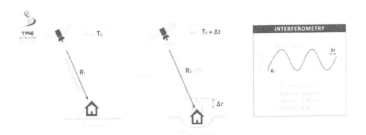

Figure 1. Schematic showing the relationship between ground displacement and signal phase shift.

second image acquisition, and this results in a phase shift from which ground deformation is derived.

This can expressed in the following relationship:

$$\Delta\phi = \frac{4\pi}{\lambda}\Delta R \tag{1}$$

where $\Delta\varphi$ is the change in signal phase, λ is the SAR wavelength and ΔR is the ground displacement (Gabriel *et al.* 1989). It is important to note that the displacement is measured in the Line of Sight of the satellite sensor (LOS).

SqueeSAR^{TM} is a proprietary multi-interferogram technique patented by TRE ALTAMIRA, which exploits statistically an entire SAR image stack and singles out measurement points (MP) on the ground that display stable amplitude and coherent phase in every image of the dataset (Ferretti *et al.* 2011).

This work involves the analysis of high resolution TSX data, which shows ground deformation across London, UK, between 2011 and 2017. It validates SqueeSAR^{TM} data with ground-based monitoring and shows which information is uniquely gained from satellite monitoring. The results demonstrate how complex deformation histories can be traced over several years; the most significant ground deformation in the study time period is related to the construction of Crossrail, which is a major new underground line crossing London in an east-west direction. The SqueeSAR^{TM} results reveal several phenomena related to this engineering project, including variation in the tunnel's settlement trough depth and width over time, and the regional impact of construction-related dewatering around Limmo Peninsula.

The dataset presented here is a SqueeSAR^{TM} analysis of 150 TerraSAR-X images regularly acquired between May 2011 and April 2017. The images were acquired in descending (west-looking) geometry with an incidence angle of 37.3°.

2 COMPARING SQUEESARTM AND CONVENTIONAL LEVELLING DATA

The SqueeSAR^{TM} measurements were validated using BRE (Building Research Establishment) levelling data obtained from Crossrail Ltd. The analysis focused on Bond Street Station in Central London. Work at this site involved the construction of two 10 m diameter platform tunnels between 20 and 24 m depth within the London Clay Formation (Abbah *et al.* 2016). The construction site is located in Mayfair underneath historical buildings that are largely supported by shallow footings, which meant that ground deformation during construction was closely monitored (Abbah *et al.* 2016).

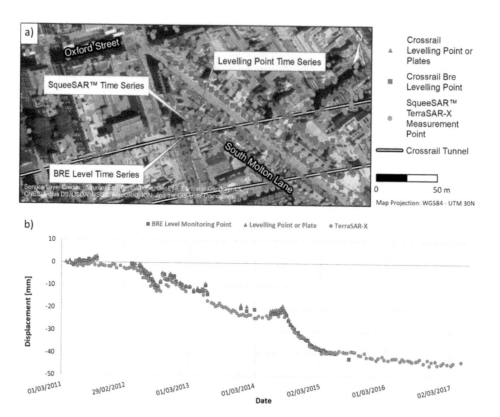

Figure 2. Comparison of SqueeSAR^{TM} measurement values with BRE Levelling and Levelling Points around Bond Street. a) Map of Bond Street showing point locations; b) time series data for location at Northern end of South Molton Lane. The SqueeSAR^{TM} time series is in excellent agreement with ground levelling data.

Figure 2 shows a comparison between three types of deformation monitoring techniques: BRE levelling (blue squares), levelling points/plates (red triangles) and SqueeSAR^{TM} MPs (green circles). The time series shown on the graph in Figure 2b are from measurement points of the three techniques, which are located within less than 10 m of each other, as indicated on the map in Figure 1a. To enable a comparison with the remote measurements, the ground-based measurements were averaged over 11 days. It is clear that the time series are in excellent agreement. This includes tracking relatively sudden movement such as the peaks in September 2012 and in August 2014. The time series also demonstrate an advantage of SqueeSAR^{TM} in long-term monitoring: this technique allows for a more consistent and longer time series than ground-based methods.

An excellent agreement in deformation measurements is found not only at the single location shown Figure 2b, but across the extent of the ground-based measuring points (Figure 3), which shows a time slice of 6th October 2014 for each of the three monitoring techniques.

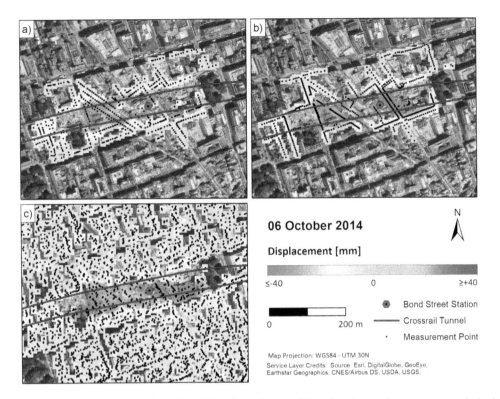

Map Projection: WGS84 - UTM 30N

Service Layer Credits: Source: Esri, DigitalGlobe, GeoEye,
Earthstar Geographics, CNES/Airbus DS, USDA, USGS.

Figure 3. Interpolated maps for a time slice of 6th October 2014, showing settlement patterns derived using satellite and ground-based methods: a) BRE levelling; b) Levelling point or plate; and c) MP derived from SqueeSAR^{TM}. The SqueeSAR^{TM} results are consistent across the entire site, not just at the single location shown in Figure 2.

3 UNDERSTANDING TEMPORAL PATTERNS IN LONG-TERM GROUND MONITORING

The timing of monitoring is important, especially to establish a baseline ahead of any construction in order to characterise normal background levels of ground movement and to accurately capture the start of the ground movement caused by construction. Furthermore, a longer monitoring period improves the understanding of any time lag between tunnelling and settlement, as well as the temporal shape and pattern of the ground response. Remote sensing, especially InSAR monitoring, is ideal for long term monitoring in urban areas, because no preparation on the ground is required before the start of construction and the monitoring can be continued at comparatively low-cost after construction is completed. The spatial coverage of InSAR monitoring is naturally more extensive than that of ground-based monitoring, and each MP has a unique displacement time series associated to it, corresponding to the image acquisition dates in the data stack that it is derived from.

To demonstrate how the ground-based and remotely sensed measurements can be integrated, Figure 4 shows two examples of time series comparisons between SqueeSAR^{TM} and BRE levelling data. Figure 4a is a map showing locations A and B. In location A, the SqueeSAR^{TM} measurements still agree with BRE levelling measurements, even though there is a gap of several months in the BRE levelling time series. Site B is interesting (Figure 4), because the ground-based monitoring at this site was apparently started too late to catch a significant amount of the ground movement related to pre-construction dewatering. SqueeSAR^{TM} monitoring can reliably track this ground movement.

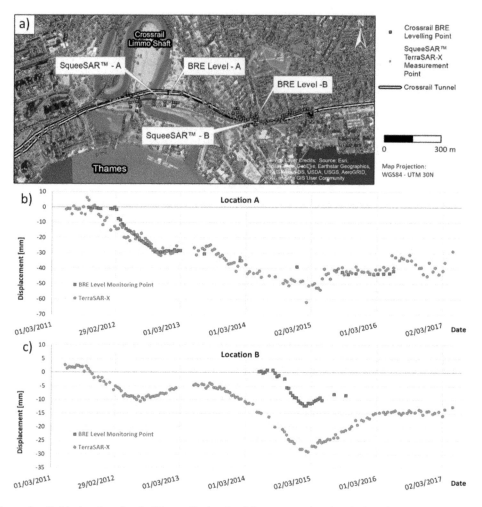

Figure 4. Validation data for the Limmo Peninsula: a) image map showing the locations and distribution of SqueeSAR^{TM} measurement points and BRE levelling points; b) and c) time series data for locations A and B.

4 REVEALING SPATIAL HETEROGENEITIES IN GROUND MOVEMENT

The regional coverage of SqueeSAR^{TM} furthermore reveals the highly complex pattern of settlement over the Crossrail tunnels (see Figure 5). The settlement appears to be controlled by a variety of factors including the effect of local geology, distance between the two running tunnels, construction method, tunnel depth and diameter, and local dewatering measures. Accurately calculating the combined influence of these factors in an empirical formula is fraught with difficulty, not least the potential unknowns in a historical city such as London. Measuring the ground deformation along the entire tunnel length with conventional, ground-based monitoring methods is neither practical nor feasible, making it an ideal application for InSAR monitoring. No other monitoring technique can provide necessary spatial and temporal detail.

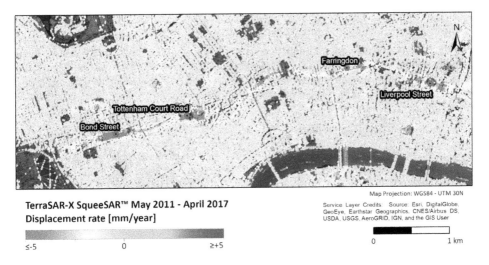

TerraSAR-X SqueeSAR™ May 2011 - April 2017
Displacement rate [mm/year]

≤-5　　　　　　0　　　　　　≥+5

Map Projection: WGS84 - UTM 30N

Service Layer Credits: Source: Esri, DigitalGlobe, GeoEye, Earthstar Geographics, CNES/Airbus DS, USDA, USGS, AeroGRID, IGN, and the GIS User

0　　　　　　1 km

Figure 5. Ground deformation map of central London, for the period between May 2011 and April 2017, illustrating the spatial 'heterogeneity' in the tunnel settlement trough along the Crossrail route.

5　DETECTING THE EFFECTS OF DEWATERING AT GROUND ENGINEERING SITES

In the eastern part of the tunnelling alignment for Crossrail, dewatering of London's regionally important, lower aquifer was necessary at several sites during construction (Semertzidou 2016), for example the main Limmo shaft, which is 30 m in diameter and 55 m deep, and that is 30 m below the pre-construction groundwater level (Roberts *et al.* 2015). London's major lower aquifer is confined by London Clay and comprised of the Upnor Formation, Thanet Sand and Chalk strata (Roberts *et al.* 2015). The most significant dewatering took place at the Limmo site and over Cross Passage 13 (CP13) (see Figure 6a) (Semertzidou 2016). At the Limmo Shaft site, water abstraction reached *ca.* 175 l/s from March to October 2012 and *ca.* 90 l/s from November 2013 to July 2015 (Semertzidou 2016). At CP13, water abstraction gradually fell from *ca.* 265 l/s in December 2013 to *ca.* 90 l/s in July 2015 (see Figure 6b).

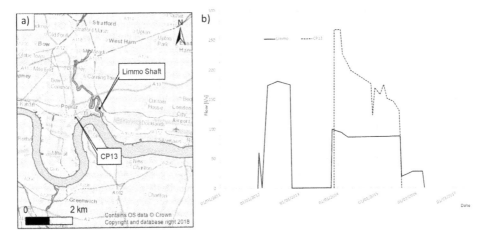

Figure 6. a) Map showing the location of dewatering sites Cross Passage 13 'CP13' and 'Limmo Shaft'. b) Graph showing abstraction rate in litres per second between 2011 and 2017. Adapted from Semertzidou 2016.

The ground displacement maps (in LOS), shown in Figure 7, are correlated with the water abstraction patterns shown in Figure 6. The first map Figure 7a shows subsidence corresponding to the first period of water abstraction at the Limmo site, shown in Figure 6b. The map Figure 7b shows the rebound that followed when water abstraction ceased. The renewed stronger subsidence shown in Figure 7c coincides with the higher amount of water abstraction, shown in the second peak in Figure 6b, and followed by the proportionately stronger ground rebound shown in Figure 7d.

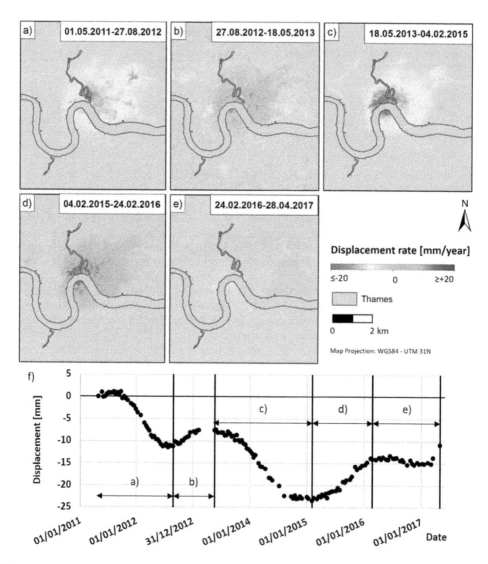

Figure 7. Two cycles of subsidence followed by ground rebound which are correlated to the rate of dewatering for the construction of Crossrail around the Limmo Peninsula. Maps a - e are showing the average displacement rate for time intervals shown in f and at the top of each map. f) shows an average time series of MPs at the centre of the ground movement around Limmo Peninsula.

6 CONCLUSIONS

SqueeSAR™ data have here proven reliable for monitoring ground movements associated with underground works of a variety of types. The validation using ground-based monitoring data clearly demonstrates that measurements provided by SqueeSAR™ are precise, accurate and reliable. Such data analysis should, and can easily be, integrated into monitoring programmes of large scale engineering projects as a matter of best practice. While PS InSAR lacks the temporal resolution of traditional surveying - and is therefore not a replacement for it - it does provide unprecedented spatial coverage at high resolution (1400 MP/km^2 equates to a point approximately every 25 m), in a way that is impractical using convention surveying.

Implementation of remote-monitoring can be complex, but guidance has been published, for example the ITAtech Guidelines (Schneider *et al.* 2015) demonstrates the benefits of including such analysis. What is remarkable in this analysis is the variety of other movements that are detected, not least ground level changes caused by groundwater level fluctuations.

These data are becoming available not just to large construction projects, but to everyone. PS InSAR results are intuitive to look at but not always easy to properly understand and interpret. It is therefore essential that construction projects acquire these data, both to better understand their associated ground movements, but also to demonstrate to the potentially anxious homeowner that settlements are isolated and nothing to do with the project (*or not!*). Coupling PS InSAR time series data with traditional monitoring provides an optimum solution for real-time construction monitoring in a wider (spatial *and* temporal) context.

ACKNOWLEDGEMENTS

We would like to thank TRE ALTAMIRA for providing the SqueeSAR™ dataset and the Crossrail Ltd. for providing the ground-based monitoring data.

REFERENCES

Abbah, J., Lazarus, D. & Jiwanji, T. 2016. Response of Buildings Supported on Shallow Footings to Tunnelling Induced Ground Movements: A Case Study of Selected Buildings at Bond Street Station. *Crossrail Project: Infrastructure design and construction*, 181–209, https://doi.org/10.1680/cpid.61293.181.

Barla, G., Tamburini, A., Del Conte, S. & Giannico, C. 2016. InSAR monitoring of tunnel induced ground movements. *Geomechanik und Tunnelbau*, 9, 15–22, https://doi.org/10.1002/geot.201500052.

Bozzano, F., Esposito, C., Mazzanti, P., Patti, M. & Scancella, S. 2018. Imaging multi-age construction settlement behaviour by advanced SAR interferometry. *Remote Sensing*, 10, https://doi.org/10.3390/rs10071137.

Capes, R. 2009. The Terrafirma Atlas.

Crosetto, M. & Monserrat, O. 2009. Persistent scatterer interferometry: Potentials and limits. *Proceedings of the ISPRR*.

Crosetto, M., Monserrat, O., Bremmer, C., Hanssen, R.F., Capes, R. & Marsh, S. 2008. Ground motion monitoring using SAR interferometry: quality assessment. *European Geologist*, 26, 12–15.

Ferretti, A., Savio, G., et al. 2007. Submillimeter accuracy of InSAR time series: Experimental validation. *IEEE Transactions on Geoscience and Remote Sensing*, 45, 1142–1153, https://doi.org/10.1109/TGRS.2007.894440.

Ferretti, A., Fumagalli, A., Novali, F., Prati, C., Rocca, F. & Rucci, A. 2011. A new algorithm for processing interferometric data-stacks: SqueeSAR. *IEEE Transactions on Geoscience and Remote Sensing*, 49, 3460–3470, https://doi.org/10.1109/TGRS.2011.2124465.

Gabriel, A.K., Goldstein, R.M. & Zebker, H.A. 1989. Mapping small elevation changes over large areas: Differential radar interferometry. 94, 9183–9191.

Garcia Robles, J., Salvá Gomar, B. & Arnaud, A. 2015. Non-Linear Motion Detection using SAR Images in Urban Tunnelling. *In: SEE Tunnel: Promoting Tunneling in SEE Region*. Dubrovnik, Croatia, ITA WTC 2015 Congress and 41st General Assembly.

Giannico, C., Ferretti, A., Alberti, S., Jurina, L., Ricci, M. & Sciotti, A. 2013. Application of satellite radar interferometry for structural damage assessment and monitoring. *Ialcce 2012 Proceedings*, 2094–2101.

Prats-iraola, P., Nannini, M., et al. 2016. Sentinel-1 Tops Interferometric Time Series Results and Validation. 3894–3897.

Roberts, T.O.L., Linde, E., Vicente, C. & Holmes, G. 2015. Multi-aquifer pressure relief in East London. *In: Crossrail Project: Infrastructure Design and Construction - Volume 2*. ICE Publishing, 135–144.

Schneider, O., Ag, A.T., Leader, S., Beth, M., Sa, S., Edelmann, T. & Ag, H. 2015. ITA Tech G Uidelines for R Emote M Easurements M Onitoring S Ystems ITAtech Activity Group Monitoring.

Semertzidou, K. 2016. *Crossrail Project Dewatering Works - Close-out Report*. CRL1-GCG-C2-RGN-CRG03-50059, https://doi.org/10.13541/j.cnki.chinade.2012.03.011.

Tunnels and Underground Cities: Engineering and Innovation meet Archaeology,
Architecture and Art, Volume 11: Urban
Tunnels - Part 1 – Peila, Viggiani & Celestino (Eds)
© 2020 Taylor & Francis Group, London, ISBN 978-0-367-46899-6

Construction methods, monitoring and follow-up of TBM tunnelling beneath an operating metro station in Bucharest

B. Bitetti, F. Valdemarin & C. Leguet
Systra, Paris, France

M. Masci & G. Martire
Astaldi, Rome, Italy

A. Stematiu
Metroul, Bucharest, Romania

ABSTRACT: Bucharest Metro Line 5 will connect the city-centre to its Western side. It consists of a 6km twin-tunnel alignment and 9 underground stations. Tunnelling works are performed with two 6.6m diameter EPBs machines. Station Eroilor 2 will be an interchanging station connecting line M5 to the existing M1 and M3 and has been used also as launching shaft for tunnelling operations toward Opera Station. The passage of EPBs about 1m below the operating Eroilor 1 bottom slab represents the most critical point of the project, thus some specific excavation procedures and preliminary mitigation works have been foreseen and implemented by the Contractor. Furthermore, special EPB follow-up procedures, contingency plan, counter-measures in case of adverse structure behaviour and rail misalignments, have been developed in agreement with the Site Supervision. This paper discusses the issues encountered on site during the works, the monitoring results, the follow-up procedure and the resulting effectiveness of the whole tunnelling process.

1   THE BUCHAREST METRO LINE 5

The plan of the urban development of Bucharest promotes the achievement of a superior level for the metro network, through the construction of several metro lines serving all the city districts: Metro Line 4 extended to the north-western suburbs, Metro Line 5 mainly East-West, and Metro Line 6 that will create a connection between Bucharest Henri Coanda Otopeni airport and stations Gara de Nord and Gara Progresu.

The Bucharest Metro Line 5 will lead an easier access to the Bucharest city-center from the Western districts. The travel time from the central existing Eroilor 1 Station to the western area of the city will reduce to about 21 minutes, which represents an important benefit for the 335,000 residents of Drumul Taberei District and surrounding areas who will benefit of an easier access city-center effectively reduce congestion on the roads.

The line consists of about 6km of twin-tunnel alignment and 9 underground stations. The new Eroilor 2 Station will play an important role for the traffic network, assuring the connection of the new Metro Line 5 to the existing metro lines 1 and 2, serving the existing Eroilor 1 Station. Civil construction works of the line started in 2011 and were completed in 2017. To date all the stations and tunnel excavation works have been completed; finishing works of stations and railways are ongoing.

The paper deals with the tunnelling works of the northern stretch of the line, between the new Eroilor 2 Station and Opera Station. It is mostly focused on the first 50m of this stretch, where tunnels pass beneath the existing Eroilor 1 Station (Figure 1). This initial part of the

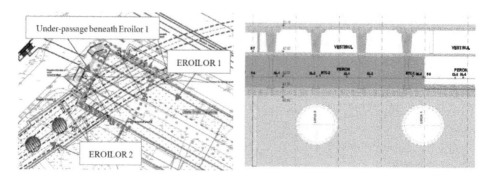

Figure 1.   Tunnels beneath the existing Eroilor 1 Station: plan view and transversal section.

stretch represents one of the most critical point of the whole project because of the physical interferences between the alignment of tunnels and existing structures of the old Eroilor 1 Station, dating back 1976.

The Eroilor 1 Station is located along Boulevard Eroilor, perpendicularly to the newly built Eroilor 2 Station, forming with it, an inter-exchange node for passengers between the new Line 5 and the existing Line 1 of the Bucharest underground transport network.

Several preliminary interventions (soils injections, dewatering wells, monitoring plan) have been required to reduce the risks of damaging the existing structures and to carry on the tunnelling operations in a safe way.

This paper especially deals with the follow-up of tunnelling operations beneath Eroilor 1 Station and with the preliminary works designed and implemented by the Contractor in total agreement with the Client and the Site Supervision, so as to reduce any impact on the station structures.

2   TUNNELLING UNDER EROILOR 1 STATION AND ITS CHALLENGES

The passage beneath Eroilor 1 station is made in twin-tunnels configuration, the distance between tunnels axes being around 16.6m. Tunnels have an internal diameter of 5.70m. They are performed by two 6.60m excavation diameter Herrenknecht EPB machines (Figure 2). The final lining is made of 5+1 precast segments, 30cm thick and having an average length of about 1.50m.

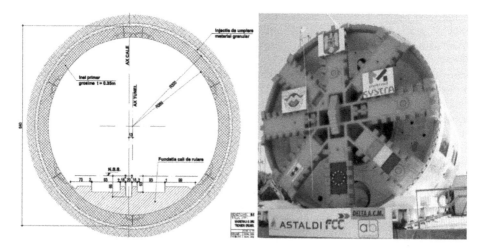

Figure 2.   Tunnel section and used Herrenknecht EPB machine used for tunnelling operations.

The intervention area is located in an alluvial plain of the river Dambovita, today regimented through an artificial canalization, characterized by lenticular stratification, sometimes interrupted, with large granulometry variations, both in vertical and horizontal. In particular, in the area where the Eroilor 1 Station was built, the geological investigation campaign indicates a geological sequence beneath the station bottom slab made of alternating clayey and sandy soils:

- Medium/large sands, with ris gravel formations between 3m and 8m (intermediate level between the floor of the vestibule and the foundation floor of the station of Eroilor)
- Clays and silty clays between 8m and 13m up to about 1–1.5 m below the Eroilor 1 Station invert
- Sands of Mostistea (fine sands), between 13m and 29m, at tunnel depth TBM.

Moreover, two aquifers have been identified during the hydrogeological investigation campaign. The deepest aquifer is located in the Mostistea complex, it is confined by the two clay layers and it interferes with the tunnels alignment. The piezometers installed in the area of the project indicate the water table located approximately 2.9m deep. According to these information, tunnelling operations result to take place in a mixed-face conditions, and beneath the water level.

2.1 The main challenges

Station Eroilor 1 is an old exiting station, dating back around 1976. The station is about 10m deep and 42m wide (Figure 3) and has two underground levels (Figure 4).

The tunnels sub-cross the existing station of Eroilor 1 in a curve, with a radius of about 260 m, passing through a station body isolated from the rest of the structure by means of

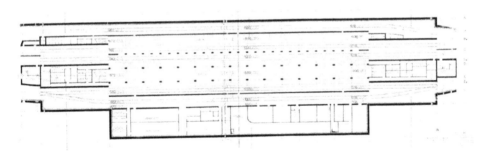

Figure 3.   Eroilor 1 station plan view.

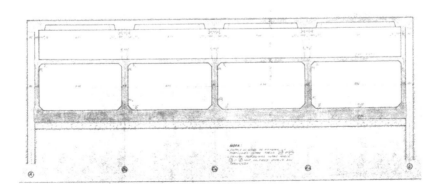

Figure 4.   Eroilor 1 station cross section.

expansion joints. The TBMs will pass beneath the station bottom slab at about 1.2m distance between the tunnels crowns and the station lower level (Figure 1).

Considering this short distance, the uncertainties on conditions of station structural elements and on the geological properties of the encountered soils, the risk of impacting the station structure represented the main challenge of the whole excavation process, even more so the Eroilor 1 Station has been kept operating during tunnelling works, because of its crucial role in the Bucharest metro system. Indeed, five operating train lines were kept working during the whole tunnelling operation.

Hence, specific tunnelling procedures and preliminary mitigation measures have been required and designed by the Contractor, to manage the risk of impacting the functionality of the station and its structures. As well, an accurate follow-up of the tunnelling operations has been required and carry on.

3 THE DESIGN SOLUTION

The design solution foreseen by the Contractor for the excavation works beneath Eroilor 1 Station, included preliminary interventions to minimize the risk of local instabilities and large deformations which could affect the station functionality:

- Vertical and horizontal soil injections to reduce the risk of water inflow during the TBM break-through
- Dewatering wells to lower the water level beneath the tunnel invert and execute the demolition and excavation works in dry conditions
- Monitoring plan for ground the station structures to control the tunnelling induced effects;

3.1 The construction sequence and the mitigation of the construction risks

In order to manage any risk related to the tunnelling operations in a so complex and risky context, the Contractor defined a precise construction procedure, resumed in the following construction steps:

- Execution of dewatering wells: 6 exploitation wells and further 6 "emergency" wells were planned, the latter to be activated in case of malfunctions of the first wells
- Execution of vertical injections from the ground level to consolidate the 0.8m to 1.0m thick strip of soil between the diaphragm walls of Eroilor 2 and Eroilor 1 Stations and of Eroilor 1 Station and the Pumping Station. The purpose of vertical injections was to ensure waterproofing and to increase safety during diaphragm walls demolition works
- Execution of drilling for horizontal injections: drillings have been done before the diaphragm wall demolition and after the water level has been lowered through the activated dewatering wells. In this way, drillings have been performed in dry conditions and any waterproofing system (type "preventer") required. Anyway, the system was conceived also to prompt react to any emergency in exceptional cases
- Execution of horizontal chemical and cement injections after the deactivation of the dewatering system
- Diaphragm walls demolition: after reactivation of dewatering, demolition has been carried out by cutting the diaphragm wall concrete by means of diamond wire. A total area of about 48m² was cut step by step isolating singular portions of the diaphragm wall . After each demolition, a 50cm thick glass fibres reinforced concrete layer was applied
- Concrete filling of the TBM starting area to ensure a perpendicular position of the machine;
- Pumping shaft (S.P.A.I.) filling with lean concrete (Figure 5)
- Construction reaction frame and starting block in Eroilor 2 Station for the TBM launch
- Start of tunnelling operations from Eroilor 2 Station toward the S.P.A.I. pumping shaft
- Deactivation of dewatering wells.

3.1.1 *The dewatering*

As previously mentioned, two aquifers have been identified in the Eroilor 1 Station area and the water level is measured about 3m deep. To reduce any risk related to water inflow during excavation works, a dewatering system made of large diameter dewatering wells (D = 700mm) and horizontal drains has been installed by the Contractor. The filtering section was positioned in the lower aquifer to lower the water level at least to 1m below the tunnels invert.

The main scope of the dewatering wells was to perform the further works (i.e. soils injections, diaphragm walls demolition and tunnels excavations), in dry conditions. This allowed to prevent the risks of water inflows during the diaphragm wall demolition works and of uplift of Eroilor 1 structures caused by the TBM face pressure. Moreover, the reduction of the water level below the tunnel invert permitted to advance during the tunnelling operation with lower face pressure.

3.1.2 *The soil injections*

Vertical and horizontal injections were foreseen to consolidate the soils between the structures diaphragm walls and beneath the Eroilor 1 Station bottom slab (Figure 6).

Vertical injections were made from the ground surface for a length of about 22m, in the 0.8m to 1.0m thick strip of soil between the Eroilor 2 and Eroilor 1 diaphragms walls and between the Eroilor 1 and the pumping shaft diaphragm walls (Figure 7). These injections allowed to execute the diaphragm wall demolition works in safer conditions, by reducing the risk of water inflow in the stations.

According to the results of the test field, two different types of injections have been chosen: the first type to strengthen the front and the second type, of thicker perimetric compartmentation, to

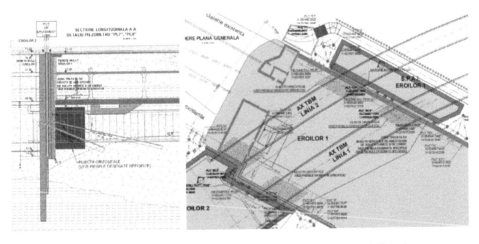

Figure 5.   Eroilor 1 Station: preliminary mitigation measures. Cross section and plan view.

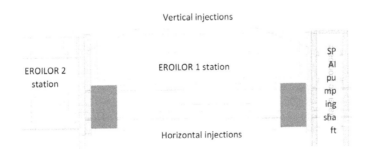

Figure 6.   Vertical and horizontal soil injections.

create a consolidated and compact core of volume behind the bulkheads so as to reduce possible dispersions in the sandy material of the former injections. 12m long cement and chemical injections have been used at this scope beneath the Eroilor 1 bottom slab.

3.2 *The monitoring plan*

As essential part of the risk management strategy, an extensive monitoring plan has been defined in order to detect all possible movements induced by tunnelling in Eroilor 1 and to minimize and to prevent all the risks dues to construction activities.

Execution of tunnelling at narrow distance from the metro operating station may obviously impact the stability of the station structures causing water inflows, erosion of the soils, with the inevitable consequences of settlements, possible deformations of the railways, and thus impacts to the metro operating lines.

The monitoring scheme conceived specifically to prevent adverse movements included:

- Structural monitoring of Eroilor 1 Station, including railways and station structure
- Monitoring of water table, especially during dewatering, consolidation works and during the TBM drive
- Monitoring of the tunnel lining
- Monitoring of TBM parameters during the passage beneath the Eroilor 1 Station bottom slab.

The existing structure of Eroilor 1 Station has been monitored through a complete monitoring system to determine the possible displacements of the internal, horizontal and vertical structures at platform and vestibule level.

Following the station structure path, a near-rectangular net has been created to install the monitoring devices inside the station (Figure 8).

The platform level has been instrumented providing the monitoring of the main three diaphragm walls, using hydraulic levelling equipment (HL) divided in three lines perpendicular to the alignment of the tunnels, with a reading station outside the zone of influence (HL-R).

The railway level has been instrumented by four devices Rail Deformation Systems (RDS) to check the deformation of the operated rail lines.

Inclinometers (CL) have been installed to the lateral walls of the station, whilst crackmeters (F) have been installed at the structural joints.

For measuring displacements in the same areas an additional topographic levelling network (RTC) has been provided to use as a double check of the above-mentioned monitoring.

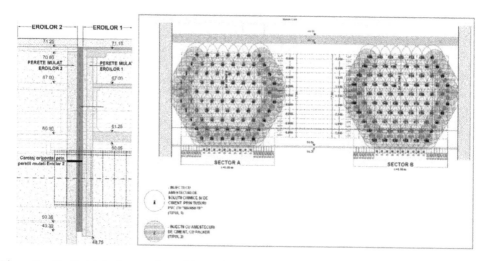

Figure 7.   Vertical injections performed from the ground level (left) and horizontal injections performed from the Eroilor 2 Station (right).

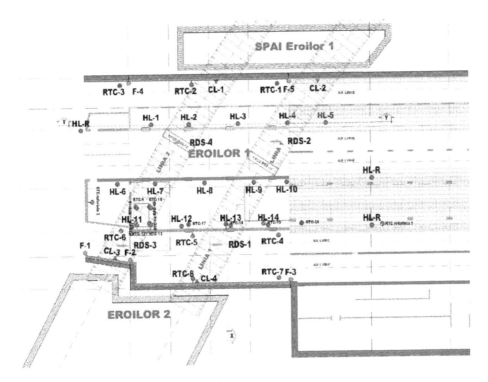

Figure 8. Eroilor 1 Station: monitoring plan view.

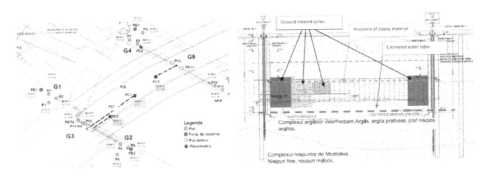

Figure 9. Installed piezometers at Eroilor 1 Station (left), predicted water level (right) during works.

The transition region located between the existing station Eroilor 1 and the underground structure of Line 1 (Eroilor 2) has been also monitored through 2 topographic levelling points, one hydraulic levelling sensor and a clinometer, installed in such area.

The water level (explained in detail in §3.2.1) has been continuously monitored during the consolidation and demolition works under the station Eroilor 1 with the purpose of checking the effectiveness of the dewatering system during the water table draw-down, the demolitions works and the TBMs drive beneath Eroilor 1 Station (Figure 9). For such monitoring the following instrument have been installed from the surface and from inside the Eroilor 2 Station:

- A total of 6 vertical piezometers, 3 on each side of the Eroilor 1 Station, 25m and 28m deep respectively
- A total of 2 horizontal piezometers equipped with electric cells for measuring the pore water pressure at the end of drillings (12m and 17m long), to check the evolution of the hydrodynamic level under the body of the station.

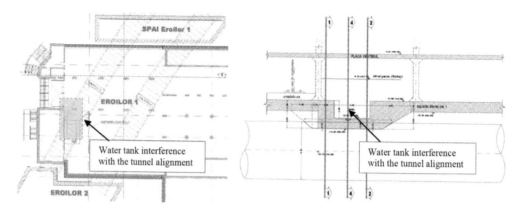

Figure 10.   Water tank below Eroilor 1 Station: plan view (left) and transversal section (right).

Figure 11.   View from the TBM excavating chamber: water tank structure and water tank steel rebars.

4   THE NEED OF A RE-DESIGN PHASE

Before the starting of the tunnelling operations a technical expertise has been carried out to check the presence of any obstacle beneath the Eroilor 1 Station invert slab. This expertise revealed the presence of water tank made by steel concrete and whose foundations interfere with the Line 2 tunnel alignment (Figure 10) and had to be demolished to permit the passage of the Line 2 TBM.

The identification of the water tank forced the Contractor to modify the initial design solution because of the need to remove the obstacle before the starting of the Line 2 tunnelling operations.

The excavation procedures initially foreseen by the Contractor need to be adapted after the water tank findings occurred before the starting of the tunnelling operations. After the analysis of several solutions, the Contractor proposed to add to the construction sequence (§3.1) the following steps to demolish the water tank base slab and to execute the tunnelling operations:

- PHASE 0: activation of the dewatering system and realization of 6m long horizontal drillings from inside the Eroilor 2 Station in the soil beneath the water tank, in order to minimize the risk of water inflow in the station in case of malfunctioning of the dewatering system

- PHASE 1: drillings and injection of expansive mix in the water tank base slab from inside the Eroilor 1 Station, to facilitate the demolition of the water tank base slab carried out from the Line 2 TBM hyperbaric chamber; casting a new concrete slab of the water tank
- PHASE 2 -5: approaching the TBM to the water tank base slab, partial emptying of the excavation chamber, executing drillings in the water tank base slab cutting and extraction of the steel rebar from the hyperbaric chamber, 15 to 20 cm advancing of the TBM (Figure 11).

5 THE CONSTRUCTION PHASE

The tunnelling operations beneath the Eroilor 1 Station, including all the operations foreseen for the demolition of the water tank, took about 20 days. They were executed between mi-June 2017 and the beginning of July. The dewatering wells were kept active during the whole demolition works and tunnelling operations. Data collected from the piezometers (§3.3.2), showed that the water level reached the designed value, keeping about 1m below the tunnel invert during the whole construction process (Figure 12).

Before the start of tunnelling operations, the diaphragm wall demolition works were executed from inside the Eroilor 2 station and the S.P.A.I. pumping shaft (Figure 13). Line 1 tunnelling works started with the TBM break-in from Eroilor 2 Station on June 16th and lasted about one month up to July 14th (Figure 14). The tunnel advancement ratio kept lower during the first two weeks of the excavation process (about 3.2 m/day), when the water tank was demolished (from June 21st to 27th), and increased up to 14.5 m/day once the water tank passage was finished, since the beginning of July up the end of the excavation works.

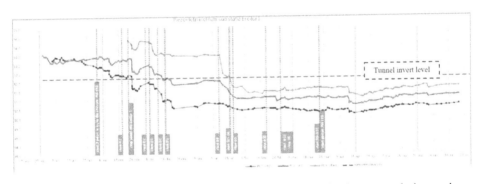

Figure 12.   Horizontal electric piezometer (PC7, PC8 and PC9) water level measures during works.

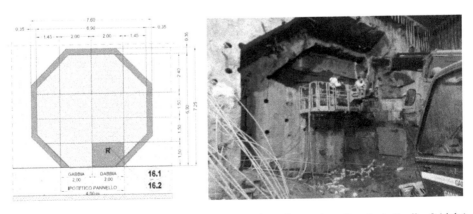

Figure 13.   Diaphragm wall demolition design sketch (left) and execution of works in Eroilor 2 (right).

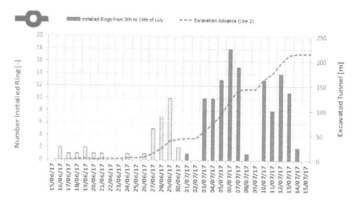

Figure 14. Advancement of Line 1 tunnelling operations beneath Eroilor 1 Station.

5.1 *The results of the construction*

The TBM parameters follow-up has been carried out during the tunnelling operations to check the parameters considered as determinant to control the induced deformations (i.e. ground settlements, station deformations…). The main recorded values of the TBM highlighted quite well the zones in which the boring have been done in ground treated zones, indeed:

- After Eroilor 1 Station break-in and during the water tank demolition the mean thrust values kept on average around 5 MN. They increased afterward and stabilized around 15 MN
- When boring through the ground treated zone a torque of about 0.6 MNm to 1.3 MNm was required, lower than the one required for the remaining stretch up to the S.P.A.I. pumping shaft (2.5 MNm to 3.5 MNm)
- As expected, the registered advancing speed was larger while boring trough ground treated zones (20 mm/min)
- The rotational speed was kept constantly around 2.5 rpm
- The registered penetration ranged between 4.0mm/rev to 9.0mm/rev

At the recorded data of the face pressure values showed that:

- As expected, during the demolition of the water tank the values corresponded to a partially fill excavation chamber. Indeed, the demolition works occurred in the "atmospheric conditions"
- The face pressure increased from 0 to 0.6 bar at the TBM axis when excavation was performed in real "closed-mode"
- A progressive decreasing was recorded when boring through S.P.A.I. pumping shaft.

Concerning the monitoring results of the structures of the Eroilor 1 Station, except some displacements increments (indirect effects of dewatering and ground treatments works performed between November 2016 and January 2017), no significant displacements, rotations and openings of structural joints have been recorded during tunnelling operations. The Rail Deformation Systems (RDS) installed on the railways confirmed such conclusions.

The most remarkable evolutions were shown by the three Hydraulic Levelling lines (almost perpendicular to the tunnel alignment). They showed a maximum settlement of about 2mm due to the TBM advancement beneath Eroilor 1 Station (Figure 15) and highlighting the stabilization of residual movements.

6 CONCLUSIONS

The construction works of the Bucharest metro Line 5 have been characterized by the presence of a very complicated stretch consisting in the tunnels passage beneath the old operating Eroilor 1 Station. Along such stretch the tunnels pass at very short distance from the station

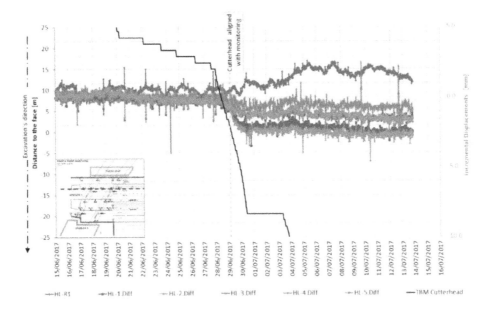

Figure 15. 1st HL line displacements from 3rd to 14th of July.

bottom slab and in complicated geological and hydrogeological conditions (mixed face conditions, below water level).

Considering the complexity of this passage, the Contractor, in total agreement with the Client and the Site Supervision put in place a whole risk management plan defining a comprehensive excavation procedure and designing several preliminary interventions, with the scope of prevent and mitigate the risk of impacting the station structures and the operating metro circulations. Moreover, further investigations carried out just before the commencement of the excavation works, revealed the presence of an existing structures (water tank) below the station bottom slab, whose foundation interfere with the Line 1 tunnel alignment.

Because of this finding the Contractor revised the design procedures and anticipated further preliminary interventions permitting to execute the demolition works of such structure in safe way before and during the passage of the TBM machine.

Dewatering system, soil injections and controlled demolition procedures have been defined and implemented. Meanwhile, to control the effects induced on the operating station, a continuous monitoring has been implemented, together with a follow-up procedure of the TBM operating parameters. This procedure allowed to check the TBM excavation parameters and to correlate them to the monitored induced deformations.

The collected data revealed that all the design parameters were respected during the excavation works and that any significant effect is induced at the station bottom slab (irrelevant displacements).

REFERENCES

B. Bitetti, E. Chiriotti, P. Monosilio, 2015. Tunnelling of Bucharest Metro Line 5: hard-points and performances. *World Tunnel Congress 2015.*
Guglielmetti, V. Grasso, P. Mahtab, A. & Xu, S. 2008. Mechanized tunnelling in urban areas. *Taylor and Francis.*
ITA WG2 (2004), Guidelines for Tunnelling Risk Management, Tunnelling and Underground Space Technology, N.19z

*Tunnels and Underground Cities: Engineering and Innovation meet Archaeology,
Architecture and Art, Volume 11: Urban
Tunnels - Part 1 – Peila, Viggiani & Celestino (Eds)
© 2020 Taylor & Francis Group, London, ISBN 978-0-367-46899-6*

Microtunneling machine completes a 110m radius curve in dense urban context

T. Blanchard & J. Bruneton
Bessac, St Jory, France

ABSTRACT: A 2m-ID microtunneling machine has completed this unprecedented performance. The machine developed was adapted to the specific constraints of the Livry-Gargan project (Paris area) with the addition of a backup tube hydraulically articulated and driven from
the surface. The microtunneling machine is unique in its short length, which allows for its
recovery from shafts with a smaller dimension but also to fit into curves of a very tight radius.
The reinforced concrete jacking pipes used on the Livry-Gargan project were also subject to a
specific curvature. The length of the pipes is reduced to 1.5 m with ends designed to ensure the
watertightness of the pipe connections in the curve. Furthermore, the thrust is efficiently distributed over the section of the pipe by the installation of a hydraulic joint interposed in
between the pipes (JackControl process).

1 INTRODUCTION

During the heavy rains, the plain of Rouailler in which the city of Livry Gargan (suburbs of
Paris, France) is located, is subject to flooding directly affecting residents. Local public institutions want to improve the management of storm water to prevent floods, but also to protect
rivers and canals from discharges from existing stormwater pipes.

The county of Seine-Saint-Denis and its "Direction de l'Eau et de l'Assainissement"
(DEA93) have launched the construction of a stormwater collector main with an inside diameter of 2000mm and a length of 620m. This tunnel will carry rainwater beneath the city of
Livry-Gargan, to relieve an existing undersized stormwater main and direct the flow to a
basin with a storage capacity of 25,000m^3.

The methodology of excavation chosen for this project was highly technical, where the
MTBM was chosen to create a tunnel with a curvature of 110m radius in the horizontal plane.

The company Bessac was entrusted with the execution of this challenging works with the
ultimate objective of unloading the existing regional stormwater main that had an insufficient
capacity during heavy rains.

2 DESCRIPTION OF THE TECHNICAL CHALLENGE – WHY MAKE SUCH A RISKY CURVE?

2.1 Dense Urban Context

The project is to be carried out in the center of the city of Livry Gargan, in predominantly
pavilion sectors where the space on the surface is very restricted and the streets are narrow.

In addition, a railway line of the SNCF (French railway public company) passes through
the city where tram-type trains are scheduled frequently.

During the technical design of this project, the DEA93 has a very precise specification. It
must consider the location of the existing stormwater collector upstream (located at the corner

Figure 1. Layout new stormwater tunnel.

of Allée Ledru Rollin and Avenue Turgot) as well as the location of the storm basin in which the new collector will end (this large basin is in the public park Beregovoy).

This information influenced the DEA93 to design for the new tunnel to be completed in 3 drives. This would include the construction of 3 shafts.

The drives 1 & 2 are relatively simple and ordinary:

- Drive 1 is installed from the shaft 1 to the basin, it is 72m long and straight
- Drive 2 is installed from the shaft 1 to the shaft 2, it is 130m long and straight

Drive 3 is the one that presents the greatest difficulties of the project: indeed, the point of departure and arrival are set by the specifications of the project (shaft 2 & 3), it must also pass under the Avenue de la Convention and under the Allée Ledru Rollin.

The angle between these two streets is important and it would have been necessary to make an intermediate shaft between these two streets to be able to complete the drive with a MTBM.

This was not possible, because these two streets are separated by the railway. The DEA 93 decided to design the project with the installation of a third tunnel with a 2200mm diameter tunneling boring machine which installs segments as it advances.

2.2 The technical choice of the pipe jacking technique with Mtbm

The tunnel-boring machine technique made it possible to achieve this section in one go with a 110m radius curve and remove the intermediate shaft in a dense urban environment (including the presence of rail infrastructure at the theoretical site of a possible intermediate shaft).

Moreover, the storage space needed around shaft 3 (corner Allee Ledru Rollin and Avenue Turgot where the space is severely restricted) for the descent of the segment, the rise of materials or the installation of bulky equipment, would have had an impact too heavy on residents and road traffic.

Finally, the DEA 93 accepted Bessac's proposal for a bold variation: install the tunnel with a MTBM of 2000mm internal diameter.

In fact, the micro-tunneling technique is the only viable alternative to reduce the constraints induced on the surface which also takes into account the urban context and the nature of the work to be carried out.

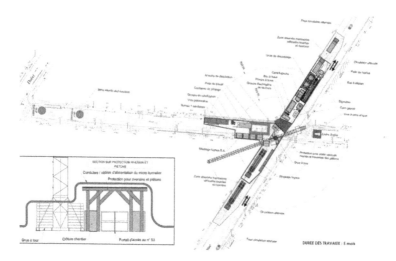

Figure 2.   Tentative site layout for TBM with segments.

Figure 3.   MTBM site layout - Reduced footprint area.

This solution generated the main challenge of the project: the creation of a 110m radius curve tunnel with a MTBM DN 2000 and concrete jacking pipes.

The Bessac micro-tunneling equipment has been custom built and adapted front this challenge.

3   GEOLOGY AND TECHNICAL CHOICE OF EQUIPMENT

3.1   *Geological context*

The route of the new stormwater pipe evolves between levels 44 and 47.1 NGF (an average depth of 11m below the ground level) and mainly intercepts the horizon of marl infra-gypsums.

This geology is known for its high clay content, which can be clogging and reduces the effectiveness of the MTBM. In addition, boulders can also be found within the marls.

A shallow water table is present at an average depth of 5m below the ground level.

3.2 *Choice of thrust equipment*

Located in an urban environment, the project had to deal with restricted size working shaft to limit surface impact. The use of a compact 850-tons thrust frame allows to reduce the dimensions of the drive shafts.

Bessac's methods department has also produced 3D drawings check that the equipment fits into the shafts.

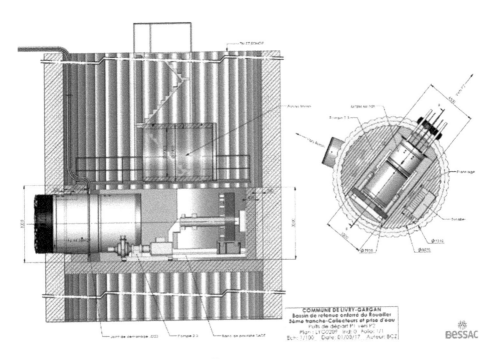

Figure 4.   3D modelling shaft 1 (drives 1 & 2).

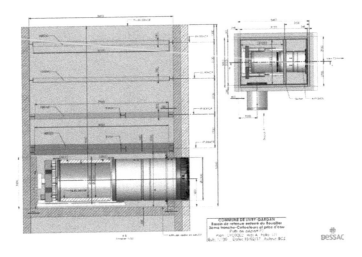

Figure 5.   3D modelling shaft 2 (drive 3).

3.3 Choice of cutting wheel

The assumptions which oriented our technical choices of the cutting wheel are the following:

− Encounter ground with clays that could clog the cutting wheel
− Risk of boulders of various sizes
− 110m radius curve route

It leads to the following characteristics to ensure a good excavation:

− Mixed wheel with disks and cutters
− 30% opening ratio of the wheel

It is the best compromise between the clogging risk in cohesive ground and the necessity to have disk cutter for large boulders.

This new cutting wheel was specially designed by Bessac and manufactured for this project.

4 FOCUS ON THE 110M RADIUS CURVE

The third drive of 420 m length included all the greatest challenges for the project:

− A passage under the railways of the SNCF (Tramway T4) with a constraint schedule requiring operation of the MTBM 24H/24.
− Unusual alignment for this technique with a 110m radius curve.

4.1 Technical choices and strategy

From the site preparation phase, Bessac knew the technical challenge represented by such a curve with a MTBM and jacking pipes which was a first on French territory.

Thus, the risks inherent to the realization of this tunnel were anticipated through the design of adapted equipment and methods:

• Use of a short micro tunneling boring machine specially adapted for this project and a specific module at the rear equipped with articulation cylinders to double the articulation and facilitate guidance,
• Use of shorter reinforced concrete jacking pipes (1.5m) to reduce the rigidity of the tunnel in the curve,
• Use of a specific joint system between each jacking pipe with computer monitoring of thrust forces and the openings between pipes
• Computer modeling of the tunnel upstream to optimize the alignment and take this curve in a more viable geometric layout.

Figure 6.   Short jacking pipes (1.5m long).

4.2 Jacking pipes

The jacking pipes were subject to a specific preparation for this curve:

- The length of the pipes is reduced to 1,5 m.
- The ends of the pipes are designed to guarantee the tightness of the slots in the curve.
- The thrust is distributed uniformly over the section of the pipes using a hydraulic joint inserted between the pipes (Jack Control® process).
- The reinforcement of the pipes is modified at the joint area (most sensitive part of the pipe in the curve).

4.3 Jack Control® process

The Jack Control process consists in setting up a hydraulic joint between each pipe to equally distribute the jacking pressure on the pipe end.

In our project, it makes possible to guarantee the reliability of the pipes, the watertightness, and the joints for radius of curvature of up to 93m!

It is enough to permit a safe guidance of the MTBM in the 110 m radius curve.

For a curvature of such a tight radius, the maximum deflection angle calculated for 1.5m pipes will be approximately 0.95 °.

We can calculate a maximum opening between pipes of about 53mm on the outside of the curve (considering a residual opening of 20mm of the inner side of the curve).

With such an opening, our system still ensures the tightness of the tunnel!

4.4 Micro Tunneling Boring Machine

A key characteristic of the MTBM is its short length that allows it to be extracted from smaller exit shafts and to fit in tighter curves.

The power pack, the airlock, and the compressed air control panel necessary for the operation of the system are housed in the first concrete pipes that follow the MTBM.

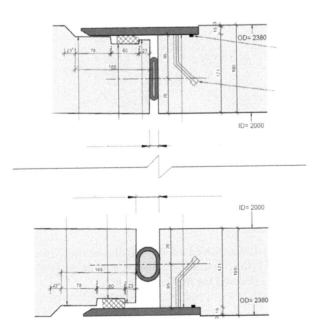

Figure 7.   Modelling tight curve pipe union with hydraulic joint.

Figure 8.   Jack Control hydraulic joint system.

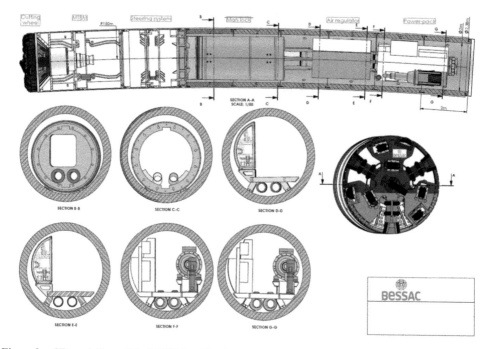

Figure 9.   3D modelling of the MTBM and back up equipment in a very tight curve.

We also add a second articulation module at the rear of the MTBM.

4.5   *Second articulation module*

The second articulation module provided an additional possibility of making an active articulation at the rear of the MTBM to meet the intended alignment. It also avoids rigidity at the connection between the MTBM and the first concrete pipe.

The second steering joint is equipped with hydraulic cylinders suitable for the project requirements.

This second articulation is actively remote controlled by the pilot.

The data can be accessed in a drive protocol.

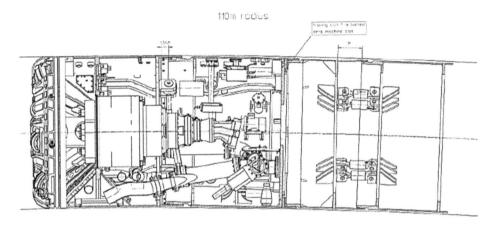

Figure 10.   MTBM and second steering joint module in 110m radius curve.

4.5.1 *General Performance*

The cutting wheel, which is fixed to the shield, excavates the ground, and defines the direction of the excavation. The machine Can is situated behind the shield and is connected to the shield by means of an active articulation. This active articulation includes steering cylinders, that allow the control of the joint, helping to excavate curves until a determined curve radius. In cases of tight curves, the second steering joint is helpful giving an additional possibility to remain on the theoretical alignment.

4.5.2 *Pipeline Route*

For drives without curves the steering cylinders elongation remain equal (e.g. at 10mm).

Figure 11.   Access to the MTBM through the second steering joint module.

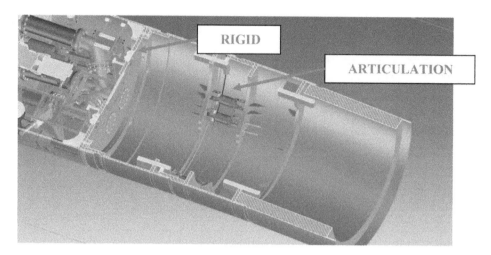

Figure 12. 3D modelling MTBM connection with second steering joint module - Curve R=65m.

For curved drives, the position of the steering cylinders shall remain at its position until the beginning of the planned curve is reached. Then the cylinders must be extracted to the required position.

In practice, it is not possible because the machine cannot instantaneously change from a straight alignment to a curved one with a radius of 110m. For this reason, such a drive has been done as a clothoid. This means the curve starts earlier, then arrives at the target curve radius and finishes past the theoretical end.

To help the machine operator a checklist has been prepared. This list shows depending on the position, the radius and direction of the curve, as well as the theoretical matching strokes of steering cylinders at the 1st and 2nd steering joints.

4.5.3 Connection MTBM and second steering joint module

The connection between the MTBM and the second steering joint module is rigid. It is made using bolts that must be tightened completely to ensure the recovery of the roll.

This rigid connection has no impact on the tight curve of the alignment. The front Can of the articulated module and the rear Can of the MTBM constitute a single rigid piece after installation.

The 3D modelisations and the calculations carried out in collaboration with Herrenknecht allow to define the minimum curve to be taken by the system [MTBM-Second steering joint]: R= 65m!

5   GYROSCOPE AND SURVEY INTERVENTIONS

The MTBM was guided by a gyroscope system (planimetry) coupled to an electronic water level (altimetry).

The accuracy of the position of the MTBM at the end of the drive was very good: largely within the guidance tolerances and our objective with less than 2 centimeters difference observed in altimetry and planimetry.

This satisfaction was obtained thanks to the computer modeling upstream of the works and a rigorous monitoring of the position of the MTBM throughout the drive.

6 CONCLUSION

Bessac has taken up the challenges of this complex project.

This project demonstrates of the importance of anticipating the hazards inherent to these specific works, through the implementation of specific equipment and methods.

The evolution of the MTBM technique in tight curves has produced an outstanding result and satisfaction for the department of Seine Saint Denis. This showed that the equipment and methods chosen by Bessac were adequate and essential for the completion of the project.

The technical challenge of a 110m radius curve with a MTBM ND2000 and jacking pipes is now performed.

Tunnels and Underground Cities: Engineering and Innovation meet Archaeology,
Architecture and Art, Volume 11: Urban
Tunnels - Part 1 – Peila, Viggiani & Celestino (Eds)
© 2020 Taylor & Francis Group, London, ISBN 978-0-367-46899-6

Risk management for tunneling-induced deformations in relation to the Eurocodes

W. Bogusz
Building Research Institute, Warsaw, Poland

ABSTRACT: Tunneling-induced ground movements are counted as a major factor guiding the design of tunnels in urbanized areas. Due to high level of third-party exposure associated with their construction, at least a simplified risk management approach should be used from the early stages of the project, to ensure its success and to protect interests of various stakeholders. Depending on expected accuracy of prediction at a given stage, qualitative as well as quantitative risk assessment methods can be used. Generally, they all involve the assessment of predicted extent of the influence zone, magnitude of expected ground deformation, and definition of limiting criteria for specific structures. Furthermore, the predicted impact is traditionally assessed in a limit state design framework, by simple comparison of predicted deformations with their assumed limiting values. However, to accommodate more risk oriented attitude, it can be extended to performance based design, where prediction is related to expected risk levels as well as potential consequences of failure for each structure. The paper presents the preliminary proposal of qualitative risk assessment approach and guidance that may be applied specifically for the purpose of assessing impact of tunneling-induced ground deformations on adjacent structures.

1 INTRODUCTION

A construction of a tunnel can have a significant impact on ground surface deformations. The extent of its zone of influence often reaches beyond the area of the construction site, affecting existing structures due to the disturbance of in-situ conditions. For shallow tunneling in highly urbanized areas, this is a subject of significant concern and often a major geotechnical risk inherent in the execution of the project, involving a significant level of third-party exposure. External limitations imposed on the designer may be a major factor governing the choice of construction methods, specific design solutions, or organization of construction activities. Despite the significance of this issue, risk assessment and design is not sufficiently standardized in regard to tunnels and other underground structures. For example, according to ISO 4356 (1977) standard, the impact on adjacent structures is for national regulations to be specified and no specific guidance is given. Similarly, current version of the Eurocode 7 (2004) does not offer any guidance on this issue beyond the need for its consideration. Not much more guidance can be expected from its upcoming new version (prEN 1997-1: 2018), as well, as tunnels will still remain beyond the scope of the Eurocodes.

The paper presents main information, challenges, and some proposals regarding risk assessment of tunneling-induced ground deformations and their impact on adjacent structures. These ideas are presented for potential implementation in a risk management framework and in reference to upcoming changes introduced by the next generation of Eurocodes (prEN 1990: 2018, prEN 1997: 2018).

2 PROJECT RISK CLASSIFICATION

Risk management techniques have been widely implemented in different industries as a comprehensive measures of assessing potential dangers, and more importantly, to balance reliability with cost effectiveness. Furthermore, they may provide a common language for engineers of different specializations, and non-engineers, often occupying decision-making positions. This process has to begin at the feasibility study stage, and then continue throughout design, construction, and the entire lifetime of the structure.

Risk can generally be divided into foreseen and unforeseen. As the latter is usually outside the influence of stakeholders, the former can be determined in a risk identification process and explicitly accounted for in a design; this includes the impact of tunneling on ground deformations and adjacent structures. Shallow tunnels are designed with the limitation of the subsoil deformations in mind, in order to:

- avoid excessive strains and additional forces in the adjacent structures, which can threaten their bearing capacity – considered as ultimate limit state (ULS);
- avoid or limit the occurrence of damage or displacements to the adjacent structures, which can worsen the state or serviceability conditions of these structures in a noticeable way – considered as serviceability limit state (SLS).

The serviceability conditions of existing structures are especially important for elements of critical infrastructures, e.g. existing railway tunnels, metro lines, bridges, etc. Such structures often have strict serviceability criteria imposed by authorities responsible for their maintenance and operation; any significant disruptions caused by construction activities may result in disproportional consequences for their owners or even the society.

In this context, risk assessment cannot be separated from consideration of potential consequences of failure. This leads to the necessity of qualitative risk classification of the project itself, as well as the potential consequences classes for each structure that may be affected. Furthermore, using a single reference level of reliability for various types of structures, under various conditions of uncertainty, is not cost-effective. Reliability discrimination, based on the expected consequences of failure and the complexity of the soil-structure interaction problem, is an effective way to balance safety with economy. Whereas, high aversion to risk may result in unnecessarily conservative assumptions and expensive design solutions, often with negligible increase of overall reliability.

EN 1997 (2004) standard already introduced the concept of three Geotechnical Categories (GC), qualifying the complexity of soil-structure interaction in a simplified manner. This approach is similar to distinction of three project classes presented by Rowe (1972), where three main levels are distinguished as:

- Class A – Important and risky – complex geology which necessitates extensive investigation, great deal of information is required for design (GC3 in EC7);
- Class B – Modest project risk and tolerable uncertainties (GC2 in EC7);
- Class C – Low risk and relatively straightforward ground conditions – little investigation is required (GC1 in EC7).

In practice, with few exceptions, all underground structures in urbanized areas should be assigned to either class A (GC3) or B (GC2). For use in rock engineering, Stille & Palmstrom (2018) suggested GC1 primarily for simple tunnels in a ground of good quality, where underlying uncertainty is relatively low and design can be based on prescriptive measures alone. In urban areas, such conditions seldom exist.

Moreover, with the upcoming changes in the Eurocodes, the concept of Geotechnical Categories will be further improved (prEN 1997-1: 2018). Especially, the consequence classes (CC)

Table 1. Consequence Classes – based on prEN 1990 (2018).

Consequence class, Description		Loss of human life	Economic, social or environmental	Factor K_F
CC4	Highest	Extreme	Huge	Outside the scope of Eurocodes
CC3	Higher	High	Very great	1,10
CC2	Normal	Medium	Considerable	1,00
CC1	Lower	Low	Small	0,90
CC0	Lowest	Very low	Negligible	Outside the scope of Eurocodes

Table 2. Geotechnical Categories – based on prEN 1997 (2018).

Consequence Class (CC)	Geotechnical Complexity Class (GCC)		
	Lower (GCC1)	Normal (GCC2)	Higher (GCC3)
Higher (CC3)	GC2	GC3	GC3
Normal (CC2)	GC2	GC2	GC3
Lower (CC1)	GC1	GC2	GC2

proposed in prEN-1990 (2018), as qualitative risk classification system, will have a direct impact on design assumptions, thus the required scope of risk assessment, as well as quality assurance procedures. The newly proposed division into five CC is presented in Table 1. Although the lowest (CC0) and the highest (CC4) classes are theoretically beyond the scope of the proposed new Eurocodes, their inclusion in the broader scope of risk management framework may be beneficial. Together with Geotechnical Complexity Classes (GCC), representing qualitative assessment of complexity of geotechnical conditions, this leads to the choice of a Geotechnical Category (Table 2).

3 PROJECT QUALITY MANAGEMENT

The reliability discrimination can be achieved not only by the modification of required margin of safety but also by differentiation in regard to quality assurance procedures. When established, a risk profile of the investment can be easily translated into specific requirements to be fulfilled throughout subsequent stages of the project. This concept has been proposed in the new version of the Eurocodes as quality management system.

For example, projects associated with higher potential consequences of failure (CC3) and complexity of ground conditions (GCC3) are therefore qualified as GC3, and often require the implementation of more advanced calculation methods, more strict quality assurance procedures, as well as additional supervision and control by an objective third-party.

Guidance concerning quality control, from design stage, through construction, and up to maintenance phase, rarely is presented in a formalized manner, however; it is often left to be specified by national regulations or guidelines for a given type of structures. Lambe (1985) distinguished different levels of control, which can be considered as precursors to quality management system proposed in prEN 1997-1 (2018), in which the level of quality control in design, construction and maintenance is governed by a risk profile of the investment, based on the assigned geotechnical category. By merging the abovementioned ideas, a proposition concerning general quality control requirements, which may be used in tunneling projects, is presented in Table 3.

The main assumption behind the quality management is that strict quality assurance procedures should reduce the probability of human error occurrence as the complexity of the project and the potential consequences increase. The presented general considerations can be further improved and refined depending on the scale and complexity of the project. For example, concerning project verification, Shirlaw & Wen (2005) distinguished between reviewing and checking a design, in the context of applying numerical methods. As the former is implemented to verify the compliance of the design with client's requirements, regulations, and the design codes, the latter is implemented to verify the reliability of a design. However, the extent to which a design should be checked can vary, depending on the complexity of the problem at hand. For projects with a high risk profile, even providing an extended independent third-party verification can be recommended. Its extent may include verification of design assumptions, based on expert judgment alone, or even additional calculations conducted by the independent engineer.

When design is conducted on the basis of calculations, any prediction of ground deformations is made before construction, only based on the data available at the time; this kind of prediction is defined by Lambe (1973) as a type A. Obviously, such prediction is conducted under conditions

Table 3. Design, verification, construction supervision, inspection, and performance control requirements.

GC	Design and verification	Construction supervision and inspection	Performance and maintenance
GC3	By qualified geotechnical engineer; Based on measured site-specific data; Complete quantitative assessment of geotechnical conditions. Independent extended verification by a third-party Detailed evaluation of critical design assumptions and predictions	Full-time supervision by a qualified engineer; Field measurement control; Independent extended inspection by a third-party.	Complete performance program; Continuous maintenance.
GC2	By qualified geotechnical engineer; Qualitative assessment of geotechnical conditions. Independent normal verification. Evaluation of critical design assumptions and predictions	Part-time supervision by a qualified engineer. Independent normal inspection.	Periodic inspection by a qualified engineer Few field measurements; Routine maintenance.
GC1	Design based on prescriptive measures Self-checking	Informal supervision. Self-inspection.	Annual inspection by a qualified engineer; Maintenance limited to emergency repairs.
-	No rational design. No verification	No supervision. No inspection.	No inspection or occasional inspection by a non-qualified person.

of significant uncertainty. Risk cannot be separated from it, which should motivate designers to seek information. This should include investigation conducted at pre-design phase, as input for the analysis, then during design, to decrease uncertainty concerning critical assumptions, and finally, at the construction stage, to validate the appropriateness of the chosen design solution. Furthermore, Lambe (1973) highlighted the interrelationship of prediction methods and data, as their quality affects the obtained results. The data in use should match the sophistication of the implemented method (i.e. calculation model). The data obtained from monitoring are as much important, not only for purpose of verification of design assumptions, but also to safe-guard one-self from unjustified claims of third parties. In the simplest cases, monitoring can be limited to visual inspections of adjacent structures; whereas, for critical infrastructure (e.g. a metro line under constant operation), continuous measurements should be conducted to limit the time from any event occurrence to the implementation of mitigation measures.

4 IMPACT ASSESSMENT

In general, the verification of the impact on the neighboring structures should be treated as a part of the design, and it is composed of following main steps:

- Assessing the extent of the zone of influence and identifying structures located within it;
- Assessing the impact on the ground displacements within the influence zone;
- Investigating the type and the state of the structures in the influence zone, as well as establishing acceptable and allowable limiting values of their deformations;
- Assessing predicted deformations of adjacent structures due to ground displacement, and the impact of those deformations on the condition and serviceability of each structure;
- Verifying any significant limit states for adjacent structures;
- Documenting the current technical state of the existing structures in the zone of influence;
- Design and preparation of the remediation measures, if necessary.
- Recommendations for the monitoring program for execution and maintenance phase.

For prediction of ground deformations due to the construction of tunnels, various calculation models can be used. These models can be divided into three main categories: analytical, empirical, and numerical models (Eurocode 7). All such calculation models are just idealizations of expected soil-structure interaction; they involve varying levels of simplifications and resulting model uncertainties (Lesny *et al* 2017). Due to that fact, simpler calculation methods are often more conservative and robust, as to ensure sufficient reliability of displacement prediction. At the same time, higher accuracy of more advanced models results in the increase in numbers of required parameters. Therefore, the most appropriate calculation method to be implemented may be different, depending on the stage of the project as well as the complexity of the problem.

When considering the extent of the influence zone of underground construction activities on neighboring area, the exact mechanism of soil-structure interaction is often complex and it depends on a combination of various factors. Therefore, empirical estimation based on the previous practice and experience is often preferred as the first approximation; more sophisticated numerical methods, e.g. finite element method (FEM), are often used at later stages of design or when more complex soil-structure interaction problems have to be considered.

In order to assure reliability of prediction obtained from FEM analysis, by application of constitutive models utilizing non-linear stiffness, a representative number of sophisticated laboratory test results are usually necessary. In comparison, most empirical models based on prediction of settlement trough are sufficiently robust to provide a prediction with simpler and relatively easier to obtain input parameters. Therefore, at the preliminary stages of design (i.e. feasibility study stage), application of simplified models may be more beneficial. From practical perspective, the impact assessment of tunneling on ground deformation can be conducted with a three step approach, where the method to be applied is selected based on two most important factors, the stage of the project (i.e. availability of data and required accuracy of prediction) and the complexity of the analyzed problem (Table 4). Generally, these steps can include:

- Step 1 – feasibility study – simplest, very conservative assumptions, using simplified calculation models or prescriptive measures, e.g. based on ITA-AITES (2014);
- Step 2 – preliminary design – empirical models, as well as numerical methods, e.g. with FEM 2D calculations for selected cross-sections, which were identified as critical, based on previous assessment;
- Step 3 – detailed design – numerical methods, FEM 2D calculations for all characteristic cross-sections, as well as FEM 3D for selected locations, when 2D model cannot provide sufficiently reliable prediction.

In some cases, when a complex spatial distribution of underground structures and soil layers is present, usually outside the range of applicability of simpler calculation models, performing spatial (3D) FEM analysis is often the only viable solution to obtain a realistic mode of deformation and assessment of possible risk.

The application of all those methods may require additional sensitivity analysis if uncertainty of the input parameters is relatively high. In situations when even the advanced analysis may not provide sufficiently reliable deterministic prediction, lower- and upper-bound solution can be provided.

Finally, when highly reliable prediction of the behavior of adjacent structures is difficult even with the use of numerical modeling and extensive sensitivity analysis, the observational method still might be used (Peck 1969, EN-1997-1: 2004).

Table 4. Levels of complexity of prediction models.

Design stage	Conditions		
	Greenfield	Normal	Complex
Feasibility study	1. Simplified		
Preliminary		2. Standard	
Detailed			3. Advanced

5 RISK MANAGEMENT BY DESIGN

In the Limit State Design framework, for any design, a verification of all relevant ultimate limit states (ULS) and serviceability limit states (SLS) have to be conducted. In the case of serviceability, usually irreversible SLS are of most concern; this includes the tunneling impact on adjacent structures. However, the choice of limiting criteria for SLS analysis is often problematic. Existing standards usually state minimum required safety levels, and they seldom offer guidance on matters of performance. There is no guideline for choosing a limiting values for all types of structures and foundations. Therefore, the designers are forced to consult different reference sources. Furthermore, the choice of reference criteria is often a conundrum for many practitioners. Structural engineers often treat geotechnical SLS limits as a responsibility of geotechnical engineers. However, in most cases, the susceptibility of the structure to the foundation deformations is the factor guiding the choice of limiting parameters.

The limiting values are usually evaluated on a case-by-case basis for each structure. Generally, the choice of the limiting criteria for a specific structure or an area should depend on (Table 5): the stage of the design (i.e. expected accuracy of prediction) and the complexity of soil-structure interaction for this structure (i.e. geotechnical category). However, an additional factor, namely the possibility of observing the actual behavior of the structure may act in favor of using simpler reference criteria, i.e. vertical displacements, under the condition of assuming sufficiently conservative values. Detailed analysis of deformations may include vertical displacements (settlements), angular distortion, tilt, as well as horizontal strains. Under some circumstances, even more detailed analysis might be justified, which may include detailed structural analysis of internal forces in the structural members due to imposed deformations. The choice of limiting values of deformations for buildings should be based on the results of technical assessment, depending on following factors:

- Type of the building and its load-bearing structural elements (susceptibility to differential deformations, as well as their impact on bearing capacity and serviceability of the structure);
- Current technical condition and preexisting damage (e.g. cracks);
- Consequences of failure and estimated value to the society;
- Comparable experience with similar buildings subjected to excavation-induced deformations.

In most cases, a detailed investigation concerning the type of a structure and its load-bearing elements, as well as documentation of existing damages, should be performed for all

Table 5. Possible choices of the limiting criteria.

	Geotechnical category of a structure or an area		
Design stage	GC1 (i.e. greenfield)	GC2 (i.e. most buildings)	GC3 (i.e. monuments)
Feasibility study	Vertical displacements		
Preliminary design			Detailed deformations
Detailed design			Forces

Table 6. The limiting values of displacements for buildings (Bogusz & Godlewski 2017).

Types of buildings and their load-bearing elements	v_{SLS} [mm]	v_{ULS} [mm]
Masonry buildings without roof and floor bands, with wooden or steel-framing floors	5 – 7	15 - 18
Masonry buildings with suspended beam-and-slab or reinforced concrete floors, or buildings constructed with precast concrete elements	7 – 9	20 - 25
Cast-in-place concrete or steel buildings	9 – 11	25 - 35

Table 7. The limiting values of displacements for installations (Bogusz & Godlewski 2017).

Types of installations	v_d [mm]
Single cables	200
Cable bands (i.e. electrical, telecommunication)	150
Water pipes Φ 200 mm	100
Natural gas pipes Φ 100 mm	150
Natural gas pipes Φ 400 mm	50
Sewage pipes	10 to 25

buildings located in the active zone of influence, as well as the buildings in the vigilance zone which are in poor technical condition or with preexisting visible cracks.

Some conservative limiting values may be found in various publications and recommendations (e.g. Boscardin & Cording 1989, Bogusz & Godlewski 2017). The example of limiting values for verification of both SLS and ULS, for buildings in good or average technical condition, can be found in Table 6. Underground installations and utilities may be subjected to the ground movement caused by excavation, as well. To avoid the disruption in their service, their displacements should be limited to a value which will depend on their function. Examples of limiting values for utilities are provided in Table 7. In comparison, the new prEN 1990 (2018) standard proposes a more general approach to serviceability criteria by distinguishing five structural sensitivity classes (Table 8).

As excavation-induced SLS will usually precede the occurrence of ULS, meeting serviceability criteria often makes detailed verification of the ULS not necessary. However, complete avoidance of the risk that can affect serviceability might not always be justified financially when the cost of remediation measures would be higher than the cost of potential repairs. If consequences of SLS failure are acceptable, when there is no possibility of loss of life and only limited economic losses due to such failure, its occurrence has to be evaluated against expected cost of preventing it. Providing sufficiently accurate analysis to allow for a certain level of deformation and possible damage may be considered as performance-based design (PBD). With the agreement of the owner of a neighboring structure, a temporary worsening of serviceability or the aesthetic condition of the structure may be permitted, during the construction phase, as the structure can be restored to the previous state, afterwards. In such case, sufficient margin of safety in regard to ULS should

Table 8. Serviceability criteria for different structural sensitivity classes - prEN 1990 (2018)

Structural sensitivity class	Description of sensitivity	Design serviceability criteria C_d		
		Maximum settlement s_{Cd}	Maximum angular distortion β_{Cd}	Maximum tilt ω_{Cd}
SSC5	Highest	10 mm	0.05%	0.1%
SSC4	Higher	15 mm	0.075%	0.2%
SSC3	Normal	30 mm	0.15%	0.3%
SSC2	Lower	60 mm	0.3%	0.4%
SSC1	Lowest	100 mm	0.5%	0.5%

Table 9. Possible acceptance levels in performance based design.

Consequence class	Limit state		
	Limiting damage	Significant damage	Near collapse
CC1	Acceptable		
CC2		Unwanted	
CC3			Unacceptable

Table 10. Risk matrix for assessment of the impact on adjacent structures (modified based on the general proposal by Eskesen et al (2004)).

	Consequences of exceeding SLS				
	Insignificant (CC0)	Considerable (CC1)	Serious (CC2)	Severe (CC3)	Disastrous (CC4)
Probability	GC1	GC1/GC2	GC2/GC3	GC2/GC3	GC3+
Very likely	UW	UW	UA	UA	UA
Likely	A	UW	UW	UA	UA
Possible	A	A	UW	UW	UA
Unlikely	N	A	A	UW	UW
Very unlikely	N	N	A	A	UW

UA – Unacceptable - risk mitigation necessary (e.g. improvement and strengthening necessary);
UW – Unwanted - risk mitigation should be considered (e.g. improvement, strengthening);
A – Acceptable - risk mitigation unnecessary, but some repairs may be required afterwards;
N – Negligible - no action necessary.

still be maintained. This is a valid approach when a possibility of damage to existing structures is not prohibited by regulations, a contract or due to low risk tolerance. Until now, the concept of PBD has been associated mostly with seismic design and is related to the assumption that some extent of damage may be allowed or even expected due to external action. This concept is presented in ISO 13824 (2009) as well as prEN 1998-5 (2017) standards. For example, prEN 1998 distinguishes three separate limit state, which together may be defined as PBD framework, namely: limiting damage, significant damage, and near collapse. In the context of PBD application when assessing the risk of damage to adjacent structures, the possible acceptance levels for occurrence of these limit states are presented in Table 9.

The risk tolerance for a specific structure is usually related to its ownership. Generally, significant third-party exposure translates to stricter limitations, often prohibiting or significantly limiting the probability of exceeding of serviceability criteria (i.e. occurrence of any damage, even those affecting the visual appearance).

For structures owned or managed by the same party, less restrictive limitations can be used, even allowing significant damage for more flexibility in design optimization, as long as a loss of stability (near collapse limit state) is still sufficiently improbable.

When limiting criteria are established and a reliable prediction is obtained, the final step is to evaluate the risk in the context of likelihood (probability) of its occurrence and the expected consequences for the specific structure. In deterministic limit state design framework, the likelihood can be associated with the expected limit state occurrence, where "very likely" is associated with exceeding the assumed limiting value. Risk can be qualitatively classified as: unacceptable, unwanted, acceptable, or negligible.

Furthermore, the risk is often managed in a proactive manner, either by its prevention (i.e. reducing the impact on the structure) or reduction (i.e. by underpinning and strengthening of the structure) to an acceptable level. When assessing a risk of a specific limit state occurrence, a simple risk matrix can be utilized, as presented in Table 10. Based on the assigned qualitative risk classification, appropriate response can be chosen regarding the need for mitigation measures.

6   DISCUSSION

Risk management approach gains favor among investors and contractors involved in development of underground structures. However, simplicity and intuitive understanding of risk management ideas are necessary to advance their use in practice. Therefore, qualitative risk assessment procedures, which are in line with current design practices, are the most optimal way to promote risk management way of thinking among investors, designers, and contractors.

The design of tunnels in Europe is still not standardized to a sufficient extent. However, the use of European standard (Eurocodes), beyond their formal range of applicability, shows the need for a more harmonized approach. Due to the specificity of tunnel design practice, a connection has to be made between a design practice and the qualitative risk management approach used in Eurocodes, i.e. consequence classes, geotechnical categories, as well as the newly proposed (prEN 1990: 2018) quality management system.

Furthermore, due to high third-party exposure, managing risk is not only limited to technical aspects of a design. It is also important to account for potential perception of the construction activities. Some safety measures offer not only technical advantages but they might be used also for psychological purposes, to assure other stakeholders that an unlikely scenario has been considered. Similarly, introducing additional monitoring measures can also have a beneficial effect on the people living next to the construction site.

7 CONCLUSIONS

As one of the most important risks inherent in underground construction activities, the impact on adjacent structures has to be explicitly considered in a design. Implementing proper risk assessment procedures is necessary to provide reliable design solutions balanced with the cost optimization. The level of assumed simplifications should depend on the complexity of the analyzed problem as well as the potential consequences of failure.

The paper presents a relation between tunnel design practice and the risk management tools available in the European standards, in the context of assessing the risk of tunneling-induced ground deformations and their impact on adjacent structures. The presented approach should not be considered as a strict requirements but as suggestions of a common framework for designs performed in accordance with European standard. Presented differentiation examples can be easily modified based on national practices and the specific requirements of a given project.

REFERENCES

Bogusz, W. & Godlewski, T. 2017. Geotechnical interaction in underground space – theory and practice. *Proc. of the 13th Intern. Conf. on Undergr. Infr. of Urban Areas*, pp. 19–31, Wrocław: CRC Press.
Boscardin, M.D. & Cording, E.J. 1989. Building response to excavation-induced settlement, *ASCE J. of Geotech. Eng.*, Vol. 115, No. 1, 1–21.
EN1997: 2004 Eurocode 7: Part 1 – General rules.
Eskesen, S.D. Tengborg, P. Kampmann, J. & Veicherts, T.H. 2004. Guidelines for tunneling risk management: International Tunneling Association, Working Group no. 2. *Tunneling and Underground Space Technology*, Vol. 19, pp. 217–237.
ISO 4356: 1977. Bases for the design of structures – Deformations of buildings at the serviceability limit states.
ISO 13824: 2009. Bases of design for structures – General principles on risk assessment of systems involving structures.
ITA-AITES 2014. ITAtech Guidelines on Monitoring Frequencies in Urban Tunneling. ITAtech Activity Group MONITORING, 20975 ITA Report No 3.
Lambe, T.W. 1973. Predictions in soil engineering. *Geotechnique* 23, No. 2, 149–202.
Lambe, T.W. 1985. Amuay landslides, *Proc. of 11th Int. Conf. on Soil Mech. and Found. Eng.*, A.A. Balkema: 137–158, San Francisco.
Lesny, K. Akbas, S. Bogusz, W. Burlon, S. Vessia, G. & Zhang, L. 2017. Evaluation of the Uncertainties Related to the Geotechnical Design Method and Its Consideration in Reliability Based Design. *Geo-Risk 2017: Reliability-Based Design and Code Developments*, GSP 283, ASCE.
Peck, R.B. 1969. Advantages and limitations of the observational method in applied soil mechanics. *Geotechnique* 19, No. 2, 171–187.
prEN 1990: 2018. Eurocode: Basis of structural and geotechnical design (final draft, 2018-05-09), CEN.
prEN 1997-1: 2018. Eurocode 7: Part 1 – General rules (final draft of 2018-05-04), CEN.
prEN 1998-5: 2017. Eurocode 8: Earthquake resistance design of structures (1st draft, 2018-05-22), CEN.
Rowe, P.W. 1972. 12th Rankin Lecture: The relevance of soil fabric to site investigation practice, *Geotechnique* 22, No. 2.
Shirlaw, J.N. & Wen, D. 2005. Checking and reviewing the output from numerical analysis. *Proc. of Underground Construction in Soft Ground*, Singapore.
Stille, H. & Palmstrom A. 2018. Practical use of the concept of geotechnical categories in rock engineering. *Tunneling and Underground Space Technology*, Vol. 79, pp. 1–11.

Tunnels and Underground Cities: Engineering and Innovation meet Archaeology,
Architecture and Art, Volume 11: Urban
Tunnels - Part 1 – Peila, Viggiani & Celestino (Eds)
© 2020 Taylor & Francis Group, London, ISBN 978-0-367-46899-6

Assessment of geotechnical capacities of spread footings due to settlements induced by tunnelling and excavation

C.W. Boon & L.H. Ooi
MMC-Gamuda KVMRT (T) Sdn Bhd, Kuala Lumpur, Malaysia

Y.C. Tan
G&P Professionals Sdn Bhd, Kuala Lumpur, Malaysia

ABSTRACT: This paper presents a framework for the impact assessment of structures on spread footings, taking into account the potential reduction of contact pressure where settlement occurs and the re-distribution of foundation loads due to differential settlement. The models demonstrate that statically determinate structures will likely settle more than the ground, and are capable of accommodating ground settlements with only modest increase in support reaction. The models for stiff statically indeterminate structures suggest that (i) there is greater load re-distribution to neighbouring supports, (ii) there is likely a lag in the first appearance of building settlement in relation to the ground settlement, and (iii) the building settlement as a percentage of ground settlement lags behind but increases with ground settlement. The models here are discussed through two case histories. They function well with observational methods, as both structural stiffness and foundation response are considered, resulting in more realistic settlement predictions.

1 INTRODUCTION

The impact of underground works to adjacent structures consists of both settlement and differential settlement, and the latter is normally considered more critical. Differential settlements also induce load re-distribution between supports but this is normally ignored to-date.

For buildings with deep foundations, the loss in mobilized resistance along the pile due to ground settlement could be recovered by mobilizing more resistance from the soil layers which are not affected along the pile body. The displacements required to remobilize the imposed load on the pile can be assessed based on load transfer analysis, i.e. *t-z* and *q-z* analysis (Boon & Ooi, 2016).

For buildings with shallow foundations, the loss in mobilized resistance due to ground settlement can also be recovered but requires the foundation to move with the ground profile, as the resistance has to be generated through the contact pressure at the spread footing. If the structure is stiff and the unloading of the foundation created by the settling ground cannot be closed, the support reactions have to be re-distributed.

However, to-date, to obtain a first assessment of the impact of tunnelling and excavation to structures on shallow foundations, the solutions for masonry structures based on Burland & Wroth (1974) and Boscardin & Cording (1989) are still used widely, and the damage is quantified using tensile strains. The calculations to estimate the tensile strains assume that the deformations of the structure are compatible with the ground, and the bearing capacity of the ground is not affected. For framed structures supported on spread footings, these calculations are less representative, as the loads applied on the ground are discrete rather than uniformly distributed. Attempts to predict the settlements of discrete footings of a framed structure have been reported in Franza & DeJong (2017). Nevertheless, the study of load re-distribution between supports is limited. Although the phenomenon of load re-distribution and loss in mobilized resistance of shallow foundations are generally considered less critical than the consequence of structural deformations, this depends on a few factors. The appreciation of the mechanism of load re-distribution would be useful in practice to facilitate engineering judgement.

The objective of this paper is to set out a framework which helps engineers to evaluate in a more systematic manner whether the bearing capacity of the foundations are compromised, and evaluate if more rigorous analyses or protective works may be required potentially based on the predicted settlements.

2 PROPOSED MODELS

In this paper, it is assumed that the two main parameters affecting the performance of the spread footings are the contact stiffness and bearing capacity, and it is important to identify if these are affected at the outset of the assessment. The models here are developed based on statics. The advantage is that it can be adapted easily to similar problems, without the concern of inaccurate calibration factors due to changes in boundary conditions, and can be solved immediately by commonly available structural program with familiar input parameters.

2.1 Structure-Foundation Response

The models for a statically determinate and indeterminate structure in response to ground settlement are distinguished here, as the mechanism of load re-distribution is different.

2.1.1 Statically Determinate Structure

For a statically determinate structure, the bending stiffness of the structure has negligible influence to the foundation reaction loads, because the structure will have to accommodate the ground movement. Ground settlements inducing differential settlements are capable of inducing eccentricity and this can cause an increase in support loads at the edge where the structure is leaning forward. A model to approximate this mechanism is shown in Figure 1.

The building settlement is a function of both the ground settlement and the contact compression at the foundation. If the distortion of the ground due to settlement is defined as i, and the additional foundation rotation to develop more contact pressure is defined as θ, then the total building rotation is $i + \theta$. The calculation procedures are discussed in Appendix A. The model here suggests that the building settlement is likely to be greater than the ground settlement for a determinate structure, and will be apparent if the contact stiffness of the foundation is compromised due to ground settlements induced by migration of fines.

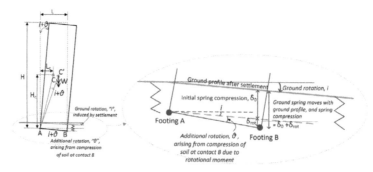

Figure 1. Model for a statically determinate structure with two spring supports, A and B, where the moment due to eccentricity of the structure is resisted by additional compression at support B.

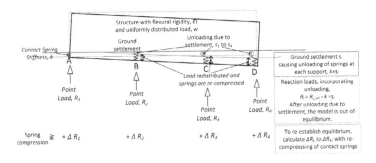

Figure 2. Adopted model for a statically indeterminate structure with multiple supports.

2.1.2 *Statically Indeterminate Structure*

For a statically indeterminate structure, there are redundant supports, and the loss in mobilized contact pressure at a foundation, can be compensated through load re-distribution to neighbouring foundations.

For a stiff structure, the structure is more efficient in re-distributing loads to neighbouring supports, and the movement of the structure will be less compatible with the ground. For a more compliant or flexible structure, the structure is less efficient in re-distributing loads, and the structure will accommodate the ground movement resulting in more compatible movement with the ground. The support reactions are not likely to experience large changes in magnitudes for a compliant structure.

The model adopted is shown in Figure 2, where the reaction loads, after taking into account of the unloading due to soil settlement, are assigned as upward point loads. To recover the reaction loads to re-establish equilibrium, the contact springs are compressed, based on the flexural rigidity of the equivalent beam. The calculation procedures are discussed in Appendix B, but common structural programs can also be used by assigning similar boundary conditions. It is noted that conventional p-Δ frame analysis may overestimate building strains, because either incomplete boundary conditions are specified, or the existing reaction loads are ignored.

3 CASE HISTORIES

Two case histories comprising statically determinate and indeterminate structures are discussed.

3.1 *Pylon Structure*

A pylon structure was located nearby an excavation in the karstic Kuala Lumpur Limestone Formation (see Figure 3). The foundation of the pylon consists of pad footings and they were monitored prior to the adjacent excavation (Figure 4a). During excavation, settlement was

Figure 3. Nearby pylon structure adjacent to a deep excavation.

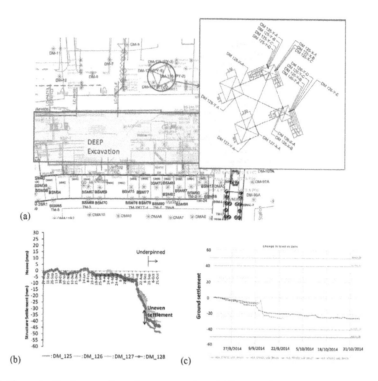

(a)

(b)

—— : DM_125 ---- : DM_126 ····· : DM_127 —●— : DM_128

(c)

Figure 4. Settlement of pylon nearby a deep excavation: (a) layout plan, (b) settlement of structure, (c) settlement of ground.

measured. The difference in settlement between supports was increasing up to approximately 15 mm (Figure 4b).

The implication of differential settlement to the foundation forces can be approximated and simplified in 2-D as a rigid beam with two spring supports. For a beam with two supports, the structure is statically determinate. The estimated ultimate bearing capacity of the pad footing

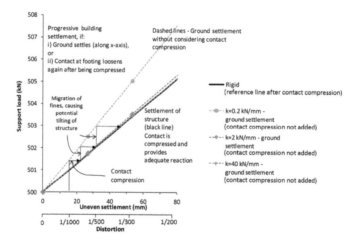

Figure 5. Influence of uneven settlement on support load, using the solution in Appendix A. It was assumed that each support was originally carrying 500 kN. The contact stiffness of 40 kN/mm was estimated using Young's modulus 10000 MPa for a pad footing of 4 × 4m. The contact stiffness of 0.2 kN/mm shows the influence of loss of contact stiffness. Span of structure is 15.3 m.

(4 × 4 m) for an undrained shear strength of 20 kPa for the soil was estimated to be around 1600 kN, or an allowable load of 533 kN. This is coherent with the estimated foundation design, with the pylon weight estimated to be around 2000 kN.

Using the model discussed in Section 2.1, it is shown in Figure 5 that an uneven settlement between the supports of 15 mm induce only marginal additional loads (y-axis). The model shows that the support loads of determinate structures are not very sensitive to small magnitudes of differential settlements especially below the typical allowable tolerance ranging between 1/500 and 1/300. Figure 5 shows how progressive movement of the structure may happen due to loosening of the contact pressure (staircase lines in Figure 5), if there is migration of fines underneath the footing. As the settlement measured at the structure was greater than the pylon (Figure 4 (b) and (c)), this was believed to be a plausible mechanism of settlement.

In the case history here, Tube-A-Manchette as well as compaction grouting was carried out. This was followed by the installation of underpinning micropiles socketed 4.5 m into limestone rock. The pylon however was subsequently relocated to accommodate a future underground entrance to a commercial development.

3.2 One-Storey Structure

A one-storey structure was in the path of the tunnel boring machine (TBM) in the karstic Kuala Lumpur Limestone Formation. There were two incidents. In the first incident, a depression was detected, and settlement was recorded at the edge of the building, as the TBM stopped for intervention, as shown in Figure 6. In the second incident, after tunneling, ground penetration radar scanning was carried out, and it was found that separation developed between the ground and the slab midspan. This was confirmed also through coring.

The ultimate bearing capacity was estimated to be 900 kN for a footing size of 3×3 m for a ground with an undrained shear strength of 20 kPa. The original support loads are estimated to be in the range of 320 – 420 kN depending on the location of the supports.

In this study, two cross sections were analysed (Figure 7). The first cross section (A-A) analyzed with ground settlement at the first edge foundation shows that the support immediately

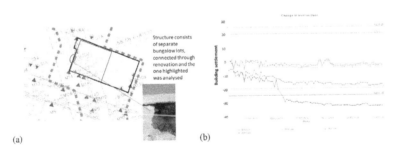

(a) (b)

Figure 6. Settlement of a one storey structure due to tunnelling: (a) layout plan, (b) settlement trend of 3 building settlement markers, (c) settlement trend of the settlement marker with the largest magnitude.

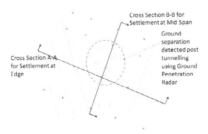

Figure 7. Analysed Cross-Sections. Red dots are foundation positions.

adjacent to the edge may increase by approximately 300 kN, with a total reaction load of 670 kN (Figure 8 (b) and (c)). This happens when the ground settlement is large enough until separation occurs and no load is transferred to the ground from the structure at the edge support. Unloading was calculated at the opposite far end, due to the flexural rigidity of the structure. However, the measurement of heave is unlikely to occur in practice. The predicted settlement of 16 mm difference in magnitude between the two end supports was calculated, and this is in the same order of magnitude compared to field measurements (17mm and 32mm at the two ends respectively). The larger magnitude of measured settlement may be possibly due to the greater extent of ground settlement affecting more than one foundation support (related to the second incident).

Using the model in Section 2.2, the building settlement as a fraction of ground settlement is plotted in Figure 9. The results show that the first 10 mm of ground settlement will result in

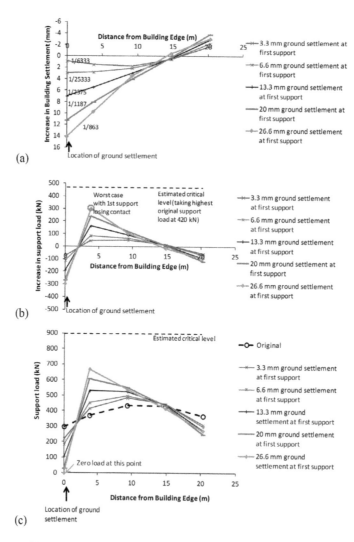

Figure 8.   Impact of loss of foundation support stiffness at the edge of the building (Cross Section A-A): (a) increase in building settlement, (b) increment in support loads, and (c) support loads. Solution was obtained using solution in Appendix B, with spring support stiffness of 30 kN/mm assuming that the ground Young's modulus was 6.5 MPa and footing size of 3 × 3 m. The building stiffness was estimated using a 200 mm thick slab and with a parallel axis theorem using the midheight of the structure (3m).

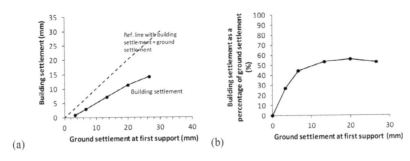

(a)   (b)

Figure 9. Building settlement as a function of ground settlement in terms of (a) magnitude, (b) percentage.

less than half of building settlement. At even smaller magnitudes, i.e. approximately less than 5 mm of ground settlement, the building settlement may not manifest with the tolerance of measurement accuracy. The building settlement as a percentage of ground settlement increases with larger magnitudes of ground settlement up to approximately 60%. This lag of building settlement compared to greenfield ground settlement predictions has been also observed in other project sites (Boon et al., 2016).

In the cross section (B-B) analysed for the second incident, the loss of support stiffness at the midspan led to an increase of edge support by 280 kN, and it also led to en-block settlements of the buildings (Figure 10), due to greater soil compression at the foundations (now taking larger loads).

The influence of flexural rigidity of the structure was studied using the first cross-section (A-A) with the edge foundation compromised, and the results are presented in Figure 11. The flexible structure is more prone to distortion, as it is more sensitive to ground settlements. The stiff structure experiences less settlement and distortion, exhibiting more linear settlements across the building, but distributes loads to adjacent supports.

In Figure 12, the response of the structure with decreasing stiffness at the edge support is shown, for different magnitudes of structural flexural stiffnesses. The flexural stiffness was estimated using the parallel axis theorem, taking the neutral axis at the midheight, for different ground slab thicknesses and assuming that the roof truss offers little rigidity. The case where the neutral axis is taken at the ground level was also compared. The results show that the ground slab thickness has little impact to the structural response, but the location of neutral axis has a major impact. The location of the neutral axis depends on the framing of the reinforced concrete structure, and is beyond the subject of study in this paper. The contribution of column stiffness to the flexural rigidity is discussed in Goh & Mair (2014). The results in Figure 12 show that flexible structures are more responsive to changes in ground stiffness and settlements.

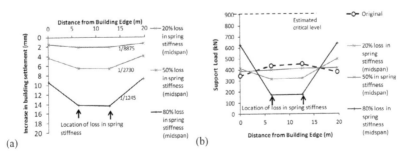

(a)   (b)

Figure 10. Impact of loss of foundation support stiffness at the midspan of the building (Cross Section B-B): (a) increase in settlement, and (b) support reaction loads.

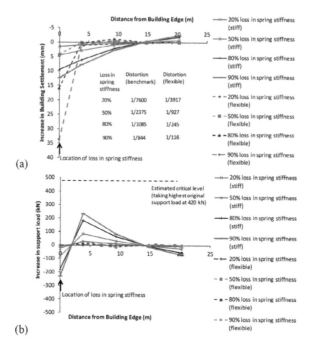

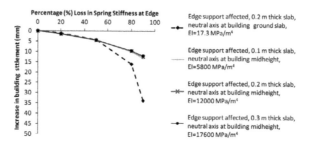

Figure 11. Influence of flexural stiffness of structure with EI = 12000 MPa/m^4 and EI=17.3 MPa/m^4 in terms of (a) settlement and (b) reaction loads, due to settlement at edge foundation.

Figure 12. Influence of flexural stiffness of structure to the building settlement.

4 CONCLUSION

The impact of ground settlement to structures with discrete loads on spread footings was studied, and two case histories were discussed. The response of the structure depends on the mechanism of the ground settlement, the flexural rigidity of the structure, and whether the structure is statically determinate or indeterminate.

For statically determinate structures, the building settlement will likely exceed the ground settlement when there is distortion of the structure. Most statically determinate structures, depending on the location of centre of gravity and span, are likely able to accommodate differential settlements, within the typical allowable tolerance ranging between 1/500 and 1/300 adopted in most projects, with modest increase in support reaction force. This range of differential settlements are two orders in magnitude smaller compared to those discussed in Burland et al. (2003), with an angle of rotation of ~5.4° or ~1/10 for the leaning tower of Pisa. However, as the structure is statically determinate, the loss of bearing capacity and stiffness at any support due to migration of fines underneath the foundation has to be reviewed. Progressive settlements may occur as a result of continual loss of contact pressure due to the migration of fines.

For statically indeterminate structures, the response of the structure depends on the flexural rigidity of the structure. For compliant structures, the settlement of the structure will be more or less compatible with the ground settlement, and the structure is more prone to distortion. For typical reinforced concrete structures with sufficient flexural rigidity, the structure is less prone to distortion. However, some of the support loads can almost double in magnitude as a result of load re-distribution for plane-strain problems, when the contact at the neighbouring foundation is lost. However, this may be within the bearing capacity, provided an original factor of safety of 3 was available. This is likely the reason why incidences of bearing capacity problems associated to underground works are uncommon. The model here can capture the delay in the appearance of building settlement in relation to settlement. It was found that the building settlement as a fraction of ground settlement increases and goes up to 60% for a moderate one-storey reinforced concrete structure. Attempts were also made to model the impact of a depression where contact pressure is lost in one of the foundations.

The calculations in this paper assume that both the structure and foundation respond elastically. The influence of moment and horizontal loads on the bearing capacity (Nova & Montrasio, 1991; Houlsby, 2016), and the influence of plasticity and creep have not been studied. Nonetheless, the models discussed here have been able to reflect typical observations encountered in practice, and will function better with observational approaches during construction.

5 APPENDICES

5.1 *Appendix A: Load-settlement response of determinate structure*

In the model in Figure 1, i is the distortion induced by the ground settlement, and θ is the additional rotation for soil compression to develop the additional reaction. Taking moment at support A, the driving moment M_D for a body with weight W can be expressed as:

$$M_D = W[H_c sin(\theta + i) + L_c \cos(\theta + i)] \tag{1}$$

where H_c and L_c are the horizontal and vertical distance to the centre of gravity. The resisting moment M_R can be expressed as:

$$M_R = R_{new}[L \cos(\theta + i)] \tag{2}$$

where R_{new} is the reaction force to maintain equilibrium. For a rigid foundation, R_{new} can be calculated easily with θ being nil. For a compliant foundation, R_{new} can be calculated as:

$$R_{new} = R_{ori} + kL(sin(\theta + i) - \sin \theta) \tag{3}$$

where k is the support spring stiffnesses. To satisfy equilibrium, i.e. $M_D = M_R$, the unknown θ can be calculated using the MS Excel Solver tool.

5.2 *Appendix B: Indeterminate structure subject to changes of support reactions*

An example for a beam with three spans (l_1, l_2 and l_3) and four supports (R_1, R_2, R_3 and R_4) with uniform load, w, is discussed (see Figure 2). The equations can be modified to incorporate more spans and more supports. From Castigliano's theorem, the energy, U, can be expressed as (Boresi et al., 1993):

$$U = \int_0^{l_1} \frac{M_1^2}{2EI} dx + \int_{l_1}^{l_1+l_2} \frac{M_2^2}{2EI} dx + \int_{l_1+l_2}^{l_1+l_2+l_3} \frac{M_3^2}{2EI} dx + \frac{R_1^2}{2k_1} + \frac{R_2^2}{2k_2} + \frac{R_3^2}{2k_3} + \frac{R_4^2}{2k_4} \tag{4}$$

where M is the bending moment at each span, EI the bending stiffness, and k is the contact spring stiffness. It is assumed that there are no moments at the two ends of the beams, i.e.

$$\frac{w(l_1 + l_2 + l_3)^2}{2} = R_1(l_1 + l_2 + l_3) + R_2(l_2 + l_3) + R_3 l_3 \tag{5}$$

$$\frac{w(l_1 + l_2 + l_3)^2}{2} = R_4(l_1 + l_2 + l_3) + R_3(l_1 + l_2) + R_2 l_1 \tag{6}$$

Eq. (5) and Eq. (6) allows R_1 and R_4 to be substituted into Eq. (4). The remaining supports are redundant supports. This leads to the following equations $\frac{\partial U}{\partial R_2} = 0$ and $\frac{\partial U}{\partial R_3} = 0$, solving which gives R_2 and R_3. Once R_2 and R_3 are obtained, R_1 and R_4 can be calculated from Eq. (5) and (6). These values are the original reaction loads R_{i_ori} before the presence of settlement.

To model the influence of ground settlements (see Figure 2), point loads are assigned at the supports $R_i = R_{i\_ori} - k_i \times s_i$ where $k_i \times s_i$ are the unloading of springs due to ground settlement. If R_i is negative, it indicates the presence of a gap that has to be closed before compressive reaction forces can be developed. The point loads R_i are treated as external loads in the subsequent calculation. To obtain equilibrium, the contact springs have to be compressed by ΔR:

$$U = \int_0^{l_1} \frac{M_1^2}{2EI} dx + \int_{l_1}^{l_1+l_2} \frac{M_2^2}{2EI} dx + \int_{l_1+l_2}^{l_1+l_2+l_3} \frac{M_3^2}{2EI} dx + \frac{\Delta R_1^2}{2k_1} + \frac{\Delta R_2^2}{2k_2} + \frac{\Delta R_3^2}{2k_3} + \frac{\Delta R_4^2}{2k_4} \tag{7}$$

At equilibrium, when any support load $R_i + \Delta R_i$ is negative, the analysis is repeated by replacing the spring stiffness with a very small number, and the point load (simulating the foundation reaction) at the support is removed from the calculation. In this paper, a symbolic mathematical toolbox Sympy operating in the Linux system was used to calculate the solutions. Eq. (4) has been benchmarked with StaadPro and the same solutions in Figure 11 were obtained.

Alternatively, the effects of settlement in Figure 2 can also be calculated using commonly available structural programs, in which case the springs are active at the outset, and are also interacting with the assigned reaction loads (in the opposite direction) and imposed load, w, in contrast to the formulation in Eq. (7). The differences in the solutions were found to be small.

REFERENCES

Boon, C.W. & Ooi, L.H. 2016. Tunnelling past critical structures in Kuala Lumpur: insights from finite element analysis and t-z load transfer analysis. Geotechnical Engineering Journal of the SEAGS & AGSSEA 47(4): 109–122.

Boon, C.W., Ooi, L.H., Tan, J.G., Low, Y.Y. 2016. Geotechnical Considerations of Deep Excavation Design for TBM Bore-Through with a Case History in Kuala Lumpur. Geotec Hanoi 2016: 375–384.

Boresi, A.P., Schmidt, R.J., Sidebottom, O.M. 1993. Advanced Mechanics of Materials. 5th Ed. John Wiley & Sons, Inc,

Boscardin, D.M. & Cording, E.J. 1989. Building response to excavation induced settlement. ASCE Journal of Geotechnical Engineering 115:1–21.

Burland, J.B. & Wroth, C.P. 1974. Settlements of buildings and associated damage. Conference on Settlement and Structures: 203–208

Burland, J.B., Jamiolkowski, M., Viggiani, C. 2003. The Stabilisation of the Leaning Tower of Pisa. Soils and Foundations, Japan Geotechnical Society 43(5): 63–80.

Franza, A & DeJong, M. J. 2017. A Simple Method to Evaluate the Response of Structures with Continuous or Separated Footings to Tunnelling-Induced Movements. Congress on Numberical Methods in Engineering, Valencia Spain, 3–5 July 2017: 919–939.

Houlsby, G.T. 2016. Interactions in offshore foundation design. Géotechnique 66(10): 791–825.

Goh, K.H. & Mair, R. J. 2014. Response of framed buildings to excavation-induced movements. Soils and Foundations 54(3):250–268.

Nova, R. & Montrasio, R. 1991. Settlements of shallow foundations on sand. Géotechnique 41(2):243–256.

Tunnels and Underground Cities: Engineering and Innovation meet Archaeology,
Architecture and Art, Volume 11: Urban
Tunnels - Part 1 – Peila, Viggiani & Celestino (Eds)
© 2020 Taylor & Francis Group, London, ISBN 978-0-367-46899-6

Geotechnical monitoring at Forrestfield Airport Link Project in Perth (WA)

M. Bragallini, S. Ganguly & D. Coulthard
FGLS JV (Field Monitoring Services, Geomotion, Land Surveys), Perth, Australia

A. Zampieri
Sisgeo Srl, Milan, Italy

D. Frontini
SI-NRW JV, Perth, Australia

ABSTRACT: The Geotechnical Monitoring of the Forrestfield Airport Link started at the end of 2016 before the construction of the new rail service (with two underground tunnels) 8.5 km long from the existing Midland Line to the eastern suburbs of Forrestfield – with three new stations at Redcliffe, Airport Central and Forrestfield. The most of the instruments are in-stalled underneath or in proximity of very sensitive buildings and sites, being in an airport area, with nearly real time system for monitoring data with wireless technology combined with telemetry. All the monitoring data are available on web base MIMS (Monitoring Information Management System)

1 INTRODUCTION

The 1.86 AUD billion Forrestfield Airport Link is jointly funded by the Australian and Western Australian governments and will deliver a new rail service 8.5 km long from the existing Midland Line to the eastern suburbs of Forrestfield – with three new stations at Redcliffe, Airport Central and Forrestfield.

The rail link will connect with the existing Midland line near Bayswater Station and will run to Forrestfield through two underground tunnels realized by two 7m diameter Mix shield TBMs, under the Swan River, Tonkin Highway and under the Perth Airport (which remains in service) to ensure minimal impact on them.

Early in 2016, the major contract for the Forrestfield-Airport Link project was awarded to Salini Impregilo – NRW Joint Venture (SI-NRW JV), the designer is Coffey.

There were two main challenges for the monitoring system. First one was that most of the instruments are installed underneath or in proximity of very sensitive buildings and sites, being in an airport area. The second one was to implement a nearly real time system for monitoring data.

Two main solutions have been therefore implemented, the first one related to the instruments installed on the stations (piezometers and strain gauges), the second one to the incline-settlement columns.

The paper describes in details how the applied solutions can be technically suitable and cost-effective for nearly real-time automatic monitoring system in sensitive sites.

2   DESCRIPTION OF THE PROJECT

2.1   *General layout*

The new Perth metro line is 8,5Km long, starting at the Bayswater junction (connection to the existing Bayswater Line) the tunnels gradually deepen towards the Swan River and re-emerge into the Redcliffe Station box; the tunnels dive again to reach the airport area, running under the Runways and Taxiways until the Central Station box close to the Airport Control Tower. In the last part of the track, the two tunnels run under bush area before crossing the Brook-field Rail line and come out into the Forrestfield Dive where the line ends.

2.2   *Geological condition*

The most superficial FAL alignment is into fluvial deposits characterized by clayey sands and silty sands (Guilford Formation GF), carbonate sandy gravel by inner shelf of nearshore marine environment deposited during multiple marine transgressions (Ascot Formation AF) and Silty sandstone with shallow-marine origin in the deepest part of the alignment.

Sea level has fluctuated significantly over geological time, with the result that the ancient Swan River has experienced several very significant fluctuations in water level. These changes in water level have created a number of erosion and deposition events resulting in paleochannels which have been infilled in a marine, estuarine and freshwater environment. These paleochannels and erosional features have been identified in close proximity to the current Swan River alignment.

2.3   *Design of the geotechnical monitoring*

The design follows the Scope of Works and Technical Criteria (SWTC) created for this FAL Project. This document defines the monitoring and instrumentation requirements, reading instrumentation frequencies, critical instruments, monitoring review levels and MIMS

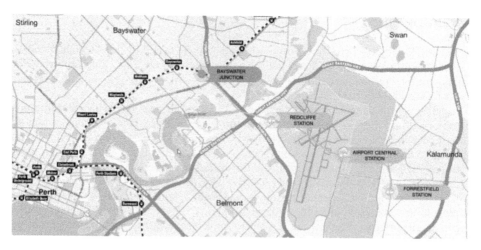

Figure 1.   Caption of FAL Project plan.

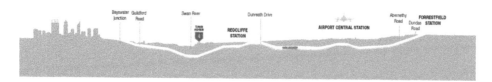

Figure 2.   Forrestfield Rail route.

Figure 3. An example of installed instrument: borehole rod extensometer.

specifications. Six different type of Array from A to F have been designed and each one has different kind of instrumentation inside. The instrumentation and Monitoring Plan (IMP) documents describe how SI-NRW JV approaches, implements and manages the geotechnical and structural instrumentation and monitoring (I&M) for the Project. The IMP conforms to the requirements of the Scope of Works and Technical Criteria (SWTC) and Project Deed (PD).

For each instrument the designer Coffey, provide the exactly position in X, Y and Z coordinate and are referenced to absolutely coordinates and all the instruments have been installed in the design position. For this reason, e.g. the length of the mono and multi anchors extensometers has been set on site during the installation.

Monitoring Review Levels (MRL) are assigned to every monitoring instrument installed for the Project. The Public Transport Authority of Western Australia (PTA) has specified three levels of MRLs for the Project, 'trigger', 'design' and 'allowable'. SI-NRW JV assigns those three MRLs on the following basis:

Trigger – a predetermined level prior to the design and/or allowable level, e.g. 70% of the expected maximum design value or measurement parameter;

Design – the best estimate of the maximum value of a measurement parameter based on design calculations using design parameters. In some instances, this may be coincident with the allowable level;

Allowable – the maximum or minimum (as appropriate) allowable reading consistent with the requirements of the SWTC or as specified by the Stakeholder.

3 INSTRUMENTATION INSTALLED

3.1 *List of Installed equipment.*

A list of all installed instrumentation is reported in the chart below:

Item	N	Casing (m)	Drillings (m)
Ground Settlement Point – Levelling	1367		
Ground Settlement Monitoring (Read with RTS)	184		
Ground Settlement Point – Prism	276		
Ground Settlement Reflector	84		
Rail Monitoring Point	888		
Structural Monitoring Point – Levelling	100		
Structural Monitoring Point – Prism	590		
Topographical	3489		
Extensometers single anchor	49		715

(Continued)

(Continued)

Extensometers double anchors	16		233
Inclinometer/Extensometer combined (Automated)	6	227	226
Inclinometer/Extensometer combined (Manual Survey)	37	957	938
Inclinometer in Diaphragm Wall	31	787	
Load cell	10		
Piezometer	243		695
Electro level beam – Monoaxial	5		
Strain Gauges Vibrating Wire Embedment (Tunnel segment)	372		
Structural Load Cell	9		
Strain Gauges Vibrating Wire	42		
Tiltmeter – MEMS – Biaxial	15		
Vibrometer (7 moved in different locations)	24		
Geotechnical instrumentation	861	1911	2838

Instrumentation installed till August 2018

3.2 Instrumentation for structures

Tiltmeters are installed e.g. into the PAPI approach laser light in airside for recording the biaxial tilt long two orthogonal axes of a building, biaxial. The two different sensor assembled permit horizontal or vertical installation

Monodirectional electro-level beams on 3m base, with digital sensor are installed on sensible buildings orthogonally the direction of excavation.

Vibrating wire crackmeters have been utilized into the project for checking the displacement between two structural elements and cracks.

The construction of the station box with a depth 15m under the ground level, with the base slab below 14m below the groundwater average level requested the control of the deformation during the de-watering stop: on this purpose, vibrating Strain Gauges directly welded on the reinforced bars have been installed.

3.3 Instrumentation for groundwater

In order to measure the pore pressure, two different ranges of vibrating wire piezometers have been supplied, most of them installed around the boxes stations (Array Type C) and along the tunnel section (Array Type F). For the installation, fully grouted method has been utilized. Standpipe piezometers equipped with submersible transducers have been utilized instead for the measurement of the water level during the pumping test.

Figure 4.   Installed horizontal tiltmeter (left) and vibrating wire piezometer (right).

Borehole extensometers with single and double anchors have been installed, the depth of which is designed in accordance to the depth of the tunnel, as defined by Array E and F in SWTC, 1 – 2m higher than the extrados of the tunnel. The length of the fiber glass rods is actually defined and set on site during the installation, considering the ground elevation in that point.

The measure reference is the head, where displacement transducers are also installed, and it is usually georeferenced topographically for correlate absolutely the settling of the anchor into the borehole. Inclinometer casings have been installed vertically into the D-Walls of the stations and dives, after the construction of the wall and capping beam. These ones are read manually with inclinometer system. The inclinometer/extensometers are installed in Array Type C close the stations and dives and Type F along the tunnel alignment; to the two orthogonal and horizontal components, with the use of the magnetic rings has been added the vertical (settlement) component. This instrument can be read manually with inclinometer system and extensometer probe or automatically with DEX-S probes. Most sensitive casings have been equipped with DEX-S probes in order to have nearly-real time information regarding horizontal displacement as well as vertical settlements.

Spot weldable strain gauges have been installed on station and dive struts (sections with n.3 instruments 120° each other's). Some struts are also instrumented with load cell interposed into the junction for understand what the load during the deepening of the excavation was and considering temperature variations.

The sections at cross passages into the tunnels are instrumented with segment with strain gauges embedded into the concrete. For each segment the cables are collected into a box that has been free after the installation and connect to dataloggers for Near Real Time (NRT) readings Vibrometers have been installed for the sake of recording the vibration of the site, during the construction of the stations, dive and of the tunnel.

In this project as specified into SWTC, many instruments have to be read in Near Real Time (e.g. piezometers during station box dewatering, load cells and strain gauges in the struts), therefore several dataloggers have been utilized. The recorder data are sent to an FTP and uploaded into MIMS for visualization and elaboration. Miniloggers are installed at the ring tunnel for reading the strain gauges into the segments, for reading the load cells and digitally the electro level beams. Miniloggers have been selected due to their small dimensions which are suitable for installation in tunnel lining.

Whenever a bigger number of instruments with different signal output has to be read, standard datalogger SISGEO Omnialog has been considered as a good solution. Installed at the Array F in airside, these ones are connected the DEX-S probes, VW Piezometers and Extensometers sensors.

In total up to 96 sensors are read and, through telemetry, the data are uploaded on MIMS.

WR LOG is wireless data acquisition network that combines state-of-the-art wireless monitoring and advanced software tools. This solution was implemented in the Stations. This kind of dataloggers are connected to the Vibrating Wire Piezometers and extensometer transducers e.g. in the Airport Central Station Gateway, where more than 15 units simultaneously

Figure 5.   Installed Inclino/extensometer DEX-S probes (left) and struts load cell (right).

Figure 6. Minilogger (n.29), standard datalogger (n.6) and WRLOG installed (n.35).

connected in to one Gateway thanks to LoRa (Long Range Radio Transmission) technology. Battery lifespan over 10 years of unattended runtime and the embedded radio technology permits therefore long-range communication up to 15km, becoming a cost-effective solution by avoiding people to interview for measures, changing batteries, change configuration, etc.

3.5 Topography Monitoring 1D and 3D

The approximate breakdown of monitoring required on the Forrestfield-Airport Link Project includes 8 ATS Systems (14x Leica TM50 & 4x Leica TS16 instruments), 2500 1D ground settlement points, 700+ 3D monitoring prisms for buildings, structures and other critical infrastructure, 650+ 3D monitoring prisms for rail monitoring, 250+ 3D tunnel convergence prisms. The readings have been undertaking using a combination of manual and automated systems.

3.5.1 Ground Settlement and Prism monitoring points

Manual readings taken using a Leica digital Level and barcode staff/Leica TM50 or TS16. The ground settlement monitoring points consist of a concrete encased star iron picket for unimproved ground conditions, while the building and structure monitoring points consist of levelling pins or 3D Prisms. Reflectorless surface technology have been used for airside Runway, Taxiways and Aprons. The errors usually associated have been significantly reduced through methods and automated process designed and developed in-house specifically to meet the projects requirement.

All the measurements are linked back to the survey control which is located well outside of the zone of influence with plenty of redundancy mark build into the system. This control is check regular throughout the works. Manual monitoring data is post processed with infield adjustment or MircoSurvey StarNet. The final results exported in *.csv format and vertical settlement comparisons computed/verified in excel. Once validated the data is then uploaded to MIMS

3.5.2 Automatic Total Station (ATS)

The ATS are commonly used on critical assets to achieve 1D or 3D monitoring in real-time. The ATS are connected to a cellular modem installed in the power and communication logger box which is powered by either solar or mains power. The measurements made by the ATS are managed by the Leica GeoMoS. With an integrated TCP/IP connection, GeoMoS is able to establish a bi-directional communication with the ATS over the modem The GeoMoS software is responsible for control of the ATS, data acquisition, data storage and automatic processing of the instrument's resection (free station computation) or orientation. Once the data is verified it is then sent to the MIMS/Maxwell Geosystem.

4 DATA MANAGEMENT AND PROCESSING

The monitoring information management System (MIMS) is delivered for the project by the collaboration of Maxell Geosystem Limited (MGS). MIMS uses MGSs MissionOS which utilizes AWS elastic web server to manage increasing data volume and uninterrupted processing requirement. A separate data server is used to accept incoming raw data from dataloggers.

Figure 7. Picture of RTS installation in airside.

All automated data is collected and stored on the dataloggers, transferred to a cloud-based data server using telemetry and then imported in to the MIMS database. The backup of the data server is taken at a stipulated time every day. For data processing of automated readings each file that is automatically transferred to the system has a configuration setup for it. This allows the acceptance of any text file type in the MissionOS so that different dataloggers and instrumentation can be used. The Critical instruments of the FAL Project are monitored using automated dataloggers and telemetry. The system also accepts manual data from any type of instruments. Manual data can be uploaded through the portal directly in the MGS template format. During manual data upload the user is prompted if there are any alarms that will be triggered from the readings. This helps in stopping human error being introduced to the manual recording of data.

5 DATA ANALYSIS

This chapter describes some of the most important information obtained by the instrumentation in correlation with processing of the works.

5.1 Extensometers data.

Extensometers data which depend from several variable circumstances (e.g. ground condition, pressure on the face, feed rate…) were in most instances extremely accurate and precise in detecting TBM-induced sub-surface ground movement, with a precise and immediate ground response to the arrival of the cutter head directly below.

The data analysis showed also the perfect comparison and the natural integration between geotechnical instrumentation and the topographical points. The surface settlements into the subsidence basin had the same trend.

5.2 Inclinometers data

The data coming from inclinometers shows that there was no significant lateral ground displacement recorded after passage of TBM. The inclinometers casing installed into the diaphragm wall recorded the same displacement during the excavation of the stations box.

5.2.1 Struts load cells and strain gauges data

The load recorded on the strut is basically affected by the excavation progress and the variation of the temperature. The comparison between the total load recorded by the load cells and the average load of the vibrating wire strain gauges spot welded on the strut explain the

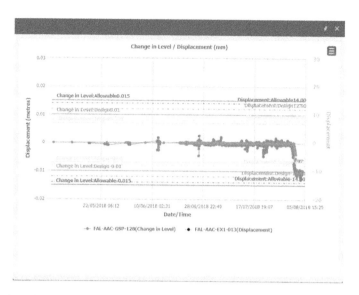

Figure 8. Caption of comparison between settlement point settlement and extensometer anchor settlement.

good feedback recorded by the instrumentation installed. The variation of the temperature has a great influence on the structural load.

5.3 Comparison between reflector and reflector less topographical reading in Airside

In Perth Airside the two tunnels run under taxiways and the two runways 06/24 and 03/21, which have to remain live during the excavation. The Near Time Monitoring has been designed for fulfill the safety requirement of this sensitive part of the project.

All sealed pavement surfaces (runways, taxiways and aprons) are proposed to be monitored using a 'reflectorless' methodology that has been designed inhouse.

Elsewhere in the grassed areas, the ground will be monitored via survey prisms being fixed onto concrete structures set into the ground.

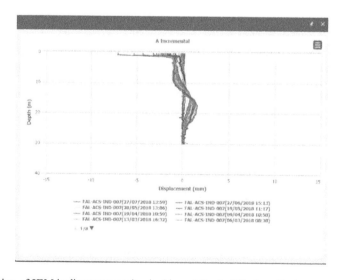

Figure 9. Caption of IEM inclinometer casing in Airport Central Station diaphragm wall.

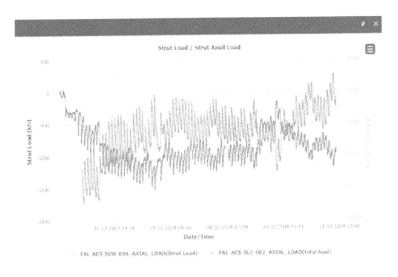

Figure 10. Caption of strut load in Airport Central Station between load cell and vibrating wire strain gauges spot-welded on strut.

The reflectorless surveying methodology requires the RTS to be elevated sufficiently above the ground in order to improve the accuracy of the measurements. A lattice tower structure approximately 2.2 m above ground level is proposed to support the RTS. The RTS is approximately 350 mm tall. The lattice structure, from Millard Towers Ltd, meets or exceeds the relevant ICAO standards for airfield frangibility

Power for each RTS has been provided by solar panel with battery, with frangible stand. A manual survey of the pavement is also done to verify the accuracy of the reflector-less data

The data recorded show the good quality of the reflector less reading with a variability of 2–3mm, understanding the trend of the settlements. The daily manual reading done this levelling confirm the trend.

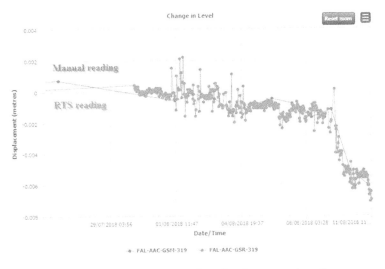

Figure 11. Caption of comparison graph between Manual and automated readings.

Figure 12. Pictures of RTS masts in airside.

5.4 *Correlations between the theoretical analyses and the measured results*

Normally expected values and the subsequent threshold levels are determined based on design assumption. A very interesting topic of this project is that expected ground deformation values have been assumed and the re-calibrated based on:

– Analysis of baseline data;
– Actual settlement measured during the excavation just before position of monitoring points;
– Correlation between geotechnical data and TBM's parameters.

As a result, expected values and subsequent thresholds have been adapted in nearly -real-time and no critical situation have been detected during the progress of the works.

6 CONCLUSIONS

Considering surrounding conditions and the specific environment on which FAL project is developed, geotechnical and structural monitoring is an essential part of latter. In order to ensure a quick and reliable flow of information, automation of the instruments has been proved to be the most cost-effective solution. Thanks to the solution adopted, main parameters are indeed measured in nearly real-time. This allows all stakeholders (Designer, Owner, Consultants, etc.) to be always updated in respect to the site situation.

Applied solutions such as wireless data transmission for piezometers in airport areas and implementation of specific instruments (incline-settlement column) have been proved to be a good solution to provide reliable data without involving people in measurement. By this way, safety and reliability have been both increased. Data obtained are consistent and described properly what is happening during the construction phases.

REFERENCES

Dunnicliff, J. (1988). Geotechnical Instruments for Monitoring Field Performances.
ISO STANDARD 18674-1 2015. Geotechnical Investigation and testing – geotechnical monitoring by field instrumentation – Part 1: general rules
METRONET https://www.metronet.wa.gov.au/projects/forrestfield-airport-link
Mikkelsen, P.E. (2003). Piezometer in fully grouted boreholes. Proceeding of Field Measurements in Geomechanics.
SISGEO Srl Technical Data Sheet www.sisgeo.com

Tunnels and Underground Cities: Engineering and Innovation meet Archaeology,
Architecture and Art, Volume 11: Urban
Tunnels - Part 1 – Peila, Viggiani & Celestino (Eds)
© 2020 Taylor & Francis Group, London, ISBN 978-0-367-46899-6

The "CNIT La Défense" railway station in Paris: A large scale underpinning in a constrained environment

A. Burdiel, J. Pinto, L. Canolle & M. Pré
Setec tpi, Paris, France

N. Coquelle & V. Rigoux
Vinci Construction, Rueil-Malmaison, France

ABSTRACT: The extension of the E line between the Haussmann-Saint-Lazare and Mantes-la-Jolie stations, one of the Paris regional express train network lines, includes the construction of a large underground railway station in the heart of the La Défense district. It is located directly beneath the iconic concrete shell structure of CNIT (Centre for New Industries and Technologies). Buildings of 5 to 10 floors are sheltered under the vault. The central volume of the station, with a capacity of 120 000 m3, must be excavated directly under these buildings. The construction sequence includes the load transfer (approximately 75,000 tonnes) on temporary foundations, the construction of columns supporting a thick transfer slab and the final transfer to this slab. So far, the first load transfer and separation of the existing shallow soles have been performed, as long as a first excavation step.

1 INTRODUCTION

The new CNIT-La Défense station on the extension of the RER E towards the West must ensure fast and fluid interconnections with the existing transport infrastructures, particularly the so-called "Heart Transport" complex, which is today the busiest multimodal station of France, and the stations of the Transilien lines. The station is located under the vault of the CNIT, which is an iconic structure, and all the buildings which are sheltered under this vault (Figures 1–2).

The volume of the station is made up of a central part located right under the last level of CNIT car park, representing a large volume of 15 m high, 108 m long and 33 m wide, and two lateral underground parts in extension of this central volume to cover the rest of the platforms with a total length of 225 m (Figure 3) (Thuaud et al. 2017, Vigneron et al. 2017, Canolle et al. 2017).

The Owner is SNCF-Réseau. The Engineer in charge is SETEC-EGIS – Cabinet Duthilleul (architect) joint venture. The construction was entrusted to the group Vinci – Solétanche – Bachy – Dodin Campenon Bernard – Spie Batignolles and began at the end of 2016.

The purpose of this paper is to present the design and the implementation of the underpinning for construction of the central part of the station. It has reached an important stage with the completion of the first step of taking over the overlying structures and structures, and the beginning of the excavation of the underground volume.

Underpinning of existing buildings is not new. We can thus cite some recent operations in France:

– Restructuring of the Gare Saint-Lazare in Paris (2010 – 2012): for the creation of 3 additional levels, full replacement of the existing masonry foundation structure with a reinforced concrete column-beam structure (Baumann 2010).

Figure 1. External aspect of the CNIT.

Figure 2. Internal aspect of the CNIT.

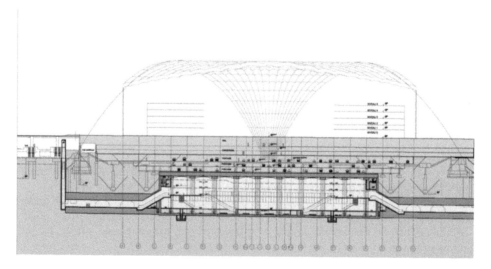

Figure 3. Longitudinal section of the CNIT station.

– Peninsula Hotel in Paris (2011 – 2014): addition of 2 basement levels to the 2 existing levels under a R + 6 building, by creation of beams on micropiles then replaced by a level of transfer floor.

- Couvent des Jacobins in Rennes (2014 – 2017): for the creation of an auditorium and a car park lot in the basement of a historical monument, the entire cloister was temporarily founded on micropiles and piles to allow the excavation over a depth of 15 m and prepare for the final underpinning structure (Bertucci 2015).

The underpinning of the CNIT is largely inspired by these examples. However the dimensions and the context in which it fits make it an exceptional work.

2 CONTEXT

The objective assigned to the underpinning work is to allow excavation of a volume of about 120 000 m3 in which will fit the central volume of the station of internal dimensions: length 108m, width 40m, depth 25m.

The overlying structures the integrity of which must be maintained consist of:

- the vault of the CNIT: the position of the EOLE station was chosen in order that no modification or underpinning of the bearings of the vault have to be implemented. Critical movements to be controlled during construction are mainly the rotational movements of the massive abutments on which the vault is supported. The acceptable value of such rotational movements was deduced from a full numerical simulation of the vault behavior: above a limit of 4mm/m, crack zones are spreading throughout the vault, beginning by regions close to the abutments (Canolle et al, 2017).
- interior buildings consisting of five basement levels, three of which are car parks, and buildings in elevation below the vault, of up to five storeys. The foundations of these buildings are individual footings under each column, with an average grid of 9 m by 9 m, but with many singularities, as it is the case in any multipurpose structure. With estimated loads ranging from 200 t to 1,600 t per column, 120 columns are to be borne by the underpinning system (total weight about 75,000 t.

The CNIT is located in the heart of the business district of La Défense: this strategic position, as well as the importance of sheltered economic activities (offices, hotel, restaurants, department stores), makes it impossible to imagine to stop the existing operation even for a short period. Only a partial immobilization was accepted:

- stopping the activities of temporary fairs and exhibitions, mainly concentrated on the first basement level, most of which could thus be devolved to the works facilities. These facilities could then enjoy the existing infrastructure capacity, which includes possibility for heavy trucks to reach the facility.
- partial interruption of the car park operation, with a total interruption limited to the last basement level.

At the end of works, a partial restructuring of the basements is planned to restore as much as possible the previous functions and capacities, but also to take advantage of the opportunity of the arrival of new passengers' flows.

The construction conditions of the underpinning are also largely dependent on the geotechnical and hydrogeological context, as well as on the constitution of the existing structure to be borne.

In terms of geotechnics, the context is generally favorable: the existing foundations are based on a thick layer of so-called "Marnes and Caillasses" made up of an intercalation of stiff layers with some slightly softer ones, which lies above a layer of coarse limestone rather weakly fractured. Both horizons present only few void-like or open-vein-type defects, as is quite frequently observed in other areas of the subsoil of the Paris Basin. However, the excavation must go down almost to the base of the coarse limestones, and the underlying grounds (Sands, Marls) are much less favorable. The aquifer, in its natural state, bathes at 6 to 8 m height the base of the coarse limestone. A de-watering system is set up for the works.

The structure of the five basement floors which are to be borne is made of cast in place reinforced concrete columns/beams/slabs, the columns being founded on insulated superficial footings. The whole structure is relatively homogeneous, as long as it was built at one time (with only a few subsequent adaptations) at the end of the 1980s. In spite of this relatively recent age, rather well documented with as-built drawings and reports, a number of issues deserved to be clarified. Thus the load borne by each column had to be recalculated before the start of work. In the same way, it was not possible for operational reasons to carry out beforehand a complete campaign of reconnaissance of the in situ concrete qualities. This was one of the unfortunate surprises of this site to note at the beginning of works, whereas the archives indicated concrete qualities ranging from C30 for the beams and slabs to C40 for the columns, qualities which are consistent with the values of the computed loads, that in reality the concrete strengths were much lower, which justified a number of the precautions which will be detailed below.

3 PRINCIPLES OF THE UNDERPINNING

The principle for the underpinning (Figure 4) is a two steps sequence:

- a first step consisting in supporting a total of about 120 columns on temporary support structures made of strong steel frameworks tightened to the columns and supported on micropiles (Figure 5). This makes it possible to separate the columns from their current footings;
- a second step consisting in constructing the permanent support structures made of a total of 60 pillars on which a transfer slab is supported (Figure 6); once this structure is completed, the column loads are transferred onto it (Figure 7).

Eventually, after this second transfer, the structure of the station is completed, secured to the underpinning structure and de-watering can be stopped.

The phenomena that can affect the CNIT structures and consequently create damage are:

- Movements during the initial loading, when it is necessary to be able to raise the current structure in order to cut the footings; an excess of load compared to the existing load can be feared.

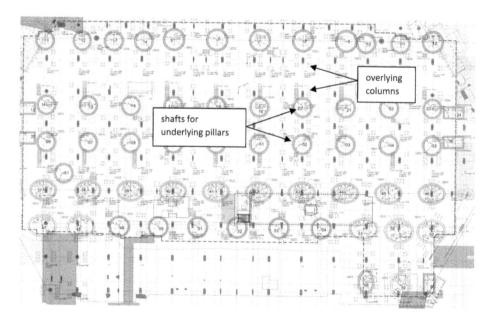

Figure 4.   Underpinning overview (horizontal cut at transfer slab level).

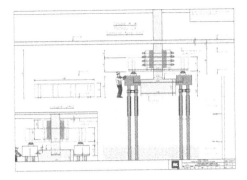

Figure 5.   First underpinning footings.

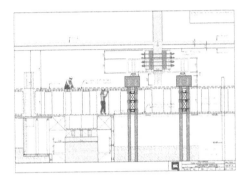

Figure 6.   Construction of the thick transfer slab.

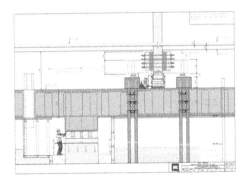

Figure 7.   Final load transfer.

– Settlement or uplift of the micropile foundations during earthworks to be carried out for the construction of the final underpinning structure (earthwork on the thickness of the transfer slab and execution of the wells necessary for construction of the future pillars), which are carried out near the ground layers which provide foundation for the temporary underpinning structure.
– Settlements of deep layers under the effect of punching by the bases of the pillars bearing the transfer slab, after the second stage of transfer of loads (load per pillar of about 3,000 tons).
– Consecutive uplift due to stop of de-watering and application of a hydrostatic pressure on the basis of the completed structure.

The basic provision to control these phenomena, at least to prevent them causing damages in the existing structure, is to provide for each step the possibility of compensating the movements that can affect the lower part made of new structures, for the upper part which is made of the existing structures to be preserved, by jacking systems which can be actuated at any time. These systems should be largely dimensioned in terms of load and with strokes deemed sufficient compared to the prediction. About this latter issue, it should be noted that a deficiency can be compensated by a temporary wedging operation allowing a catch-up stroke.

Here below provisions and procedures put in place for the first stage of underpinning, which is now almost 100% completed, are detailed.

4 IMPLEMENTATION OF THE FIRST STEP OF UNDERPINNING

4.1 *Means*

The means implemented for the first stage of underpinning, from top to bottom, for each of the columns, are the following (Figure 8):

- concrete girdles tightened by short prestressed bars to the existing column in the height of the last basement;
- steel frameworks on which the girdles are supported;
- jacks for the column uplifting;
- concrete blocks crowning groups of micropiles.

The construction sequence is as follows: first the micropiles are built, then the crowning concrete blocks; the girdles come afterwards and eventually the steel framework between the blocks and the girdles after placing the jacks on the concrete blocks. The load to be borne is known only through computation, all these elements, including the jacks, are dimensioned with a high coefficient (40%) over the calculated load, which is the coefficient usually recommended in this kind of operation. Due to the weakness of the concrete constituting the columns with respect to the expected resistance, for about 60% of the columns, these devices had to be completed by a first complementary concrete girdle around the columns in order to create sufficient embracing stresses in order to resist to crushing by prestressing.

All these devices had to be implemented inside the last basement of the car park with a very limited height (1.90 m under beams). That's why it took nearly 14 months to achieve the required 16,000 m micropiles.

4.2 *Monitoring*

Monitoring (Figure 9) is focused on:

- loads applied by jacking and existing loads;
- possible movements of the existing structure.

Figure 8. Devices for underpinning.

Regarding the loads, the following measurements are available:

– a direct measurement of the pressure in the jacking system;
– an indirect measurement of load variation in existing columns. This measurement is obtained by means of vibrating-cord extensometers arranged vertically on the columns. For sake of ease of installation, but also reliability of readings, they are located, not in the last basement, but at the next higher level. Any load variation in a column will result in a vertical strain that will be detected by the extensometer. Multiplied by the Young modulus of concrete, it will give an idea of the stress variation in the columns. The Young modulus is not well known, but it is of no real importance insofar as the objective is to stay as close as possible to an absence of variation.

Regarding the movements of the existing structure, the measurement of settlements is carried out by means of water-level gauges, which, for the same reasons as the extensometers, have been implemented in the penultimate basement level.

All of these instrumentation devices are continuously monitored in real time.

The recordings, at least for the extensometers and gauges, were begun with an advance of several months on the underpinning works. This allowed noticing daily variations in leveling as

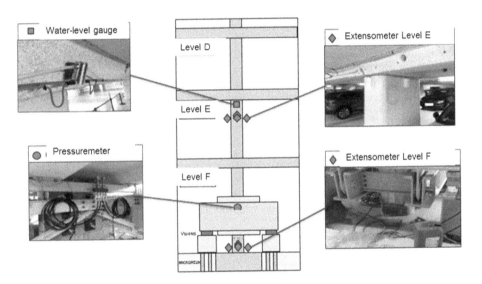

Figure 9. Monitoring devices.

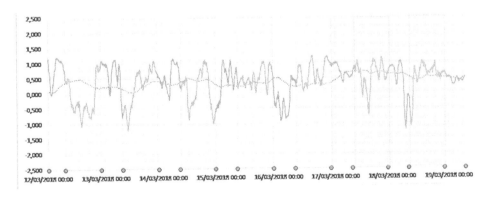

Figure 10. Settlement variations during one week (mm).

well as load, between day and night, between working days and the others, not negligible, due to the operation "life" (arrival of vehicles in car parks, deliveries, etc ...) (Figure 10). This natural "noise" makes it more difficult to perceive the effect of the underpinning, and had to be taken into account during monitoring. On the opposite, the vault abutments are very stable.

4.3 The underpinning procedure

The main difficulty of the first underpinning operation is that the main purpose of being able to release the column from its footing implies unloading of the latter, and then may be accompanied by ground uplift. It may therefore be necessary to raise the column significantly, with the consequences that this may produce in the existing structure. The sensitivity to this phenomenon is exacerbated by the weakness of the concrete columns. On the other hand, it should be remembered that at the beginning of this stage, only an estimate of the actual load currently going through the columns is available.

This led to a fairly complex procedure for dealing with a large enough number of parameters to monitor with the aim of determining the time when the footing of the column can be safely sawn. This time is reached when the column begins to be actually raised. After a number of tests (on columns considered as non-critical), a list of criteria could be established:

– the loading of the jacks of a given column has a significant effect on pressure of the jacks of neighboring columns, say 1%; this criterion is especially relevant when neighboring columns are themselves already supported;
– the curve (load in the jacks) vs (strain in the column) presents an inflection, supposed to represent the moment when the column "takes off";
– the settlement measured by the water-level gauges is significant, say 1.5 mm;
– the strain variation in the column which is measured by the extensometers is significant, say 2.10–5;
– the relative displacement between the column and the surrounding ground measured through a micrometer is significant, say 0.5 mm.

During all underpinning process, the parameters are continuously monitored (Figure 11), in order to make sure that the decision is made at the right time.The decision to separate the footing from the column is taken as soon as at least two of the criteria are met. Most of the time, the first met criteria were the effect on neighbouring columns and the strain variation in the column.

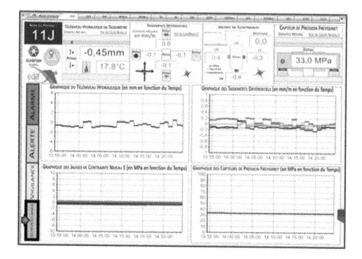

Figure 11.   Monitoring screen for one column.

After cutting the footing, the entire load of the column passes in the jacks; this allows knowing precisely this load. The follow-up of the settlement then makes it possible to readjust the column to its initial position.

4.4 *The follow-up procedure during subsequent work*

Upon completion of the underpinning procedure, the single hydraulic circuit that controls all the jacks assigned to a given column is closed: the jacks remain locked at a constant volume, the load remaining measured via the sensors of hydraulic system pressure. Throughout the rest of the work, the variations of this pressure, as well as the movements of the gauges and extensometers are monitored and any anomaly is rectified if necessary by reactivation of the concerned hydraulic circuit.

5 CONCLUSIONS

To date, almost all the existing columns to be underpinned have been put on their jacks and separated from their footings, and the first earthworks are well advanced (Figure 12).

A first assessment can be made on comparison between actual measured loads and estimated loads. It turns out that, probably because of the relative simplicity of the existing load-bearing system, either globally, or column by column, the measured loads are not that far from the estimated sum of the permanent loads and 30% of the variable loads (according to Eurocode regulations). Moreover, movements due to the works remained minimal, well below the usual limits, including for very sensitive buildings. Till now, any damage (checked by visual inspection) was avoided, in spite of the observed weakness of the column concretes.

To date, the corrections that had to be carried out remained quite limited, showing a good stability of the system. It is expected that it is secured in the long term, until the second operation of underpinning. For this purpose, preventive maintenance measures are taken.

However, it is worth remembering that these results were not acquired in advance, and are the result of a fairly long maturation (estimated at around six months) during which various tests and adjustments were made.

"Last but not least", the works could proceed till now without any disturbance of the activities maintained in the CNIT. Only some transient problems of noise due to vibration during various demolitions operations not directly related to the implementation of the proper underpinning could affect these activities.

At the end of the day, the finalization of the underground works, with a final aspect as expected in the here below (Figure 13) architectural view is under good way.

Figure 12.   First stage earthworks.

Figure 13.   Architectural view of the finalized works.

REFERENCES

Baumann O. 2010. Restructuration lourde, la gare Saint-Lazare restructurée. *Le Moniteur 4 June 2010*: 60–63
Bertucci A.-E. 2015. Un couvent en lévitation à 15 mètres du sol. *Le Moniteur, 20 February 2015*:36–39
Thuaud O., Canolle L. 2014. EOLE line western extension: survey works beneath the CNIT in La Défense business district. *Tunnels et Espace Souterrain 245, Sept-Oct 2014*: 445–454.
Vigneron J., Riché C., Pré M. 2017 A new RER station under the CNIT in Paris-La Défense: multiple connexions in a dense and complex urban surrounding. *AFTES Congress Paris 2017*.
Canolle L., Marlinge J., Berend L., Lanquette F., Pré M. 2017 Technical design of the RER station under the CNIT in Paris-La Défense: control of constructive risks in a complex surrounding. *AFTES Congress Paris 2017*.

Tunnels and Underground Cities: Engineering and Innovation meet Archaeology, Architecture and Art, Volume 11: Urban Tunnels - Part 1 – Peila, Viggiani & Celestino (Eds)
© 2020 Taylor & Francis Group, London, ISBN 978-0-367-46899-6

Bucarest Metro Line M5: Underpassing of existing Eroilor1 subway station during regular train transit

V. Capata
S.G.S. Studio Geotecnico Strutturale S.r.l., Rome, Italy

M. Repetto
Astaldi S.p.A., Rome, Italy

F.L. Bircolotti
S.G.S. Studio Geotecnico Strutturale S.r.l., Rome, Italy

ABSTRACT: The realization of M5 metro line in Bucarest has foreseen the underpass of the existing semi-shallow Eroilor1 station at a distance of about 1 m from the bottom slab. The excavation was carried out using two TBMs-EPB departing from the new station Eroilor 2. The most critical aspects were the demolition of the concrete DW for the part below the existing station bottom slab and the extreme excavation conditions due to a 20 m hydraulic head in presence of sand. It was necessary to realize preventive soil improvement under the foundation slab of the existing Eroilor1 station with vertical and horizontal injection using both cement and chemical grouting. All activities were carried out in presence of a dewatering system, lowering the water level more than 20 m. An automatic monitoring system of the hydraulic and tenso-deformative state of the area allowed the continous check of safety conditions during the excavation.

1 INTRODUCTION

1.1 *Type area*

The project of the new line 5 "Magistrala" in Bucharest, between station Raul Doamnei - Hasdeu (Opera), has a length of approximately six kilometers and consists of about four kilometers of circular cross section tunnels, using two TBM EPB (Earth Pressure Balance) of diam. 6.6 m with 30 cm thick reinforced concrete lining, and for about two kilometers of top down built stations and shafts, distributed as follow:

– Raul Doamnei
– Brancusi
– Romancierilor
– Parc Drumul Taberei
– Drumul Taberei 34
– Favorit
– Orizont
– Academia Militara
– Eroilor 2

In correspondence of the limit between new Eroilor II station and the old station Eroilor I the new line underpass the old one, with a gap between the existing bottom slab foundation and the top of the TBM shield of just 1.20 m. The underpass has been realized using an EPB TBM. This was necessary to prevent the dismantling of the existing station diaphragms walls made of steel reinforced concrete.

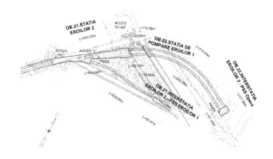

Figure 1.   Plan of connection area between New and Old station.

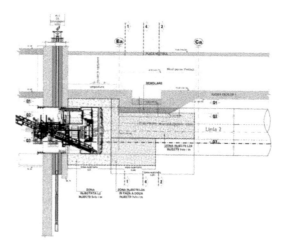

Figure 2.   Longitudinal section of connector tunnel.

2   GEOLOGICAL AND GROUND WATER CONDITIONS

The geotechnical characterization of soil in this area derives from several investigations surveys carried out by boreholes and laboratory tests.
    The soil profile consists of:

Unit 1: Backfill 0.40–2.60 m thick;
Unit 2: Soft clay and silty clay, grey, maximum depth 5.5 m;
Unit 3: Medium sand and gravel;
Unit 4: Clay and silty clay, brown,consistency medium, maximun depth 16.0 m;
Unit 5: Medium or fine sands with silty sand interbedded, maximum depth 33 m, lacustrian, high plasticity.

Table 1.   Geotechnical parameters for soil involved in excavation.

Soil	γ [kN/m^3]	ϕ' [°]	c' [kPa]	Ic	K0
Medium sand	20	30	0	-	0.57
Silty clay	20	24	20	0.75	0.66
Fine sand	20	35	0	-	0.49
Hard clay	19.5	24	15	0.85	0.66

Figure 3.   medium sand and gravel, Unit 3.

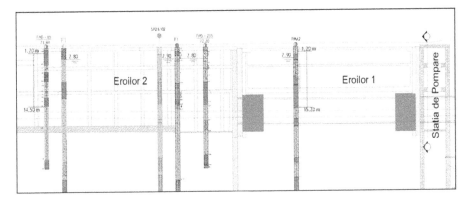

Figure 4.   Longitudinal profile with borehole showing presence of sand and clay interbedded.

According to the depth of the sandy layer, the soil interested from TBM cross is character-ized by a strong heterogeneity and can be definitively defined as fine medium silty sands with interbedded silt and clay.

However, sand has strongly influenced the demolition works taking also into account the presence of groundwater. The water indeed is located in the sand layers at about 1.5 m below ground level, thereby in the intervention area during the cross passage all digging and demoli-tion operations would have been done under high hydraulic loads up to about 20 m.

3   INTERVENTIONS TO SECURE THE EXCAVATIONS DURING THE DISMANTLING WORKS AND BREAK OUT TBM OPERATION

In order to avoid soil instability and structure failure two different improvement works were designed: soil injection with cement and chemical grouting and use of dewatering wells to lower the hydraulic pressure.

Grouting was done by injecting both in vertical and horizontal direction for a length range from 6 to 12 meters for the horizontal, and of about 22 meters for the vertical: the last ones were positioned in the narrow space available between the diaphragms of the Eroilor 1 and Eroilor 2 station.

All the activities were performed once the effect of wells have been reached, through the use 22 wells (diameter of wells was 700 mm), and 12 horizontal drains equipped with vacuum pumps and realized from the already excavated Eroilor 2 station and SPAI shaft.

Once the lowering of water table was reached as designed, the injection works have been carried out without using preventer system during horizontal drilling.

Taking into account the grain size of soil to be improved it was decide to use cement and chemical grouting. A special field test was performed to define the optimal grouting composition.

Regarding horizontal consolidation and its injection methods (cement and chemical), according to the results of the field test, two different types were chosen:

a first type of injection to strengthen the soil behind the diaphragms walls to be cut in order to create a solid and compact volume core;
a second type around the perimeter of excavation with the purpose to limit possible dispersions into the sandy soil of the grouting injection.

In addition, horizontal consolidation has been designed and subdivided into three different phases:

A. Horizontal drilling 12 m long, insertion of a PVC valved pipe coupled with VTR lamellar with anchor function, injection of the first 6 meters behind the diaphragms by cement (first injection) and then chemical (second injection);
B. Injection between 6 to 12 m in length by injection stept of 1 meter;
C. Consolidation of the space between 0–1 meters and 5–6 m by injections of acrylic solutions: this action was done to have the effect of confining the consolidated soil in step A as a plug, also considering the preexistence of compartment injections around the dismantling zone.

Once executed all the demolition operations, it was positioned the equipment thrust for TBM departing from Eroilor 2 station.

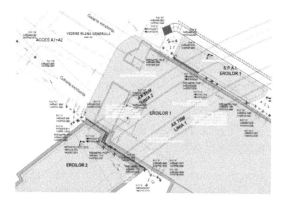

Figure 5.   Well location around Eroilor I.

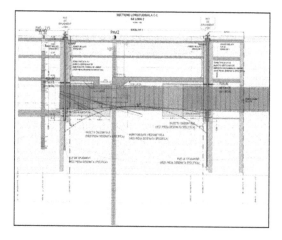

Figure 6.   Dewatering effect during pumping.

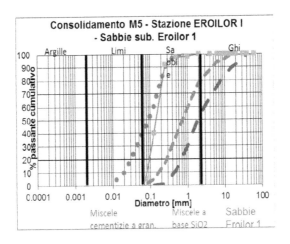

Figure 7. Cambefort curves.

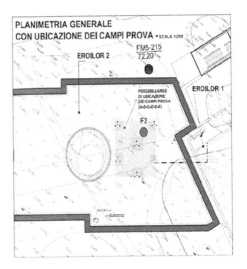

Figure 8. Trial test location.

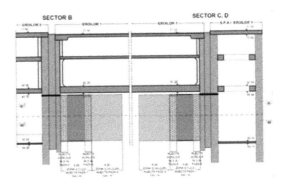

Figure 9. Longitudinal profile of grouted zone.

Figure 10.   Injection phase.

Figure 11.   Diaphragm wall after cutting.

Figure 12.   Diaphragm wall during cutting.

4   DESIGN STUDIES AND SETTLEMENT PREVISION

A series of analysis by 3D Fem model were performed to evaluate the stress and deformations status induced by TBM passing on the existing structures. The stress strain state of the whole area interested by the TBM excavation was verified using 3D model in order to simulate the interaction effect during cross passage of TBM.

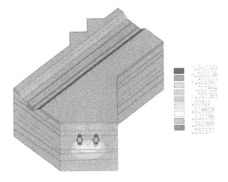

Figure 13.   Analysis results.

5   MONITORING SYSTEM

In the same time a sophisticated ad intense monitoring system was placed at level of platform of Eroilor 1 station and on the rails.

The monitoring system has provided essentially

– Measures of deformation of structures using Optical targets, miniprism and crackmeters
– Biaxial clinometers
– Levelling of rails
– H-level monitoring system
– Monitoring of ground water

The result were very positive and showed maximum settlements lower than 1 cm (8mm). The two TBMs have crossed the station in 21 days.

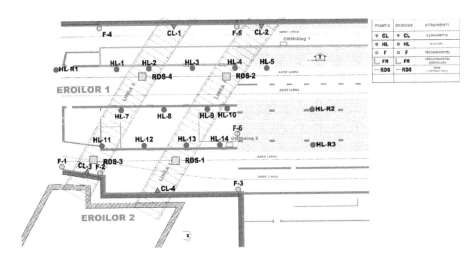

Figure 14.   Monitoring plan.

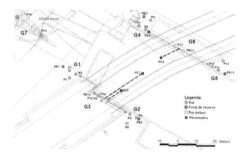

Figure 15.   Dewatering wells.

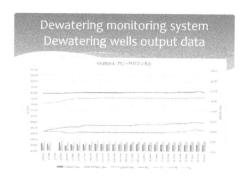

Figure 16.   Ground water lowering (TBM invert az z = 52).

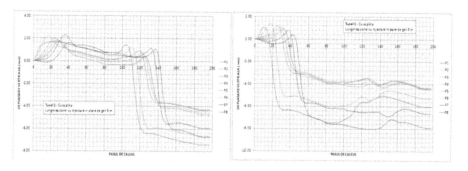

Figure 17.   Maximum settlement and recorded stress.

6   CONCLUSION

The underpass crossing of the station Eroilor I was a classic example of geotechnical engineering that only through the improvement of the soil it was possible to complete. Injections of cement and chemical grout and an efficient dewatering have allowed to carried out the complex work in safe conditions and by the scheduled deadline.

REFERENCES

Anagnostou G. & Kovari K. 1996. Face stability conditions in earth pressure balanced shields. *Tunnelling and Underground Space Technology*, 11(2):165–173.

Burland J. B. & Wroth C. P. 1974 Settlement of buildings and associated damage. Proceedings of a *Conference on Settlement of Structures*, Cambridge, 611–654.

Burland J. B. 1997 Assessment of risk damage to buildings due to tunnelling and excavation. *Earthquake Geotechnical Engineering*, 1997 Ishihara, 1189–1201.

Chambon P. & Corte J. F. 1994 Shallow tunnels in cohesionless soil: Stability of tunnel face. ASCE *Journal of Geotechnical Engineering*, 120, 1148–1165.

Rankin W. J. 1988 Ground movements resulting from urban tunnelling: predictions and effects. *Engineering geology of underground movements*, Geological Society Engineering Geology Special Publication, 79–92.

Tunnels and Underground Cities: Engineering and Innovation meet Archaeology,
Architecture and Art, Volume 11: Urban
Tunnels - Part 1 – Peila, Viggiani & Celestino (Eds)
© 2020 Taylor & Francis Group, London, ISBN 978-0-367-46899-6

Milan Metro Line 4: The Dateo underground station

L. Carli, P. Galvanin & G. Peri
Alpina S.p.A., Milan, Italy

A. Carrettucci & M. Lodico
MetroBlu S.C.r.l, Milan, Italy

ABSTRACT: The Metro Line 4 is the new underground line under construction in Milan, connecting the east to the west of the city centre. The underground Dateo station is one of the boldest along the whole line. Its depth is the maximum never reached by the subway lines built in Milan and it is determined by the exigency to pass under the existing underground Bovisa-Dateo Railway Link. The excavation is 31 m deep, 20 m below the water table, 60 m long and 23 m wide, having a distance from adjacent buildings less than 3 m. Retaining structures consists of 50 m high concrete diaphragms, 1.20m thick, with a 20 m injected bottom plug. It was excavated using a combined bottom-up/top-down method. The paper focuses on two key aspects of the design: settlement control of the nearby existing buildings and construction countermeasures to avoid drainage and counteract high underground water pressure.

1   INTRODUCTION

The Line 4 - also called Blue Line - is a new subway under construction in Milan, financed by M4 S.p.A, a holding formed by a public shareholder - the Municipality of Milan - and some other private ones, among which Salini Impregilo SpA and Astaldi SpA, which are also the main Contractors for the whole infrastructure.

Once completed, the line will connect the western and eastern sides of the City – from Linate airport up to the border with Buccinasco Municipality – becoming a strategic infrastructure of the mass-rapid transport system in Milan. The 15 km of the new Line will connect 21 stations and 30 intermediate shafts; the Line 4 will be the second driverless line of Milan. Underground stations are excavated in loose sandy soils below the water table, alongside the existing buildings and infrastructures in a densely populated urban area. The metro stations are built using "bottom-up" methods below the water table or "combined bottom-up/top-down" methods for deeper excavations. The level of tracks is placed in a range between 15 and 28 m below the existing ground level: the maximum depth is reached in correspondence of the Dateo Station. This depth is the maximum never reached by the subway lines built in

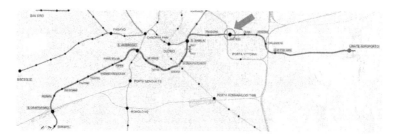

Figure 1.   Alignment of the new Milan Metro Line 4 – Dateo Station in the highlighted circle.

Milan and it is determined by the exigency to pass under the existing underground Bovisa-Dateo Railway Link. The excavation depth, the high hydraulic head due to the water table and the presence of a great number of masonry buildings, some of which date back to the early 1900s, required an accurate analysis of the construction methodology, followed by a detailed design of the construction choices to fulfil the severe safety requirements and tough construction schedule fixed by the Municipality.

2 STATION LAY-OUT: SHALLOW VERSUS DEEP SOLUTION

The existing Dateo FS station, the railway tunnel, together with the sewage culvert called "Nosedo" nearby the ground surface were the key elements to be faced at the beginning of the design process: as a matter of fact, the existing underground railway intersects the new Metro 4 alignment almost orthogonally at a depth of 17 m, preventing the possibility to built the new Metro Line and the Dateo Station. Two different solutions were studied to overcome this problem: the first ("A" solution hereinafter) envisaged to cross over the existing railway (Fava & Galvanin, 2012), while the second ("B" solution) entailed to pass below the existing railway by means of twin tunnels excavated with two TBM machines. In the "A" solution, the Dateo station and the whole stretch of the Line passing over the existing railaway was conceived as "cut and cover" tunnel, executed with a "bottom-up" procedure, inside concrete diaphragms: existing roads should have been diverted on alternative routes, to gain the working areas along the Corso Plebisciti, creating great impacts on the traffic in the city center. This solution was very simple to be built; nevertheless, it would have created some problems to the altimetric alignment of the new line, forced to pass below the Nosedo and over the near existing tunnel with a partial demolition of the concrete revetment, as shown in Figure 3. Moreover, the "A" solution would have been not suitable for the excavation of Line 4 tunnels using the TBMs: the machines should have been dismantled before the Dateo station and remounted in the next one, with a remarkable loss of time and not negligible extra-costs. The "B" solution was developed to supersede the limits of the previous one, achieving a consistent reduction of

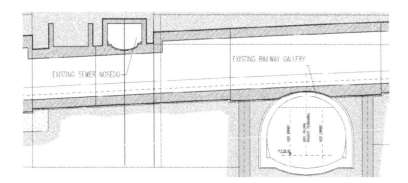

Figure 2.   Metro Line 4: longitudinal profile solution "A". The Nosedo sewage culvert is on the left, while the existing railway tunnel is inside the diaphragms on the right.

Figure 3.   Metro Line 4: longitudinal profile solution "B".

Figure 4. Dateo Station -solution "B": 3D Model of the internal lay-out.

the impacts on traffic, a significant saving of the working areas and an improvement in the construction time schedule for tunnel excavation by means of TBMs.

In the "B" solution, the Dateo Station is placed underneath Corso Plebisciti, on the East side of the existing Dateo Railway Station, but the axis of the new station is rotated roughly 8° with respect to Corso Plebisciti axis road to avoid the interference between some deep foundation piles of the existing railway which would have obstacle the TBMs.

The Dateo Station lay-out in solution B is divided into two parts:

– a shallow portion (11.70 m from the ground level) which links the existing railway station to the main part of the new Metro 4 station;
– a deep portion, having planimetric dimensions 60 x 23 m and a maximum depth from the existing ground level equal to 32 m.

Due to the depth of the railway tracks, the Dateo functional lay-out is quite different from the standard one used for the majority of the Metro 4 shallow stations. The main portion of the station (see Figure 4) is hugely characterized by architectural choices studied to create a wide and open internal volume, without columns or intermediate slabs, useful to place the internal access stairs and escalators, which present a remarkable length, due to the vertical distance between the existing ground and the platform level. The architectural choices, as well as the above mentioned phisical constraints, required a carefull preliminary analisys of the construction methods, followed by an extensive structural analysis, as described in the following paragraphs.

3   DESIGN SOLUTIONS FOR THE DEEP STATION

3.1   *Geological and geotechnical setting*

The Dateo station, located in the eastern part of the Milan historical city center, interferes - from a geological point of view - with the so called recent "Diluvium", formed by loose silty-sandy granular soils having high permeability (10-5/10-6 m/s) and characterized by a random variability of their granulometric composition, from silt to cobbles. Some more frankly silty clayey strata can be present, 2-3 meters thick. The geotechnical surveys executed before starting the design stage and deepened up to 60-70 m from the ground surface – well below the depth usually investigated for standard buildings in Milan - confirmed the first hints from technical literature and some typical characteristics of the Milan underground soil, for depths greater than 30-35 m (Cancelli, 1967).

It is worth to note that the deeper surveys highlighted the presence of two silty clayey layers, the first at 37-38 m from the ground surface - having a thickness of 3.50 m - and the second, with a similar thickness, at a depth of 57 m from the surface, with a higher content of silt and sand.

The presence of a clayey layer at a depth next to 40 m is a characteristic "marker" of the underground soil in Milan, as evidenced in the past by several geological studies (i.e. Desio '53, Niccolai et al. '67). This layer, almost continuos beneath the whole city center, marks certainly a inter-glacial phase during which glaciers withdrew up to Alps. This layer marks also a transition from mainly gravelly-sandy soils in the first 30-35 m, to silty-sandy soils in which the gravel component is negligible. This granulometrcic variation, well highlighted by

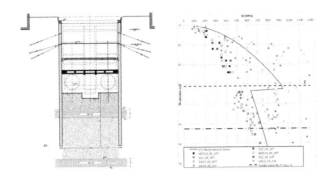

Figure 5. Dateo Station: simplified stratigraphy (brown strata: silty clay) and distribution of small strain E_0 modulus with depth.

geotechnical surveys, impacts also on geotechnical properties of soils as well shown in Figure 5: as a matter of fact, soil below 35 m is less stiff than the upper portion and the v_s shear velocities, as well as the small strain E_0 modulus, exhibit a significant discontinuity: all this information, above all the E modulus values for displacement and settlement calculation, was taken into account in the geotechnical and structural analysis and in the study of injections to be used for excavation waterproofing.

3.2 Choice of the construction method

Once the station lay-out was defined according to the planimetric and altimetric alignment of the Line as for "B" solution, the design focused on the choice of the most suitable construction procedures: station lay-out, its position and depth, the presence of a powerful aquifer and of numerous buildings at the ground surface as well as the construction time-schedule deeply affected the design choices: after several analysis and different proposals, a combined *bottom-up/top-down* excavation strategy was considered the most suitable in order to:

- guarantee the maximum excavation safety, limiting the settlements underneath the existing buildings placed in a range from 3 m to 10 m from the perimetric retaining walls of the station;
- comply with the construction time schedule, allowing the breakthrough of the TBMs used for the tunnel excavation without interferences with the construction stages of the station: in other words, a complete separation of the two working processes (one for the station and the other for tunnels) was possible thanks to the chosen solution, thus creating a remarkable time saving in the construction schedule;
- execute the ground injection for excavation waterproofing as fast as possible.

Different solutions of combined *bottom-up/top-down* excavation sequences (Peri et al., 2011) were analysed to achieve the above goals:

- the "B1" solution envisaged two intermediate reinforced concrete slabs, having a thickness 50/60 cm, using the top down procedure: these two slabs should have been withstood by vertical columns, executed in advance from the ground level by means of deep drilling, as shown in Figure 6 on the left;
- the "B2" solution entailed a different approach, based on a system of temporary steel struts in lieu of the upper slab of the "B1" solution, plus a unique intermediate slab. This slab, 140 cm thick, is conceived to counteract the perimetric wall thrust working as a one-way slab, having 23 m span, hinged on the diaphragms, without any kind of intermediate support. To achieve this result the slab was designed as a latticework, lightened with polystyrene (Figure 6, on the right).

The "B2" solution was preferred to the "B1" because it allowed to eliminate the support columns, removing any kind of internal constraint for the TBMs breakthrough, without waste

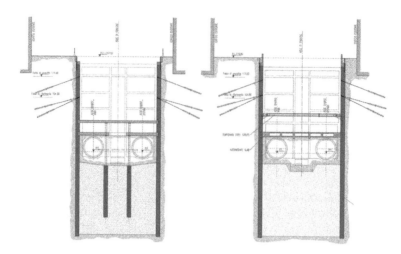

Figure 6.   Dateo Station – Excavation sections and retaining structures – Solution "B1" (left) vs. Solution "B2" (right).

of time for the waterproofing and final revetment of the temporary support columns. In a nutshell, the "B2" solution envisaged the following constructions stages:

1. concrete diaphragms execution, 120 cm thick and 50 m long, using a clamshell bucket;
2. excavation up to the ground water table: the concrete walls are restrained by two pre-stressed anchor levels, the first placed roughly 5 m below the ground level, inclined to avoid interferences with the existing foundations, and the second placed half a meter above the water table, 10 meters below the ground surface;
3. once reached the working level for the second anchors level, the injection of the soil below the foundation slab is carried out;
4. after the injection completion, the excavation is deepened up to 18 m below the existing surface and the steel trusses are installed: this system is composed by waling beams (2 HEM 900) and circular hollow steel struts (diameter 1100 mm, 20/26 mm thick), spaced with an interval in the range from 4.5 m to 6 m;
5. the excavation is deepened up to the level of the intermediate slab, casted in situ on the bottom of the excavation, roughly 20 m below the existing ground surface: this slab marks the limit between the bottom up and the top down procedure: underneath the slab, the excavation continues as in a top-down mode; some hatches were provided in the slab to allow the excavation of the remaining portion of the station, up to the foundation level, and the lifting and lowering of the construction materials,
6. excavation is carried on below the slab executed in the previous stage and the foundation slab is completed;
7. after the TBMs breakthrough, the platform level is casted, while the bottom-up construction of the upper part of the station continues, dismantling the temporary steel struts, without interferences with the tunnelling process.

The combined procedure *bottom-up* and *top-down* proved to be the right choice for the following reasons:

– the use of steel struts and - above all - the intermediate slab casted before reaching the foundation level were able to reduce the horizontal displacements of the diaphragms, limiting the settlements of the adjacent buildings, improving the safety of the excavation;
– the removal of anchors below the water table entailed to reduce construction time and to eliminate the problems related to drilling operations below the water table in loose soils (which usually require preventers and accurate procedures to avoid dangerous water leakages inside the excavation);

– the complete separation between the station construction procedures and the tunnel excavation process avoided interferences with TBMs break in and out: also in this case the time saving was remarkable and the twin tunnels were completed within the scheduled time.

3.3 *The excavation retaining walls*

The deep excavation inside the perimetric walls was executed using reinforced concrete diaphragms having a thickness of 1.20 m and a length equal to 50 m, considered in the structural design like permanent retaining structures.

Two different construction options were taken into account before making the final choice: the first one envisaged the use of an hydromill, while the second the use of a clamshell bucket. Hydromills are normally used for diaphragms having a length grater than 30-35 m and they guarantee optimal performances in terms of verticality tolerances, execution speediness and joint-water tightness. Notwithstanding the aforesaid advantages, they usually require, while operating in loose soils and if compared to other systems, a lower density bentonite slurry to maximize the performance of the excavation tools. Moreover, they require a quite huge working area, larger than the one available in Dateo: due to these constraints the daily production could have been reduced below the optimum standard.

Having in mind the aforesaid limitations, a second option was examined based on a heavy (20 ton) clamshell bucket, with an hydraulic closure system type "BAYA", in the version stretched by TecSystem for this kind of applications: the bucket verticality is controlled by an electronic device to assure the compliance with the vertical design tolerances (0.5%). The satisfying experiences with this kind of equipment - gained during construction of Line C in Rome and High Speed Railways in Florence, encouraged to propose this solution as the final one: as then experienced on site, the bucket was able to guarantee the same performances of the hydromill in terms of diaphragms quality, continuity and verticality: the good final results can be appreciated also looking at Figure 8, where the absence of water leakages through diaphragm joints and from the bottom of the excavation is well remarkable.

The weathertightness of joints between diaphragms was studied during the design stage: the joint executed on site is a male-female type and was created in the diaphragms using "stop-ends" sheetpiles. These joints can be shaped while casting the concrete diaphragms, executing at first the primary panels, installing on its two lateral faces the trapezoidal sheetpiles, having an appropriate length if compared to the depth of the excavation (36 m for the Dateo Station).

Once completed the primary panels, the secondary ones are excavated and, as soon as the excavation is completed, the sheetpiles are removed by means of a vibrator. The extraction time is normally 2-5 mins for each sheetpile without damaging the adjacent panels. For Dateo station the male-female joint water tightness was improved using two different water stop: the

Figure 7. Top down excavation stage.

first is the standard PVC profile, while the second is a continuous and injectable pipe to be used in case of failure of the PVC profiles, allowing the injection of polyurethane resins.

3.4 *Soil injection for the waterproofing of the excavation*

The presence of granular and permeable soils below a high water-table and the impossibility to execute a dewatering of the excavation by means of pumps, to avoid unacceptable settlements of the adjacent buildings, imposed to inject the soil below the bottom of the foundation slab, before removing the soil below the water table.

During the design stage, three different options were evaluated. The first solution ("J1" on the left in Figure 8) envisaged a massive injection of the soil for a total height equal to 20 m below the foundation slab (Balossi Restelli, 1981). This injection was executed using a TAM methodology (tubes à manchettes injections) with cement and silicates controlled by volume and pressure. In this solution the length of the diaphragms is optimized since the hydraulic pressure acting on the bottom part of the excavation is counteracted by the self-weight of the injected soil and by the "arch effect" of the massive treated soil. The "J2" solution (in the centre of Figure 8) entailed the execution of a thin waterproof layer placed at a depth where the self-weight of the soil above the waterproof layer equals the water pressure. The waterproof treated soil, in this case, is composed by a first layer of cement injected soil, thick 0.50 m, a second layer 1 m thick, injected with high quantities of sodium aluminate and a final capping layer injected as the lower one. At the bottom of the excavation a layer injected with micro-fine cements was added to improve the soil passive thrust acting on the diaphragms, to reduce the horizontal displacements of the perimetric walls. This solution would have been able to optimize the total injected volume and also the execution time; conversely it would have required a greater length of the perimetric walls (up to the depth where the self weight of the treated soil equals the water pressure) and a greater accuracy in drilling procedures to assure the correct verticality, having a length more than 40 m.

The "J3" solution, (on the right in Figure 8) was based on a TAM injection of cements and silicates as in J1, but the treated soil was designed as a "sandwich" to optimize the total injected volume. The diaphragms - as explained for the J2 solution - should have been stretched up to the pressures equilibrium point.

The three different solutions were object of careful cost-benefit analysis, at the end of which it was decided to operate with the J1 solution described above: the key points in the final choice turned out to be:

– the verticality check during drilling phases - much less stringent in solution J1 tanks to the massive soil treatment;
– the impossibility - due to technological constraints - to guarantee the continuity of the male-female joint below a depth of 36 m (maximum length of the sheet piles), with possible water leakages from joints in the diaphragms, which could have had compromised the stability of J2 and J3 solutions.

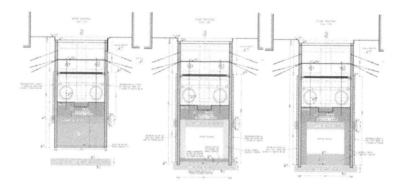

Figure 8.   Different injection solutions studied for the waterproofing of the bottom of the excavation.

4 DESIGN CRITERIA AND ANALYSIS OF THE RETAINING STRUCTURES

4.1 *Assessment of settlements induced by the excavation*

While excavating, the loosening of the ground behind a retaining wall creates a subsidence basin, at the ground surface, potentially able to induce unacceptable settlements of buildings next to the excavated area and their potential damaging: therefore, it is necessary to carry out an estimate of the possible damages caused by settlements to verify their admissibility. Analyses of potential damages, caused to buildings or other facilities next to an excavation, is usually divided into the following phases:

1. assessment of vertical subsidence and consequent maximum values of vertical settlements and distortions potentially induced by the excavation on the buildings;
2. evaluation of category of the potential damages on buildings to be compared with the allowable ones.

The starting key point for each estimate of expected potential damages is the definition of the foundation settlements induced by the excavation. Considering the importance of the project and the proximity of existing buildings to the excavation, 2D FDM numerical analyses were carried out, assuming plane strain conditions, following an excavation procedure and a gradual staged construction, to have a realistic prediction of settlements and damages. The calculation code used was FLAC 2D (Fast Lagrangian Analysis of Continuous), developed by ITASCA Consulting Group. The material behaviour in the models was described by means of linear elastic constitutive laws for structures and elasto-plastic laws, based on the so called "CY-Soil" model implemented within the calculation code, for granular soils. The "CYSoil" model was used to take into account the different elastic properties of soils along unload-reload conditions. In particular, the compressibility modules were evaluated using the equation:

$$E_s = m \, Pa \, (\sigma_h/P_a)^n \tag{1}$$

where P_a = reference pressure = 1 atm; m=1000; n=0.5 and assuming a coefficient equal to 3 within the model "CYSoil" to simulate unload and reload conditions avoiding an overestimate of the induced settlements.

The numerical models were able to highlight some elements useful for excavation monitoring during construction:

– the maximum horizontal displacement of the concrete perimetric walls is equal to roughly 15 mm at a depth 20 m from the ground surface;
– maximum vertical settlements observed on the buildings are in the range 8-10 mm; in the numerical analysis a building was simulated at a medium distance equal to 5 m from the diaphragms;

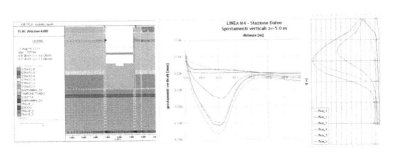

Figure 9. FDM 2D model, maximum vertical displacements beneath building foundation and wall horizontal displacements for different excavation steps (mm).

- examining the subsidence curves, the vertical displacements induced at the ground surface are pretty greater than those induced immediately below the buildings foundation level. This effect is due to the foundation stiffness and to the stress-release in the soil, caused by the excavation;
- the subsidence basin is quite huge, due to the excavation depth (30-40 m), so buildings inside this basin are potentially involved by the excavation in the so called "sagging" or "hogging" zones;
- the shear strains in the soil behind the concrete walls are approximately 1‰; thus they are in good accordance with the choice of the operative E_{op} modulus used in the numerical analysis, and assumed equal to 1/3 of the small strain modulus E_0 shown in previous § 3.1.

4.2 *Structural analysis*

Design and verification of the whole structure, partly "top down" and partly "bottom up", were developed starting from the identification of its characteristics in terms of functionality, construction requirements, expected performances and applied loads to form the database to be used in structural analysis.

The various structural typologies used for construction were modelled by means of different FEM codes (Midas Gen and Sap 2000), to specialize, in the most rigorous and efficient way, the necessary checks, taking into account the differences in the structural typologies and construction procedures. The numerous construction phases and static schemes, together with the need to anticipate the execution of the intermediate slab, suggested to develop different models, with increasing degrees of complexity, as listed below:

- 3D models for the system of temporary steel struts, counteracting the concrete walls;
- 2D shell models for the dimensioning and verification of the intermediate concrete slab;
- plane frame models, for analysis of the effects caused by possible differential settlements of the base slab and by the staged construction and loading;
- a full three-dimensional model to check the final station lay-out and to carry out detailed analysis of the foundation structures and intermediate floors.

4.3 *Monitoring strategy*

Before starting the excavation of the Dateo station, a monitoring plan was prepared considering the proximity of the excavation to the existing masonry buildings, some of which date back to the early 1900s, as usual in the central areas of the city centre.

The installed instrumentation allowed to carry out surveys on buildings while excavating, according to construction sequences, to detect in advance the extent of possible excessive settlements or rotations of buildings produced by the excavation. The following quantities were subjected to periodic monitoring:

- vertical displacements of buildings;
- rotations of buildings;
- movement of existing structural joints on buildings;
- horizontal displacements of the diaphragms at various depths from the ground surface;
- stress and strains inside the steel struts and the intermediate concrete as well as temperature variation;
- width of pre-existing on new cracks present on buildings.

The numerical analyses carried out during the design stage allowed to set the attention and alarm thresholds for each measured value, together with the countermeasures to be implemented, if the thresholds were reached or exceeded: monitoring data made it possible to ascertain the correspondence between the behaviour of diaphragms and buildings predicted by numerical models and the observed one during the construction phase.

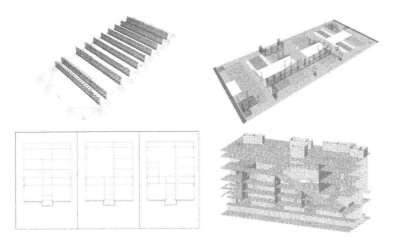

Figure 10. Different structural models used for station design. From left, clockwise: 3D models for the steel truss behaviour analysis, 2D model for the top-down slab, 2D simplified models for analysis of effects induced by construction stages, 3D complete model.

5 CONCLUSIONS

Solutions provided for construction of Dateo station, described in the present paper, are the result of a design process which required a careful analysis and comparison of different solutions, to optimize the whole lay-out and construction process of this strategic station, placed in a complex urban context. Many recent experiences in the construction of similar structures show that design choices are strongly affected by conditions present on site, construction time-schedule and specific exigencies of the stakeholders involved in the project. The desired optimizations of the construction process can be attained only through an appropriate preliminary analysis and a coherent design development, which usually require an appropriate time frame to be fully detailed: this time frame should be taken into account during the drafting phase of the construction program. If these analyses and structural modelling are adequately detailed, they allow a real improvement of the construction process, minimizing the "adjustments on site" to solve contingent and unforeseen problems.

The experience gained in the construction of Dateo station could be a useful reference for future further refinement of combined top-down and bottom-up excavation procedures for deep under-ground structures in urban areas, to achieve even more faster and reliable construction processes.

Balossi Restelli A. 1981. *"Iniezioni in terreni sciolti"*. X Ciclo di Conferenze di Meccanica dei Terreni e Ingegneria delle Fondazioni, "Metodi di Miglioramento dei Terreni", Politecnico di Torino, Turin.

REFERENCES

Balossi Restelli, A. 1981. *"Iniezioni in terreni sciolti"*. X Ciclo di Conferenze di Meccanica dei Terreni e Ingegneria delle Fondazioni, "Metodi di Miglioramento dei Terreni", Politecnico di Torino, Turin.

Cancelli, A. 1967 *"Caratteristiche geotecniche dei terreni di fondazione del territorio milanese."* Atti del VII Convegno di Geotecnica – Associazione Geotecnica Italiana – Cagliari 6-7 Febbraio 1967.

Fava, A & Galvanin, P et al. 2012. *"Design and construction aspects of Milan underground Line 5 – passing over the Bovisa-Dateo railway tunnel"* - Geomechanics and Tunnelling, Volume 5, Issue 3, pages 243–260, Jun. 2012.

Peri, G & Galvanin, P. & Socci A. 2011. *"Metropolitana di Milano Linea 5: Aspetti progettuali ed esecutivi inerenti la realizzazione della stazione Garibaldi con procedura "top-down"*, Milan, MadeExpo 2011.

Tunnels and Underground Cities: Engineering and Innovation meet Archaeology,
Architecture and Art, Volume 11: Urban
Tunnels - Part 1 – Peila, Viggiani & Celestino (Eds)
© 2020 Taylor & Francis Group, London, ISBN 978-0-367-46899-6

Grand Paris Express Ligne 15 sud T2C: TBM driving parameters and complex passing of critical zones

S. Cavagnet, A. Giammarino, L. Mancinelli & C. Crémer
Lombardi Ingegneria S.r.l., Milan, Italy

ABSTRACT: The paper describes relevant design aspects for the realization of the southern portion of Line 15, structure that represents a part of Grand Paris Express Project in the surroundings of Villiers Sur Marne. The lot includes two double-track circular single tunnels (main line and SMR stretch), a large one excavated with conventional method, underground stations, shafts and connecting passages. The line tunnels, 8.70 m in (internal) diameter, are realized by using EPB TBMs, with the contemporary disposition of a precast lining composed by 5+1 r. c. segments. The focus of the contribution is mainly related to the studies carried out to define advancing procedures in terms of head pressures, back-filling pressures, and to estimate the expected settlements. Thresholds for particular crossings in correspondence of low covers, buildings, interferences etc. are presented and discussed, also with a strong link to numerical studies; the approaches followed for the definition of alert and alarm situations are also presented.

Technological aspects and solutions are then described, summarizing the impact of the works in terms of surface settlements based on monitoring system feedback.

1 INTRODUCTION

1.1 *Grand Paris Express project*

Grand Paris Express is an ambitious program of extension of the Parisian underground network that will be completed by 2030. The purpose of the intervention is to provide the city with a peripheral circle that intercepts the numerous radial lines already present between metro, RER, railways, in order to create an extremely efficient public transport network .

The project foresees the creation of about 200 km of automatic line which develops mainly in an underground tunnel, 68 stations, up to 150 ventilation or emergency shafts and 7 technical maintenance centres.

1.2 *Focus on Line 15 Sud T2C*

The work package of Line 15 named "T2C" is situated near Villiers Sur Marne in the southeastern outskirts of the city. It includes:

- two mechanised tunnels, principal and SMR lines, respectively 4.7 km and 2.2 km long; the principal tunnel connects Noisy Champs and Bry-Villiers-Champigny stations while the SMR tunnel connects principal tunnel and maintenance site;
- an interconnection shaft approximately 180 m long (O807), 92m of which are characterized by a traditional tunnel;
- two TBMs launch shafts (O802P and O813P);
- seven circular shafts used for evacuation and/or ventilation.

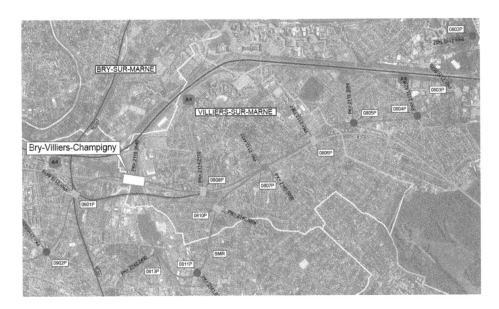

Figure 1. Overview on work package Line 15 T2C.

2 GEOTECHNICAL CONTEXT

The area under study is constituted by a stratigraphic succession that includes recent formations of "Remblais (R)" and "Eboulis (Eb)" (Quaternary formations) and sedimentary layers of Tertiary, from the "Calcaire de Brie (TB)" until the layer of "Marnes et Caillasses (MC)". In particular the site can be divided into two macro areas of study:

- at East, on the Noisy-le-Grand plateau, with a preponderance of TB and of the flush zone of "Argiles vertes (GV)" and "Limons des plateaux (LP)";

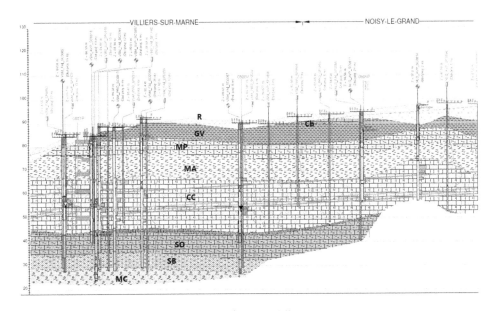

Figure 2. Typical geological tunnel profile on Ligne 15 T2C.

– at West, with a zone characterized by the flush formations of "Marnes de Pantin (MP)" and "Marnes d'argenteuil (MA)", resting on the layer of the "Calcaire de Champigny (CC)" and "Calcaire de St. Ouen (SO).

Except for some areas close to break-in/break-out, the TBMs digs mainly in the formation of CC (Figure 2). The thickness of this soil horizon is about 15–20m on the whole sector and its heterogeneous character has posed multiple problems in terms of geotechnical parametrization. In fact, this formation alternates layers with typical characteristics of a heavily fractured rock mass, mostly in its upper part, with others marly layers in the lower part at contact with SO. In addition, although representing the non-gypsiferous facies of Ludien, the presence of gypsum levels in some areas has been observed. The main geotechnical parameters defined for the different horizons are presented in Table 1.

Among the geological formations in the Paris region, several are permeable and constitute aquifer reservoirs. Concerning the T2C, four different water tables can be distinguished:

– the Oligocene aquifer of TB resting on the substratum of GV (very shallow and without any impact on the tunnel);
– the Eocene aquifer of MP resting on the substratum of MA, which impacts the tunnel at East launch shaft 813;
– the Eocene aquifer of CC, which impacts the tunnel on most of the track but with limited hydraulic charge;

Table 1. Geotechnical parameters for T2C.

Horizon	γ_h kN/m^3	E_m MPa	P_{lm} MPa	α -	E_{tunnel} MPa	c' kPa	φ' °	c_{tr} kPa	φ_{tr} °	c_u kPa	φ_u °	k_0 -	ν -
R	18	5–8	0.5–0.8	0.5	24	0	25	0	25	15	20	0.5	0.33
Eb	19.5	9–16	0.8–1.2	0.5	44	0	25	0	25	5	20	0.5	0.30
LP	21	07–12	0.9–1.5	0.5	28	0	25	0	25	40	15	0.7	0.33
TB	19	15–33	1.1–1.9	0.67	60	10	25	10	25	10	25	0.5	0.30
GV	18.5	10–13	0.7–0.9	1.0	24	20	15	80	0	80	0	0.8	0.35
MP	19	15–18	1.4–1.6	0.67	48	20	25	70	15	120	5	0.6	0.35
MA	18.5	29–36	1.8–2.0	0.67	96	40	15	90	10	140	0	0.6	0.35
CC(1)	19	18–35	2.3–3.4	0.5	92	15	30	50	20	100	15	0.5	0.30
CC(2)	19	37–59	3.7–4.6	0.5	184	30	30	80	20	150	20	0.5	0.30
CC(3)	19	90–131	5.0–5.5	0.5	440	40	30	100	20	170	20	0.5	0.30
SO	18	137–181	5.9–6.4	0.5	624	30	30	80	20	150	15	0.5	0.30
SB	21.5	110–157	6.1–6.7	0.5	516	25	30	40	25	80	20	0.5	0.30
MC	20	123–200	6.2–6.6	0.5	680	40	35	120	30	180	30	0.5	0.35

(1) Altered; (2) Degraded; (3) Healthy

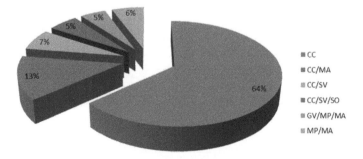

Figure 3. Geology at tunnel face.

– the Eocene aquifer of SO, always separated from the CC aquifer, considering the presence of strongly marly horizons between these two soil layers; tunnel meets this aquifer exclusively at the lowest point of its profile.

An automatic reading system of piezometers is always active during the entire duration of the works, in order to confirm the water levels assumed during the design phase.

3 MECHANIZED TUNNEL

3.1 *Excavation kinematic*

The tunnel is characterized by a double-railway track monotube structure and will be bored with a **TBM-EPB** (Earth Pressure Balance). A concrete reload that includes the equipment is is provided on the entire layout.

The kinematics of excavation envisage a first **TBM** (the so-called TBM-1) launched from the shaft 813 and digging up to interconnection shaft 807. A second **TBM** (TBM-2) is launched from the shaft 802 near Noisy-Champs Station towards Bry-Villers-Champigny Station, crossing shaft 807.

A particularity concerning the crossing of shaft 807 is that the section across the traditional tunnel will be excavated only when the upper part of the gallery will be finalized.

3.2 *Principal characteristics of the TBM*

The TBM has a maximum diameter of 9840 mm at the excavation face with a minimum of 9800 m at the end of the shield. It is equipped with the joints necessary to ensure the seal between shield and extrados precast segments and between shield and excavated soil. Three injection devices are provided for the control of confining pressures at the face and along the machine:

– a bentonite injection system at the front of the cutting wheel;
– a bentonite injection system along the shield in order to reduce the soil deconfinement and the friction of machine;

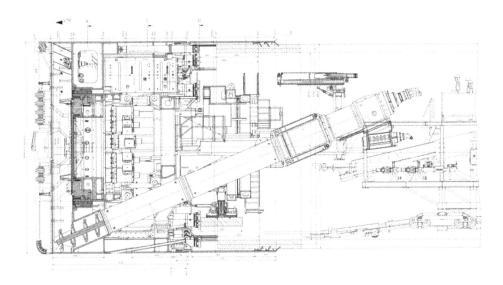

Figure 4.   TBM's profile.

Figure 5.   Ring seal (left) and back view of TBM during assembly at shaft 813.

– a stuffing mortar injection device at the back of the shield to fill the annular void between precast segments put in place and excavated ground profile.

The ring of the lining is composed by 6+1 precast segments 1.5m or 2.0m long, to respect the radius of curvature provided for the tunnel.

3.3   Evaluation of operational parameters

The excavation sites are heavily anthropized and affected by the presence of buildings and infrastructures; this aspect require a specific design of operational parameters with utmost caution with regard of possible impact of works. As known, in fact, the underground blind-bore works have a direct influence in terms of subsidence on the surface.

The design activity was therefore developed providing numerical elaborations with different analysis detail levels, according to the expected risk scenarios. The approach has been oriented towards a proposal to be confirmed and/or adjusted according to the first monitoring data. Specifically, the calculation of confining pressure is based on the following procedure:

1) preliminary screening on the line, based on results of settlements obtained by semi-empirical method of Peck (1969), assuming literature values of volume loss and k parameter; potential singular and critical points were identified;
2) calculation of the minimum confining pressure by analytical method of Anagnostou and Kovari (1994), to ensure the stability of the excavation face;

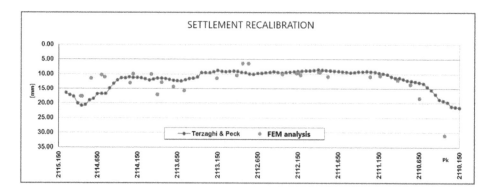

Figure 6.   Vertical settlement recalibration.

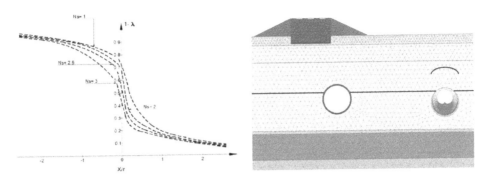

Figure 7.   Panet curves (left) and confining pressure simulation in the 2D FEM model.

3) 2D finite element calculations on up to 40 sections along the tunnel profile, to define the recommended confining pressure and to assess the impacts on the surrounding areas.
4) recalibration of the Peck analysis using the values of volume loss and k parameter extrapolated from FEM analysis (Figure 6).

The numerical computations with finite elements, in addition to the explicit simulation of the confining pressures, involved the adoption of some hypothesis concerning the convergence of the cavity while the excavation proceeds.

The deconfinement factor has been taken for two situations, at face (λ=0.3) and at the end of shield (λ=1.0), with reference to the Panet curves shown in Figure 7 for an elastic configuration of the face. In fact, the assumption is that the application of the confining pressure ensures de stability of the face, limiting the appearance of plastic zones.

Each 2D computational section has been implemented by schematically reproducing the stratigraphy of the site and hydro-geological configuration. The analysis has been conducted according to the Staged Construction and using a plastic drained condition of soil. The principal steps of analysis are listed in Table 2.

The evaluation of the settlements is done according to a parametric variation of the applied pressure, in a range defined as follows:

$$Pmin \geq max \text{ (hydrostatic pressure; pressure ensure face stability)} \quad (1)$$

$$Pmax \leq min \text{ (pressure ensure acceptable damage class on buildings;}$$
$$\text{blow} - \text{out pressure; TBM operational limit)}$$

A verification of the contact between ground and TBM shield has been made, in order to validate the calculated pressure and verify the compatibility between convergence of the cavity and shield diameter. For each calculated section, a systematic control of the convergence

Table 2.   Principal steps of 2D FEM analysis.

Step	Description
#0	Initialization of stress according to k_0 procedure (geostatic condition)
#1	Activation of elements representing the surrounding building foundations; Application of distributed equivalent loads
#2	Excavation of tunnel: face condition (λ=0.3) and application of confining pressure
#3	Excavation of tunnel: back shield condition (λ=1.0) and application of confining pressure
#4	Installation of final lining*

* This step has shown very low influence on deformation regime

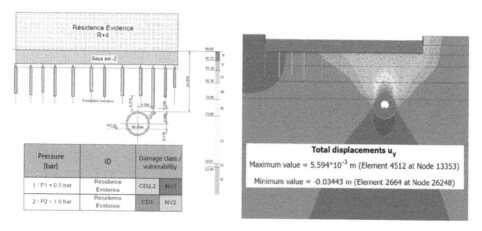

Figure 8. Example of building vulnerability evaluation with 2D FEM model.

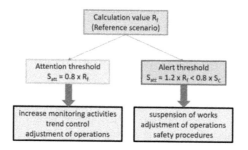

Figure 9. Definition of thresholds and procedures during monitoring activity.

value has been done, with a maximum 4 cm. In the case of this value was exceeded, even if the control parameters were verified, an overpressure has been imposed.

The damage class of the buildings has been evaluated considering three parameters: maximum vertical displacement, maximum deflection and horizontal deformation. A vulnerability matrix allows to evaluate the type of monitoring to be implemented and the acceptability of deformation values calculated, with reference to a contractual threshold S_C; this threshold is reached if a very vulnerable structure is detected and it represents the limit beyond which consolidation interventions are necessary.

Prior to the start of excavation activities, for each of the examined deformation criteria, attention and alert thresholds were defined according to the calculated values, which therefore represents the reference scenario. The flowchart represented in Figure 9 summarizes the process of thresholds definition that, associated with monitoring activities, allows to accurately manage possible critical issues during the working phase.

4 MONITORING ACTIVITY

Several monitoring sections have been set up along the tunnel route to be able to evaluate the correct progress of the excavation activities and verify the level of reliability of the estimated theoretical deformations. Five reinforced sections on the first excavated section (813–807) were equipped with 7 monitoring surface points, two inclinometers on the sides of the tunnel and an extensometer in correspondence of tunnel vault (Figure 10).

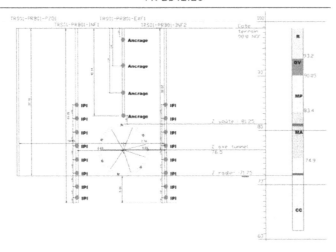

Figure 10. Typical reinforced monitoring section.

The TBM-1 was launched in June 2018 with a reduced back-up due to the small size of the shaft 813. After the first 100m of excavation, the machine was stopped below an area with no buildings to proceed with the complete installation of the back-up. The excavation operations then resumed at the end of July 2018.

During the excavation section with the reduced back-up, the measurements carried out reported values in line with the numerical estimations or slightly higher in the first segment, immediately following the break-out; this trend is mainly due to the calibration of the driving parameters and the rather slow progress of the machine, also due to the clogging phenomena with excavation face mainly in the MP.

At the re-start of July, the excavation speed gradually increased accompanied by a decrease of surface settlements. Upon reaching the first reinforced section, the monitoring data generally showed values which are lower than predicted (Figure 11 and Figure 12), despite a maintenance operation carried out precisely at this section.

These initial monitoring data allow to recognise that the quality of the ground in which the TBM excavates has potentially better characteristics than expected, leading to a positive trend in view of the passage in the urbanized area within which many sensitive structures have been

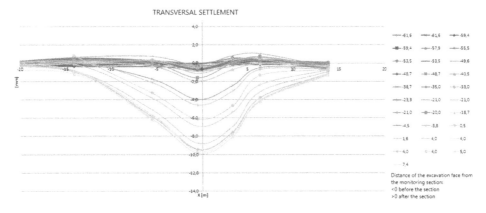

Figure 11. Transversal settlements on reinforced monitoring section along the first TBM's drive (813–807).

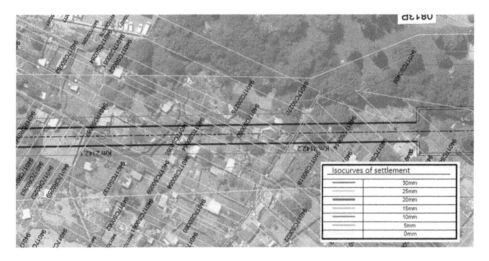

Figure 12. Isocurves of settlement estimated by calculation and position of the first reinforced section.

identified. The TBM-2 was instead launched from shaft 802 in September 2018, digging in a different geological context: the excavation face for the first section is in the GV formation on which the very deformable layer of TB persists. Currently, the first monitoring data are not yet available.

5 CONCLUSIONS

This paper presents the main aspects related to the definition of the driving parameter of the TBMs, concerning the tunnelling works of Line 15 T2C of the Grand Paris Express.

After a description of the geological and hydrogeological context, the approach and the adopted calculation methodology has been exposed in detail, with particular reference to the management of critical points and the definition of the reference thresholds for monitoring activities.

The excavation of the first tunnel section started in June 2018 from shaft 813, which allowed for a first feedback on the impact generated by the excavation operations on the surface.

In general, the measurements showed a trend of deformation values lower than the theoretical estimations. The progress of the work, currently underway, and the acquisition of new monitoring data, will make it possible to verify more in detail the reliability of the theoretical predictions also in different geological conditions, coverage or in presence of sensitive buildings.

ACKNOWLEDGEMENTS

The Authors want to thank all Lombardi Group and Alliance teams for their support and collaboration.

REFERENCES

Anagnostou G., Kovári K. 1996. Tunneling and Underground space Technology. *Face stability condition with Earth-Pressure-Balance shields*.
Cassani G., Mancinelli L. 2005. IACMAG Torino. *Monitoring surface subsidence for low overburden TBM tunnel excavation: computational aids for driving tunnels*.

Damiani A., Crova R., Mancinelli L., Avitabile E. 2018. Strade e Autostrade – n° 128. *I lavori di prolungamento della Linea 1 della Metropolitana di Torino verso Bengasi.*

Mair R. J., Taylor R. N., Burland J. B. 1996. Proc. Int. Symposium on Geotechnical Aspects of Underground Construction in Soft Ground, London (eds R. J. Mair and R. N. Taylor), pp. 713–718. *Prediction of ground movements and assessment of risk of building damage due to bored tunneling.* Balkema.

Mancinelli L. 2005. Geotechnical and geological engineering – Vol. 3, n. 3 – pp. 263–271. *Evaluation of superficial settlements in low overburden tunnel TBM excavation: numerical approaches.*

Tunnels and Underground Cities: Engineering and Innovation meet Archaeology,
Architecture and Art, Volume 11: Urban
Tunnels - Part 1 – Peila, Viggiani & Celestino (Eds)
© 2020 Taylor & Francis Group, London, ISBN 978-0-367-46899-6

A new analytical method to define the confinement pressure operation range for a closed TBM

G. Champagne De Labriolle & L. Pavel
ARCADIS ESG, Le Plessis Robinson, France

G. Teulade & O. Givet
ARCADIS ESG, Labège, France

ABSTRACT: Definition of operational range for the design and execution of TBM built tunnel is essential, in order to guarantee personal safety, neighbourhood integrity, and top quality at the end. In this article is firstly introduced a methodology, emphasising on the relation between standards context and justifications to produce. Several improvements of existing analytical methods are presented, particularly about front stability. Regarding this existing bibliography on this thema, we show that our new approach is better adapted to realistic situations (circular tunnel, multilayer case, drained or undrained conditions), theoretically more robust (new choice for lateral coefficient pressure, comparison with well-known limit analysis that provide theoretical bounds), and is geometrically coherent with failure mechanism observed in situ or in laboratory. Then, we present an application to real historical case.

1 INTRODUCTION

The control of tunnel face stability during tunnel excavation is essential to guarantee worker's safety, neighboring integrity, and good execution of support and lining.

After talking about normative issues, we will present the improvement we suggest to analyze face stability during tunnel excavation with a closed TBM (Slurry, EPB or VD-TBM).

2 NORMATIVE GUIDE TO DEFINE OPERATIONAL RANGE OF CONFINEMENT PRESSURE OF A CLOSED TBM

It is needed to evaluate 5 different kinds of confinement pressure at TBM axis:

- Minimal pressure to seal the tunnel: $p_{front,min,ELU\ HYD}$. A simple analytical method can be used: the water pressure pw must be overwhelmed.

$$p_{front,min,ELU,HYD} = \max\left(p_w(crown) + \frac{\gamma_M h}{2} ; p_w(crown) - \frac{\gamma_M h}{2} \right) \quad (1)$$

where γ_M = unit weight for material in the chamber of TBM (depend on confinement mode); h = height of the tunnel.

- Minimal pressure to avoid collapse: $p_{front,min,ELU\ GEO}$. This aspect will be studied in chapters 2, 3 and 4, through analytical method
- Maximal pressure to avoid blow-out: $p_{front,max,ELU\ GEO}$. The pressure at the crown mustn't be greater than initial vertical stress. Other considerations can be found in (Champagne de Labriolle, 2018)

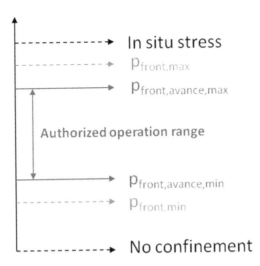

Figure 1. Operation range definition – inspired from (ITA/AITES/DAUB, 2016).

- Minimal pressure to respect settlement criterions: $p_{front,min,ELS}$. Not discussed here: numerical methods are needed.
- Maximal pressure to avoid heaving of the surface or a closed structure: $p_{front,max,ELS}$. Not discussed here: numerical methods are needed.

Finally:

$$p_{front,min} = \max\left(p_{front,min,ELS}; p_{front,min,ELU}\right)$$

$$p_{front,max} = \min\left(p_{front,max,ELS}; p_{front,max,ELU}\right)$$

(2)

where γ_M = unit weight for material in the chamber of TBM (depend on confinement mode); h = height of the tunnel; and for $p_{front,min,ELU}$, it depends on hydrogeological conditions:

$$p_{front,min,ELU} = \begin{array}{ll} p_{front,min,ELU,HYD} + p_{front,min,ELU,GEO} & drained\_conditions \\ \max\left(p_{front,min,ELU,HYD}; p_{front,min,ELU,GEO}\right) & undrained\_conditions \end{array}$$

(3)

To define the operation range of confinement pressure for a closed-TBM, it remains only to add mode variability to minimal pressure and substract it to maximal pressure. The frequent value is v = +/- 10 kPa in slurry mode and v = +/- 20 to 30 kPa in full EPB mode (GEO report No. 249, 2009; GEO report No. 298, 2014).

$$p_{front,avance,min} = \max\left(p_{front,min,ELS}; p_{front,min,ELU}\right) + v$$

$$p_{front,avance,max} = \min\left(p_{front,max,ELS}; p_{front,max,ELU}\right) - v$$

(4)

It is then possible to draw a typical range like following scheme:
In Eurocode 7, we choose 3rd approach (partial coefficient « A2 + M2 + R3 »), in which the drained ground parameters are divided by 1.25 and undrained parameters are divided by and 1.4. In the following article, we suppose that all parameters are already included.

3 EXISTING APPROACH TO ANALYSE FACE STABILITY

A well-known way to analyze the face stability of a tunnel is to use analytical closed-form founded on silo mechanism (Janssen, 1895) pressing down on a dihedral wedge (Horn, 1961).

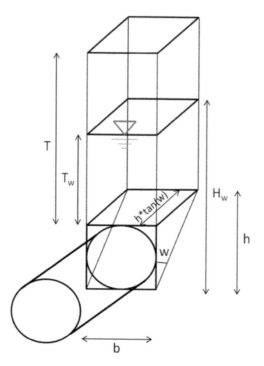

Figure 2.   3D failure mechanism.

It was firstly developed in the 90's by (Jancsecz & Steiner, 1994; Anagnostou & Kovári, 1994). The problem is illustrated on Figure 2.

The stages are described below:

- Use the well-kown approach of (Terzaghi, 1951) allowing to calculate the vertical pressure applied on a tunnel in 2D scheme and extend it in 3D in order to estimate the vertical pressure at the bottom of the silo
- Establish forces equilibrium for a given angle, and find the needed confinement force thanks to several assumptions
- Find the maximum needed confinement force under varying angle. It means that we finally know the most probable failure mechanism

Then, this method was improved successively by (Anagnostou & Serafeimidis 2007; Anagnostou, 2012; Anagnostou & Perazzelli 2013, Perazzelli & Anagnostou 2017), to reduce progressively the number of simplifying assumptions of the original method, and allow to take in account an unsupported span.

The remaining assumptions are listed below:

- Unique and homogeneous ground layer
- Empirical choice for lateral aerth pressure coefficient
- Water table in the silo (not crossing the tunnel) or without aquifer
- Rectangular excavation shape
- Linear distribution of confinement pressure on the tunnel face

4   IMPROVEMENTS OF EXISTING APPROACH

By addition to the existing approaches we discussed before, we are going to add three improvements, using the same symbols.

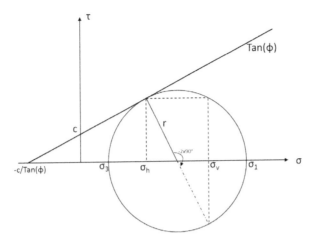

Figure 3. Mohr circle at collapse plane.

4.1 New theoretical earth pressure coefficient

Instead of using an empirical value for the lateral earth pressure coefficient K (= 1 for (Terzaghi, 1951; Anagnostou & al, 2012, 2013, 2014), =0.4 or 0.8 in (Anagnostou & Kovári, 1994, 1996), …), we prefer to use a purely theoretical expression.

That's why we suggest using the approach of (Sequeira Nunes Antao, 1997), using Mohr circle at the boundaries of the silo and the wedge. As the Mohr-Coulomb criterion is reached on theses boundaries, we can obtain the link between vertical an horizontal pressure by a rotation of 2x90° (see Figure 3).

We finally obtain:

$$\sigma_h = \sigma_v \frac{\cos^2\varphi}{1 + \sin^2\varphi} + c\frac{\sin 2\varphi}{1 + \sin^2\varphi} = \sigma_v\lambda + cK_c \tag{5}$$

where σh and σv = horizontal and vertical stress.

In this approach, we have a rigorous result which needed no more assumption than those we already had. When $\varphi = 0°$ (purely cohesive ground), we have $\sigma_v = \sigma_h$.

4.2 Explicitly elliptic cross section, whose width is β, height is D

To take in account this geometry, the width b used in differential equations has to vary with z, with the following boundaries:

$$b(0) = 0 \quad b\left(\tfrac{D}{2}\right) = \beta \quad b(D) = 0 \tag{6}$$

That means we have h=D, b=β, and:

$$b(z) = \beta \cos\left(\sin^{-1}\frac{2z - D}{D}\right) \tag{7}$$

4.3 Random shape of confinement pressure on the tunnel face

Take this in account allows to calculate explicitly the face stability in the following situations:

- Water table level below the crown
- The TBM bores mixed ground
- The TBM cutterhead is maintained (partial air pressure, see Gugliemetti & al, 2008)
- Use of face reinforcement for conventional tunneling (Anagnostou & Serafeimidis, 2007)

The face will be separated in small slices (~ 0.1 m) where we consider that every parameter is a constant. The confinement pressure becomes an input data and not a result: we will check that the chosen shape of confinement pressure on the tunnel face is sufficient.

5 NEW SOLUTION FOR THE STUDY OF FACE STABILITY

5.1 Solution in the silo

After writing the balance between stresses at the boundaries of the silo, following the approach of (Terzaghi, 1951), we have:

$$P\tau dz + (\sigma_v + d\sigma_v)S = \sigma_v S + S\gamma dz \tag{8}$$

where S is the surface and P the perimeter of the transversal section of the silo.

$$\frac{P}{S}\tau + \frac{d\sigma_v}{dz} = \gamma \tag{9}$$

We state that R=S/P. This solution is working, even a not rectangular silo is chosen. If we have an unsupported span « e »:

$$R = \frac{b(h\tan w + e)}{2(b + h\tan w + e)} \tag{10}$$

After the equation was resolved, we must be sure that the vertical pressure is note negative:

$$\sigma_{clef} = \sigma_v(T) = \max\left(0; \sigma_{v\infty}\left(1 - \exp\left(-\frac{\lambda T \tan\varphi}{R}\right)\right) + q\exp\left(-\frac{\lambda T \tan\varphi}{R}\right)\right)$$

$$\sigma_{v\infty} = \frac{R\gamma + c(K_c \tan\varphi - 1)}{\lambda\tan\varphi} \quad \lambda = \frac{\cos^2\varphi}{1+\sin^2\varphi} \quad K_c = \frac{\sin 2\varphi}{1+\sin^2\varphi} \tag{11}$$

To prepare the next chapters, we state the two following definitions, V_{silo} represeting the vertical force transmitted by the silo to the wedge, and $\sigma_z(h)$ being the vertical stress at the top of the wedge

$$V_{silo} = \beta D \tan w\left(1 + \frac{e}{D\tan w}\right)\sigma_{clef} \quad \sigma_z(h) = \left(1 + \frac{e}{D\tan w}\right)\sigma_{clef} \tag{12}$$

5.2 Solution in the wedge

The method of slice, firstly suggested by (Anagnostou, 2012) is represented on Figure 4 below.

Using this method, added to the improvements we presented before, we obtain the following expression for vertical pressure at the bottom of the slice k (see the demonstration and details in Champagne de Labriolle, 2018):

$$V_k = s_k\beta^2 E_{sk}(z_k) + c_k\beta^2 E_{ck}(z_k) - \gamma_k\beta^3 E_{\gamma k}(z_k) + V_{k-1}E_{Vk}(z_k) \tag{13}$$

Where E_{sk}, E_{ck}, $E_{\gamma k}$ and E_{vk} are analytical expressions which depend on z_k.

We note that V_k only depends on V_{k-1}. As V_0 at the bottom of the wedge equals 0, we can then have every V_k until the top of the wedge.

If $V(D)$, vertical force at the top the wedge, is superior to V_{silo}, vertical force obtained at the bottom of the silo, the equilibrium is granted. If it's not, it means that the shape of the confinement pressure of tunnel face is not high enough.

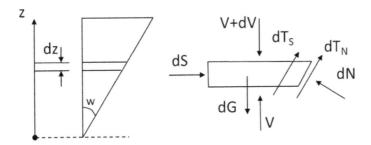

Figure 4.   Method of slices (Anagnostou, 2012).

5.3 New simplified analytical closed-form solution: cohesive-frictional ground

To compare our new solution with existing approaches, we present below a simplified solution where we suppose that ground parameters are constants all over the tunnel and the overburden, with a linear shape for confinement pressure, like in (Anagnostou & Perazzelli, 2013):

$$s(z) = s_{axe} + \delta s\left(\frac{z}{D} - \frac{1}{2}\right) \tag{13}$$

Where δs is confinement pressure gradient between bottom and top of the face, and s_{axe} is the axis confinement pressure. Using the same symbols by deleting the "k" reference, we rewrite the solution. In this simplified case, we can have directly the confinement pressure (see the demonstration details in Champagne de Labriolle, 2018):

$$\frac{S_{axe}}{\gamma D} = f_1^D - f_2^D \frac{c}{\gamma D} + f_3^D \frac{\sigma_z(D)}{\gamma D} + f_4^D \tag{14a}$$

with:

$$f_1^D = \frac{M_\gamma J_2}{P_s J_1} \quad f_2^D = \frac{P_c + M_c \frac{DJ_3}{2\beta J_1}}{P_s} \quad f_3^D = \frac{\tan w \exp\left(-\frac{\chi\pi}{2}\right)}{-4 P_s J_1} \quad f_4^D = \frac{1}{2} - \frac{J_2}{J_1} \tag{14b}$$

with:

$$J_1 = \frac{2 - 2\exp\left(-\frac{\chi\pi}{2}\right)}{16\chi + \chi^3} \quad J_2 = \frac{2\exp\left(-\frac{\chi\pi}{2}\right)\left(72 + 28\chi^2 + \chi^4 - 12\exp\left(\frac{\chi\pi}{2}\right)\left(6+\chi^2\right)\right)}{\chi(4+\chi^2)(16+\chi^2)(36+\chi^2)}$$

$$J_3 = \frac{-\exp\left(-\frac{\chi\pi}{2}\right)\left(10 + 6\exp\left(\frac{\chi\pi}{2}\right) + \chi^2\right)}{64 + 20\chi^2 + \chi^4} \tag{14c}$$

and:

$$\chi = \frac{D}{\beta}\Lambda \quad \Lambda = \frac{2\lambda \tan\varphi}{\cos w - \sin w \tan\varphi} \quad M_c = \frac{\Lambda \tan w(1 - K_c \tan\varphi)}{\lambda \tan\varphi}$$

$$P_c = \frac{\Lambda}{2\lambda \tan\varphi \cos w} \quad M_\gamma = \tan w \quad P_s = \tan(\varphi + w) \tag{14d}$$

5.4 New simplified analytical closed-form solution: purely cohesive ground

It is well-known that the silo mechanism is not adapted to purely cohesive ground, where all theories linked with upper bounds and lower bounds are clearly preferred. That's why in this case, we don't use the silo pressure, but the expressions of (Perazzelli & Anagnostou, 2017), page 5 et 6, equations 21 to 25, that allows to combine the classical approaches in purely cohesive ground and the wedge mechanism, thanks to (Gunn, 1980) results. We have:

$$\sigma_{clef} = q + \gamma T - Q_s(w)C_U$$

$$Q_s(w) = \max\left(2\ln\left(\tfrac{2T}{\beta}\right); 2\ln\left(\tfrac{2T}{D\tan w}\right); 4\ln\left(\frac{2T}{\beta\sqrt{1+\left(\tfrac{D}{\beta}\tan w\right)^2}}\right)\right) \tag{15}$$

In the previous results in the wedge, we bring φ down to 0 in the coefficients $f_1{}^D$ à $f_4{}^D$:

$$f_1^D = \tfrac{1}{2} \quad f_2^D = \tfrac{1}{\cos w \sin w} + \tfrac{4D}{\pi\beta\cos w} \quad f_3^D = \tfrac{4}{\pi} \quad f_4^D = 0 \tag{16}$$

We note that the linear variation of confinement pressure as no impact here, because $f_4{}^D=0$. For a circular cross-section (as $w \geq 45°$ for a purely cohesive ground: $Qs_2 < Qs_1$):

$$S_{axe} = \tfrac{4}{\pi}q + \gamma D\left(\tfrac{4T}{\pi D} + \tfrac{1}{2}\right) - \left(\tfrac{1}{\cos w \sin w} + \tfrac{4}{\pi\cos w} + \tfrac{4}{\pi}Q_s(w)\right)C_U$$

$$Q_s(w) = \max\left(2\ln\left(\tfrac{2T}{D}\right); 4\ln\left(\tfrac{2T}{D}\cos w\right)\right) \tag{17}$$

It is then possible to write the corresponding « Stability Number »:

$$N_{s,cible} = \left(\frac{1}{\cos w \sin w} + \frac{4}{\pi\cos w} + \frac{4}{\pi}Q_s(w)\right)\frac{\pi}{4} \tag{18}$$

5.5 Comparison and theoretical boundaries

5.5.1 Cohesive-frictional ground

We draw below an abaqus (Figure 5 on next page) for a constant confinement pressure on the tunnel face, where we show the results of (Anagnostou, 2012) and the minimum theoretical bound of (Mollon & al, 2010), for T/h = 1 and various friction angles (20° to 35°).

We note that our new solution respects the theory for every angle, which is not the case for the results of (Anagnostou, 2012).

Then, we compare our new approach with static and cinematic solutions of (Atkinson & Potts, 1977) for a purely frictional ground with φ = 30° ou 35° (respectively at left and at right on Figure 6 next page) (NB: the static approach of (Panet & Leca, 1988) is better for T/h < 0.8). Our solution must be between these two boundaries: this is true for T/h > 0.5 (very low overburden), but not for (Anagnostou, 2012). By the way, we show that the approaches of (Piaskowski & al, 1965) and (Carranza-Torres, 2004 & 2013) are not adapted to purely frictional ground.

5.5.2 Purely cohesive ground

The best way to demonstrate the efficiency of an approach in purely cohesive ground is to show the « Stability Number » associated to. We compare our new approach for square or circular on (Figure 7), where varying values for relative overburden are used, with 3 well-known methods (Broms & Bennermark, 1967; Davis & al, 1980; Carranza-Torres, 2004 & 2013). We present also the results obtained with (Piaskowski & al, 1965), to show that it's not adapted to purely cohesive ground.

This confirms that the improvement we made to silo & wedge failure mechanism for a circu-lar tunnel is more robust than existing methods.

Everywhere on (Figure 7), our new approach is closed to the well-known approaches. The main point is that our approach can be extended to much more complicated cases (multilayer case, elliptic excavation, arbitrary shape for confinement pressure ...).

We add that with this method, on obtain optimum failure angles between 45 and 60°, when silo theory for cohesive-frictional ground gives angles between 20 and 40°. This is coherent with laboratory test and observed in-situ historical cases (Leca, 2007).

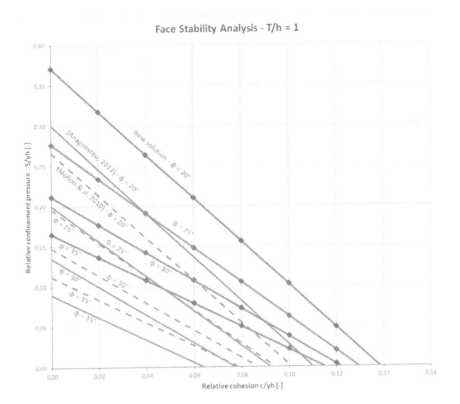

Figure 5. Comparison between the new simplified solution (full line & dots), (Anagnostou, 2012) with λ=1 (full line), and (Mollon & al, 2010) (dashed line).

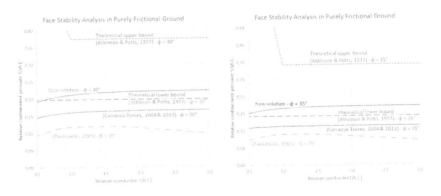

Figure 6. Purely frictional soil ground – Left: φ = 30°; Right: φ = 35°. Comparison between the new simplified solution, upper and lower bound from (Atkinson & Potts, 1977), and formulas from (Piaskowski & al, 1965) and (Carranza-Torres, 2004 & 2013).

6 HISTORICAL CASE

On the figure 8 below, we present, for typical geological condition of than can be encountered in Grand Paris Express (200 km of new metro), an example of pressure operation range for a 10 m diameter EPB TBM (unit weight of 15 kN/m^3 for conditioned ground)

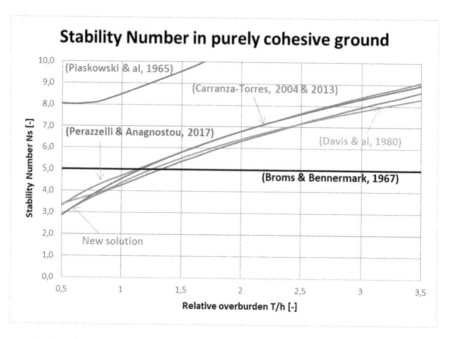

Figure 7. Purely cohesive soil – targeted Ns.

We used a classical criterion for blow-out: the pressure at the crown cannot be superior to initial total stress with a factor of safety of 1.1. We show the uplift limit, in case of a water table equal to the ground level.

We also added a minimum value if the full EPB mode is imposed: the chamber at the front, behind the cutterhead, must be filled with conditioned ground. It has an impact on the

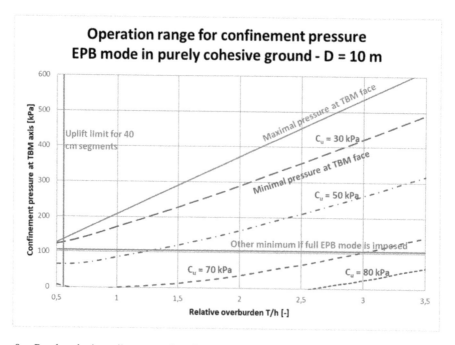

Figure 8. Purely cohesive soil – range of confinement pressure required.

minimum value at the axis. This mode is essential if it is necessary to use bentonite all around the tail shield to control the settlements.

REFERENCES

AFTES 2013. Recommandation: Nomenclature simplifiée des tunneliers. *Tunnels et Ouvrages Souterrains, Mai/Juin*, pp. 141–161.

Anagnostou G. 2012. The contribution of horizontal arching to tunnel face stability. *Geotechnik, vol. 35, 1*, pp. 34–44.

Anagnostou G., Kovári K. 1994. The face stability of slurry shield-driven tunnels. *Tunnelling and Underground Space Technology 92*, pp. 165–174.

Anagnostou G., Kovári K. 1996. Face stability in slurry and EPB shield tunneling. *Geotechnical Aspects of Underground Construction in Soft Ground*, pp. 453–458. Rotterdam, Balkema.

Anagnostou G., Perazzelli P. 2013. The stability of a tunnel face with a free span and a non-uniform support. *Geotechnik 36, 1*, pp 40–50.

Anagnostou G., Serafeimidis K. 2007. The dimensioning of tunnel face reinforcement. In: Barták, J., Hrdina, I., Romancov, G., Zlámal, J. eds.: Underground Space – the 4th Dimension of Metropolises. *ITA World Tunnel Congress*, Prague, 2007, pp. 291–296.

Aristaghes P., Autuori P. 2001. Calcul des tunnels au tunnelier. *Rev. Fr. Geotech, vol 97, 2*, pp. 31–40.

Atkinson J. H., Potts D. M. 1977. Stability of a shallow circular tunnel in cohesionless soil, *Géotechnique, 27, 2*, pp. 203–215.

Broms B. B., Bennermark H. 1967. Stability of clay at vertical openings. *J. Soil Mech. Fndn Div. Am. Soc. Civ. Engrs 93, SM1*, pp. 71–94

Carranza-Torres C. 2004. Computation of factor of Safety for Shallow Tunnels using Caquot's Lower Bound Solution. *Itasca Consulting Group*. Minneapolis.

Carranza-Torres C., Reich T., Saftner D. 2013. Stability of shallow circular tunnels in soils using analytical and numerical models. *In Proceedings of the 61st Minnesota Annual Geotechnical Engineering Conference*. University of Minnesota, St. Paul Campus. February 22, 2013.

Champagne de Labriolle G. 2018. Détermination de la fenêtre de pilotage de la pression de confinement d'un tunnelier fermé dans un sol cohérent-frottant ou purement cohérent. *Rev. Fr. Geotech, vol 155, 3*.

Davis E.H., Gunn M.J., Mair R.J., Seneviratne H.N. 1980. The stability of shallow tunnels and underground openings in cohesive material. *Géotechnique, vol. 30, n°4*, pp 397–416.

Eurocode 7 2005. *NF EN 1997–1, Calcul géotechnique, Partie 1: Règles générales*. AFNOR.

GEO report No. 249 2009. Ground control for slurry TBM tunnelling. *Geotechnical Engineering Office*, Hong Kong, 57 p.

GEO report No. 298 2014. Ground control for EPB TBM tunnelling. *Geotechnical Engineering Office*, Hong Kong, 82 p.

Gugliemetti V., Grasso P., Mahtab A., Xu S. 2008. *Mechanized tunneling in urban areas*. Taylor & Francis, London.

Gunn M. J. 1980. Limit analysis of undrained stability problems using a very small computer. *Proc., Symp. on Computer Applications to Geotechnical Problems in Highway Engineering*, Cambridge Univ., Cambridge, U.K., pp. 5–30.

Handy L. H. 1985. The Arch in Soil Arching. *Journal of Geotechnical Engineering, vol. 111*, No. 3, March.

Horn M. 1961. Horizontaler Erddruck auf senkrechte Abschlussflächen von Tunneln. *Landeskonferenz der ungarischen Tiefbauindustrie Deutsche Überarbeitung durch STUVA*, Düsseldorf.

ITA/AITES/DAUB 2016. *Recommandations for Face Support Pressure Calculations for Shield Tunnelling in Soft Ground*, p. 64.

Jancsecz S., Steiner W. 1994. Face support for large Mix-Shield in heterogeneous ground conditions. *Tunneling 94*. London.

Janssen H.A., 1895. Versuche über Getreidedruck in Silozellen. *Zeitschrift des Vereines Deutscher Ingenieure*, p. 1045.

Leca E. 2007. Settlements induced by tunneling in Soft Ground. *Tunnelling and Underground Space Technology, vol. 22*, pp. 119–149.

Leca E., Panet M. 1988. Application du calcul à la rupture à la stabilité du front de taille d'un tunnel. *Rev. Fr. Geotech, vol. 43*, pp. 5–19.

Maidl B., Herrenknecht M., Maidl U., Wehrmeyer G. 2012. *Mechanised Shield Tunnelling*. Ernst & Sohn, A Wiley Company.

Perazzelli P., Anagnostou G. 2014. A limit equilibrium method for the assessment of the tunnel face stability taking into account seepage forces. *World Tunnel Congress 2013*, pp. 715–722. Geneva.

Perazzelli P., Anagnostou G. 2017. Analysis Method and Design Charts for Bolt Reinforcement of the Tunnel Face in purely Cohesive Soils. *J. Geotech. Geoenviron. Eng.*, 2017, p. 1439.

Piaskowski A., Kowalewski Z. 1965. Application of thixotropic clay suspensions for stability of vertical sides of deep trenches without strutting. *Proc. Of 6th Int. Conf. On Soil Mech. And Found. Eng. Montreal*, Vol. 111.

Sequeira Nunes Antao A. M., 1997. *Analyse de la stabilité des ouvrages souterrains par une méthode cinématique régularisée*. Thèse, ENPC.

Serafeimidis K., Ramoni M., Anagnostou G. 2007: Analysing the stability of reinforced tunnel faces. *Geotechnical engineering in urban environments, XIV* European *conference on soil mechanics and geotechnical engineering, Madrid. Volume 2*, pp. 1079–1084. Millpress Science Publishers Rotterdam.

Terzaghi K. 1951. *Mécanique théorique des sols*. Paris: Dunod.

Tunnels and Underground Cities: Engineering and Innovation meet Archaeology,
Architecture and Art, Volume 11: Urban
Tunnels - Part 1 – Peila, Viggiani & Celestino (Eds)
© 2020 Taylor & Francis Group, London, ISBN 978-0-367-46899-6

TBM Selection for tunneling beneath urbanized areas at shallow depth. A practical case: The machine selection for the excavation of Metro Gran Paris Line 15 Section 2C

M. Concilia
Impresa Pizzarotti & C. S.p.A., Parma, Italy

ABSTRACT: The paper presents arguments and describes a method for the selection of TBM when the excavation has to be performed at shallow depth in highly urbanized areas. As a practical application an innovative approach, based on multi-criteria analysis, for the selection of the machines to be employed for the excavation of the Line 15 section 2C of the Grand Paris Metro between Noisy Champs and Bry Villiers Champigny. The aim of the paper is also to stimulate thoughts regarding the importance of the geotechnical information needed for a comprehensive risk assessment analysis and the related actions requested to mitigate the residual risk below the acceptable level.

1 INTRODUCTION

For the excavation of tunnels at shallow depth in highly urbanized areas, the selection of the machine to be adopted has to be performed in accordance with the expected, or requested, average advance rates and in compliance with the admissible surface settlement.

When tunneling in soft or unstable ground under the water table, the only method to confine the settlement under the desired limit is to support the tunnel face and the periphery of the tunnel behind the tail shield, by applying an active pressure to counterbalance the stress generated by the ground and the water on the excavation surfaces.

The knowledge of the stress conditions requires an assessment of the geotechnical characteristics of the soils expected along the tunnel alignment that is essential to predict the level of risk associated with the type of machine; the correct selection should guarantee the lowest level of residual risk.

While the knowledge of the geotechnical conditions, particularly soil type, plasticity and permeability are clearly considered fundamental for the analysis of the geotechnical related risks, it should not be underestimated how important, in the selection process, are those factors related to the logistics, the environmental constrains and the availability of experienced personnel which we can define as production related risks.

The only machines capable to apply on the face and on the excavation periphery behind the shield an active pressure to counterbalance the pressures generated by the terrain and the water, are the Earth Pressure Balance machines (EPB) and the Slurry Shields (SS).

To emphasize the importance of the selection of the type of machine to be adopted, we should bear in mind that once the selection is made, this is irreversible.

This paper presents the multi-criteria analysis for the selection of the most suitable machine and a recent application of this approach for the excavation of the tunnels of the Lot GC01 of the Line 15 of the Grand Paris Metro Line 15 in Paris.

2 THE DIFFERENT TYPES OF FACE SUPPORT

The support, necessary to prevent an excessive stress reduction of the face and on the periphery of the tunnel, and in the worst cases the collapse of the tunnel face, can be obtained by providing a fluid pressure at the face, as for the Slurry Shields, or by pressurizing the excavated material directly at the face, as for the Earth Pressure Balance Machines.

In Slurry Shields application, the fluid, typically frictionless, consists of a suspension of bentonite in water, which, acting as support medium, is capable to transmit the required level of pressure to the tunnel face.

The bentonite suspension is pumped into the excavation chamber in which the differential pressure Δp, between the boring fluid and the neutral stress, induces, in case of relatively low permeability soils, the penetration of the suspension into the soil and the formation at the face of a filter cake acting as a sort of impermeable membrane though which the confinement pressure σ_{tot} is applied.

The water migration through the membrane, despite is very limited in quantity and prolonged for a short time duration, induces an increase of the water pore pressure Δu.

The dissipation of Δu, if the total stress σ_{tot} remain constant, generates an increase of the effective stress and the equilibrium, at which corresponds an instantaneous transmission of the stabilizing pressure, is reached when:

$$\sigma_{tot} = \sigma' + \Delta\sigma' + u + \Delta u \tag{1}$$

In case of coarser grained soil, the membrane could not always be formed, but the penetration of the bentonite suspension into the soil, bind the grains together, providing shear resistance to the soil even though the supporting capability rapidly decreases with the penetration distance.

In Earth Pressure Balance Shield application, the face stability is granted by the effective support pressure acting on the face by the excavated material, opportunely conditioned by foaming agents eventually mixed with specially developped polymers to reduce the ground permeability or the stickiness behaviour of the muck.

It has to be mentioned that for values of the permeability coefficient higher than 5×10^{-3} ms^{-1} the bentonite slurry can easily flow through the soil without any supporting action to the face as well as, on the other hand, for terrains containing more than 30% of fine grained soils (<0.06 mm) the application of EPB shield is more suitable.

Because of these limitations, the most relevant factor for the selection of the type of shield to be adopted for the excavation of a tunnel at shallow depth through soil behaving materials is the grain size distribution of the soils through which the excavation has to be carried out.

The EPB application is therefore possible in those terrains where the grain size allows the pressure applied in the excavation chamber to be dissipated through the screw conveyor, without flowing in an uncontrolled manner, and when the percentage of fined grained soils enables the transmission of the pressure generated by the machine advance to the tunnel face.

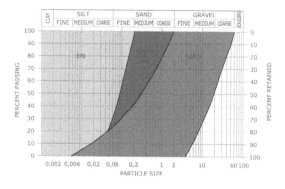

Figure 1.   Application ranges for Slurry and EPB Shields.

The extension of the range of application of the EPB machines towards the terrains characterized by a large portion of granular soils has been recently proved successful through the injection of carbonaceous fillers or bentonite in the excavation chamber.

3 ADVANTAGES AND DISADVANTAGES OF SLURRY AND EPB MACHINES

Slurry machines are well suited for tunneling through water bearing granular soils and heterogeneous deposits that can be easily separated from the slurry.

The main advantage is represented by the possibility to apply a positive fluid pressure at tunnel face that can counteract to both groundwater and soil pressure; the fluid pressure can be rapidly and accurately controlled, thus the application of this type of machine results very well adapted when the excavation has to be performed through loose running soils under high water pressure.

Since the utilization of a frictionless fluid, instead of the excavated soil for the pressure application to the face, the power consumption and the torque required for the cutterhead rotation, results lower than the values requested by an EPB machine; the wear of the cutterhead results also reduced by the lubricating action of the bentonite fluid.

In boulder forming grounds, the application of a Slurry Shield could be generally preferred since these machines are equipped with stone crusher and the cutterhead can be equipped with disc cutters, hence the frequency of downtime and hyperbaric intervention is sensibly reduced.

The size of the treatment plant can be considered a disadvantage for this type of machines, as well as the requested continuous monitoring and maintenance to prevent any clogging of the slurry circuit and to guarantee the desired density of both outgoing and incoming slurry.

Problems with the management of the treatment plant can be amplified when abrasive soils are encountered during the excavation, since the increased wear may damage the slurry pumps and pipelines resulting in prolonged stoppage for maintenance and repairs.

A technical limit for the application of a Slurry Shielded machine can be represented by cohesive and sticky clays that could cause, unless appropriately conditioned, clogging of the cutterhead and severe problems to the slurry treatment plant.

High content of particles with grain size < 0.06 mm may represent a limit for the successful application of Slurry machines.

Tunnelling with EPB machines, typically performed through fine grained and cohesive soils, has the major advantages in the elimination of the slurry separation plant.

With respect to the risk of face instability, the consequences of such event in case of EPB machine is less severe if compared with the consequences resulted in case of Slurry machine application.

When the excavation is performed through stable ground EPB machines can also be utilized in open mode, resulting in higher production rates, but, as shown in Figure 2, they require more power and torque than Slurry Shields.

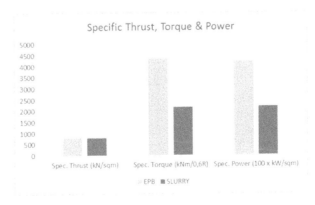

Figure 2. Typical values of torque and Thrust for medium-large diameters machines.

If we consider the interaction with buildings and structures above ground, in case the ground conditions enable the use of both technologies, a Slurry Shield would provide less level of risk because the easier control of the pressure but, with respect to the risk of face instability the EPB application would provide a higher safety level since the excavation chamber is filled up with excavated muck, while a sudden drop of the support pressure in a Slurry Shield could generate unacceptable surface settlements.

From the above-mentioned advantages and disadvantages appears clear that the selection of the most suitable machine for the excavation of a tunnel at shallow depth through soil behaving materials is rather complex and critical and an approach based of Multi-criteria analysis should be always implemented.

4 THE RELEVANT RISK FACTORS: MULTI-CRITEREA ANALYSIS

As previously stated, one of the most relevant factor for the selection of the machine type to be adopted for the excavation of a tunnel through unstable formations at shallow depth under urbanized areas, is the grain size distribution of the materials expected along the tunnel alignment.

This is clearly linked with the soils properties associated with the soil grain size, such as the water permeability and the plastic behavior for the fines fractions, but they are not the only factors to be considered for the correct selection of the machine.

Some of the machine technical characteristics requested from the tunnel design, such as the machine nominal diameter and the confinement pressure, have a determined influence on the machine type selection too.

For nominal excavation diameters above 12 metres, the torque requested for an EPB machine could result a technical limit and therefore the application of a Slurry Shield results more favorable than an EPB, as well as, in case of confinement pressure above eight bars, the application of an EPB machine may result not favorable.

From a sensitivity analysis, based on the type of structures above ground and on the response to the settlements expected along the tunnel alignment, the accuracy of pressure control may induce to prefer a Slurry Shield instead of an EPB or, on the contrary, the impact of the auxiliary plant may lead to prefer an EPB machine for the excavation from a shaft in a confined highly urbanized area.

An approach based on the analysis of all the relevant risk factors is therefore needed; this approach could be implemented on a quantitative or qualitative bases.

By adopting a quantitative approach, once the risks criteria have been determined, the level of risk R is obtained by multiplying the occurrence probability P with the Impact level I, of any considered risk.

The risks can be divided in groups and for each group n criteria can be established for any of the m machines considered in the analysis (in the examined case m = two, EPB or Slurry Shield).

If a relative weight W_C is introduced for each of the n groups of criteria considered, the final score, based on which the machine selection could be made, can be represented as:

$$R = \sum_1^n W_c \left(\sum_1^n p_{Ci,m} I_{Ci,m} \right) \qquad (2)$$

where $p_{Ci,m}$ represents the probability for the ith hazard of the Cth criteria and $I_{Ci,m}$ represents the consequence level estimated for the same hazard.

By adopting a qualitative approach, since the process to identify the hazards remain unchanged, the only difference is the risk classification that can be represented by the following risk matrix.

A multi-criteria analysis allow representing, in a form of a qualitative matrix, the advantages and disadvantages of the considered types of soft ground tunnelling machines.

Table 1. Risk Matrix (ITA, 2004. Guidelines for tunnelling risk management).

RISK MATRIX Frequency	Consequence Disastrous 5	Severe 4	Serious 3	Considerable 2	Insignificant 1
Very likely 5	Unacceptable	Unacceptable	Unacceptable	Unwanted	Unwanted
Likely 4	Unacceptable	Unacceptable	Unwanted	Unwanted	Acceptable
Occasional 3	Unacceptable	Unwanted	Unwanted	Acceptable	Acceptable
Unlikely 2	Unwanted	Unwanted	Acceptable	Acceptable	Negligible
Very unlikely 1	Unwanted	Acceptable	Acceptable	Negligible	Negligible

The hazards can be grouped into the following categories or criteria:

- Risks related to the geotechnical properties of the soils:
  - Grain size distribution (presence of cobbles and boulders)
  - Permeability
  - Abrasiveness
  - Plasticity for the fine particles
  - Swelling behavior

The plasticity of the fine fraction of the soil can be utilized to evaluate the potential clogging behavior that can be considered generally less problematic when the EPB is selected, instead of a Slurry Shield, for the excavation through clayey materials.

- Risks related to machine-ground interaction:
  - Clogging of the cutterhead and/or within the excavation chamber
  - Rapidly developed convergence due to swelling behavior of the encountered soil at tunnel depth
  - Settlement caused by loss of volume and/or rapid variation of the water pore pressure within permeable horizons

It is worth to mention that other than those risks listed above, the uncertainties related to the geotechnical investigation campaign, such as those related to the evaluation of the interstitial water pressure, within the impermeable horizontal layers, may represent another risk factor that could lead to select the machine improperly.

Finally, the suspected presence along the tunnel alignment of karsts cavities, despite this risk has the same score for both types of machines EPB or Slurry, required attention relatively to the auxiliary equipment necessary to investigate the ground ahead of the machine in order to minimise the residual risk under the acceptable level.

- Risks related to various operative issues and peculiar project constrains:
  - Management and disposal of the excavated muck
  - Pollution caused by the use of special polymers and or bentonite
  - Space availability for the site facilities
  - Management of the tunneling activities: special precautions to be taken for particular sections of the tunnel or because of peculiar conditions expected along the tunnel route

The above list, definitely not exhaustive, present some of the issues, not directly related to the expected ground conditions, to be considered when the machine is selected through a multi criteria analysis.

5 EPB OR SLURRY IN CRITICAL CONDITIONS

Sudden loss of face pressure could be defined as critical condition for the excavation at shallow depth in urbanized areas and, since this phenomenon is generally unexpected or suddenly encountered during the drive, it may represents a pitfall for the entire project.

Pressure loss phenomenon and, if uncontrolled, the consequent face collapse, could happen when the excavation involves heterogeneous materials in terms of permeability such as the case of excavation trough clayey soils with lenses of highly permeable soils.

With respect to the selection of the machine, it is worth to mention the behavior differences between the EPB machines respect Slurry Shield machines when such conditions are found.

By using an EPB machine, the risk of causing a large volume loss resulting from a localized face instability is limited by confinement offered by the excavated and conditioned muck inside the excavation chamber, while, in case of Slurry Shield machine it may be impossible to prevent the washing out of the slurry from the excavation chamber.

The reason can be understood by comparing the different density between the excavated material, generally ranging from 18 to 20 kNm^{-3}, and the bentonite slurry that has a density varying from 10.5 to 13 kNm^{-3}.

6 A PRACTICAL APPLICATION: THE GRAND PARIS METRO LINE 15 SECTION 2C

For the construction of the Section 2C of the new line 15 of the Grand Paris Metro project, in the East quadrant of the city of Paris, the Joint Venture between Demathieu Bard, NGE Génie Civil, Impresa Pizzarotti SpA and Implenia had been contracted.

The scope of work is the construction of the 4673 metres long tunnel, connecting the new stations of Noisy Champs and Villiers Champigny, the construction of a second tunnel from the Depot to the switching shaft, namely 807, as well as the construction of 9 emergencies and service shafts two of which to be used for launching the machines.

Approximately 60% of the length of the tunnel driven from Noisy-Champs will be excavated through altered limestones, namely Calcaire de Champigny (CC), with soil like behavior, while the remaining part will be driven through silty and clayey soil formations, namely Marnes d'Argenteuil (MA) and for a limited portion through Argilles Vertes, while for the shorter drive the geology will be predominately constituted by carbonaceous formation.

From Figure 4, it is evident that the terrains expected along the tunnel route fall, in terms of grain size distributions, within the typical range of application of the EPB technology.

It is worth to be mentioned that only the sands namely "Sables Verts", expected for a very limited portion of the tunnel length and with a maximum thickness of 2 metres, are cohesionless, therefore the risk related to running ground could be considered negligible.

The permeability coefficient for the expected soils ranges between 1.3x10^{-5} and 1.0x10^{-9} ms^{-1}, hence also for the permeability the EPB technology is the most appropriate method.

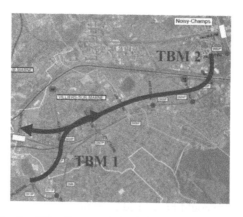

Figure 3. Line 15 South Section 2C – Construction of the tunnels from Noisy Champs to Bry Villiers Champigny and No.9 shafts.

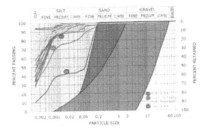

Figure 4. Grain size distribution of the expected soils along the tunnel alignment.

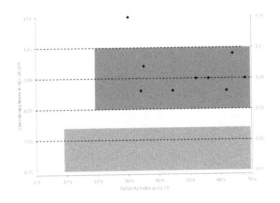

Figure 5. Clogging potential expressed in accordance with Thewes & Burger criterion.

The geotechnical risks considered for the selection of the machines for these alignments are:

- Risk of encounter cavities along the tunnel route due to the anthropic activities in the area or as a result of the dissolution of the gypsum formations below the tunnel level.
- Sensitivity of the excavated terrains respect the subsidence at shallow depth.
- The possible presence of boulders or unaltered rock formations along the alignment.
- Risk of clogging during the excavation through plastic formations.

Regarding the potential clogging the risk can be easily evaluated by adopting the method proposed by Thewes and Burger (Thewes & Burger, 2004) based on the determination of the Atterberg Limits.

Considering the expected level for the water table, the necessary face-support pressure will be, for the entire length of the tunnels, always lower than 3 bars, hence the risks related to the face pressure control will be negligible; it is worth to be mentioned that for face pressure above 5 bars, the dissipation trough the screw conveyor could result problematic, requiring special attention for the design that may lead to equip to machine with very long conveyors.

Regarding the experience in the region of Paris of other projects of similar nature, such as the recent extension of the metro lines 12 and 14, where from geotechnical point of view the conditions are very similar to those expected along the line 15 route, the selected machines are in both
cases EPB types.

The results of the multi-criteria analysis are presented in the table below.

From the table above it is evident that the most suitable type of machine is the EPB machine and for this reason two twins, state of the art EPB machines, manufactured by Herrenknecht, with an excavation nominal diameter of 9860 mm, have been selected for the excavation and lining of the tunnels.

Table 2. Multi-criteria matrix for TBM selection.

Criteria	Parameters	EPB		Slurry Shield		Line 15 Conditions	EPB	Slurry Shield
		Favorable	Unfavorable	Favorable	Unfavorable			
Geology	Grain Size	< 0,06mm (>25%) < 0,002mm (>5%)	> 2mm (10%)	> 2mm (10%)	< 0,06mm (>25%) < 0,002mm (>5%)	< 0,06mm (>25%) < 0,002mm (>5%)	Very Favorable	Slightly Favorable
	Permeabilty	<= 10^{-4} - 10^{-2} m/s	>= 10^{-3} m/s	>= 10^{-3} m/s	<= 10^{-4} - 10^{-2} m/s	>10^{-7} m/s	Favorable	Unfavorable
	Abrasiveness	Foam injection may have a positive affect on the tolls wear		Bentonite slurry have great influence on friction reduction		Low abrasiveness	Favorable	Very Favorable
Gelogical Risks	Cavities	For both types of machines the cavities represent a difficut condition				The possibility of encounter cavities cannot be excluded	Adoptable with precautions	Adoptable with precautions
	Swelling	Limited by the reduced quantity of water inside the chamber			Elevated risk due to use of water for slurry	Negligible risk	Very Favorable	Slightly Favorable
	Clogging	Machine to be equipped with facilities to inject anti clogging additives inside the chamber			For high percentage of fines particles the separation from muck could represent a practical limit	Unwanted risk; Atterberg Limits shows high risk of clogging	Very Favorable	Unfavorable
Operational Risks	Control of the consinement pressure		Not rapidly governed	Pressure can be controled in extremely rapid manner		Low face-support pressure; Negligible risk	Favorable	Very Favorable
	Control of the volume loss	Quantity of excavated material can be monitored precisely			Excavated volume cannot be monitored from the machine	Unwanted risk	Very Favorable	Unfavorable
	Management of the muck		Prolonged time for classification	Muck not polluted by additives after separation of the bentonite	In case of fine grained soils separation of the bentonite is not practicable	Silty Clayey soils	Very Favorable	Unfavorable

7 CONCLUSIONS

This paper has presented the different interaction behavior of two soft ground tunneling technologies when are used for the excavation at shallow depth under urbanized areas and, illustrated the multi-criteria analysis adopted for the selection of the first machine to be employed for the excavation of the extension of Gran Paris Metro Line 15 Section 2C.

Since the impact of the tunneling technology selection could be extremely severe, all variables possibly influencing the success of the project in terms of time and cost should be thoroughly analyzed.

Figure 6. Line 15 South Section 2C – The Herrenknecht EPB machine selected for the project.

The method presented could be adopted for any project of similar nature; however, it is also important to consider that when both types are suitable, the decision shall be taken considering the contractor experience as well as the availability of specialized personnel to perform the works.

REFERENCES

AFTES, 2005. GT4R3A1 Recommendations: Choosing mechanized tunneling techniques. *Tunnels et Ouvrages Souterrains, Hors Serie no.1*, pp. 137–163.

AFTES, 2012. GT32R2A1 Recommendations: Characterisation of geological, hydrogeological and geotechnical uncertainties and risks. *Tunnels et Espace Souterrain, no.232*, pp.315–355.

Anagnostu, G. & Kovari, K. 1994. The Face stability in Slurry-shield driven Tunnels. *Tunnelling and Underground Space Technology No.2*, pp.165–174.

Anagnostu, G. & Kovari, K. 1996. Face stability Conditions with Earth-Pressure-Balanced Shields. *Tunnelling and Underground Space Technology No.2*, pp.165–173.

Bezuijen, A. 2011. Foam used during EPB tunneling in saturated sand, description of mechanisms. *Proceedings of WTC 2011*, Helsinki.

Burger, W. 2013. Multi-Mode TBMs – State of the Art and Recent Developments. *2013 Rapid Excavation & Tunneling Conference & Exhibit*, Washington.

DAUB, 2010. Recommendations for the selection of tunneling machines.

Guglielmetti V., Grasso P., Mahtab A., Xu S. 2007. Mechanized Tunnelling in Urban Areas. *Taylor & Francis e-library*, London.

Maidl B., Herrenknecht M., Maidl U., Wehrmeyer G. 2013. Mechanised Shield Tunnelling. *Wilhelm Ernst & Sohn*, Berlin.

Thewes M. and Burger W. 2004. Clogging risks for TBM drives in clay. *Tunnels & Tunnelling International, June*, pp.28–31.

Tunnels and Underground Cities: Engineering and Innovation meet Archaeology,
Architecture and Art, Volume 11: Urban
Tunnels - Part 1 – Peila, Viggiani & Celestino (Eds)
© 2020 Taylor & Francis Group, London, ISBN 978-0-367-46899-6

Settlement reduction in clean sandy soils with EPB – good practice for backfilling grout injection and soil conditioning. The case of Warsaw Metro Line 2

E. Dal Negro, A. Boscaro, A. Picchio & E. Barbero
Underground Technology Team, MAPEI S.p.A., Milano, Italy

T. Grosso
Astaldi S.p.A., Rome, Italy

ABSTRACT: Warsaw Metro Line 2 eastern extension is developed to improve the transportation system of the Polish capital, reducing the distance between residential areas and city center. Tunnel alignment is under-passing residential areas composed by old buildings, boring through clean gravelly fine sand. The lack of fines and the groundwater in the soil, may cause difficulties in the management of the earth pressure. Soil conditioning has been a key parameter for the successful control of the pressure in the bulk chamber. Injecting small quantities of high performances polymer, has been fundamental to obtain a suitable conditioned soil, able to be managed with the screw conveyor. Another key aspect has been the use of a specific mix design of the A component of the two-components backfill grout, with the correct dosage of accelerator, and modified injection parameters, achieving the reduction of the volume loss due to the gap between soil and lining.

1   INTRODUCTION

The construction of the extension of the Warsaw Metro Line 2 is certainly a further step in the development of the public transport system in the capital of Poland and it's included in the vast expansion program of public infrastructure; the so-called "Odcinek II A" is in fact destined to become the main connection between downtown to the north-east part zones of the city and on the future it will be still extended with other 3 additional stations connected by twin tubes.

The tunnel alignment underpass populous residential areas with buildings in critical structural conditions. The particular geological conditions and the shallow overburden, required particular excavation procedure able to allow the minimization of surface settlements.

2   THE NORTH-EAST WARSAW METRO LINE 2 EXTENSION PROJECT

The construction works, awarded to Astaldi S.p.A., includes the realization of 3 Stations and 2,2 km tunnels in natural excavation for each tube. During excavations were actually involved the use of two TBMs EBP type, accurately refurbished after the excavation of the same metro line under the historical center of the city.

The section between C-18 Trocka station and C-15 Dworzec Wilenski Station runs along a curved path and with shallow overburden under the Targowek and the Praga District, where buildings that survived the Second World War are located. The critical stretch of the excavation is the last part, from C16 Szwedzka station and the dismantling chamber, temporary

Figure 1. Warsaw Metro Line 2 – North Eastern extension.

obtained, inside of station C15 holding track, already active part of the underground in operation. From the station 2 tubes run under the historical area, the right one under Strzelecka street and the left one running for roughly 500 meters under a row of old ancient building, with different elevation and with 2,5 meters basement deep. The overburden between the top of TBM excavation and the lower level of basement was in a range of 7- 10 meters, in clean sandy soil, under water table.

3 GEOLOGICAL CONDITIONS AND DESIGN APPROACH

Along the whole route the soil to be excavated and the soil above the tunnels was mainly granular soils. All unit belonged to the Quaternary period and was of sedimentary origin. The water table is 3 meters deep from the ground level. Only the bottom part of the excavated front was silty and clayey material. The geological profile is shown in figure 2.

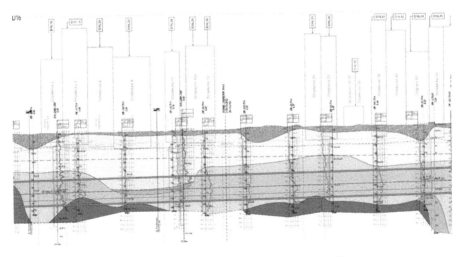

Figure 2. Warsaw Metro Line 2, North Eastern extension geological profile.

Several buildings, on Praga district, along strzelecka street area, are often in a critical structural condition and consequently more subject to possible damage.

From the Building Permit Design, was proposed a "classic" solution using sub-horizontal drilling carried out from the ground level under the buildings foundations in order to install sleeved grouting pipes for subsequent grouting.

The difficulties in obtaining the necessary working areas, as well as the issues and the required time to get the necessary permits for temporary occupation of private properties and important traffic routes, was not compatible with the general planning of the construction works. Due to the above mentioned issues, the huge impact on date of completion the works and cost impact, was decided to limit soil improvement to 5 ancient buildings, with very fragile structure condition and conservation. Furthermore, this solution minimizes the impact of the works on the urban area, especially reducing the risk of damage to the buildings located along the alignment of the tunnels. The final design approach, also connected to budget and management choice in accordance with client was to limit soil improvement under the foundation of 5 buildings, and try make an accurate excavation by TBM EPB closed mode, with some specific solution and procedures, studied with the technician of Mapei S.p.A. The precautions described on the next chapter.

4 SPECIFIC SOLUTION FOR SURFACE SETTLEMENT REDUCTION

4.1 Initial situation

Due to the peculiarity of the building above the tunnel alignment previously described, the operation in closed mode and the accurate control of the earth pressure are key parameters to be keep under strict observation to minimize the surface settlements.

Another fundamental aspect to be accurately managed to minimize the loss of volume and, therefore, settlement on the surface, in the annular gap backfilling injection. The volume and pressure control of the injection have to be carefully managed. The injected volume have to reach at least the theoretical value, with pressures of about 0.5 bars exceeding the earth pressure applied to the tunnel face.

Due to the above mentioned geological and hydrogeological conditions, the soil, properly conditioned to achieve good TBM performance and a proper pressure transmission to the tunnel face, showed, due to the lack of finer fraction and to the saturated conditions of the sandy soil, a too liquid behavior on the TBM belt. Beyond the logistic problems related to the muck management from the tunnel belt up to the truck transportation, a too liquid behavior may cause difficulties in the earth pressure control. In fact, a too liquid muck, in presence of pressure in the excavation chamber, may flow along the screw conveyor by extrusion, pushed by the earth pressure. Therefore, instead of creating a plug for the water in the screw conveyor able to keep the required pressure in the excavation chamber, may cause difficulties in the control of a constant earth pressure, as required by the design.

The backfilling injection behind the segment have been carried out properly since the beginning of the excavation, with injection pressure and volumes as per design.

4.2 Specific solution for soil conditioning

As mentioned in previous paragraph, the liquid consistency of the muck may cause difficulties in the pressure management and, therefore, in the minimization of the surface settlement. The solution adopted to solve the mentioned issue, have been the injection in the excavation chamber of Mapedrill M1, soil conditioning polymer with water absorption an lubricating properties. The use of about 0.3 – 0.4 kg of Mapedrill M1 per cubic meter of soil, brought to an effective control of the soil consistency. By the observation of the TBM data during the use of the polymer, any uncontrolled fluctuation of earth pressure was visible. Although the muck

Figure 3a.   visual aspect of the muck before the use of Mapedrill M1.

Figure 3b.   visual aspect of the muck after the use of Mapedrill M1.

was drier that before, the cutterhead torque and the TBM performances remained the same. The injection of Mapedrill M1 in the chamber and the fast reaction time, allowed the treatment of the soil close to the screw conveyor inlet, without affecting the consistency of the conditioned soil close to the cutter head. In figure 3, pictures of the soil before the use of Mapedrill M1, and during the injection.

4.3   Backfilling grout injection improvement

Due to the non-cohesive nature of the soil, every step of the excavation foreseeing an unsupported situation has to be minimized as much as possible. For this reason the standard mix design of the A component, studied in the Mapei R&D laboratory during the job site preparation have been modified. Also the proportions between the two components, have been modified in order to reduce the gel time. To reduce the gel time without affecting the performances of the grout, an increase of cement has been done on the A component recipe, together with a slight increase of Mapequick CBS System 1, the admixture to allow an A component workability of 3 days. The proportion between Mapequick CBS System 3 (accelerator) and A component, have been reduced. These modifications brought to a reduction of about 2 seconds of the gel time and to a higher viscosity of the backfilling grout. The modified mix design has been studied and tested by Mapei technicians, at the mixing plant prior to use it in production, to evaluate the most suitable mix design for the specific situation

Also the injection parameters have been modified in order to minimize the possible fall of soil towards the void and to compensate the loss of injected volume due to possible

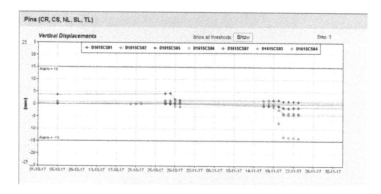

Figure 4a. vertical displacements on the monitoring section with ground pins installed on the ground level and representing the movement of the surface at the passage of first and second TBM;

permeation of the grout in the sandy soil. The injection pressure has been increased up to almost 1 bar exceeding the earth pressure. The total injected volume per ring have been also increased in order to keep the pressure level required to minimize the settlement.

5 TBM EXCAVATION PERFORMANCES

With the application of the countermeasures described in the previous chapter during excavation, earth pressure was kept regularly on right range design, aiming to have full excavation chamber and checking the volume of excavated ground never exceeded the theoretical values, which made it was possible to avoid settlement around the area of the cutter head. On the other hand, the volume of grout filled in the gap behind the lining at the tail skin was always higher than the designed value.

The first of two underground routes began on October 2017 with the TBM S644, which drilled the left tunnel towards the C15 Station and successfully completed the critical passage on December 2017. Then, following alignment under Strzelecka street, the TBM S760 passed the same route between October and January 2018. Both underground passages, about 1150 m long, included the 550 meters under ancient building and Strzelecka street with shallow overburden, were achieved with average daily advances of 18,75 m and 18,5 m respectively, with peaks of 33 m per day. The recorded volume loss was less than 0.30% instead of 0.44 % as well as less total settlement under the urban areas lower than one centimeter.

As per common approach during execution of such type of underground project in urban area, a monitoring system was installed in order to check settlements and deformation during TBMs performances. The system was able to register continuously data. The configuration of the system was set up to furnish automatic and manual measurements. The behaviors of the buildings were observed during the passage of the two TBMs were similar. The settlement occurred in a similar way, both in terms of kinematics and values.

6 CONCLUSIONS

The settlements measurement underlines the success of the EPB technology in urban areas under sensitive buildings. The high attention to the excavation parameters, such as earth pressure, muck extraction control, backfilling injection pressure and volume, have been a fundamental aspect of the settlements minimization.

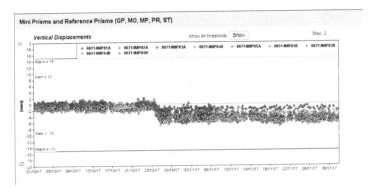

Figure 4b.   vertical displacements on the levelling pins installed on building N.214;

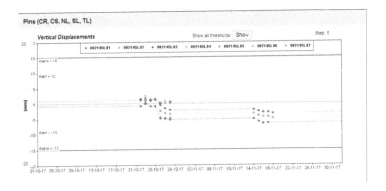

Figure 4c.   real time monitoring of miniprism installed on building N.214;

The good practice, together with the use of a specific soil conditioning products such as Mapedrill M1, and the adjustment of the backfilling grout mix design and injection procedure, combined with good TBM crew skills and careful supervision, brought to the respect of the admissible settlements limits, without the use of extensive grouting campaign below the buildings. The synergy between contractors and suppliers, have been an important aspect to share expertise with the common target of a safe and effective TBM excavation, with the expected results and good production rates.

REFERENCES

AFTES, (French National Tunnelling Association), *Choosing mechanized tunnelling techniques* (2005) Paris.

EFNARC (2005). *Specification and guidelines for the use of specialist products for Mechanized Tunnelling (TBM) in Soft Ground and Hard Rock* www.efnarc.org.

Dal Negro, E., Boscaro, A. & Plescia, E. (2014). *Two-component backfill grout system in TBM: The experience of the tunnel "Sparvo" in Italy*, Proceedings of TAC Congress 2014: "Tunnelling in a Resource Driven World", Vancouver, 26–28 October 2014.

Dal Negro, E., et al. (2014). *Two-component backfill grout system in double shield hard rock TBM. The "Legacy Way" tunnel in Brisbane, Australia*, Proceedings of ITA-AITES World Tunnel Congress 2014: "Tunnels for a better life", Foz do Iguacu, Brazil, May 2014.

Guglielmetti, V., Mahtab A. & Xu, S. (2007) *Mechanized tunnelling in urban area*, Taylor & Francis, London.

Pelizza, S., et al. (2010). *Analysis of the Performance of Two Component Back-filling Grout in Tunnel Boring Machines Operating under Face Pressure*, Proceedings of ITA-AITES World Tunnel Congress 2010: "Tunnel vision towards 2020", Vancouver,14 20 May 2010.

Thewes M., & Budach C. (2009). Grouting of the annular gap in shield tunnelling – An important factor for minimization of settlements and production performance, Proceedings of the ITA-AITES World Tunnel Congress 2009 "Safe Tunnelling for the City and Environment", Budapest, 23–28 May 2009.

Tunnels and Underground Cities: Engineering and Innovation meet Archaeology, Architecture and Art, Volume 11: Urban Tunnels - Part 1 – Peila, Viggiani & Celestino (Eds)
© 2020 Taylor & Francis Group, London, ISBN 978-0-367-46899-6

Bucharest line 5: 3D numerical modelling of tunnelling beneath operating station

K. Daneshmand, B. Bitetti & G. Ragazzo
Systra Engineering, Paris, France

ABSTRACT: The Bucharest metro Line-5 plays a strategic role to improve the urban transport system, connecting the new Raul-Doamnei station to the existing Eroilor-1 station. Tunnelling works were carried out by two 6.6m diameter EPB machines. The most critical point of the project is the passage of TBMs at about 1m below the bottom slab of the operating station. Due to the high-risk level of impacting the station stability and functionality during tunnelling, detailed excavation procedures and mitigation measures were designed by the Contractor. To verify the efficiency of the designed measures, a detailed 3D FEM analysis was developed prior to the commencement of the works. The analysis results validated the Contractor's procedures and identified the most critical aspect of the tunnelling beneath the station. Accordingly, a detailed follow-up procedure was implemented during the tunnelling works to check the compliance of the excavation procedures and station behaviour during the TBM advancement.

1 INTRODUCTION

The Bucharest metro Line 5 plays a strategic role to improve the urban transport system, connecting the new Raul-Doamnei station to the existing Eroilor 1 station. The line consists of 9 underground stations and about 6km long twin-tunnels alignment. Construction works started in 2013 and lasted until 2017. The geological sequence along the tunnels mostly consists of the alternation of sandy and clayey soils, while the water table is always located few meters below the ground surface. The tunnelling works have been carried out by two 6.6m diameter EPB machines. The most critical point of the project has been the TBMs passage at about 1m distance below the bottom slab of the operating Eroilor 1 station, where excavation has been performed in mixed-face conditions and below the water table.

Considering the proximity of twin-tunnels to the existing station structure, the whole excavation procedure has been designed to avoid any risk of impacting the station stability and functionality. Detailed mitigation measures and preliminary works have been foreseen and carried out by the Contractor: lowering the water level to approximately 1m below the tunnels invert in the area beneath the Eroilor 1 station, cement and chemical soil injections beneath the Eroilor 1 bottom slab for TBMs breaking-through and monitoring system for ground and structures.

In order to verify the efficiency of the designed mitigation measures, a numerical analysis was developed prior to the commencement of the works. Hence, a 3D FEM model was developed, using Plaxis 3D code, to check the tunnelling impact on the Eroilor 1 bottom slab (Vermeer 2001). Furthermore, during the execution of the works, a detailed follow-up procedure has been implemented in order to check the compliance of the excavation procedures, the efficiency of the designed mitigation measures and the global station behaviour during the TBM advancement.

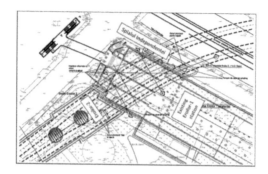

Figure 1.   Location of the existing Eroilor-1 station.

2   GENERAL DESCRIPTION OF THE DESIGN

The general design developed by the Contractor for tunnelling beneath the Eroilor 1 is described in the following sections.

2.1   General layout

The tunnel alignment is characterized by twin-tunnels configuration. Tunnels are carried out by two *Herrenknecht* EPB machines with the excavation diameter of 6.60m and intrados diameter of 5.70m. In the stretch beneath the Eroilor 1 station the distance between tunnels axes is around 16.6m. The tunnel lining is made of 5+1 precast segments, 30cm thick and average length of 1.50m. The table hereafter summarizes the general layout of the tunnels and station.

2.2   Geological and geotechnical environment

Information derived from Borehole 1 (corresponding to FM5-215, the closest borehole to the station) and Borehole 2 (about 18m far from borehole 1) has been used to define the geological sequence beneath the Eroilor 1 station (Figure 3). It consists of the alternation of sandy and clayey layers. Moreover, the piezometers installed in the area of the project indicate the water table located approximately 2.9m below the ground surface.

Table 1.   General layout.

Main bored tunnel	Depth of the tunnel axis	16.6m
	Level of the tunnel axis	55.7
	Tunnel crown to the bottom slab	1.2m
Segmental lining	Extrados diameter	6.3m
	Intrados diameter	5.7m
	Thickness	0.3m
	Ring width	1.5m
TBM	Face diameter	6.54m
	Shield length	9m
	Shield thickness	0.04m
	TBM conicity	0.3%
Station structures	Station length *	70.4m
	Station width	54.4m
	D-wall thickness	0.8m
	Bottom slab thickness	1.2m

* Considered in the 3D model

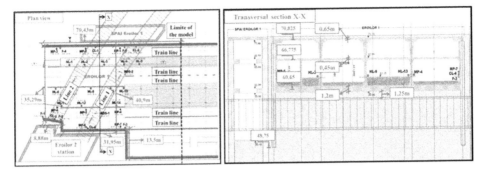

Figure 2. Eroilor-bottom slab plan view (left) and transversal section X-X (right).

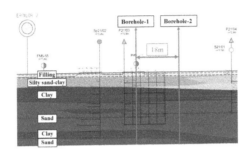

Figure 3. Geological profile and boreholes in the Eroilor 1 station area.

2.3 *Planning of works and mitigation measures*

According to the Contractor design, the tunnelling operations beneath Eroilor 1 station start from Eroilor 2 station by breaking through the Eroilor 1 retaining wall and proceed toward the SPAI shaft (Figure 2). The Line 2 is executed before starting the Line 1 excavation.

To reduce any risk of water inflow and to keep the Eroilor 1 functionality, the following mitigation measures have been designed by the Contractor and implemented on site before the TBM breaking-through:

– Dewatering: the water level is lowered down to 1m beneath the tunnels invert;
– Cement and chemical horizontal soil injections beneath the Eroilor 1 bottom slab. They are injected from inside the *"Eroilor 2"* and *"SPAI Eroilor 1"* in order to provide a sealing and compact block reducing any further risk of water ingress during the TBMs breaking-through.

Furthermore, due to the high sensitivity of Eroilor 1 station to the excavation process, an accurate control of the TBM excavation parameters (face pressure, soil conditioning, TBMs advancement forces...) has been required and implemented during the works.

3 NUMERICAL MODELLING

In order to study the behaviour of the bottom slab during the tunnelling operations as well as the compliance and effectiveness of the designed mitigation measures, a 3D FEM model has been developed using PLAXIS 3D code.

3.1 *Model size and developed mesh*

Considering the real geometry of the tunnels and existing station, the developed model includes:

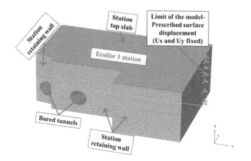

Figure 4. Schematic presentation of the developed 3D model.

- Three levels of slabs of the Eroilor 1 station;
- Five active train lines at the bottom slab, simulated by surface loads;
- The improved soil zones for TBM breaking through, consistent to the Contractor design.
- The twin-tunnels alignment passing beneath the bottom slab.

The model dimensions are defined in order to avoid any boundary condition effect:

- Model dimension along X-axis: 130 m
- Model dimension along Y-axis: 100 m
- Model dimension along Z-axis: 72.2 m

The mesh was generated to be fine enough around the tunnels and station bottom slab to obtain reliable results without affecting the calculation time. In order to decrease the number of elements in the 3D model and to reduce the calculation time, Eroilor 1 station was not entirely modelled. Only the part of the station which can be affected by the tunnelling works has been modelled. The limit of the station was modelled by impermeable imposed surface displacement element while the displacements are defined to be fixed in horizontal directions and free in the vertical direction (Figure 4).

3.2 Characteristics of the materials

3.2.1 Soil properties
The geological sequence derived from the performed investigation campaigns (§ 2.1) has been implemented in the model. However, soil layers with insignificant thickness are considered to have insignificant effect on the scope of the work and they have been neglected. Therefore, the developed model consists of 6 soil layers, as shown in Figure 5.

The ground behaviour has been simulated considering the Hardening-Soil constitutive law. Moreover, the *"undrained behaviour"* has been defined for the cohesive soils and *"drained*

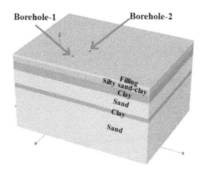

Figure 5. Developed ground stratigraphy.

Table 2. Soils geotechnical parameters.

Layer	γ	E_{50}	E_{oed}	E_{ur}	c'	ϕ'
	kN/m³	MPa	MPa	MPa	kPa	°
Filling	17.5	8	8	8	1	18
Silty sand-clay	19.0	10	10	10	0	20
Clay	20.0	10	10	10	18	20
Sand	20.0	25	25	25	0	32

Table 3. Soil parameters for injected zone.

Soil improvement	γ	E	c'	ϕ'
	kN/m³	MPa	kPa	°
Injected soil	21.0	150	18	28

behaviour" for the cohesion-less ones. The main geotechnical properties of the soil layers inserted in the model are summarized in Table 2.

For the injected soils beneath the Eroilor 1 station, the Mohr-Coulomb constitutive law and *"non-porous"* material have been considered, with the following mechanical properties:

3.2.2 Station structural elements

In the developed 3D model, *"plate"* elements have been used to model the station retaining walls, the three station slabs, TBM shield and segmental lining.

The conicity of the nine meters TBM shield was modelled by a 0.04%/m linear contraction along the first 7.5m of the shield, from face to tail, and 0.3% constant contraction for the last 1.5m of the shield (Guglielmetti et al. 2008).

Moreover, elasto-plastic *"node-to-node anchor"* elements have been used to model the columns with no tensile resistance.

The rigidity of the structural elements has been defined according to the corresponding elastic module and thickness.

In order to well simulate the soil-structure interaction, interface elements have been defined at the contact between the *"cluster"* elements, representing the soil layers, and the *"plate"* elements, representing the structural elements. The strength parameter of the soil-structure interfaces (R_{inter}) has been considered equal to 0.67 for all the layers, to consider 2/3 of the adjacent soil strength parameters.

3.2.3 TBM face and backfilling pressure

The face and backfilling pressures have been considered for the TBM modelling. The TBM face pressure was modelled by surface load increasing from crown to the invert. As indicated in the Contractor design, face pressure has been set equal to 0.5bar at the crown.

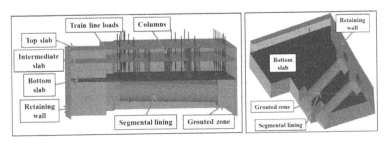

Figure 6. Schematic presentation of the developed 3D model-side and bottom view.

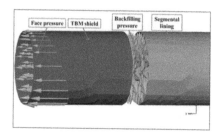

Figure 7. Schematic presentation of the developed 3D model-TBM and segmental lining.

The backfill grouting pressure was applied on the unsupported soil in between the shield tail and the last installed ring. This pressure was modelled by surface load applying vertically to the surrounding soil which is increasing from top to bottom (0.75bar at top).

4 3D CALCULATION SEQUENCE AND MAIN RESULTS

The general description of the construction sequence implemented in the model and the main obtained results are presented hereafter.

4.1 TBM advancement and excavation phases

The TBM excavation has been modelled using step-by step excavation method (Daneshmand et al. 2014). It seems that the first step-by-step methods have been introduced by Hanafy & Emery (1980) followed shortly after by Katzenbach & Breth (1981). It thereupon has been considered by Wittke (1984) and Swoboda et al. (1989). Accordingly, each phase of the TBM advancement in the developed model includes: excavation of one slice of the soil ahead of the face (1.5m width), changing the position of the face and backfilling pressures, advancement of the TBM shield, updating shield contraction values and installing the ring.

The developed 3D model includes 77 phases to simulate the complete TBM advancement and Contractor's construction sequences including all the designed technical interventions. The modelled excavation sequence is presented in the following table.

4.2 Initial condition before tunnelling

The first excavation phases (phase 0 and phase 1) consist of reproducing the existing on-site condition before starting the tunnelling operations. This includes the calculation of the geo-static condition and the excavation of the existing Eroilor 1 station. This has been done by:

Table 4. Modelled excavation sequence.

Phase ID	Phase description
0	Geo-static condition
1	Station emplacement
2	Dewatering + Soil improvement zones + Set the displacement/excess pore pressure to zero
3 to 8	Line 2: Break-in + Step-by-step TBM shield emplacement and excavation
9	Line 2: TBM totally in the soil + 1st TBM advancement
10 to 27	Line 2: TBM advancement step 2 to 19
28	Line 2: TBM advancement step 20 + Break-out (ring 1)
29 to 37	Line 2: TBM advancement step 21 to 29 (last)
38 to 44	Line 1: Break-in + Step-by-step TBM shield emplacement and excavation
45	Line 1: TBM totally in the soil + 1st TBM advancement
46 to 67	Line 1: TBM advancement step 2 to 23
68	Line 1: TBM advancement step 24 + Break-out (ring 1)
69 to 77	Line 1: TBM advancement step 25 to 32 (last)

- Activating the station structural elements (retaining walls, slabs, columns...), the related interfaces and the charges representing the train loads;
- Deactivating the volumetric soil elements (clusters) inside the station and setting them to dry in order to simulate the station excavation.

4.3 Dewatering and soil improvement

In this phase, all the induced effects by the station excavation (displacements, excess pore pressures) have been set to zero in order to reproduce the existing on-site initial condition before starting the tunnelling operations.

Afterwards, the water table below the station bottom slab is lowered down to 1m below the tunnel invert to model the preliminary dewatering operation, according to the Contractor design. The dewatering is kept active during the tunnelling operations of the both lines.

Furthermore, the horizontal injection zones were modelled by changing the soil properties of the related clusters beneath the bottom slab to the "*injected soil*" ones (Figure 6).

The vertical and horizontal displacements induced by the dewatering inside the soil and bottom slab are shown in Figure 8 and Figure 9.

It can be remarked that:

- The maximum vertical displacement on the bottom slab after dewatering reaches approximately to 13mm;
- The horizontal displacement along the X-axis (parallel to the train lines) varies between 4.8mm and 6mm along the width of the station;
- The horizontal displacement along the Y-axis (perpendicular to the train lines) varies from -1mm to +1.3mm along the length of the station.

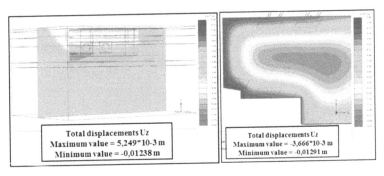

Figure 8.   Soil vertical displacements (left) and bottom slab vertical displacements after dewatering (right).

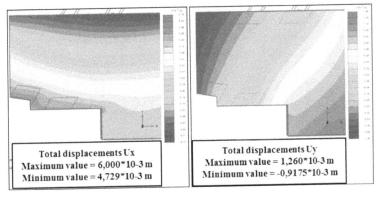

Figure 9.   Bottom slab horizontal displacements after dewatering (phase 2).

4.4 Calculation main results

In order to check the induced displacements during the tunnelling operations, 25 monitoring points have been defined on the Eroilor 1 bottom slab (Figure 10). In order to analyse the bottom slab behaviour, the induced displacements at these points have been studied for the most significant calculation steps:

- Phase 2: Dewatering and soil improvement;
- Phase 16: Line 2 excavation arrived to the middle of the station width;
- Phase 37: Line 2 excavation finished;
- Phase 45: Line 1 TBM broken-out and out of the improved zone;
- Phase 56: Line 1 excavation arrived to the middle of the station width;
- Phase 77: Line 1 excavation finished.

Figure 11 shows the obtained vertical displacement at the position of each control point for the above mentioned main calculation phases. It can be remarked that the main displacement occurs during the dewatering phase while the maximum induced displacement by the tunnelling operations is less than 2mm. Moreover, the maximum induced displacement occurs at the point 14 which is above the crown of Line 1 tunnel (the last excavated) and it reaches about 14mm, as shown in Figure 12.

The induced maximum vertical displacement and the differential settlement along the five train lines (Figure 10) are presented in Figure 13. It can be remarked that the maximum calculated settlements concern Train Line 2 and 3 (TL2 and TL3) located almost in the middle of the station width (about 14mm) and it is reached mostly during the dewatering stage. Moreover, the maximum differential settlement is about 9mm, along almost 18m span. Furthermore, the

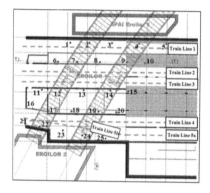

Figure 10.   Monitoring points on the Eroilor 1 bottom slab.

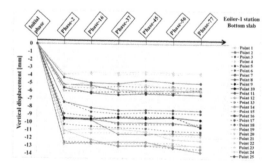

Figure 11.   Induced vertical displacement of the bottom slab during the construction for monitoring points.

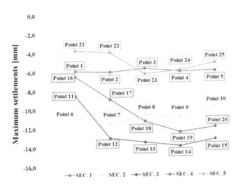

Figure 12.   Maximum induced settlement of the monitoring points.

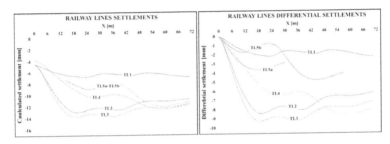

Figure 13.   Calculated total and differential settlements along the 5 existing train lines.

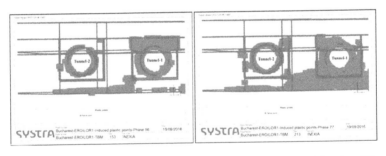

Figure 14.   Induced plastic zone around the excavated tunnels-Phase56 at left side, phase 77 at right side.

maximum slope of the differential settlement curve is about 2.5mm, along 3m span, which is respecting the defined Contractor threshold: 6.5mm along 3m span.

The induced plastic zones around the excavated tunnels for two phases (phase56 and phase77) are shown in Figure 14: there is an important risk of soil plastification between the tunnel crown and the station bottom slab. Therefore, it confirms the need of providing a detailed monitoring plan and follow-up procedure as well as an accurate control of all the TBM advancement parameters during the excavation works.

5   CONCLUSION

The tunnelling operation beneath the bottom slab of the existing Eroilor 1 station represents one of the most critical points of the project due to the high risk of inducing damages to the existing and operating Eroilor 1 station. Therefore, specific tunnelling procedures and mitigation measures have been required and designed by the Contractor in order to minimize any risk of damaging the station structures (i.e. dewatering during tunnelling, soil improvements...).

A 3D FEM analysis was carried out by *Systra* to study the efficiency of the Contractor's designed procedures as well as the effects induced by tunnelling works on the structures of the Eroilor 1 station. All the construction stages have been considered in the developed model.

The obtained results showed that:

- The main bottom slab displacements occur during the dewatering stage, when the maximum settlement of the bottom slab reaches about 13mm;
- The total settlement reaches about 14mm at the end of the tunnelling operations. The increment of settlement induced by tunnelling operations is not significant (less than 2mm);
- The induced horizontal displacements due to the dewatering phase are varying between 4.8m and 6mm along the X-axis (parallel to the train lines), whilst they vary from -1mm to +1.3mm along the Y-axis (perpendicular to the train lines);
- The observed induced plastic points around the two tunnels indicate the risk of soil plasticization at the tunnels crown. Therefore, it is necessary to well-manage the TBM parameters and the excavation procedures during tunnelling.

Moreover, looking at the calculated displacements corresponding to the 25 monitoring points, defined by the Contractor, it has been remarked that the maximum induced displacement is reached in the middle width of the station. Actually, the induced displacements along the existing train lines indicate that the largest settlement corresponds to the train lines 2 and 3, crossing the tunnel alignments at their middle length. Nevertheless, the maximum differential settlement is respecting the admissible thresholds defined by the Contractor.

In conclusion, according to the results of the numerical calculations, the whole Contractor's designed procedures seem to be effective. However, a detailed follow-up procedure is required to control the excavation parameters and the related significant negative effects.

Moreover, calculation results indicate that the dewatering procedure is the most affecting intervention (in terms of induced displacements). Indeed, this has been confirmed during the follow-up procedure, showing that the global behaviour fits well with the results of the developed FEM model.

REFERENCES

Daneshmand, K. Floria, V. Peila, D. & Pescara. 2014. Cross passages between twin tunnels. Preliminary design schemes. World Tunnelling Congress. Foz do Iguaçu. Brazil.

Guglielmetti, V. Grasso, P. Mahtab, A. & Xu, S. 2008. Mechanized tunnelling in urban areas. Taylor and Francis.

Hanafy, E.A. & Emery, J. J. 1980. Advancing face simulation of tunnel excavation and lining placement. Under Ground Rock Engineering: 119–125. V.22.

Katzenbach, R. & Breth, H. Nonlinear 3d analysis for NATM in Frankfurt clay. 10th Int. Conf. Soil Mech. and Found. Eng: 315–318. Volume 1. Rotterdam: Balkema.

Swoboda, G. Mertz, W. & Schmid, A. 1989. Three dimensional numerical models to simulate tunnel excavation. Proceedings of NUMLOG III, Elsevier Science Publishers Ltd: 536–548, London.

Vermeer P.A. 2001. On a smart use of 3D-FEM in tunnelling. Plaxis Bulletin (11): 2–7.

Wittke, W. 1984. Rock mechanics. Springer-Verlag, Berlin.

*Tunnels and Underground Cities: Engineering and Innovation meet Archaeology,
Architecture and Art, Volume 11: Urban
Tunnels - Part 1 – Peila, Viggiani & Celestino (Eds)
© 2020 Taylor & Francis Group, London, ISBN 978-0-367-46899-6*

Innovative station design for the second tramway line in Nice

C. David & S. Minec
Bouygues Travaux Publics, Guyancourt, France

ABSTRACT: Nice Côte d'Azur Metropole has initiated the construction of a second tramway line, which will ultimately link Nice Côte d'Azur airport with the port of the city.
The first segment of this line is an underground section between the Boulevard Grosso and Ségurane Street, which serves 4 stations on its route. The technical challenge lies in limiting the impact of construction on existing buildings, while digging in a variety of grounds, ranging from very raw and permeable soils to clayey soils. Another challenge was to reduce the duration of the works.
This article presents the measures implemented to limit settlement of the surrounding structures, the observational method developed to adapt measures based on the conditions encountered and the innovative solutions developed at the tunnel and station interfaces to meet the program requirements: for some stations, the tunnel was already built when D-walls were realized and excavation started.

1 INTRODUCTION

The line 2 of the Nice tramway is built partially underground below Nice historical center. One of the biggest challenges is to limit the impact of the line construction on the existing building while boring through a variable and complex geology. Thus, the management and control of settlements is a critical issue for the project. Dealing with this issue includes actions such as settlement prediction and vulnerability assessment during design stage and implementation of the observational method during construction stage.

The construction of the underground part of the line 2 was awarded to Thaumasia, joint venture lead by Bouygues Travaux Publics.

At the beginning of the contract, the Client requested Thaumasia to shorten the global work duration. Some acceleration measures were implemented, such as boring the tunnel while the underground stations are still under construction. This staging leads to strong interactions between Dwall construction and station excavation on one hand, tunnel boring and lining dismantling on the other hand.

2 PROJECT OVERVIEW

2.1 *Tramway line 2 underground section*

The underground section of the tramway line 2 is 3.2km long from the port to the boulevard Grosso. This section of the tramway line is built underground in order to limit the impact of the line on the public space and to preserve the traditional architecture of the city center. The Thaumasia contract for the line 2 includes the construction of:

- 2900 meters of tunnel bored with a slurry TBM (internal diameter is 8.5m and lining thickness is 40cm). The launching shaft is located in Ségurane street, close to the port, and the tunnels ends in the Grosso retrieval shaft.

- 4 deep parallelepipedic underground stations (Garibaldi, Durandy, Jean Médecin and Alsace Lorraine). Length of each station is about 60m for 20m width, depth varies between 21m to 25m.
- 2 sections in open cut

200meters of open cut on the western part beyond the TBM retrieval shaft of Grosso
300meters of open cut on the eastern part before the TBM launching shaft of Segurane
The project is built by Thaumasia, a joint venture lead by Bouygues Travaux Publics that includes the following contractors: Bouygues Travaux Publics Régions France, Solétanche Bachy France, Solétanche Bachy Tunnels, CSM Bessac, Snaf and Colas Midi-Méditerranée.

2.2 Geotechnical profile

The underground section is located in the alluvial plain of the Paillon River. This plain is constituted of various sedimentary formations and alluvial material. The geotechnical campaigns run before works indicated that the stratigraphy is very heterogenic and that the expected ground conditions are very variable. The tunnel started in consolidated limestone at the bottom of the castle hill, then drives through a mix face with Keuper formation (rock substratum) and clay, then encounters mix face with sand, gravel and clay, and ends through

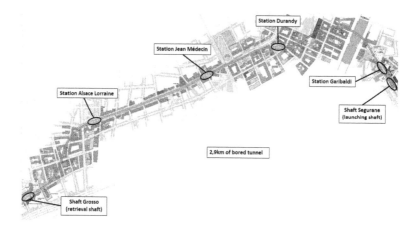

Figure 1.   General overview of the underground section.

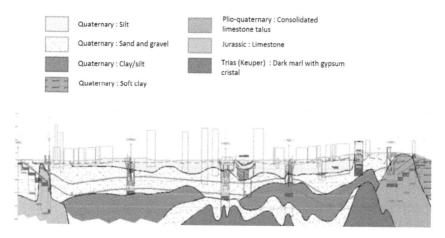

Figure 2.   Geological profile along tunnel.

limestone (rock substratum). The figure hereunder gives the main formations encountered during tunnel boring.

2.3 Tunnel structure

The tunnel is a segmental lining constituted of 8 segments, including a smaller key segment. The ring external diameter is 9.3m and its length is 1.6m. The segment current thickness is 40cm and 25cm at the pad location. Temporary bolts are set up between rings (2 per segment) and between key and counter-key segment (1 on each side). The segment formwork design includes the possibility of installing shear cones between rings (2 per segment).

The temporary tunnel parts, built inside the stations, are made of steel reinforced fiber concrete (SFRC), class 5d. The concrete grade is a C40/50.

3 MANAGEMENT AND CONTROL OF SETTLEMENTS

The tunnel is built in the very center of the city of Nice. The tunnel alignment has been chosen in order to maximize the tunnels section bored under the road avoiding the buildings. However, in several areas, the tunnel is bored below sensitive buildings. Because of high density of buildings at surface and the lack of space, the underground stations are located close to existing buildings. For these reasons, the limitation of settlements produced by excavation works on the existing buildings is a key issue. The settlement criteria under buildings is very drastic: 10mm maximum allowable settlement.

In order to cope with these constraints, the following design options have been taken:

- Use of a TBM with slurry confinement in order to be able to apply high confinement pressures in this very variable geotechnical environment
- Use of Dwall panels in T shape and ground improvement below raft level strutting Dwalls to limit the Dwall deformations
- When program allows it, boring of the tunnel through station after the lateral Dwalls are completed

3.1 Building vulnerability assessment

The vulnerability of the building impacted by the tunnel is assessed according to AFTES recommendation (French association for tunnel and underground works). This assessment deals with about 300 buildings at the vicinity of tunnel alignment and stations. Based on buildings investigation and site visits, the initial vulnerability index is assessed for every building ($\varepsilon_{initial}$). The impact of the tunnel boring is assessed thanks to finite elements modeling; taking into account the stratigraphy, soil mechanical parameters, hydraulic conditions and buildings loads. With the settlement curve (maximal settlement and slope) and the characteristics of the building (location on the settlement curves, width, type of foundation...), the impact of the tunnel boring on the building is assessed (ε_{works}). The final state of the building $\varepsilon_{final} = \varepsilon_{initial} + \varepsilon_{works}$ is used to determine if the impact on the building is acceptable or if reinforcement measures are required before tunnel boring.

3.2 Observational method

The buildings located inside the work impact area, 30m on each side of the tunnel axis and 25m around stations D-walls, were monitored. At the vicinity of the tunnel axis, topographical measures were collected in order to follow settlements. The buildings facades were equipped with target cells, which allow manual and automated measures. These measures gave information not only on settlements but also on tipping and horizontal displacements of the buildings. A measure was taken every 30minutes. This monitoring system allowed to know in real time if settlement pre-defined criteria were overpassed.

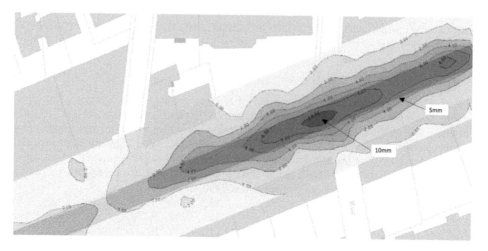

Figure 3. Monitored settlements for a tunnel section.

When the settlement criteria were about to be exceeded, that is to say when the trigger level were exceeded, mitigation measures were implemented. Settlement criteria were 10mm under buildings and sensitive infrastructures and 15mm under the road.

For the station excavation:

- Increase of the frequency of level measures
- Installattion of additional monitoring profiles (with inclinometers, settlement sensors, pore-pressure cells)
- Change of excavation staging, especially the strutting, with the possibility of adding active struts

For the tunnel boring:

- Modification of the recommended confinement pressure based on observations made in areas with similar configuration (water level, ground condition, ground cover. . .)
- Adjustment of the confinement pressure based on the real time monitored settlement

4 INTERFACE BETWEEN STATION EXCAVATION AND TUNNEL LINING

4.1 Construction sequence

4.1.1 Tunnel boring
The four underground stations are located on the tunnel alignment. The TBM had to drive through the stations while boring the tunnel. In order to optimize the construction program, the TBM start has been brought forward. Because of this early start, the TBM drove through the underground stations before excavation works were completed.

Garibaldi station is located right next to the TBM launching shaft. The TBM drove through Garibaldi station at an early stage of the station construction. Indeed, only the tympanum Dwalls panels and the ground treatment plug right below the tunnel had been realized. The lateral D-wall and the sides of the ground treatment plug (outside the tunnel area) were completed after the TBM passing through. This construction sequence allowed an early start of the TBM and thus saved time on the global program.

For Durandy and Alsace Lorraine stations, tympanum and lateral Dwalls had been built before the TBM drove through the station and excavation works had already started. In these cases, the surrounding Dwalls were acting as screens limiting the impact of the settlements induced by tunnel boring on the surrounding buildings outside the D-wall "box". This configuration is favorable as the "tunnel" settlements do not add to the "station" settlements. This is interesting in a very dense urban area, where settlement criteria are very strict. This configuration also allowed carrying out TBM maintenance operations (in confinement chamber) inside the closed box.

For Jean Médecin station, the TBM drove through it as the lateral D-walls were about to be completed. The excavation of the station had not yet started, ground and water covers were similar to outside the station.

For the four tunnel sections built inside the stations, the segments were temporary and were dismantled during station excavation. Specific segments with steel fiber reinforced concrete (SFRC) were precast for this use. This material is less expensive than classical reinforced concrete and more advantageous during tunnel lining dismantling as no reinforcement cage is be to be demolished. Moreover, with SFRC segments, the sorting of demolished material for discharge is easier.

4.1.2 Station excavation sequence

Once the D-walls were completed, the stations were excavated and the slabs were built from top to bottom. The excavation and slab sequence was decided in order to limit the D-Walls deformation and thus the impact on the surrounding structure, and also to optimize the program and be able to work safely in the tunnel during as many excavation stages as possible.

For stations Alsace Lorraine, Durandy and Jean Médecin, the slabs were built gradually from top to bottom except for the mezzanine slab that intercepted tunnel level. That slab was built after tunnel dismantling and raft construction. In such configuration, before raft completion, the station D-walls span between the technical slab above tunnel and the raft bottom of excavation (10m to 15m meters span). This configuration was made possible especially thanks to the ground improvement below raft level, acting as an underground strut. In addition to that, for Durandy, Alsace Lorraine and Jean Médecin stations that were built in especially weak ground, the D-Wall were made of T-shaped panels in order to improve the D-Wall rigidity.

Design calculations were performed to check the tunnel lining stability and to define an excavation level below which circulation inside the tunnel was not allowed anymore for safety reasons. A specific works staging was defined (see §4.2): the excavation under the technical slab started with a localized excavation down to 0.5m above tunnel crown, then localized dismantling of tunnel lining crown and then gradual excavation and lining dismantling progressing parallel to tunnel axis as represented on Figure 5.

For Garibaldi station, all the slabs were built gradually from top to bottom (included the mezzanine slab, which intercept tunnel level). As the mezzanine slab intercepts the tunnel, the

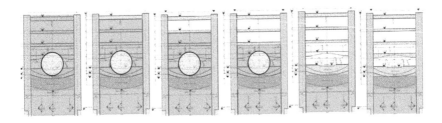

Figure 4. Construction sequence for Jean Médecin station.

Figure 5. Construction sequence under technical slab.

slab could only be cast once the tunnels lining crown was dismantled. In order to stabilize tunnel segment once the crown was dismantled and to provide sufficient stiffness of ground support to the lateral D-walls, the tunnel was filled with poor concrete. The tunnel is filled up to the level below the mezzanine slab before the tunnel crown segments are dismantled. As the excavation proceeds to reach the mezzanine slab level, the passive earth pressure that can be mobilized under this level is equivalent to the case without tunnel. The tunnel filling provides the tunnel with a stiffness that is similar to the stiffness provided by the soil in place before tunnel construction. This arrangement is required because the mezzanine slab has to be poured before excavating down to the raft level, dueto the absence of ground treatment under raft level and the greater flexibility of the D-wall (compared to the T-shape panels of others stations).

Moreover, this filling provides a flat area for concreting the mezzanine slab and no scaffolding is required to support the formwork inside the tunnel lining.

4.2   Design and mitigation measure for tunnel lining

4.2.1   Impact of lateral D-wall on tunnel lining
During excavation, especially between technical slab level and raft level, D-walls strongly mobilize passive earth pressure below excavation level, and thus induce large load on the

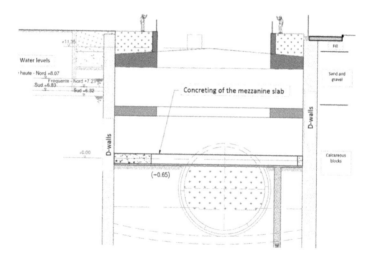

Figure 6.   Garibaldi station – concreting of the mezzanine slab.

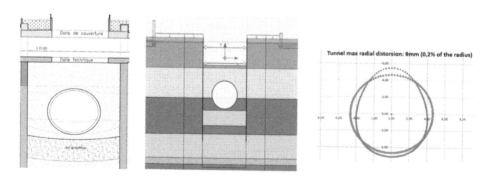

Figure 7.   Cross section perpendicular to tunnel axis - 2D finite elements model for Alsace Lorraine station.

tunnel lining. As excavation works progress, the ground cover above tunnel crown decreases and the lateral earth pressure, coming from passive earth pressure mobilized by the D-walls, increases. This leads to an increase of anisotropic loading on tunnel lining and induces lining ovalization. The acceptability of this loading for the tunnel lining, in terms of stresses and deformations, was checked using 2D soil/structure interaction models.

These models allow modeling the D-wall and tunnel structures and the whole construction sequence from D-wall construction to concreting of the last slab. From the calculation results, a minimal excavation level is defined for which stress and deformation are acceptable for tunnel lining with a reasonable safety factor. Under this level, traffic inside the tunnel is stopped and a specific longitudinal phasing is carried out. Shear cones are systematically installed between tunnel rings built inside the stations. This improves the segmental lining rigidity because the tunnel lining acts like a continuous tube regarding local loading or unloading (such as a localized excavation 0.5m above tunnel crown).

4.2.2 *Impact of tympanum D-walls on tunnel lining*

As the TBM drives through tympanum Dwall, three Dwall panels and thus the inside rebar-cages are cut in two pieces. These panels are connected to the tunnel lining. As the station excavation works progress, the cut panels deform (mostly inward the station). As long as the tunnel lining is still intact inside the station, it acts as a longitudinal strut for the tympanum panels and limits the D-wall displacement. The major part of the displacement occurs once the tunnel lining inside the station is dismantled. Because the first permanent tunnel rings are sealed in the D-wall panel, a large longitudinal displacement of the cut panels can lead to differential longitudinal displacement between adjacent rings. Consequently, between rings EPDM gaskets can decompress and lose their watertightness ability.

In order to mitigate the risk of reducing the watertightness capacity of the tunnel permanent lining, cut panels displacement were limited to 15mm. This 15mm displacement limit has been checked with calculations of 2D cross sections (parallel to tunnel axis). For some station, some corner struts were added.

In addition, reinforcing plates were installed between the five first permanent rings in order to limit the opening of the joints between rings.

4.3 *Tunnel lining monitoring*

In order to follow and control the impact of station excavation on the tunnel lining, a monitoring program is defined. This monitoring aims at following the impact of the lateral and longitudinal D-wall displacements on the tunnel and also the tunnel deformation during station

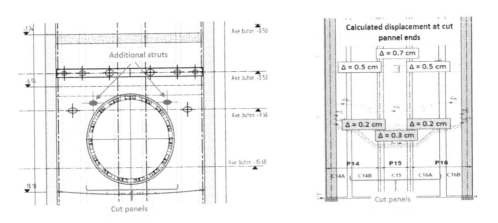

Figure 8. Tympanum D-wall front view – Predicted displacement of cut panels and additional struts.

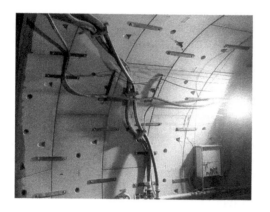
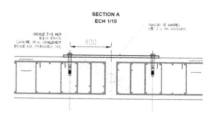

Figure 9.   Permanent rings side view – Reinforcement plate to limit joint opening.

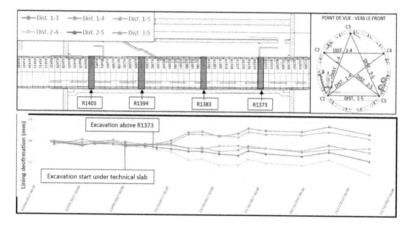

Figure 10.   Alsace Lorraine - Ring monitoring set up in order to follow lining ovalization during excavation.

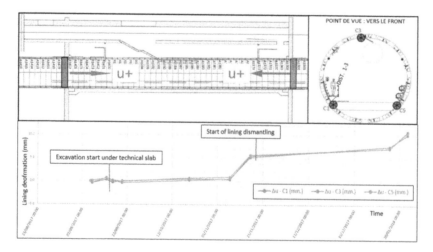

Figure 11.   Alsace Lorraine – Eastern tympanum - Ring monitoring set up in order to follow lining longitudinal displacement.

Table 1. Comparison between calculated and monitored lining ovalizations.

Calculation results

Calculation phase	Vertical ovalization	Horizontal ovalization
Excavation at -3.2 NGF	0.10%	-0.09%
Excavation 0.5m above crown level	0.11%	-0.11%
Final ovalization	0.16%/14.6 mm	-0.16%/-14.6 mm
Monitoring results		
Maximal value of ovalization*	0.11%/10.2 mm	-0.18%/-16.7 mm

* Ovalization maximal monitored on the 4 rings of Alsace Lorraine

Table 2. Comparison between calculated and monitored tympanum displacements.

Calculation results

Horizontal displacements (tunnel crown level)	Eastern tympanum	Western tympanum
Section with offset from tunnel axis	13 mm	11 mm
Section at tunnel axis	6 mm	5 mm
Monitoring results		
Maximal value of the 3 targets	2 mm	10 mm

excavation. Five targets are positioned at the lining intrados on several rings inside the stations. The distances between the targets are measured at different stages of station excavation. These measures allow estimating the tunnel lining ovalizations. The magnitude of the ovalization is related to the remaining ground cover and the intensity of horizontal earth pressure.

Three targets are positioned at the intrados of the tunnel lining on the ring sealed in the tympanum D-walls. These targets allow to monitor the longitudinal displacement (parallel to tunnel axis) of the first permanent rings (at the connection with tympanum D-walls). The major part of the longitudinal displacement occurs once the tunnel lining inside the station is dismantled.

The monitoring results in term of ovalization and longitudinal displacement were compared to calculation results. The orders of magnitude of the observed displacement were similar to what was expected. Hereunder are given the comparison for the Alsace Lorraine station.

The tympanum D-walls move toward inside the station. The longitudinal displacements of the tympanum D-wall were a bit smaller than expected.

5 CONCLUSION

At the time this article is written, the tunnel for the second tramway line in Nice is completed. Garibaldi station is currently being excavated while the other stations have been already excavated, and the internal slabs have been built.

The innovative sequencing of the works that was proposed, which consist in boring the tunnel through the stations before these were excavated and even before the D-Walls were completed, was a real asset and made it possible to reduce the overall project program by several months.

However, when adopting such a solution, one shall focus its attention on the tunnel/D-Wall interface. When the lining was dismantled, some movements of the remaining rings towards the station were observed; these displacements lead to some water leaks at the tunnel/station interface. Because these ring displacements were limited, thanks to the metallic reinforcing plates connecting the rings, the water income was limited and could be treated by injections at the interface.

Based on the results of the monitoring of the tunnel during station excavation, we can say that the preliminary assessment of the tunnel ovalization during station excavation was well correlated to the measured tunnel ovalization.

REFERENCE

AFTES recommendation – GT16R1F1 – Settlements induced by tunnelling

Tunnels and Underground Cities: Engineering and Innovation meet Archaeology,
Architecture and Art, Volume 11: Urban
Tunnels - Part 1 – Peila, Viggiani & Celestino (Eds)
© 2020 Taylor & Francis Group, London, ISBN 978-0-367-46899-6

Synthetic aperture radar interferometry applied to tunneling projects monitoring: The ATLAS processing chain

N. Devanthéry, J. González-Martí & B. Payàs
SIXENSE SATELLITE, Barcelona, Spain

ABSTRACT: One of the challenges in urban tunneling projects is to guarantee that the infrastructure assets crossing or adjacent to the tunnel alignment and other new build elements are not affected by the construction activity.

Radar Satellite Interferometry (InSAR) is a non-invasive surveying technique which provides millimetric deformation measurements of terrain structures over wide areas without any need to access site. This technique allows a comprehensive and periodic vision, with the same accuracy as manual levelling in cities for a fraction of the cost of traditional systems.

ATLAS is the Sixense's InSAR processing chain, aimed to monitor the different tunneling activities phases: access shaft excavation, tunnel construction and settlement. In this paper, various successful monitoring site examples related to different tunneling work phases will be shown.

1 INTRODUCTION

In urban tunneling projects, one of the major challenges is to guarantee that the different infrastructure assets crossing or adjacent to the tunnel alignment and other new build elements are not affected by the construction activity (in the short or long term). This can be difficult because for traditional monitoring a good deal of installation and preparation is required, often not allowing much time between the time the instrumentation is installed and commissioned and the start of the construction works, which might not give enough background data to determine and properly understand each assets' natural movement behavior.

These limitations can be overcome by the use of remote sensing techniques. The availability of historical data, the large size of the images that allow measurements over wide areas (and not only locally) and the possibility of monitoring without accessing site provide new advantages to tunneling monitoring. Radar satellite interferometry is a well-known technique which allows to derive millimetric measurements of individual terrain structures over wide areas in both urban and non-urban environments. InSAR technique is based on the exploitation of a set of radar data acquired from space. The large size of satellite images and the availability of high-resolution sensors enable to retrieve detailed information on target structures over large areas, usually both inside and outside the area of interest. For any point, satellite time series allow the follow-up of phenomena every few days during the monitoring period.

InSAR comes with no (or very limited) installation costs, minimal health and safety risks, and of course maintains a similar monitoring accuracy to other manual systems. The use of InSAR techniques provides huge cost saving whilst maintaining all the technical/design requirements, since in terms of technical capabilities InSAR is entirely comparable with precise levelling. However, where the precise levelling is carried out in one specific area, and will only provide information in that area at a rate of around 100 points monitored per day, InSAR can provide more than 20,000 measurements points per km^2 in urbanized environments.

ATLAS Interferometry chain is the Sixense's processing chain, which has been developed with the aim of monitoring geotechnical and structural deformations linked to urban construction activities. It is especially suited to monitor the different tunneling activities phases: access shaft excavation, tunnel construction and settlement. Taking advantage of the experiences in geotechnical and automatic surveying, ATLAS has been used for measuring vertical ground and structure movements, mainly related to volume loss. Volume loss control is one of the main objectives in big tunneling and excavation works on densely urbanized cities, where InSAR advantages complement more conventional approaches.

In this paper, various site examples will show monitoring of different tunneling work phases: (i) Access shaft excavation, including dewatering. Dewatering has shown on numerous occasions an impact over long distances, which are covered by ATLAS measurements due to the large size of satellite images, which also allows overcoming the problem of loss of references for the ground instruments. (ii) Tunneling phase, when satellite measurements are used to check any widening of the settlement inside and outside the planned ZOI, as a back-up and verification of the ground instruments, and (iii) Upon termination of the works, when satellite measurements provide a technically and financially efficient way of controlling long term stabilization of movements.

2 MEASURING DEFORMATION FROM SPACE

Radar satellite interferometry is a widely used technique to monitor terrain deformation over wide areas, see Bamler and Hartl (1998) and Rosen et al. (2000). It is a non-invasive technique which requires two or more Synthetic Aperture Radar (SAR) images acquired over the same area in different period of time. SAR images are complex images: they consist in the amplitude and the phase component. SAR interferometry uses the phase difference between different acquisitions, the so-call interferometric phase (Φ_{int}), to compute ground displacements measurements based on the radar waves travel time differences. The working principle is illustrated in Figure 1.

The interferometric phase consists in different components: the component due to the displacement, the topography, orbital error, atmospheric effects, and noise. If the magnitude of the deformation is much bigger than the other components the displacement can be estimated directly using the $\Delta R = \frac{4\pi}{\lambda}\emptyset_{int}$, where λ is the radar wavelength.

To obtain millimetric precision on the deformation measurements, the interferometric phase component due to the movements must be isolated, with InSAR advanced techniques. The ATLAS processing chain, developed around the core software GAMMA (Werner et al. (2003)), is based on the Persistent Scatterer Interferometry (PSI) technique (Ferretti et al. (2000 and 2001)). It requires a stack of around 18 or more SAR images to estimate the velocity of deformation, the accumulated deformation for each date and the historical time series over

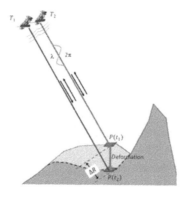

Figure 1. SAR interferometry working principle.

a set of measurements points, called Persistent Scatterers (PS). The PS are points with low level of noise and with phase stability along all the period of study. They usually are located over man-made structures, rocks or arid regions, i.e., structures which reflects well the radar signal. Its density can reach 20,000 measurements points per km^2 in urban areas.

The availability of the last generation of high resolution sensors, such as TerraSAR-X from the German Aerospace Center, and the CosmoSkyMed constellation from the Italian Spce Agency, brought improvements with respect to the medium resolution sensors. High resolution imagery, acquired with the so-called StripMap mode, offer spatial resolution of only 3x3 meters, which can be even less when using other specific acquisition modes such as Spotlight mode. When using High Resolution imagery with X band, the measurement precision of movements in the vertical axis is about 3 mm, sometimes better. The accuracy of the point positioning depends on the satellite. In High Resolution Strip Map stacks this is expected to be around 1 or 2 meters. The high resolution of the ground footprint of the images, the high sensitivity of the X-band to subtle measurements, and the large density of measurements points, makes this technique very suitable to monitor deformation on assets and infrastructures in urban construction activities, such as the activities related to tunneling works.

3 TUNNELING MONITORING

Ground movements caused by major construction projects, such as tunneling, have the potential of causing damage to overlying structures. This is even more probable when the project is done under or around heavily urbanized areas. In order for this to be measured it is necessary to understand not only the ground behavior before the start of the construction activity, but as well the structure behavior. The way that we will have this control is through monitoring techniques, and it should be carried out while the construction is being done, and after it's finished until ground settlement due to the construction activity has faced out. InSAR technique can overcome the limitations of traditional monitoring techniques by means of providing historical data measurements, wide area measurements without losing resolution over single structures, and providing measurements regularly after the end of the project.

ATLAS has been demonstrated to be useful in numerous construction works activities, in different work phase:

i. During the access shaft or station excavation phase, including any potential dewatering. The satellite measurements cover a very large area of interest, much larger than the official planned ZOI (Zone of Influence, outside which there should be no movement according to design, and outside which references are chosen for ground instruments). Dewatering activities have shown on numerous occasion an impact over long distances, sometimes very unexpectedly. ATLAS measurements easily cover such large areas and can overcome the potential problem of loss of references for the ground instruments.

ii. During tunneling, where satellite measurements are used again to check any widening of the settlement trough over and outside the planned ZOI, as a back-up and verification of the ground instruments.

iii. For litigation mitigation: by proving that areas outside the ZOI are not impacted, they are an extremely efficient tool against unjustified claims.

iv. Upon termination of the civil works, where satellite measurements provide a technically and financially efficient way of controlling long term stabilization of movements.

3.1 *Dewatering works*

Dewatering usually results in settlement as porewater pressure in the soil reduce. SAR interferometry has demonstrated to be a fundamental tool to measure the magnitude and the extend of the deformation, which can vary a lot depending on the type and properties of the soil.

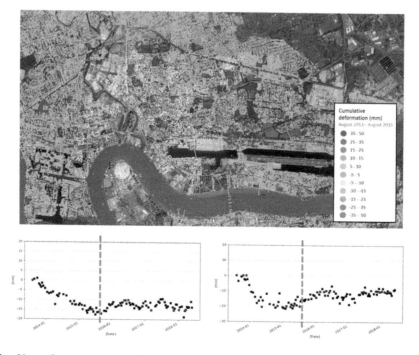

Figure 2. Upper figure: Accumulated deformation map over Limmo area (London) derived during the period August 2013 to August 2015. Stable points are represented in green, subsidence is represented in the colors ranging from yellow to red, and uplift is plotted in colors ranging from pale to dark blue. Lower figures: Deformation time series of two points located on the area affected by the dewatering.

Upper Figure 2 shows the extend and behavior of the deformation due to the dewatering in the Limmo shaft construction related to the Crossrail tunnel construction in London (Black, 2015). This accumulated deformation map and deformation time series were derived using a set of TerraSAR-X high-resolution images during the period comprised between August 2013 and August 2015. The stable areas are represented in green, while the subsidence is shown in the scale of colors from yellow to red.

The results derived with InSAR confirmed that the ZOI was at least 2.5 km with a maximum settlement up to 35 mm, much more than what was expected, reaching the opposite side of the river. Moreover, these results allowed to detect a fault in the area (left part in upper Figure 2, abrupt change of movement to stability) that affected drastically the dewatering regime on its two sides.

Lower Figure 2 shows the deformation time series over two points located on the area affected by the dewatering (orange points in the deformation map). Both time series show a strong subsidence until 2015, with different intensities and subsidence rates, and then an uplift up to 10 mm in the first one and stability on the second one, probably related to the stop of the dewatering activities. Both historical time series show the detailed behavior of the deformation, with a precision below 3 mm.

The high density of measurement points in urban areas allows the retrieval of information about the general behavior and the extent of the deformation (as shown in Figure 2), as well as the movement over single structures. Figure 3 shows the accumulated deformation map over the East India Dock Road, which spans from Burdett Road on the west, to the bridge over the River Lea on the east. This area was affected by the dewatering over Limmo area during the Crossrail tunnel construction works: see the orange and red points in Figure 3. In this case the detailed deformation behavior of the road and surrounding structures has been measured over a dense network of points, providing a perfect section and a complete analysis over individual structures.

3.2 *Tunneling construction*

One of the most important control measures during the tunneling construction phase, is the 'ground loss', i.e. the volume loss. It depends upon the depth of the tunnel, ground conditions, machine characteristics, workmanship, among other parameters. InSAR has been successfully used as a complementary source of settlement data information system, which could be considered as 'absolute' and in the same grid for the entire project, especially in big extension projects.

Figure 4 shows the deformation map over a section of the Crossrail project, the new railway line which crosses Greater London from east to west, with five tunneled sections totaling up to 42 km in length. The results were derived applying ATLAS to a set high-resolution Terra-SAR-X images, with a great density of measurement points: 821 000 points were measured with a precision of 3 mm, over an area of 1 km each side of the 42 km of tunnel alignment. The image shows deformations of more than 15 mm of subsidence fitting perfectly the area of the tunnel alignment, see colors from yellow to red in the center of the map, during the period from August 2013 to May 2018.

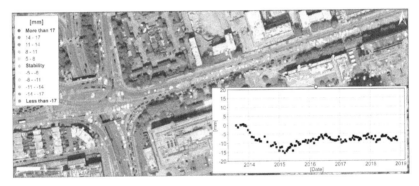

Figure 3.   Accumulated deformation map and time series over East India Dock Road (London).

Figure 4.   Accumulated deformation map over the tunnel alignment of Crossrail project, in London.

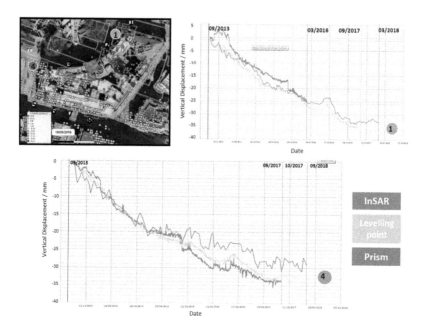

Figure 5. Upper-left figure: Accumulated deformation map over Finsbury Circus from August 2013 to June 2016, related to Crossrail project. Upper-right figure: Deformation time series over Group 1 of points. Lower figure: Deformation time series over Group 4 of points. InSAR time series are represented in pink line while manual and automatic monitoring system time series are represented in red and orange.

3.3 Long term movement stabilization

InSAR is a very suitable tool to measure the long-term movements, right after the end of the construction activity, or after many years if claims are raised in a later stage. In the case of Crossrail tunneling, ATLAS results were used to monitor the stabilization of the settlements. Figure 5 (upper left) shows the accumulated deformation map over the area of Finsbury Circus after almost 4 years of work, on June 2016. More than 25 mm in subsidence were measured (pink and red points).

The time series over the maximum deformation area are plotted in Figure 5, (pink lines) and compared against the other monitoring systems (red and orange lines). Upper right figures shows the measurements over Group 1, pink group in the top-right side of the map, and Group 4, orange group of points in the left side of the map. They show two main behaviors of stabilization: Group 4 show stabilization from the beginning of 2016, while Group 1 show that the stabilization starts in 2017. The comparison between InSAR time series and manual monitoring systems showed that the data coming from InSAR is absolutely comparable to the manual or automatic monitoring in place during the construction activity.

4 CONCLUSIONS

One of the major challenges in urban tunneling projects is to secure that the construction activity doesn't affect, in the short or long term, the different infrastructure assets crossing or adjacent to the tunnel alignment and other newly built elements.

Radar satellite interferometry (InSAR) is a non-invasive surveying technique able to measure millimetric motion of terrain structures over wide areas in both urban and non-urban environments. This technique is based on the exploitation of synthetic aperture radar images (SAR) acquired from satellite-sensors.

Sixense has developed its own treatment processing chain, ATLAS, which was developed with the goal of monitoring structural deformations linked to urban construction. In tunneling activities, it has proved to be useful as: (i) a source of information for the ground/structure behavior before the start of the construction activity, (ii) a complementary source of settlement data, which could be considered as 'absolute' and in the same grid for the entire project, especially in big extension projects, and (iii) a check for the long term movements, after the end of the construction activity, or after many years if claims are raised in a later stage.

In this paper, some examples of InSAR monitoring on different stages of the tunneling activities have been shown using high-resolution data. The accumulated deformation maps and time series related to dewatering, tunnel construction and long-term movement stabilization have been presented over a dense set of measurement points and a very high precision of 3 mm in the vertical displacement.

REFERENCES

Bamler, R., Hartl, P. 1998. Synthetic aperture radar interferometry. *Inverse Probl.14*, R1–R54

Black, M., Dodge, C., Yu, J. 2015. Crossrail project: Infrasture design and construction. *ICE Manuals*, 978-0-7277-6102-6

Ferretti, A., Prati, C., Rocca, F. 2000. Nonlinear subsidence rate estimation using permanent scatterers in differential SAR interferometry. *IEEE TGRS*, 38 (5), 2202–2212.

Ferretti, A., Prati, C., Rocca, F. (2001). Permanent scatterers in SAR interferometry. IEEE TGRS, 39 (1), 8–20.

Rosen, P.A., Hensley, S., Joughin, I.R., Li, F.K., Madsen, S.N., Rodriguez, E., Goldstein, R.M. 2000. Synthetic aperture radar interferometry. *Proc. IEEE*, 88 (3), 333–382.

Werner, C., Wegmüller, U., Strozzi, T., Wiesmann, A. 2003. Interferometric point target analysis for de formation mapping. *Proceedings of IGARSS*, Toulouse, France.

Tunnels and Underground Cities: Engineering and Innovation meet Archaeology, Architecture and Art, Volume 11: Urban Tunnels - Part 1 – Peila, Viggiani & Celestino (Eds)
© 2020 Taylor & Francis Group, London, ISBN 978-0-367-46899-6

Undercrossing live tunnels in parallel below the water table in an urban environment

V. Diwakar, T. Amsyar & N. Ramesh
Land Transport Authority, Singapore, Singapore

ABSTRACT: Singapore's Mass Rapid Transit network has been expanding rapidly as part of the country's roadmap towards developing a comprehensive public transport system. While the increased connectivity of the metro network brings about convenience to commuters, constructing additional infrastructure for the network has become increasingly challenging as a result. Contract T222 involves constructing the Outram Station and the Thomson-East Coast Line tunnels. The Outram Station will become an interchange connecting two existing lines in the central business district. The construction involves overcrossing the existing North East Line and undercrossing the East West Line (EWL) tunnels. As the new tunnels are located directly under the EWL tunnels in a parallel manner, this makes T222 a challenging project within the context of a highly dense urban district and complex geological conditions. This paper shall discuss the tunnelling methodology for the undercrossing of the EWL, challenges faced and key takeaways for future application.

1 INTRODUCTION

The Thomson-East Coast Line (TEL) is a Mass Rapid Transit line that is currently under construction in Singapore. The 43km long TEL consists of 31 new stations connecting Woodlands in the north to Marina Bay in the south which is the Central Business District (CBD) zone, and along the east coast of Singapore. The scope of work for contract T222 entails the construction of the Outram Park station and twin bored tunnels towards Havelock and Maxwell stations. The TEL involves 6 interchanges with about 5 contracts tunneling within the Railway protection zone (RPZ).

With reference to Figure 1, there are 2 launching shafts (LS1 & LS2) located within Outram Park Station. Starting from LS2, the 360m-long twin bored tunnels 1 and 2 towards Maxwell station were constructed using two Earth Pressure Balance Tunnel Boring Machines (EPB TBM). Subsequently, the TBMs were reassembled at LS1 to complete the 840m-long twin bored tunnels 3 and 4 towards Havelock station. For LS1, there was undercrossing of several infrastructures, namely the EWL tunnels, Wangz Hotel, Link Hotel, Cape Inn, SANA building and the PUB canal.

The primary focus of this paper is on the undercrossing of the EWL tunnels, which differs from previous works such as the C937 of Downtown Line Stage 3 (DTL3) under Fort Canning Hill. C937 had the alignment overcrossing the existing North East Line (NEL) rail tunnel, undercrossing North South Line (NSL) and Circle Line (CCL) for a short length and perpendicular to these existing lines (Sze et al., 2015). T222 is different because it is situated in a sensitive location, hence the alignment had to be chosen such that the new tunnels had to be built parallel to the existing live tunnels. With reference to Figure 2, the undercrossing spans over 365m, with 260 rings built within this section.

As the tunneling works of T222 were located within the RPZ, the works were required to be in compliance with the Code of Practice for Railway Protection and Development

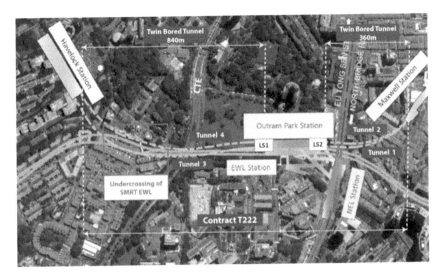

Figure 1. Tunnel drives from Outram Park Station to Havelock Station and Maxwell Station.

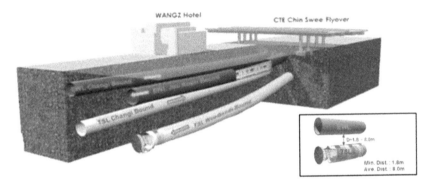

Figure 2. T222 Tunnel drives undercrossing beneath EWL Tunnels.

of Building Control Rail (DBC Rail). It is noted that the EWL line was the first MRT line to be in operation and is the oldest, hence there is little existing data available about the project.

2 GEOLOGICAL CONDITIONS

Data on geological conditions was obtained from Geotechnical Interpretative Baseline Report (GIBR) and soil investigation report by the main contractor. Figure 3 below shows the geological profile of the section where undercrossing occurs.

The geological conditions were categorized as Jurong Formation S(III) and S(IV). The Jurong Formation comprises a combination of sedimentary rocks of different strengths, ranging from stronger rocks such as cemented sandstone and siltstone, to relatively weaker rocks such as shale. The rock characteristics for S(III) and S(IV) are described in Figure 4. Groundwater was approximately 10-15m below ground level. The constructed twin bored tunnels have an overburden of more than 20m, and the vertical clearance between the tunnels and EWL tunnels is 6m.

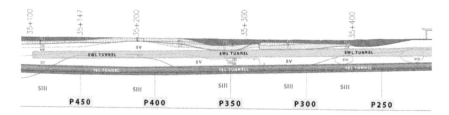

Figure 3. Geological profile of T222 Tunnel - North Bound under the EWL (green).

Grade	Characteristics
S(VI)	Bedding destroyed
S(V)	Rock weathered down to soil-like material, but bedding intact. Material slakes in water.
S(IV)	Core can be broken by hand or consists of gravel size pieces. Generally highly to very highly fractured, but majority of sample consists of lithorelics. Rock-quality designation (RQD) generally = 0, but RQD should not be used as the major guide for assessment. For siltstone, shale, sandstone, quartzite and conglomerate, the slake test can be used to differentiate between Grade V (slakes) and Grade IV (does not slake).
S(III)	Considerably weakened and discoloured, but larger pieces cannot be broken by hand. RQD is generally >0, but RQD should not be used as the major criterion for assessment.
S(II)	Slightly weakened, slight discolouration, particularly along joints.
S(I)	Intact strength, unaffected by weathering.

Figure 4. Rock Classification (taken from Singapore Standard Codes of Practice 4: 2003).

3 PRE TUNNELING WORKS

3.1 *Stakeholders*

Stakeholder management was one of the important component which ensured the successful delivery of the project. The key stakeholder for the project was Singapore Mass Rapid Transit (SMRT), the train operator owning the concession for the EWL. This was because daily train commuters might be affected if there were any disruption to the train services as a result of the construction. The risk of train service disruption was rather high due to close proximity of the tunneling works to the existing EWL tunnels. Hence, proper and timely coordination with SMRT was crucial during the undercrossing of the EWL to ensure that this risk was properly managed.

In this case, the tunnels also run beneath some of the surface structures within the Tiong Bahru and Zion Road, hence affecting the CTE (express way), Wangz Hotel, Link Hotel and SANA Building. LTA Project team had to develop a Decanting Procedure in case of emergency as both TBMs were mining between the hotel piles as seen in the below figure.

As for the EWL, an Emergency Circular (OPS) was agreed between the regulator and detailed down. This had the information pertaining to the key contact person, emergency track access slots, flow chart detailing the steps to be followed in case of any breach of levels and instance when speed restriction needs to be imposed. As for the CTE (expressway) an in

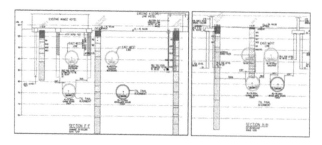

Figure 5. Undercrossing CTE viaduct (left) and Wangz Hotel (right).

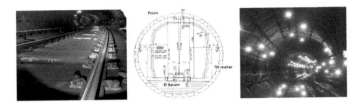

Figure 6. From left to right: Reinstated track bed, instrumentation details of live tunnel and prism layout.

depth analysis of the bearing condition and assessments were made for the CTE viaduct, as the piles toe was within 5m of the influence of the TBM (Figure 5). This was to ensure there are no structural concerns arising from the tunneling works.

3.2 Instrumentation and Coordination

Another aspect of pre-tunneling works was the installation of the instruments within the two EWL bound tunnels covering a length of 800m consisting of Prism, EL beams, Tilt and Vibration meters apart from the surface instruments. Real-time monitoring was employed so that the movement of the track bed and the structural integrity of the tunnel were monitored closely at all times (Figure 6). The key challenges during this phase was getting the approval and clearance from the authorities and operator (LTA DBC and SMRT), which took more than 4 months as the alignment was within the RPZ and had to be carefully assessed by prior to approval.

Another challenge faced was the short engineering hours when accessing the EWL tunnels. The duration of the engineering hours was only 4 hours (12am-4am), and getting the track access was subject to availability given that SMRT has their own track maintenance regime. As the tunneling works need to adhere to the CPRP. A breach of the Code can entail a train speed restriction being imposed on affected train services, or even a suspension of the train services. This was one of the key risks to be managed which involved a proper and systematic coordination during the parallel undercrossing as exposure of this particular risk was high .A detailed damage assessment was needed and it required a pre-condition survey to understand the current state of the tracks and structure. This also acted as separated document to protect the interests of the relevant authorities, contractor and operator.

The assessment showed that existing track operational levels was on the lower side as a slight movement could breach the work suspension levels for the train operation and also suspending the tunneling works. Considering this to be high risk, both LTA and SMRT had met up to come up with track replacement programme. This replacement works on completion was helpful to increase the tolerance margin for the TBM to underpass.

Close coordination between the LTA Project team, LTA DBC and SMRT was very critical during tunneling works as the procedures for track access application and safety measures had to be finalised and approved before any construction could commence. Various coordination meetings were held during the undercrossing phase. Some of the key agreements was to share the instrumentation results with SMRT on a daily basis throughout the undercrossing of the EWL tunnels. As part of a joint effort with SMRT additional track surveys were conducted on weekly basis to ensure that the EWL tunnels were structurally sound.

By maintaining a healthy professional relationship with SMRT through regular coordination meetings and site visits, the undercrossing of the EWL tunnels was executed smoothly and safely

4 TUNNELLING WORKS

This section will discuss the actual tunneling operations with reference to the TBM parameters when undercrossing the parallel sections. The first drive, TBM 3, took 158 days and the second drive, TBM 4, took 89 days. Mitigation measures and lessons learnt from the first tunnel drive had helped to understand the geology that led to reduction in time thus reducing the risk exposure under the live tunnels.

4.1 Mining Sequence and Lessons learnt

Two EPB TBM were deployed for tunneling throughout Contract T222. Due to the technical complexity of operating the EPB TBM, one key was to get familiarize with them, so as to ensure no major operating issues will arise during undercrossing. The undercrossing was in close proximity to the existing line, hence it was susceptible to major stoppages. In order to familiarise with the machines at little risk, the EPB TBM were first launched to drive towards the Maxwell Station, a relatively shorter tunnel drive. Subsequently, the EPB TBM were dismantled and reassembled at LS1 to complete the longer drive towards Havelock Station as shown in Figure 1. Some of the key specifications put in place for the cutter head was fitting them with grizzly bars to prevent damage or blockage from boulders.

The EPB TBM were also equipped with 2 pressure sensors at the crown along the front and middle shield to measure the soil collapse as shown in Figure 7. In addition, automatic face control was equipped to the TBM and shield bentonite was being injected to minimize the ground loss due to the overcut of the shield under the EWL. Two 5m^3 bentonite tanks were installed within backup as a contingency to fill the chamber if CHI had to be abandoned. As a result of these retrofitting, the risk of downtime was greatly reduced.

This had helped the team understand the mechanical issues and as well the working crew to get familiarised with the machine. There was reduction of about 50% of the total downtime by

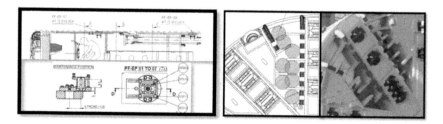

Figure 7. Diagram of 2 soil collapse sensors (left) and grizzly bars between the cutter arms (right).

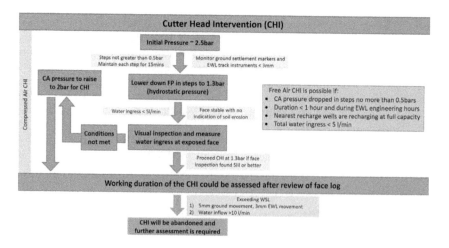

Figure 8. Flow Chart for CHI in compressed air.

this approach which in turn was one of the key to minimize the Mechanical TBM risk under the EWL.

4.2 Cutter Head Interventions (CHI) under the EWL line

One other major challenge was the cutter intervention under the EWL line. As mentioned, the geological conditions were categorized as Jurong Formation S(III) and S(IV). This formation - a combination of sandstone and siltstone – posed a challenge due to the high density.

Another challenge was managing the tunneling works where groundwater was approximately 10m below ground level. Caution had to be taken to ensure that there was no drawdown of water. Recharge wells were installed at regular intervals along the alignment as a contingency measure, and an additional set of cutter tools were stocked in place as a backup.

The flowchart above illustrates the step-by-step procedure for any unplanned CHI to mitigate the risk of settlement and water ingress, the CHIs were performed under compressed air. The initial pressure of the compressed air dive was targeted to be approximately 2.5 bar and gradually reduced after assessing the condition. The CHI duration was relatively longer, as it was completed in 6 days with a total of 25 dives. In order to reduce the duration, free air intervention was planned during non-operating hours of the live line under strict criteria. If water ingress was detected, the face pressure would be raised to 2 bar. Reducing the duration of the CHI also meant reducing the exposure time of the operation to potential risks.

The piezometers were found to be very reactive and sensitive under the EWL (Figure 9). This was of concern for the structures resting on the soft ground along the alignment. The recharge wells, installed during pre-tunneling stage as contingency measure, ensured the stability of the ground conditions to minimize the settlements.

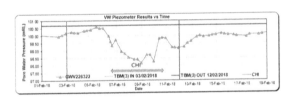

Figure 9.   Piezometer movement for TBM 3 during CHI at Ring 350.

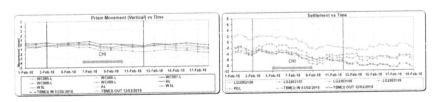

Figure 10.   Prism movement (left) and settlement markers (right) for TBM 3.

Figure 11.   Worn-out disc cutters TBM 3.

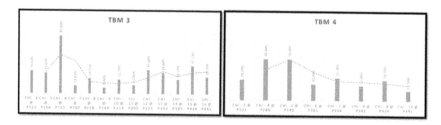

Figure 12. Comparison between TBM 3 and TBM 4 on the percentage change for tools replaced during each intervention with the ring interval.

With reference to Figure 10, it was evident that the impact on prism movement and settlement was minimal despite reactive pore pressure trends. This implied the structural stability of the tunnel. In addition, the maximum dip allowed was 0.5mm and the corresponding twist was kept below the alert level of 1:500. For straight alignment, the versine was kept below 3mm, whereas for curved alignment, the versine was kept below 4mm.

One other key observation from the tunnel drive was the tool wear trend. TBM 3 has 13 CHIs whereas TBM 4 has 8 CHIs during the undercrossing of the EWL Tunnels. The reason being a conservative approach followed for the first tunnel drive. Quick Intervention resulted in changing less % tools. This resulted in overloading of disc cutters while mining in sandstone and thus having more stoppages.

The observations of the cutter face and the EWL movements from the frequent CHI in return had proven to be advantageous for TBM 4. It is evident from the parameters were relatively more varied during TBM 3 - the first.as a result of extensive calibration to determine the appropriate ground conditions. Samples taken during the intervention were assessed and adjusted to get the right conditioning. Foam concentration was increased to have a thick concentrated bubbles reduce wear and tear of the disc cutters. Below trend plot shows CHI comparison in relation to the % tool wear, Ring interval and the numbers while undercrossing the EWL.

4.3 Tunneling parameters

The following section discuss the parameters at which the two tunnel drives were operated under the EWL and they indicate the comparison between the cutter head speed and torque

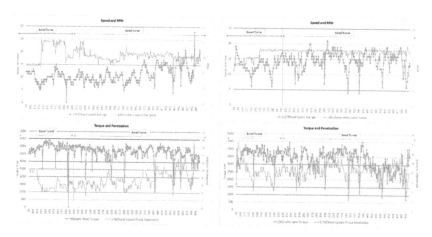

Figure 13. Parameters for TBM 3 (left column) and TBM 4 (right column).

exerted, versus the actual penetration and rounds per minute completed. In this case the torque was limited at about 4000kNm so as to avoid overheating of the cutter head motors. The rpm was adjusted between 1.5 and 2rpm to get a better penetration. These parameter was helpful to achieve better results under the TBM 4 condition and led to an increase in the penetration from 6-12mm which can be seen from the below comparison plots of the 2 drives.

It is evident that the parameters were relatively more varied during TBM 3 - the first drive compared to the other drive which was more consistent in operating the later part the torque was dropped due to a small interface of weathered silt stone. The face pressure was very well maintained within the range set and approved by the designers. An impact assessment was prepared to assess all the critical factors before arriving at the design face pressure. The challenge was to come up with the design limits considering the EWL and the surface structures sitting on soft strata.

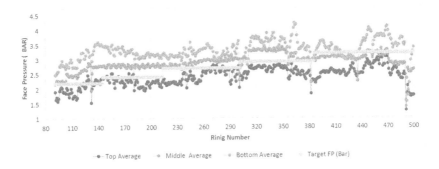

Figure 14.   Sample face pressure trend plot for TBM 4.

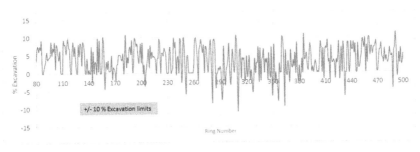

Figure 15.   Graph of Percentage (%) Excavation against Ring Number for TBM 4.

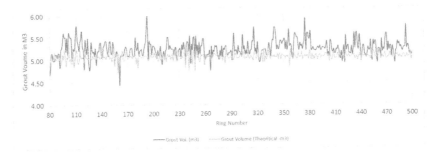

Figure 16.   Theoretical vs actual grout volume used.

Due to the shallow cover of the EWL at the initial section the face pressure was maintained slightly higher than the target/design face pressure. In addition, any over/under excavation was maintained within 10% limit and anything beyond this limit was carefully assessed in terms of density and volume (Figure 15).

Due to the close cover of the EWL line, it was ensured that the required volume of primary grout was closely monitored. Additional quantity slightly more than the theoretically required volume was injected to ensure that there were no voids. Figure 16 shows the volume intake for TBM 4.

All the above tunneling parameters was the key to control the EWL movements within the tolerance levels and complete the tunnels successfully.

5 CONCLUSIONS

The undercrossing of the EWL tunnels posed many unique challenges. Firstly, the undercrossing, stretching over 365m, was in a parallel alignment under existing live tunnels. Secondly, the construction was in close proximity to several infrastructures. Thirdly, the geology of the ground and ground water conditions were difficult. These challenges added to the complexity of the undercrossing, and presented practical implications which required effective project management throughout the entire project. The above has been elaborated in the detail in this paper.

Several key factors contributed to the successful completion of the undercrossing:

- A systematic approach was taken with regards to risk management. This entailed effective communication and close cooperation with the stakeholders- the regulators, contractor and the affected community; the earlier coordination was helpful to minimize some of the key risk due to the life of the EWL tunnel.
- Maintenance of the TBM was properly planned. This involved accurate face support pressures and ground conditioning agents from the lessons learnt from the earlier drives. The combination of the knowledge, experience and communication between the team and operators was crucial to the project
- The construction team was highly effective in analysing problems, learning from mistakes and implementing mitigation measures such as the CHI procedure which helped to manage the risk during the stoppage under the EWL. The setting up recharge wells and additional piezometer proved to be useful to assess and control the ground water table. A recommendation for future monitoring of ground water would be the inclusion of water standpipe within the zone around the piezometer, as this could help to better understand and assess the ground movements in relation to the level of water table.

The end result is an underground tunnel successfully built in difficult ground and ground water conditions. There were no major disruption to the live line, and zero loss of ground during excavation was achieved. Risks were assessed and actively managed. As a result of this project, the range of EPB TBM -related operations was effectively expanded, opening up greater opportunities for planning future tunneling projects under such conditions.

ACKNOWLEDGEMENTS

The authors wish to acknowledge the contributions from the Designers, Consultants, Contractors and other fellow colleagues involved, who made this project a success. A special mention to SMRT, who were very supportive during this critical undercrossing.

REFERENCES

O'Carroll, J.B., 2005. *A Guide to Planning, Constructing, and Supervising Earth Pressure Balance TBM Tunneling.* Parsons Brinckerhoff.

Park, C., Synn, J.H. and Lee, S.D., 2002. Estimation of net penetration rate and thrust force of a large diameter shielded TBM in soil. In *AITES-ITA Downunder 2002: 28th ITA General Assembly and World Tunnel Congress, Sydney, Australia, 2-8 March 2002: Modern Tunnels-Challenges and Solutions* (p. 494). Institution of Engineers.

Shirlaw, J.N., Hencher, S.R. and Zhao, J., 2000, November. Design and construction issues for excavation and tunnelling in some tropically weathered rocks and soils. In *ISRM International Symposium*. International Society for Rock Mechanics.

Shirlaw, J.N., Ong, J.C.W., Rosser, H.B., Tan, C.G., Osborne, N.H. and Heslop, P.E., 2003. Local settlements and sinkholes due to EPB tunnelling. *Proceedings of the Institution of Civil Engineers-Geotechnical Engineering*, 156(4).

Singapore Standard. Code of Practice for Foundations. 4: 2003, ICS 91.040; 93.020. 2003.

Sze, Y.S.E., Yee, T.C.J., Kim, I.H., Osborne, N.H., Chang, K.B. and Siew, R., 2016. Tunnelling under-crossing existing live MRT tunnels. *Tunnelling and Underground Space Technology*, 57, pp.241-256.

T222 Geotechnical Interpretative Baseline Report, September 2016.

Tunnels and Underground Cities: Engineering and Innovation meet Archaeology,
Architecture and Art, Volume 11: Urban
Tunnels - Part 1 – Peila, Viggiani & Celestino (Eds)
© 2020 Taylor & Francis Group, London, ISBN 978-0-367-46899-6

Seismic analysis of an urban subway structural system composed by shield tunnels, a cross passage and a vertical shaft for the evaluation of the structural unsafe zones

F. Durán Cárdenas
EQELS Co., Ltd. EIRL, Lima, Perú

J. Kiyono
Department of Urban Management, Division of Earthquake and Lifelines Engineering, Kyoto University, Kyoto, Japan

ABSTRACT: The seismic response analysis of a subway structural system constructed using two shield tunnels connected by a crosspassage including a shield tunnel-vertical shaft inter-section is analysed herein with the objective to identify visually the vulnerable locations of a subway structural system during a strong earthquake. Such identification is based on the evaluation of shear strains and stresses due to the torsional effects induced by the rotational motion of the vertical shaft at the tunnel-vertical shaft intersection. The zones of great vari-ation of shear strains and stresses in the compression and tension zones represent the zones of potential cracks which mean unsafe structural zones. Three models are used herein for the nonlinear finite element analysis to represent the coupling between the structural components of the subway system. The results presented herein show that the structural unsafe zones in a subway structural system are located in the vicinity of intersections.

1   INTRODUCTION

The construction of shield tunnels for subways has resulted in a good alternative to alleviate the transport problem in large cities. Subways include underground passages and under-ground stations which are crowded by people in peak hours in which the train platforms are also very congested with transit passengers. In this scenario, the occurrence of a large earth-quake may cause panic to people inside the subways and underground stations intending to rush to escape zones. In the same way, people inside the halted trains need to be evacuated safely and have opportune to exit through designated safe zones. It prompts the necessity in earthquake-prone countries to make plans for earthquake disaster prevention for subways and metros including tunnels, cross-passages connecting train platforms, ventilation shafts, and evacuation shafts. In this framework, the structural evaluation of the vulnerable and safe zones of the underground system is performed herein through a seismic analysis of a subway structural system composed by structures interconnected each to other as a whole system.

Past experiences of earthquake damage to tunnels and underground stations have shown that large ground deformations surrounding the tunnels may lead to critical structural damage and failure of subway structural components such as the collapse of central concrete columns of a box-type cross-section tunnel at the Daikai subway station in Kobe City, Japan, during the 1995 Kobe Earthquake (Mw 6.9) which was originated at a fault crossing in north-ern Awaji Island at a depth of about 16 kilometres and at distance of about 20 kilometres from Kobe City, with a peak ground acceleration of 812 gals at the surface, Durán (1995). Other earthquake damage to tunnels occurred during the 2004 Niigata ken-Chuetsu Earth-quake in Japan which caused severe damage to five mountain tunnels such as the Uonuma,

Figure 1. Spalling of concrete in the crown of Tawarayama Tunnel in Kumamoto Prefecture, Japan, Kyushu Regional Development Bureau (2016).

Myoken, Wanatsu, Tenno and Shin-Enokitoge tunnels, Yashiro & Kojima (2007). Other cases of earthquake damage occurred during the 2016 Kumamoto Earthquake of Mw 6.0 in Kumamoto Prefecture, Japan such as the considerable spalling of concrete in the crown of Tawarayama Tunnel produced by compression effects in the tunnel, Kyushu Regional Development Bureau (2016), which is showed in Figure 1.

Earthquake response of tunnels depends greatly on the ground and geological conditions as well on the type of seismic motion. Considering these factors, the seismic analysis of the subway structural system presented herein also takes into account the dynamic properties of the soil such as the shear wave velocity as well as the acceleration record of the 1995 Kobe Earthquake; which besides its long duration it has several peak variations in its frequency content.

It is also emphasised that, during strong earthquakes, the motion of vertical shafts (emergency and ventilation shafts) induce torsional effects directly the joints and connections of the segments that compose the tunnel in the vicinity of tunnel-shaft intersections and crosspassage intersections, (Durán et al, 2012), the combination of torsional moments and shear deformation at tunnel-shaft intersections may cause cracks and failure of segments that may precede the drop of blocks of concrete as well as damage to pipes attached to the tunnel, this damage put in risk, the life of passengers and the safety of trains.

2 OBJECTIVE

The objective of this paper is to identify the safety zones for evacuation and escape purposes based on the seismic response of a subway structural system composed by tunnels interconnected with crosspassages, ventilation and/or emergency shafts. Under the seismic analysis presented it is possible to analyse the structural behaviour in the vicinity of tunnel-shaft connections and tunnel-crosspassage connections determining the zones of large shear deformations and stresses which will be useful to identify the most critical zones that need additional reinforcement as well as the safe locations for evacuation purposes and for designation of safety zones for halting the trains during an strong earthquake. The identification of the safety zones for evacuation purposes is done by the examination of the shear deformation and torsional effects at the vicinity of tunnel-shaft intersections and tunnel-crosspassage intersection; it is useful to make plans for emergency routes avoiding critical zones of potential structural failure.

3 MODEL OF THE SUBWAY STRUCTURAL SYSTEM FOR THE SEISMIC ANALYSIS

The subway system analysed herein is composed by a system of two parallel precast segment lining tunnels connected each to other by a reinforced concrete crosspassage which in turn is near to a vertical shaft connected to one of the tunnels as showed in Figure 2. The vertical

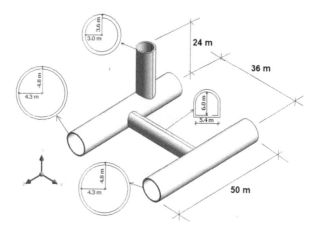

Figure 2. Tri-dimensional view of the subway structural system composed by subway tunnels, a cross-passage and a vertical shaft.

shaft represents a ventilation or emergency shaft, with a final reinforced concrete lining. Although the vertical shafts are constructed at a distance very near to shield tunnels, in this paper, for simplification of the finite element model it is assumed that a vertical shaft is not connected with an adit but the vertical shaft is connected directly to a shield tunnel, this consideration did not affect the torsional effects on the tunnels induced by the rotational motion of the vertical shaft.

3.1 Discretised Models of the Subway's coupled structural components

To perform the seismic analysis of the subway structural system shown in Figure 2, it is taken into account the construction sequence of the system by following a partial discretisation of the coupled structural components from the initial stage of construction shown in Figure 3(a), until completed underground system which is showed in Figure 3(c). The purpose of the discretisation is to evaluate the interaction effects between the structural components of the underground system. The points denoted by A, B, C and D indicated in Figure 3 represent the points where the results of the seismic response are compared. For example, for the point "A" it will be compared the results of the three models showed in the Figure. In this format, the interaction effects for each case can be evaluated. For the Finite Element models of the three cases presented, no flexible joints are considered either at the intersection of the tunnel and vertical shaft or at the tunnel with the cross-passage.

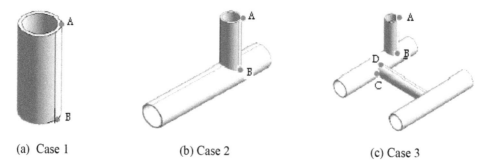

(a) Case 1   (b) Case 2   (c) Case 3

Figure 3.   Cases of sequential coupling of the structural components of the subway structural system.

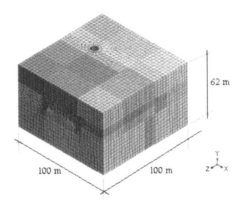

Figure 4.   Three-dimensional Finite element Model for the Subway structural system.

3.2   Three-dimensional Finite element Model for the Subway structural system

Three dimensional finite element models were used for the three model cases shown in Figure 3 by using "solid elements" for the soil and "shell elements" for the tunnels, crosspassage and the vertical shaft. The depth of the model is 62 meters below the surface as shown in Figure 4. Viscous boundaries were used at the side boundaries to avoid wave refraction at such locations. The base of the model is fixed. The shield tunnels, crosspassage and the vertical shaft are assumed with the same concrete material, thus the modulus of elasticity for all the structural components is E= 30500 MPa with Poisson Modulus $\nu = 0.2$. The ANSYS® software was used to perform the seismic analysis herein.

3.3   Geotechnical properties of the soil Stratum

The finite element model of the subway structural system analysed considered three soil stratums whose properties are indicated in Table 1. The objective of considering three soil stratums was to observe the variation of stresses and deformations of the subway system's structural components in the proximity of the division line of two soil stratums.

3.4   Input Motion

The input motion used for the seismic analysis performed herein is shown in Figure 5.

The input motion used was the acceleration record of the 1995 Kobe Earthquake which had a magnitude Mw 6.9, which was selected due to its great variation of frequencies besides of its long duration. The input motion is applied at the base of the three dimensional Finite Element Model shown in Figure 4. Although the seismic analysis was performed considering 50 seconds of duration, only time-history responses of 10 seconds of duration are shown in this paper due to limitations of printed space.

Table 1.   Geotechnical properties of the soil stratums.

Thickness of Soil stratum	Soil depth	Poisson Modulus	Friction Angle of Soil	Soil Cohesion C	Elasticity Modulus	Shear Wave Velocity
m.	m.			kPa	MPa	m/sec
17.0	2.5	0.4	35°	3.0	500	150
15.0	3.0	0.2	35°	3.0	700	180
30.0	3.0	0.2	35°	3.0	1000	210

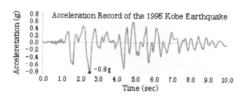

Figure 5.   Input motion used for the seismic analysis of the subway structural system.

4   SEISMIC ANALYSIS RESULTS

The seismic analysis of the subway structural system was performed without considering flexible joints; it is done recognising the importance of evaluating the stresses and strains at intersection joints of subway structural systems without flexible joints to assess the level of strains to take into account in the design of flexible joints. The displacements, shear strains and stresses obtained from the seismic analysis will be evaluated at the critical locations such as points "B", "C" and "D" for the three cases of subway structural systems shown in Figure 3.

4.1   Comparison of the seismic displacements at the intersections of the coupled structural components for each case-model analysed

The results for the lateral displacements of the four different cases of the subway structural system are shown in Figure 6. The location at top of the vertical shaft is denoted by the letter "A".

It can be seen that the maximum displacement at point "A" respect to the base of the vertical shaft is 0.0126 m. Also, it can be observed from Figure 6(b) that the largest lateral displacement at point "A" occurs when the vertical shaft is coupled with the tunnel.

From Figure 6, it can also be observed that at the Vertical shaft-Shield tunnel intersection denoted by point "B", the lateral displacement reach its maximum value of 0.2297 m which corresponds to the complete subway structural system with all the structural components interconnected. In similar form, the lateral displacement at location "C" corresponding to the shield tunnel-cross passage intersection increases according the addition of more structural components to the subway structural system as can be seen in Figures 6(b), and (c) in which the maximum lateral displacement at location "C" of 0.2182 m is reached for the whole subway system fully interconnected.

At the left border of the tunnel-cross passage intersection denoted by letter "D" the maximum lateral displacement does not occur in any of the three cases when the structural components of the subway structural system are coupled.

4.2   Comparison of the seismic shear strains at the intersections of the coupled subway structural models in the vertical plane parallel to the cross-section of the tunnels

The variations of shear strains in the whole subway structural system will be the cause of shear cracks in the crosspassages and at the entrance or exit locations of emergency or ventilation shafts.

During an strong earthquake the torsional effects produced by the rotational motion of the vertical shaft in the vertical plane XY parallel to the cross-section of the tunnels induces variations of shear strains to the other intersections of the subway structural system as shown in Figure 7(a), these variations of shear strains also extend to the whole subway structural system as it is shown in Figure 7(b). In this way, it is possible to visualise the extent of the most safe distance from emergency or ventilation shafts to cross passages in order to minimise the torsional effects of the vertical shafts on cross-passages.

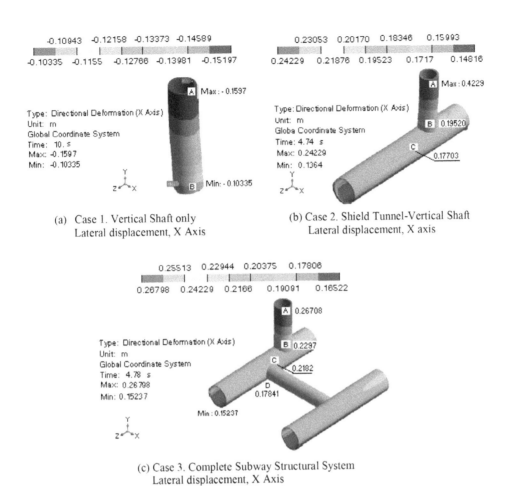

(a) Case 1. Vertical Shaft only
Lateral displacement, X Axis

(b) Case 2. Shield Tunnel-Vertical Shaft
Lateral displacement, X axis

(c) Case 3. Complete Subway Structural System
Lateral displacement, X Axis

Figure 6.   Lateral displacements of the Subway structural system for the 1995 Kobe Earthquake.

The zones in the subway structural system in which the variation of shear strains occur extensively could be designated as "critical zones" because it is a zone where shear cracks may occur during an strong earthquake and may represent an unsafe zone for people.

Also, from Figure 7(a) it can be observed that for the structural system composed only by a vertical shaft coupled to a shield tunnel, the shear strains do not exhibit a great variation.

4.3   Comparison of the seismic shear strains at the intersections of the coupled structural subway structural models in the horizontal plane

It can be observed from Figure 8(a) that shear strains in the horizontal plane do not exhibit large variations when there are intersections along one tunnel only. However, when a tunnel is connected to a vertical shaft and also to a cross-passage such as the case showed in Figure 8(b), important variations of shear strains with opposed signs occur in the subway structural system.

The rotational motion of the vertical shaft at the tunnel-shaft intersection in the vertical plane parallel the transverse and longitudinal directions of the tunnels will produce shear strains in the subway structural system as shown in Figure 8(b) which in turn will be the source of the variation of shear strains in the horizontal plane parallel to the longitudinal axis of the shield tunnels as shown in Figure 8(c).

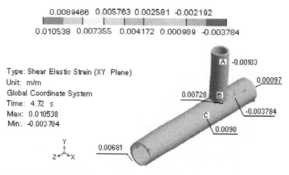

0.0089466 0.005763 0.002581 -0.002192

0.010538 0.007355 0.004172 0.000989 -0.003784

Type: Shear Elastic Strain (XY Plane)
Unit: m/m
Global Coordinate System
Time: 4.72 s
Max: 0.010538
Min: -0.003784

(a) Case 2. Shield Tunnel - Vertical Shaft
     Shear strains in the transversal plane of the tunnel

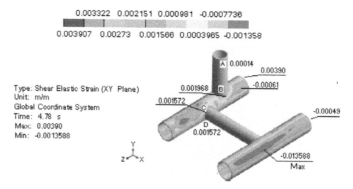

0.003322 0.002151 0.000981 -0.0007736

0.003907 0.00273 0.001566 0.0003985 -0.001358

Type: Shear Elastic Strain (XY Plane)
Unit: m/m
Global Coordinate System
Time: 4.78 s
Max: 0.00390
Min: -0.0013588

(b) Case 3. Complete Subway Structural System
    Shear strains in the transversal plane of the parallel tunnels

Figure 7. Shear strains in the vertical plane parallel the cross-section of the tunnels due to seismic effects.

4.4 Comparison of the normal stresses at the intersections of the coupled structural components for each case-model analysed

On the way to identifying the unsafe zones of subway structural systems in earthquake-prone cities, it is also very important to evaluate the stresses seismic response in the structural components of a subway. The evaluation of shear stresses in the subway structural systems is directly related to the evaluation of shear strains presented in Section 3. Thus, in this section the focus is on the evaluation of normal stresses produced in the components of the subway structural system analysed. From Figures 9(a), (b), (c) and (d) it can be seen that critical locations of normal stresses due to seismic effects are at the intersection of the shield tunnel with the vertical shaft; this result also shows that normal stresses can not be neglected in subway structural systems with the presence of vertical shafts. It can be also noticed that the normal stresses in the directions of the vertical and longitudinal axes of the tunnels reach maximum values at the intersection of the shield tunnel with the vertical shaft and also at the tunnel-cross passage intersection.

From Figures 9(c) and (d) it can be observed that the sign of the normal stresses shifts from positive to negative and viceversa at the opposed faces of the tunnels, crosspassage and vertical shaft. This implies that during a strong earthquake opposed faces of the same tunnel change from tension to compression and viceversa. This conclusion is important for strengthening the intersections of the tunnel with a vertical shaft and the tunnel with a crosspassage.

The evaluation of normal stresses at the intersection zones of shield tunnels will be also useful to gain insight about the stress levels across segment joints in the radial and longitudinal directions of the shield tunnel, based on these results, it will be possible to assess the

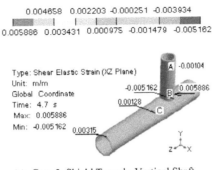

(a) Case 2. Shield Tunnel - Vertical Shaft
Shear strains in the horizontal plane

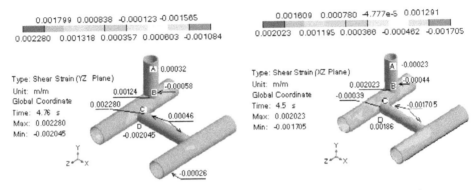

(b) Case 3. Complete subway structural system
Shear strain in the vertical plane YZ

(c) Case 3. Complete subway structural system
Shear strains in the horizontal plane

Figure 8. Shear strains in the subway system in the horizontal and vertical planes due to seismic effects.

normal forces at segment joints produced by the complex interaction mechanisms between tunnels, crosspassages and vertical shafts of a subway structural system such as the cases analysed herein.

4.5 Evaluation of Structural Unsafe zones in a subway structural system

The evaluation of structural unsafe zones in the subway structural system is presented according the earthquake response of cases of coupled components shown in Figure 3. In this way, the following cases were evaluated:

– Case 1. Tunnel-Vertical shaft system
– Case 2. Vertical shaft -Tunnel system
– Case 3 Vertical shaft -Tunnel -Crosspassage system

4.5.1 Evaluation of Structural Unsafe zones in a tunnel-vertical shaft system

During a strong earthquake, the coupled motion of a vertical shaft-tunnel system such as the case 2 shown in Figure 6(b) shows that the lateral displacement at top of the vertical shaft is the largest in this type of coupled system, thus, non-structural components such as stairs, equipment and pipes attached at top of Emergency or Ventilation shafts require more care in the fixing details. Another unsafe zone in a tunnel-vertical shaft system is located at the intersection of the tunnel with the vertical shaft in which occurs important variations of shear strains in the vertical plane parallel to the cross-section of the tunnel as shown in Figure 7(a) as well as maximum normal stresses in the vertical axis of the subway system as shown in

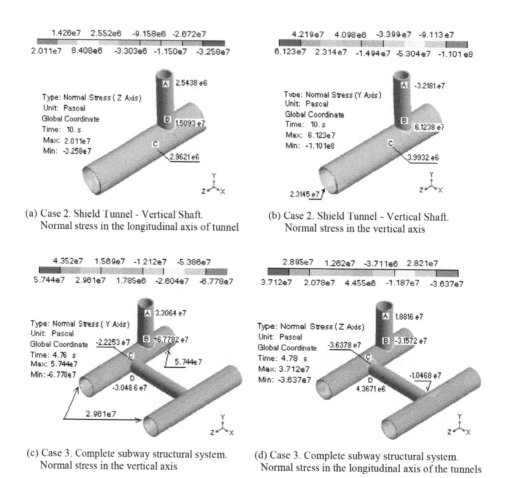

(a) Case 2. Shield Tunnel - Vertical Shaft.
Normal stress in the longitudinal axis of tunnel

(b) Case 2. Shield Tunnel - Vertical Shaft.
Normal stress in the vertical axis

(c) Case 3. Complete subway structural system.
Normal stress in the vertical axis

(d) Case 3. Complete subway structural system.
Normal stress in the longitudinal axis of the tunnels

Figure 9.   Normal stresses in the subway structural system due to seismic effects.

Figure 9(b). However, the variations of shear strains and normal stresses do not extend along the tunnel, by contrary these are localised at the tunnel-vertical shaft intersection in which it is required a proper connection such a flexible joint to ensure a safe earthquake response at this zone as well as to prevent inundation.

4.5.2 Evaluation of Structural Unsafe zones in a Vertical shaft-Tunnel-Cross passage system

The most critical case among the earthquake response of the subway structural systems analysed herein corresponds to the case of two parallel tunnels connected by a cross-passage including the presence of a vertical shaft intersecting one of the shield tunnels as shown in Figure 3(c). In this type of subway structural system variations of shear strains occur extensively as well as localised shear strains. Regarding to the earthquake response of the two parallel tunnels, it can be observed from Figure 7(b) that, maximum shear strains in the tunnels occur extensively near the intersections with crosspassages and emergency or ventilation shafts; this implies zones of potential shear cracks during strong earthquakes. Near theses zones, it would not be recommended to construct train platforms.

When examining the shear strains produced by the strong motion in the crosspassage, it can be noticed that, the crosspassage exhibits maximum shear strains in the crown of the crosspassage at the intersection zone with the tunnel as shown in Figure 8(b), as well as, maximum shear strains spreading along the wall of the crosspassage as shown in Figure 8(c). This implies that crosspassages may become structurally unsafe zones when they are located in the area of an emergency or ventilation shaft connected to a shield tunnel. In these zones, concentration of people should be prevented during strong earthquakes. From the viewpoint of the structural safety of the crosspassages it is recommended to construct them far enough from emergency or ventilation shafts connected to the tunnels.

4.6 Conclusions

Within this paper, it has been shown that the determination of the unsafe zones of a subway structural system in cities prone to strong earthquakes can be modelled appropriately through a complete analysis of the subway structural system considering the tunnel-emergency or ventilation shaft intersections as well as the tunnel-crosspassage intersections. The purpose of this paper was to identify and evaluate the unsafe zones of a subway structural system in earthquake-prone cities from the view point of its structural response. Such unsafe zones in the subway structural system were examined by means of a visual evaluation of the lateral displacements, the variation of shear strains and the variation of normal stresses obtained by a seismic finite element analysis of three cases of a subway structural system. The following are made:

- By carrying out a seismic analysis similar to the one presented herein, it is possible to evaluate the influence of torsional effects produced by the rotational motion of a vertical shaft at intersections of shafts, crosspassages and tunnels, the zones of high variation of shear strains can be used to designate the "unsafe zones" or critical zones.
- The identification of zones with great variation of shear strains in the subway structural system can be used to identify the zones that require additional strengthening in order to reduce the shear cracking induced by the vertical rotational motion of the vertical shafts.
- The identification of the shear crack-prone zones between the intersections of tunnels, crosspassages and emergency or ventilation shafts will provide a method to designate the unsafe zones for people in the area of tunnel-shaft intersections and nearby crosspassages.
- In a vertical shaft-tunnel-cross-passage system such as the one shown in Figure 7(b), the structural unsafe zone is located at the crown of the crosspassage cross-section in the intersection zone with the tunnel in which concentration of people should be prevented during strong earthquakes. For example, emergency exits should be away from these locations.
- In subway structural systems composed by two parallel tunnels connected by a crosspassage, including the presence of a vertical shaft intersecting one of the shield tunnels, it is recommended to construct the cross-passages far enough from the presence of emergency or ventilation shafts connected to shield tunnels.

REFERENCES

ANSYS® 15, *Help System, ANSYS, Inc.*

Durán, F., Kiyono, J., Maruo, Y., & Tsunei, T. 2012. Seismic Response Analysis of a Shield Tunnel Connected to a Vertical Shaft, *15th World Conference on Earthquake Engineering*: CD-ROM, Lisbon.

Japan Society of Civil Engineers, JSCE. 1999. Report on Actual Damage Caused by the Hanshin Awaji Earthquake Disaster: 231–246.

Kyushu Regional Development Bureau. 2016. Damage to the Tawarayama Tunnel in Kumamoto Prefecture, Japan, http://www.qsr.mlit.go.jp/n-kisyahappyou/h28/data_file/1481589087.pdf, (in Japanese).

Yashiro, K. & Kojima, R. 2007. Historical Earthquake Damage to Tunnels in Japan and Case Studies of Railway Tunnels in the 2004 Niigata ken-Chuetsu Earthquake, Japan: *QR of RTRI*, 48(3).

Tunnels and Underground Cities: Engineering and Innovation meet Archaeology,
Architecture and Art, Volume 11: Urban
Tunnels - Part 1 – Peila, Viggiani & Celestino (Eds)
© 2020 Taylor & Francis Group, London, ISBN 978-0-367-46899-6

Design and construction of N1 Shaft for Dudullu-Bostanci Metro Line

F. Efe, M. Kucukoglu & H. Nurnur
EMAY International Engineering and Consultancy Inc., Istanbul, Turkey

ABSTRACT: At Dudullu Station of Dudullu-Bostanci Metro Line, it was decided to construct a shaft as the fastest way to start the excavation of the station tunnels. The shaft structure was constructed with pile and conventional system that is approximately 19 meters in diameter and 42 meters in depth. Two P Type platform tunnel, one B3 Type escalator tunnel and one B5 Type connection tunnel were excavated from the end of the shaft. The shaft and the station tunnels were excavated in Filling, Silty Clay and Limestone. Because of the existing school and cut and cover structure of the line, careful measurements were taken during shaft excavation. In this paper, the main steps of the shaft design and construction will be explained and the measurements made at the site during the construction will be shown. These monitoring data will be compared with the numerical analysis results obtained during the design phase.

1 PROJECT BACKGROUND

The Dudullu-Bostanci Metro Line, planned to open in 2019, will be 14 kilometers long. The journey will take 21 minutes with the Dudullu-Bostanci Metro Line, which will have 13 stations. With the connection to be made at Kozyatagi Station, transfer to Kadıkoy-Pendik Metro Line will be provided.

This metro line starts from Bostanci Station and ends at Depo Station and extends in north-south direction by intersecting with Kozyatagi Station (with M4 Line) and Dudullu Station (with M5 Line). The route of the metro line is shown in Figure 1.

Dudullu Station is one of 13 stations. Dudullu Station is a drilling tunnel type station that will work integrated with Uskudar-Cekmekoy Metro line. The general layout of the tunnel is shown in Figure 2. The construction work of the station is carrying out with a shaft has been built from the center and there are platform tunnels, platform connection tunnels and one escalator tunnel which will be used as transfer between two lines.

Dudullu Station is a tunnel type station consisting of N1 shaft and P1, B2, B3, B5 type tunnels. In order to perform the excavation of this station, the permanently designed N1 shaft had been opened first and then the construction of platform and connection tunnels had been done. However, when considering the geological formation and surface structures of the station area, it is necessary to determine the stages of this production. In this context, the excavation steps obtained as a result of the analyses are shown in Figure 3.

The purpose of constructing the N1 Shaft is to speed up tunnel excavations of the station. It had been decided to construct the shaft at the center of Dudullu Station. After the excavation of the shaft structure, tunnel excavations have begun.

The N1 Shaft at Dudullu Station had been constructed at the position corresponding to 11+960 km of the left tube and 11+974 km of the right tube. There is a secondary school right next to the shaft construction site. During the software analyses and construction, this school was taken into consideration and has been kept under constant observation.

Figure 1. Dudullu-Bostanci Metro Route.

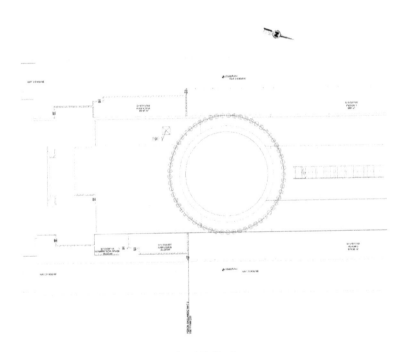

Figure 2. General layout of the Dudullu Station N1 Shaft.

After the excavation process of the shaft, 3 tunnels had been excavated, including 2 P1 type Platform Tunnels and 1 B5 type Connection Tunnel. The B3 type Escalator Tunnel will be excavated next. The 3D analysis method was chosen for the design because it is the method that will best take into account the effects of the complex structure of the project. Stability analysis had been performed by the PLAXIS 3D software using finite element method. Thus,

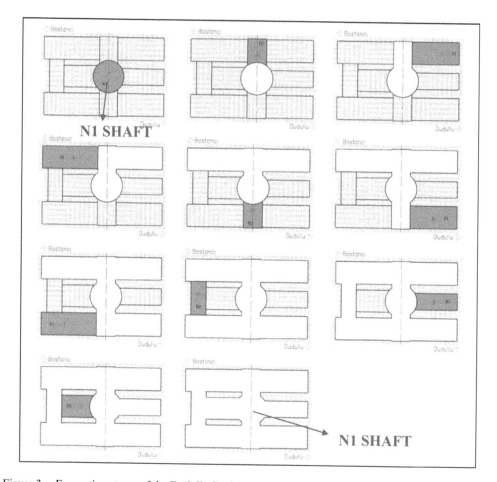

Figure 3. Excavation stages of the Dudullu Station.

the deformations that could be occur in the shaft structure and all the forces acting on it can be calculated as close to the truth.

This article focuses on the design of the Dudullu Station N1 Shaft and summarizes the points to be considered during the design phase and provides detailed information on numerical analysis.

2 GEOTECHNICAL EVALUATIONS

2.1 *Soil properties and design parameters*

The GMD11-01 drilling had been chosen and examined for Dudullu Station project. The geological units seen in the excavation area of N1 Shaft are clay and limestone. As a result of the Atterberg limit tests carried out on the samples taken from the drilling, the plasticity index value for the "silty clay" soil unit was determined as 18% on average. The Young's modulus value was obtained from the pressuremeter test.

The soil parameters determined by taking into account the drilling studies, test results and the graphics and values in the literature are summarized in Table 1. The values found during the determination of the strength parameters have been used by reducing to remain on the safe side.

Table 1. Soil parameters.

Lithology	SPT N_{av}	PI_{av} (%)	ϕ' (°)	c' (kPa)	E_s (kPa)	ν	γ (kN/m3)
Silty Clay	46	18	30	0	40,000	0.35	20

2.2 Properties of rock parameters

2.2.1 Rock mass classification systems

Rock mass classification systems are fundamental to empirical design approaches and are widely used in engineering applications. These systems should not be considered as a tool that can provide engineering design alone. These classifications should be used with observation-based and analytical solution methods considering design goals and field geology in order to be able to formulate the final design. These methods usually try to determine the quality of the rock by using the discontinuity properties of the rock, the rock strength, the state of water, the state of disintegration and the stress state around the underground opening. In Turkey mostly RMR and Q classification systems are preferred at tunnel designs. For the Dudullu Shaft, rock quality classifications were also made using Q and RMR methods. The RMR classification results can be seen from Table 2 and Table 3.

2.2.2 Hoek-Brown Method

The Hoek-Brown failure criterion is an empirical criterion, based on the experimental and theoretical interaction of rock-related fracture mechanics.

GSI (Geological Strength Index) values are needed to determine Hoek-Brown strength parameters in rock masses. GSI was first used by Hoek in 1994. GSI values do not aim to make rock mass classification exactly, but give rock quality.

In the method of the 1989 version of the RMR system;

$$GSI = RMR - 5 \tag{1}$$

Correlation will be used to determine the GSI value.

The rock mass characteristics used to find the material parameters according to the Hoek-Brown method are shown in Table 4.

Using the values given in the table above, the geomechanical parameters of the Limestone were calculated. The results are shown in the Table 5.

Table 2. RMR classification results of the Dudullu Shaft.

Lithology	Strength of intact rock	RQD	Spacing of discontinuities	Condition of discontinuities	Groundwater	RMR_{basic}
Limestone	4 (33 MPa)	13 (%71)	8	15	10	50

Table 3. RMR final results of the Dudullu Shaft.

Lithology	RMR_{basic}	Adjustment for discontinuity orientations	RMR_{final}
Limestone	50	-5	45

Table 4. Rock mass characteristics.

Lithology	σ_{lab} (MPa)	GSI	m_i	D	MR	γ (kN/m³)	h (m)
Limestone	33	40	10	0.8	400	26	40

Table 5. Geomechanical parameters of the Limestone.

Lithology	mb	s	a	c (MPa)	ϕ (°)	E (MPa)
Limestone	0.281	0.0001	0.511	0.141	37.96	673.51

3 DEVELOPMENT OF THE FINITE ELEMENT MODEL FOR SHAFT

3.1 Geometry and construction sequence

Dudullu-Bostanci metro line Dudullu station will be integrated with the existing Uskudar-Cekmekoy metro line Dudullu station. It had been found suitable to build a shaft at the center of the planned station so that new excavation works would not affect the existing station. It had been decided to continue the project with tunnels to be excavated through this shaft structure. The N1 Shaft structure can be seen in Figure 4.

The N1 Shaft is designed as a pile system measuring 19 meters in diameter and 42 meters in depth. According to the geotechnical evaluations, it has been determined that the shaft structure could be constructed with 29 m long 80 cm diameter piles and 16 m long conventional excavation. It has been decided that these piles will be constructed 100 cm apart and ring beams will be built every 3 meters in the vertical direction. The first 7 rows of beams were designed as 55x75 cm and the last row of beams are dimensioned to be 75x120 cm due to tunnel excavations to be made below. Conventional excavation has been carried out using 30 cm of shotcrete till the end of the shaft. Also with 50 cm intervals HE200B steel beams and

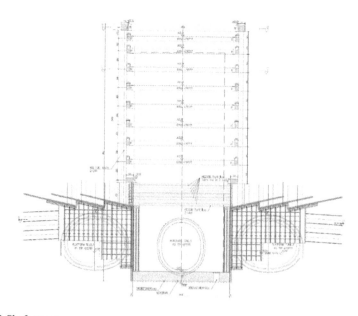

Figure 4. N1 Shaft structure.

Figure 5. Excavations of N1 Shaft and the tunnels.

soil nails up to the upper limit of B2 type tunnels have been used for the construction. After the completion of the shaft excavation, the tunnel excavation process had been started.

3.2 Model description

In the PLAXIS 3D analysis model, embedded beam material for piles, beam material for ring beams and steel beams, plate material for shotcrete had been used. The analysis model can be seen from Figure 6.

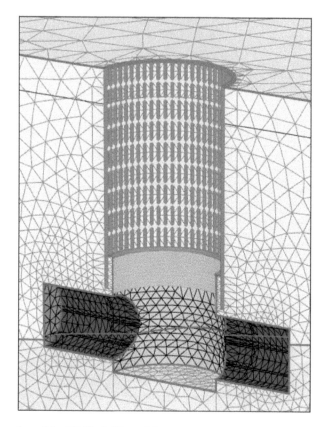

Figure 6.   Cross section of the N1 Shaft 3D model.

3.3   *Results of numerical analysis*

The details of the results of the PLAXIS 3D analysis using finite element method are given below. When the results are examined, it can be seen that the deformations are stabile during

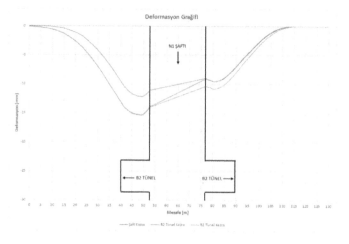

Figure 7.   Diagram of vertical deformation after excavation of shaft and B2 tunnels.

Figure 8.   Chart of measurement results in the field.

Table 6.   Maximum values of the measurements in the field.

Date	Y(m)	X(m)	Elevation(m)	Δ Elevation(m)	Δ X(m)	Δ Y(m)
24.10.2017	429,647.4270	4,542,530.1570	149.7870	-21.00	-17.00	10.00
31.10.2017	429,647.4260	4,542,530.1580	149.7820	-26.00	-18.00	11.00
06.11.2017	429,647.4240	4,542,530.1610	149.7810	-27.00	-20.00	14.00
13.11.2017	429,647.4230	4,542,530.1580	149.7820	-26.00	-21.00	11.00
20.11.2017	429,647.4230	4,542,530.1620	149.7800	-28.00	-21.00	15.00
27.11.2017	429,647.4240	4,542,530.1600	149.7750	-33.00	-20.00	13.00

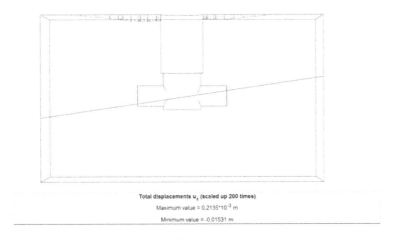

Figure 9.   Surface deformations during the shaft excavation.

the shaft excavation. It is observed that deformation increases when tunnel excavation starts after shaft construction.

As the depth of the shaft excavation increases, the deformations and forces affecting the piles increase. The deformation graph generated as a result of the analysis is shown in Figure 7.

The shaft construction had been continued by comparing the PLAXIS 3D results with the field measurements. In case of any incompatibility or problematic situation, it has been notified that the construction should be stopped and get support from the designers. Measurements in the field are carried out regularly and presented to the project engineer. The data obtained from the measurements on the field are shown in Figure 8. Some of the values from this chart in the last month are summarized in Table 6.

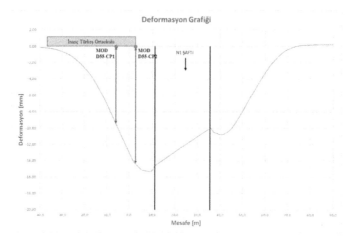

Figure 10. Deformation graph formed under the school building during shaft excavation.

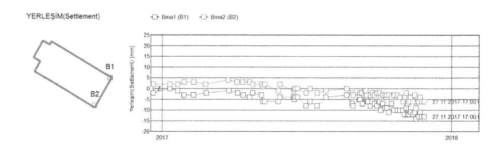

Figure 11. Chart of measurement results on the buildings.

Table 7. Maximum values of the measurements on the buildings.

Date	Y(m)	X(m)	Elevation(m)	Δ Elevation(m)	Δ X(m)	Δ Y(m)
24.10.2017	429,605.2420	4,542,550.9660	164.7980	-10.00	13.00	-3.00
31.10.2017	429,605.2440	4,542,550.9680	164.7980	-10.00	15.00	-1.00
06.11.2017	429,605.2440	4,542,550.9670	164.7960	-12.00	15.00	-2.00
13.11.2017	429,605.2460	4,542,550.9660	164.7960	-12.00	17.00	-3.00
20.11.2017	429,605.2450	4,542,550.9670	164.7950	-13.00	16.00	-2.00
27.11.2017	429,605.2460	4,542,550.9670	164.7950	-13.00	17.00	-2.00

Figure 9 shows the vertical displacements at the surface during the shaft excavation. Deformation graph formed under the school building during shaft excavation can be seen from Figure 10. The data obtained from the measurements under the school are shown in Figure 11. The maximum values from this chart in the last month are summarized in Table 7.

4   CONCLUSIONS

In this paper, 3D analysis was performed using the parameters found from geotechnical methods and it was determined that there would be no problems if the specified supports were used. The results of the analysis were compared with the results of field measurements. This comparison does not reveal any significant difference. The results also show that there is not any risk for the school building either.

REFERENCES

EMAY International Engineering and Consultancy Inc. 2017. *Design Report of Dudullu Station N1 Shaft.*
General Directorate of Highways. 2013. *Highway Technical Specifications.*
PLAXIS BV. 2017. *PLAXIS 3D User Manuals.*
Turkish Standards Institution. 2000. *TS 500 – Requirements for Design and Construction of Reinforced Concrete Structures.*
U.S. Department of Transportation Federal Highway Administration. 2009. *Technical Manual for Design and Construction of Road Tunnels – Civil Elements.*

Tunnels and Underground Cities: Engineering and Innovation meet Archaeology,
Architecture and Art, Volume 11: Urban
Tunnels - Part 1 – Peila, Viggiani & Celestino (Eds)
© 2020 Taylor & Francis Group, London, ISBN 978-0-367-46899-6

Technical solutions when designing the new bus terminal in Slussen, Stockholm, Sweden

A. Ehlis
WSP, Luleå, Sweden

ABSTRACT: The old bus terminal at Slussen in central Stockholm has been an eyesore for the area for decades. Plans to replace the terminal have finally come to fruition, but because it is located in such an urban area, there has been a shortage of space. So where to build the new bus terminal? Underground of course! But the underground space there is already in use. The existing metro and other facilities are directly adjacent to the planned terminal with very low rock cover. The newly designed departure hall for the bus terminal will pass through an abandoned emergency shelter cavern and the ticket hall will replace an old parking garage. The old parking garage is a bedrock cavern and additional rock excavation is needed in existing caverns to accommodate elevators, escalators and information boards. Unique designs have been applied to solve the challenges in each of these cases.

1 INTRODUCTION

Slussen lies in central Stockholm. It is an urban area with complex infrastructure, bridges for cars, cyclists and pedestrians, trains, metro, ferries and housing all have a place in Slussen. There is also an old bus terminal that is heavily used. Many of Slussens facilities have reached their technical life span and need replacing. One of these is the old bus terminal. Therefore a new bus terminal is being designed in bedrock. In Figure 1 the area of Slussen can be seen.

But the bedrock is not unused; there is an old parking garage, a Metro station and an emergency shelter cavern already excavated there.

2 THE NEW BUS TERMINAL

The buses will enter the new terminal from an existing road along the waterfront via an entrance ramp and a 29 meter wide portal into the existing rock cliff that runs along the road. After passing through the entrance ramp they will reach the arrival hall. In the western part of the terminal the buses will turn underneath the entrance hall. And from there the buses will enter the departure hall before again passing through the entrance ramp to head out onto the streets of Stockholm. Between the arrival hall and the departure hall the biggest hall is located, the waiting hall. Figure 2 shows the three main halls, the arrival hall, the departure hall and the waiting hall. These main halls are 200 meters long and have spans of up to 24 meters.

The entrance hall is located in the north-western part of the terminal; it connects the new terminal with the metro station of Slussen and will be the main entrance for all passengers. The three main halls are connected with 18 cross tunnels.

Rock pillars will remain between the cross tunnels and the main halls as the load bearing system of the terminal. These have been carefully designed using analytical and 3D numerical methods.

Figure 1. overview from Slussen in 2009.

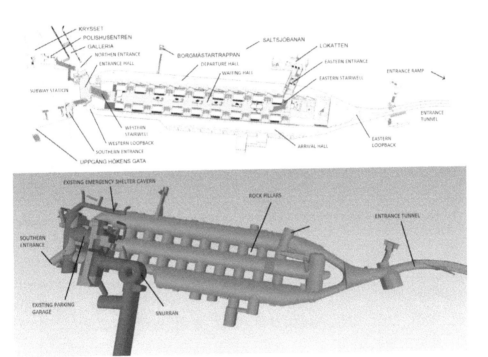

Figure 2. Overview over the bus terminal. Top: Features of the new Slussen bus terminal. Bottom: View from the BIM-model showing designed rock caverns. Existing caverns are shown in blue, and new-ly designed tunnels are shown in grey.

3 INTEGRATION OF THE NEW BUS TERMINAL INTO EXISTING CAVERNS

3.1 The entrance hall

The entrance hall will be located in the existing Katarina parking garage. The new entrance hall needs to accommodate all new installations like information boards, escalators, ventilation, etc. The existing parking garage contains concrete structures, and the existing space in the parking garage is not enough to fit all these new installations and at the same time allow the flow of people. Therefore, existing concrete structures needs to be removed and more rock needs to be excavated. Excavation in the parking garage will mostly be done as side drifts, top head excavation, and significant benching. It is uncertain if the existing concrete structure is carrying any load from the rock or not. Therefore the processes will run parallel, tearing down concrete, excavating rock and installing shotcrete and rock bolts as part of the permanent rock reinforcement.

3.2 Connection to the Metro station

The new bus terminal will be connected with the existing metro station. One of these connections will be at the existing entrance to Slussen metro station at Götgatan. This will be the southern entrance to the bus terminal. The new bus terminal will connect to the existing entrance via an existing cavern. Clos up of the area can be seen in Figure 3. Because the bus terminal and the metro station are at different elevations, and the existing cavern needs to fit an escalator and associated engine room, further excavation of rock needs to be performed. This requires rock to be excavated very close to the existing metro train tunnel. So close that theoretically a thin slab of rock could be kept between the two caverns, but the stability of the slab

can't be guaranteed without extensive reinforcement. Maybe that wouldn't even be enough and therefore the rock will be excavated. This needs to be done while the metro traffic is running.

During construction a wall will be built in the existing tunnel, and behind this wall the rock will first be excavated using wire sawing and then it will be carefully fragmented and removed.

3.3 KF's emergency shelter

After passing through the western loop and just before heading into the departure hall buses will pass under the existing emergency shelter cavern of KF, seen in Figure 4. The location of the shelter is partly in the new tunnel, therefore the existing shelter is to be integrated into the new tunnel. A mould will be built in the emergency shelter replicating the theoretical contour of the tunnel belonging to the future bus terminal. Systematic installation of bolts in the roof of the cavern that extend out in the cavern and reaches almost to the mould are installed. The emergency shelter will then be filled with steel reinforced cement.

The emergency shelter will be filled because if it isn't the room will have to be accessible for inspections throughout the lifespan of the new terminal.

After the cavern has been filled the excavation of this part of the bus terminal can be carefully performed. The excavation will be done in short and limited rounds and between every round permanent rock reinforcement will be installed using reinforced shotcrete and rock bolts.

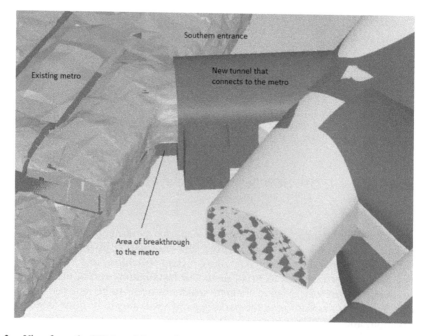

Figure 3. View from the BIM-model over the southern entrance to the bus terminal. The light grey area is a scanned area of the existing metro. The darker grey is newly designed tunnels and the yellow area is the existing Katarina parking garage.

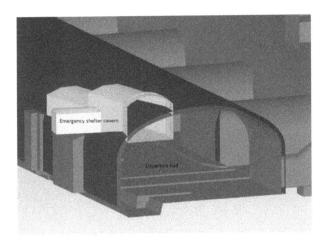

Figure 4. Section from the BIM-model showing the existing emergency shelter cavern (Yellow), the planned bus terminal (grey) and the departure hall(blue).

4 PASSAGE UNDERNEATH SNURRAN

After passing through the arrival hall the buses will use the western loop back to get to the departure hall. At the start of the loop back, the tunnel passes underneath and close to an existing cavern called Snurran. Snurran is part of the Katarina parking garage. The tunnel of the terminal has a span of twelve meters and the rock coverage with Snurran is as low as 2,8 meters. Snurran cavern has a donut-shape, Figure 5, and is twenty five meters in diameter, it is eighteen meters high and has a central pillar with a diameter of 8–9 meters.

A tube umbrella has been designed to guarantee the stability of both the existing Snurran and the new tunnel while excavating, Figure 6 is showing two sections from passage of Snurran. Because of the length of the passage and span of the tunnels a traditional spiling solution isn't

enough. A traditional spiling solution usally consists of rock bolts, instead this is tubes that are drilled into the rock and then grouted both internally and around the tubes. Also it is a sensitive area, the stability of the existing cavern cannot be compromised. The pipe umbrella is designed using analytical methods and beam theory (John & Mattle 2002). With the assumption that the pipes acting as reinforcement of the rock beam above the tunnel of the bus terminal. Figure 7 is showing a profile from the passage under Snurran.

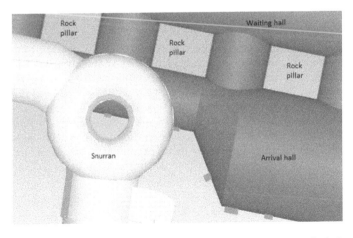

Figure 5. Overview over Snurran from the BIM-Model. The grey areas are newly designed tunnels, and the yellow areas are existing caverns.

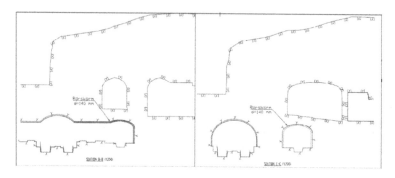

Figure 6. Sections from Snurran. The pipe umbrella (rörskärm) can be seen along the theoretical rock contour.

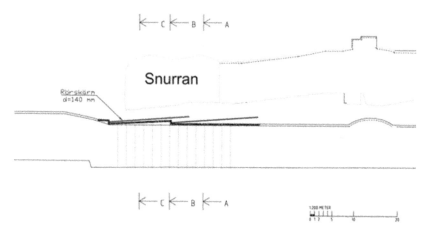

Figure 7. Profile showing the pipe umbrella principle.

In production the pipe umbrella will first be installed under Snurran, then the rock will be carefully excavated in short and limited rounds. Rounds of excavation will be maximum two meters long. After two rounds have been excavated, the first round will be reinforced using shotcrete and rock bolts for permanent reinforcement.

The stairs of Lokatten

The old stairs of Lokatten connect the two streets Stadsgårdsleden and Katarinavägen with each other and are classified as heritage sites that should not be removed. However, they are in disrepair and have been closed off for public used for some years. Within this project, they will be restored, and an entrance to the terminal will be built through the pillars of the stairs. This will be done by removing every stone of the stairs, piece by piece and marking them up for recognition. While the stairs are dissembled a new tunnel will be excavated, this tunnel will be the eastern entrance. When the rock has been excavated the job of restoring the stairs will be started. Piece by piece, exactly as it was the stairs will come back to life.

5 CONCLUSIONS

These are just a few of the challenges when designing and constructing large infrastructure facilities underground in urban areas. Every existing building, cavern, road, etc. is owned by someone. The municipality, the government, individuals, companies and all of these have an opinion or wish of how to execute the design and construction of the new bus terminal in Slussen. Our design has managed to incorporate most of these wishes.

REFERENCES

John, M. & Mattle, B. 2002. Design of tube umbrellas. *Magazine of the Czech tunneling committee and Slovak tunneling association ITA/AITES underground construction (Development, research, design realization)* 3/2002: 4–11.

Tunnels and Underground Cities: Engineering and Innovation meet Archaeology,
Architecture and Art, Volume 11: Urban
Tunnels - Part 1 – Peila, Viggiani & Celestino (Eds)
© 2020 Taylor & Francis Group, London, ISBN 978-0-367-46899-6

Design of a deep metro station for complex train operations under high passenger demand with different construction stages, case study: Gayrettepe Metro Station, Istanbul

K. Elmalı, H.T. Tunçay, A.O. Toksöz & H. Akın
Yüksel Proje Uluslararası A.Ş., Ankara, Turkey

ABSTRACT: 3 Deck Great Istanbul Tunnel Project aims to connect Europe and Asia under the Bosporus via Metro and Highway systems in a single tunnel. At the intersection of different metro lines, a unique station had to be designed to overcome many challenges such as; high number of passenger transfer demand, rolling stock transfer between different metro lines during operation, high volume of passenger movement, excessive depth of platform level due to the alignment design limitations of the Bosporus crossing, construction of the station under two separate contracts in different time frames, construction of complicated tunnels and station in dense urban patterns and coordination with the municipality and other contractors with respect to planned and ongoing underground projects. This paper will present its readers the encountered problems and developed solutions during the design process of Gayrettepe Station.

1 INTRODUCTION

The 3-Deck Great Istanbul Tunnel Project is a combined highway and metro system connecting two continents under Istanbul Strait (the Bosporus) in Istanbul, Turkey. The project consists of 16.5 km highway and 31 km metro line with 14 stations. The Bosporus crossing section of the project is 4.3 km long and through a single TBM tunnel with 16.8 m outside diameter. The tunnel contains 2×2 lane highway and double track metro.

Marmaray commuter rail system immersed tube tunnel under the Bosporus was holding the record of the deepest tunnel in İstanbul below sea level with a depth of 60 m until Eurasia Highway system TBM tunnel constructed at 106 m below sea level. The 3-Deck Great Istanbul Tunnel Project Bosporus crossing tunnel designed at 130 m below the sea level is expected to break the record not only in Istanbul but also many in the world.

The metro section of the project will be located between İncirli (European side) and Söğütlüçeşme (Asian side). In addition, metro line will provide connection to the Istanbul New Airport at the European Side and Sabiha Gökçen Airport at the Asian Side. The highway section of the project will be located between Hasdal (European side) and Çamlık (Asian side). With the realization of the project, travel time between the European Side and the Asian Side will be 14 minutes by using the highway and 6 minutes by using the metro system.

It is expected that 1.5 million passengers will be using the metro line on daily basis with the completion of the project. At the intersection of different metro lines a unique station had to be designed to overcome many challenges such as; high number of passenger transfer demand, rolling stock transfer between different metro lines during operation, high volume of pedestrian movement, excessive depth of platform level due to the alignment design limitations of Bosporus crossing, construction of the station under two separate contracts in different time frames, construction of complicated tunnels and station in dense urban patterns and coordination with the municipality and other contractors with respect to planned and ongoing underground projects.

Figure 1. Image showing the Metro lines (author).

Gayrettepe station has unique characteristics with respect to architectural and engineering perspectives. This paper presents design concept of the station to overcome architectural and engineering challenges.

2 AIM OF THE PROJECT

2.1 Passenger mobility in istanbul and projected numbers

The Ministry of Transport and Infrastructure General Directorate of Infrastructure Investments (UAB-AYGM) conducted a study to improve transportation capacity between two continents in Istanbul and developed a mega transport project with combined metro and highway systems. This study was announced to the public in February 2015 (Republic of Turkey Ministry of Transport and Infrastructure 2015). In this study, existing and future transportation demand between the two continents was assessed. According to the transportation survey, demand for the crossing, which was 1.3 million in 2015, was estimated to reach 3.8 million in 2023. Total capacity provided by the existing bridges and tunnels will be insufficient in 2020. As a result, it was realized that a new high-capacity transportation system is required.

A new high capacity transportation system was developed by the Ministry of Transport and Infrastructure consisting of 16 km long 2x2 lane highway system with a daily capacity of 120,000 vehicles and 31 km long double track metro system having 15 stations with a daily capacity of 1,5 million passengers.

Table 1. Population of Istanbul and daily transportation demand for intercontinental crossing (author).

	2000	2014	2023
Population*	10	14	17
Daily crossing demand*	0.8	1.3	3.8

* millions.

Table 2. Existing daily capacity for intercontinental crossing (author).

	2014	2023
15th July Martyrs Bridge	355,000	355,000
Fatih Sultan Mehmet Bridge	475,000	475,000
Yavuz Sultan Selim Bridge	-	475,000
Eurasia Tunnel	-	235,000
Marmaray	125,000	900,000
Metrobus	160,000	200,000
Maritime	235,000	250,000
Total	1,350,000	2,890,000

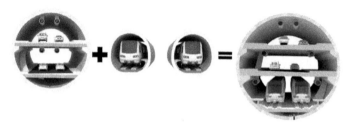

Figure 2. 3 Deck Great Istanbul Tunnel Bosporus crossing (author).

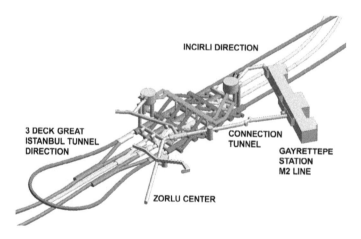

Figure 3. Gayrettepe Station (author).

Bosporus crossing section of this project was anticipated as a three-deck TBM tunnel to combine both transport systems in a single structure, which is unique for many aspects in the world.

3 DESIGN DEVELOPMENT PHASE, CHALLENGES AND SOLUTIONS

3.1 *High volumes of passenger and rolling stock transfer between two metro systems*

The planned İncirli-Söğütlüçeşme metro line follows the existing metrobus route and passes through high travel demand points of Europe and Asian sides of Istanbul. Furthermore,

Table 3. Alternative-1 three transfers (author).

	Travel Time (minutes)	traveled distance (kilometers)
SAW – Ayrılıçeşme (M4 Line)	43	30.4
Transfer at Ayrılıkçeşme Station	7	-
Ayrılıkçeşme – Yenikapı (Marmaray)	13	10
Transfer at Yenikapı	7	-
Yenikapı – Gayrettepe (M2 Line)	14	10
Transfer at Gayrettepe	10	-
Gayrettepe – Istanbul New Airport	26	33.4
Total	120	83.8

Table 4. Alternative-2 three transfers (author).

	Travel Time (minutes)	traveled distance (kilometers)
SAW – Ünalan (M4 Line)	40	27.9
Transfer at Ünalan Station	15	-
Uzunçayır – Zincirlikuyu (Express Bus)	24	9.5
Transfer at Zincirlikuyu	15	-
Gayrettepe – Istanbul New Airport	26	33.4
Total	120	70.8

Table 5. Alternative-3 two transfers (author).

	Travel Time (minutes)	traveled distance (kilometers)
SAW – Pendik (M10 Line)	15	10.5
Transfer at Pendik Station	7	-
Pendik– Halkalı (Marmaray)	90	70
Transfer at Halkalı	7	-
Halkalı – Istanbul New Airport	27	33.8
Total	146	114.3

Table 6. Alternative-4 direct with no transfer (author).

	Travel Time (minutes)	traveled distance (kilometers)
SAW – Istanbul New Airport	60	71

necessary arrangements were made to ensure passenger transfer to existing and/or planned metro systems network. The passenger capacity of the metro system is 75,000 passengers per hour per direction. In addition to 11 passenger transfer points along the metro system, Gayrettepe station is the main transfer point for passengers and rolling stocks between Istanbul New Airport metro line and Incirli-Söğütlüçeşme metro line.

Existing transfer means between Sabiha Gökçen Airport (SAW) and Istanbul New Airport have difficulties in terms of travel time and the number of transfers. There are three alternative routes, which requires travel times as listed below.

Gayrettepe district is located on the European side of Istanbul and it is regarded as downtown areas combined with commercial, residential and even industrial zones. Gayrettepe station is oriented at such a special point where İncirli-Söğütlüçeşme and Gayrettepe-Istanbul New Airport metro lines intersect. Since the passenger movement demand between the

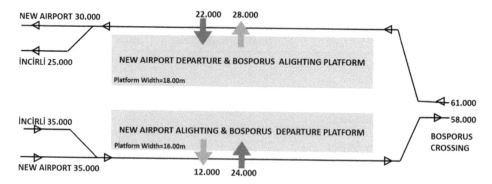

Figure 4. Platform requirements for platform widths according to TS12127(TSE, 1997) and NFPA 130 (National Fire Protection Association., 2017) (author).

continents is high, passenger transfer between metro lines would create numerous problems such as; extensive passenger circulation areas, long travel times and congestion points of passenger flow. To overcome these, it was decided to transfer passengers between two continents by continuous and direct train operations on İncirli-Söğütlüçeşme and Gayrettepe-Istanbul New Airport metro lines. By this way, the passenger movement and consequent excessive passenger circulation areas were eliminated. A similar approach was also applied for Altunizade station on the Asian side to connect Sabiha Gökçen Airport and İncirli-Söğütlüçeşme metro line.

By integrating three metro lines, it is expected that 6,5 million passengers per day will be using the whole metro system network.

Through this connection;

– A train departing from Istanbul New Airport will be able to go directly to Sabiha Gökçen Airport or Söğütlüçeşme district.
– A train departing from Sabiha Gökçen Airport will be able to go directly to Istanbul New Airport or Incirli district.
– A train departing from Söğütlüçeşme district will be able to go directly to Istanbul New Airport or Incirli district.

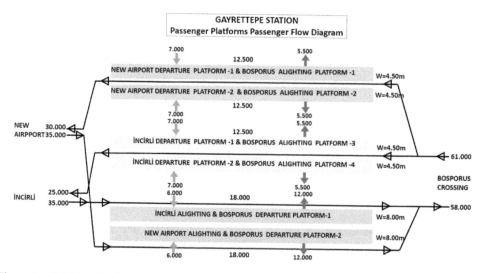

Figure 5. Platform solution to distribute passenger load (author).

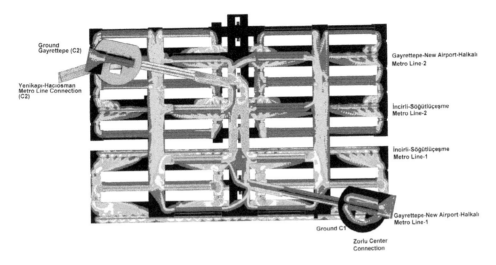

Figure 6. Pedestrian modelling of Gayrettepe Station (author).

With realization of 3-Deck Great Istanbul Tunnel Project, Sabiha Gökçen Airport and Istanbul New Airport connection will take one hour without any transfer.

Based on the passenger demand obtained from Istanbul Metropolitan Municipality capacity calculations of the station was conducted.

The calculations revealed that under the given passenger loads width of platforms had to be wider than 16 m, which is not a practical dimension for passenger access to vehicles and for tunnel construction due to its huge dimensions.

It was decided to dedicate separate platforms for both lines and both directions. As a result, passenger loads for separate platforms could be balanced between platforms.

To validate the station design, an agent-based pedestrian simulation was also conducted. Simulation was executed using year 2040 peak hour passenger density. Pedestrian velocity was taken as 60 m per minute. As a result, it was seen that maximum passenger density in the station did not exceed 34-49 people per minute at any location.

3.2 *The depth of the platform level*

In the Bosporus crossing, tunnel was located at a depth of 127 meters below sea level where the deepest rock gravel is located. Maximum gradient for the metro section is 5%. Due to the limited gradient, platform level of the station was calculated as 70 m below the ground level.

Passengers can access the station from the ground level via concourse structures which are located at both sides of Büyükdere Street. From concourse levels, passengers are directed to the main transfer tunnel. Main transfer tunnel is connected to passenger transfer tunnels with 8 stair tunnels. Passenger transfer tunnels are used to transfer passengers between two metro lines without any need to go upper levels. Each transfer tunnels are connected to platform level by 8 stair tunnels.

3.3 *Construction planning of the station*

Gayrettepe Station is a complex station which combines two separate metro lines namely Gayrettepe-Istanbul New Airport metro line and Incirli-Söğütlüçeşme metro line.

Although this is a combined station to be constructed at once, Gayrettepe-Istanbul New Airport metro line has already been tendered and is under construction. As a result, concourse structures, stair tunnels, main transfer tunnels, passenger transfer tunnels and platform tunnels of Istanbul New Airport metro line had to be constructed by the existing contractor,

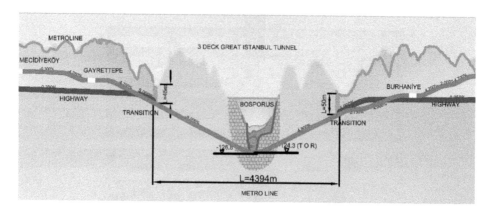

Figure 7. Bosporus crossing elevation (author).

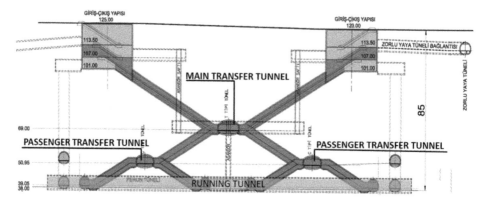

Figure 8. Section view of the station (author).

Figure 9. On the left; structures to be constructed under the scope of Gayrettepe-Istanbul New Airport metro line contractor. In the middle; structures to be constructed under the scope of Incirli-Söğütlüçeşme metro line contractor. On the right; completed station *(author)*.

whereas Incirli-Söğütlüçeşme metro line platform tunnels with its stair tunnels have to be constructed by another contractor in the future.

To achieve staged construction explained above, a construction planning study was conducted and handed over to the existing contractor. This plan would enable the Incirli-Söğütlüçeşme metro line contractor to complete the construction of its scope

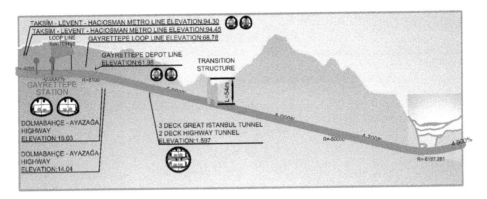

Figure 10. Gayrettepe Station and nearby infrastructures (author).

Figure 11. Artistic Rendering of the station (author).

without interrupting the operation and/or construction of Gayrettepe-Istanbul New Airport metro line.

3.4 *Design limitations due to the urban and infrastructural context*

As mentioned above, Gayrettepe district is highly populated residential and commercial area containing dense urban pattern and infrastructure network. Around the station area, there are existing metro lines, highway tunnels, utility lines and high-rise buildings with numerous basement floors, which had a direct impact on the orientation of the station and approaching tunnels. Potential clashes with other existing or planned underground and aboveground structures were avoided by close coordination with the municipality, infrastructure authorities, state highway department and other contractors.

In addition, extensive settlement analysis was carried out considering the construction stages of the station. Special measures were taken to minimize potential impact to existing structures during construction of the tunnels.

4 CONCLUSION

The 3-Deck Great Istanbul Tunnel Project connects Europe and Asia under the Bosporus with metro and highway. The project stands out with its unique approach to its deep stations with high passenger volumes. Gayrettepe station sets as a groundbreaking example not only in Turkey but also in the world in terms of its design solutions to integration of different metro lines, platform level design, pedestrian movement solutions and construction planning.

REFERENCES

National Fire Protection Association 2017, *NFPA 130 : Standard for fixed guideway transit and passenger rail systems*.

Republic of Turkey Ministry of Transport and Infrastructure 2015, *İstanbul'un yeni mega projesi, 3 Katlı Büyük İstanbul Tüneli*.

Turkish Standards Institution 1997, *TS 12127: Şehir içi yollar-Raylı taşıma sistemleri bölüm 1: Yer altı istasyon tesisleri tasarım kuralları*: 32.

Tunnels and Underground Cities: Engineering and Innovation meet Archaeology,
Architecture and Art, Volume 11: Urban
Tunnels - Part 1 – Peila, Viggiani & Celestino (Eds)
© 2020 Taylor & Francis Group, London, ISBN 978-0-367-46899-6

Surface settlement prediction of a high visibility urban tunnel in soft soil by combined 2D and 3D finite element analysis

S. Fauriel & N. Dupriez
CSD Ingénieurs SA, Lausanne, Switzerland

ABSTRACT: The paper presents the methodology and results from the construction design analysis of the Route des Nations 491m tunnel in Geneva, Switzerland. This high visibility 12m diameter urban tunnel, is excavated using conventional methods under 6.5 to 18m cover, within heterogeneous glacial till containing multiple confined aquifers, requiring cautious predictions of settlements. The method consists in excavating adits, header and subsequently, bench and invert. Soil reinforcement is ensured by a pipe umbrella and face bolting. The complexity and stakes involved implies the use of a 3D numerical model, to calibrate the 2D models, and to extrapolate the short-term trough settlement predictions as a function of the tunnel excavation face advancement. The paper evaluates the influence of the adits being excavated with differing productivity. It investigates the Peila approach to face bolting, and assesses its impact on settlement predictions. Finally, implementation of the determined thresholds in the monitoring procedure is discussed.

1 INTRODUCTION

Surface settlement is one of the main issues of tunneling in an urban environment. An accurate prediction of the tunneling-induced displacements is hence a key element of the design studies of any urban tunnel.

Surface settlement is difficult to predict using conventional methods of analyses owing to the complex soil behavior and construction sequence. Numerical analysis such as finite element is therefore often employed.

Most of the analysis used to-date are mainly two-dimensional. However, such analysis may not accurately account for many of the events which occur during the tunneling, such as the excavation of twin lateral adits with different productivity and face bolting. Furthermore, deconfinement of the ground before primary lining installation cannot be precisely reproduced in two-dimensional analyses. Such aspects of the construction phase are mimicked through other means.

Using the design undertaken for the Route des Nations tunnel in Geneva, this paper identifies the important assumptions made in the two-dimensional analyses undertaken for the Route des Nations tunnel in Geneva. Furthermore, the impact of these assumptions on the short term settlement predictions is assessed.

1.1 General description

The Route des Nations tunnel in Geneva, Switzerland is a 491m long urban tunnel. The longitudinal profile is characterized by a 4.88% constant slope. The intrados radius is 5.26m, centered at 1.59m above the axis of the road surface. The tunnel is to provide two 3.75m wide traffic, as well as two 1.20m wide utility corridors.

1.2 Site conditions

The Route des Nations tunnel is excavated under a cover above the tunnel keystone ranging from about 6.5m at the Northern starting point (PR 723) to about 18m at the passage under

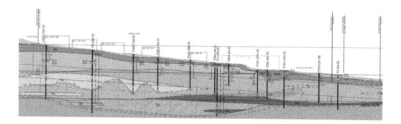

Figure 1. Longitudinal hydro-geotechnical profile of the Route des Nations tunnel.

the Voie-du-Coin surface road (PR 480), then decreasing back to 6.5m at the Southern starting point (PR 232).

The tunnel is entirely located in heterogeneous glacial till containing multiple confined aquifers. Their geotechnical nature consists in compact gravelly silts, compact clays silts and sandy-silty gravels.

The tunnel is located in a dense urban environment consisting mainly of brick or concrete houses. Importantly, there is also a brick church from the beginning of the 20th century with a non-negligible historical value. The displacements induced by the tunnel excavation were thus an important issue in the design studies.

1.3 Construction and excavation sequence

Given the presence of multiple confined aquifers in the area of the tunnel, dewatering of the main aquifers to 5m below the base of the tunnel was planned.

Then, the general construction method consists of initially excavating two lateral adits, used both for exploration and for the foundations of the header's temporary support. The header is then excavated before undertaking the bench and the invert excavations. To serve as pre-supports and maintain the stability of the working area, pipe umbrella systems and face bolting are implemented.

The temporary support of the adits and the tunnel consists of steel profiles, bolts and shotcrete.

Then, the final lining is concreted after waterproofing and lastly, the interior structures are constructed.

2 METHODOLOGY

2.1 3D finite element models

2.1.1 Geometry and mesh
The finite element predictions of the lateral adits excavation and of the whole tunnel were performed using the software Plaxis 3D. Two distinct three-dimensional models were used:

- Model A: full model of the two lateral adits to investigate on the effect of twin tunneling with and without different productivity and face bolting modelling.
- Model B: half-symmetrical model of the complete tunnel excavation cycle to investigate on deconfinement and face bolting modelling.

Model A is 60m long (y axis), 60m wide (x axis) and 45m high (z axis). The mesh is refined in the vicinity of the lateral adits. The mesh is composed of 84'654 elements and 143'184 nodes.

Model B is 80m long (y axis), 45m wide (x axis) and 45m high (z axis). The mesh is refined in the vicinity of the lateral adits. The mesh is composed of 84'230 elements and 137'905 nodes.

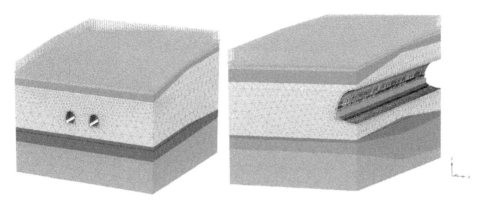

Figure 2.   3D finite element model A - Adits excavation (to the left) and B - Complete tunnel excavation (to the right).

A reference section located at the center of each of the two models was determined. This section corresponds to PR 640 of the tunnel.

2.1.2 *Modelling of temporary support*
Soil reinforcement ahead of the face by a pipe umbrella is not modelled.
Face bolting is modelled in two alternative ways as illustrated in Figure 3:

– Explicitly by "embedded piles".
– Implicitly by an equivalent face retaining pressure determined according to Peila (1994).

The "embedded piles" are able to take into account the soil-bolt interface to reproduce closely the behavior of the reinforcement.
The equivalent face retaining pressure is calculated as the sum of the normal forces mobilized in the bolts over the section of the face.
Finally, temporary support consisting of centering and shotcrete is modelled by means of "plate" elements of equivalent inertia and rigidity. The bolts are not considered.

2.1.3 *Calculation phases*
The calculation phases are the following:

– Initialization of stress state.
– Aquifers dewatering.
– Excavation of the lateral adits with a delay of 30m, respectively 0 m between the two fronts:
    – Face bolting.
    – Excavation by steps of 1 m.
    – Installation of primary support by steps of 1 m.

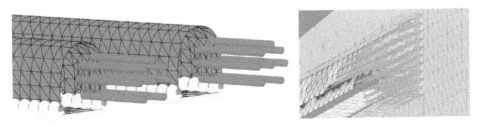

Figure 3.   Modeling of face bolting in model A (to the left) and in model B (to the right).

- Construction of the lateral foundations for the header's temporary support.
- Excavation of the header:
  - Face bolting.
  - Excavation by steps of 1 m.
  - Installation of primary support by steps of 1 m.

- Excavation of the bench and the invert:
  - Face bolting.
  - Excavation by steps of 1 m.
  - Installation of primary support by steps of 1 m.

2.2 2D finite element models

2.2.1 Geometry and mesh
The two-dimensional finite element predictions of the tunnel excavation were performed using the software Plaxis 2D in plane strain conditions.

Several two-dimensional finite element models were used in the Route des Nations tunnel design studies. One of these is representative of the mid-section of the three-dimensional models (PR 640) and only this one will be presented for the purpose of the present paper.

The total width of the model is 90m and its height approximately 50m. The mesh is refined in the vicinity of the tunnel. It consists of 2'920 elements and 24'251 bolts.

2.2.2 Modelling of temporary support
Temporary support consisting of centering and shotcrete is modelled by means of "beam" elements of equivalent inertia and rigidity. The bolts are not considered.

2.2.3 Calculation phases
The calculation phases are the following:

- Initialization of stress state.
- Aquifers dewatering.
- Excavation of the lateral adits:
  - Excavation using a stress relaxation factor.
  - Installation of primary support.

- Construction of the lateral foundations for the header's temporary support.
- Excavation of the header:
  - Excavation using a stress relaxation factor.
  - Installation of primary support.

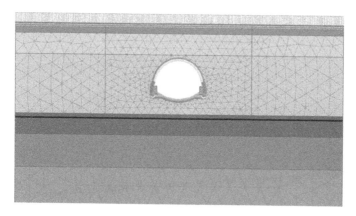

Figure 4.   2D finite element model in correspondence to the 3D finite element models.

– Excavation of the bench and the invert:
  – Excavation using a stress relaxation factor.
  – Installation of primary support.

3 RESULTS AND DISCUSSION

3.1 *Twin tunneling of the lateral adits*

As part of the design studies of the Route des Nations tunnel, the prediction of tunneling-induced ground movements during excavation of the twin adits was carried out with numerical methods. The study allowed assessing the effect of one adit being excavated faster than the other. Two simulations were performed:

– Simulation A1. Excavation of adit 2 with a delay of 30m with respect to adit 1.
– Simulation A2. Excavation of adits 1 and 2 in parallel.

The profiles of surface settlement at section PR 640 obtained with model A simulations A1 and A2 are shown in Figure 5. The surface settlement troughs are given as the adit front passes the reference section and at the end of the excavation. Figure 6 shows the longitudinal surface settlement profiles for both simulations A1 and A2. Finally, the maximum surface settlement as a function of the advancement of the excavation front with respect to reference section PR 640 is shown in the Figure 7 for simulations A1 and A2.

From the above, it can be seen that the surface settlement troughs at the end of excavation are not notably influenced by a difference in productivity of the two adits and the maximum surface settlement reaches 20.4 mm in both simulations. There is however a slight translation of the surface settlement trough towards adit 2 in simulation A1.

3.2 *Face bolting*

The impact of the modelling approach for face bolting was investigated on by performing the following types of simulations with models A and B:

– Simulation A2. Face bolting modelled explicitly by "embedded pile" elements.
– Simulation B. Face bolting modelled implicitly by an equivalent face retaining pressure determined according to Peila (1994).

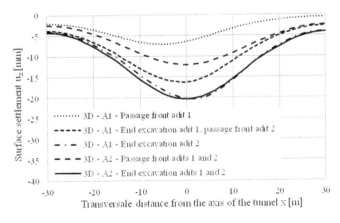

Figure 5. Transverse profiles of surface settlement at PR 640 obtained with model A (simulations A1 and A2).

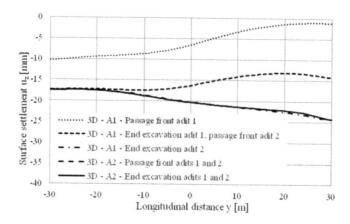

Figure 6. Longitudinal profiles of surface settlement obtained with model A (simulations A1 and A2).

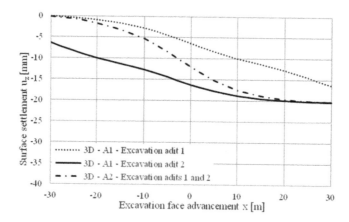

Figure 7. Maximum surface settlement as a function of excavation face advancement obtained with model A (simulations A1 and A2).

Comparison of the results between models with "embedded beam rows" and face retaining pressure was performed by face extrusion.

The impact of face bolting modelling on extrusion of the tunnel face is shown in the figures below.

The modelling approach for face bolting has a notable impact on the front extrusion.

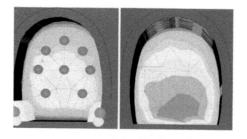

Figure 8. Front extrusion comparison in model A2 (to the left) and model B (to the right).

Table 1. Comparison of maximum face extrusion.

Model	Maximum extrusion [mm]
Model A, simulation 2	10.8
Model B	5.8

3.3 Two-dimensional model calibration

The reference section three-dimensional settlement trough was simulated in a two-dimensional analyses.

To simulate the section in two-dimensional analyses, the so-called beta-method was used where the lining is installed after a prescribed amount of unloading. After installation of the lining, initial stresses are completely reduced and this induces loading of the lining.

The beta-value in the two-dimensional analyses is assigned by assuming that the section area dA of the two-dimensional settlement trough is equal to the volume per meter dV/d of a given excavation step, calculated using the longitudinal settlement trough.

Figure 9 shows the two-dimensional transversal settlement profiles using beta-values of 0.40 for the excavation of the adits and 0.20 for the excavation of the header, bench and invert.

From the above, it can be noted that the obtained three-dimensional and two-dimensional settlement troughs are almost identical using the beta-method. The obtained beta-value is however dependent on the excavation phase and it was found to be significantly higher for the excavation of the lateral adits due to the temporary support being more flexible.

3.4 Monitoring procedure

Based on the previously discussed analysis, the settlement troughs were determined for 6 critical two-dimensional sections over the length of the tunnel. The obtained results were used in determining thresholds in the surface settlement monitoring procedure.

In fact, numerous studies (Peck 1969 and others) proved that the settlements can be described with a good approximation using a normal probability Gaussian function. This function was calibrated on the numerical settlement troughs by optimization with the least squares method as per Peck's approach. The Peck approach was proven to correctly reproduce the numerical settlements troughs.

The obtained total and differential displacements at each critical building were used to determine their risk level and the thresholds for the monitoring procedure.

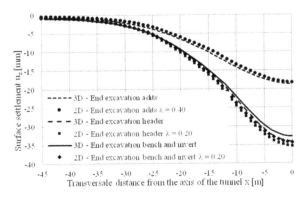

Figure 9. Transverse profiles of surface settlement at PR 640 obtained with three-dimensional model B and the two-dimensional model.

During the excavation works, continuous monitoring of the displacements in the tunnel and at the surface is carried out. The observed ground response is compared to the predicted one and the necessary modifications are planned to guarantee the excavation stability and limit the surface settlements.

4 CONCLUSION

From the numerical modelling performed during the design studies relating to the Route des Nations tunnel in Geneva, the following conclusions can be drawn:

- The specific simplifying assumptions relating to the adits being excavated with differing productivity did not have any notable influence in the surface settlement trough.
- The Peila approach to face bolting did have an impact on the short-term settlement predictions. The modelled front extrusion obtained using the approach was lower than that obtained modelling explicitly the face bolts.
- The beta-method was successfully used to calibrate the two-dimensional models.
- Implementation of the determined thresholds in the surface settlement monitoring procedure was finally discussed.

The authors would like to thank Andrew Bourget for his guidance and contribution to the design studies of the Routes des Nations tunnel. They would also like to acknowledge The Joint Venture NATIF and the Master of Work, République et Canton de Genève, Département des infrastructures (DI), Office cantonal du génie civil, Project manager Jorge Canameras.

REFERENCES

Anagnostou, G. & K. Serafeimidis 2007. The dimensionning of tunnel face reinforcement. *Underground space – The 4th dimension of metropolises, ITW World Tunnel Congress 2007.*
Panet, M. 1995. Calcul des tunnels par la méthode convergence-confinement. *Ponts et Chaussée, Paris.*
Peck, R.B. 1969. Deep excavation and tunnelling in soft ground. *7th ICSMFE State of the art Volume: 225–290.*
Peila, D. 1994. A theoretical study of reinforcemnet influence on stability of tunnel face. *Geotechnical and Geological Engineering 12 (3):145–168.*
Ramoni, M. & G. Anagnostou 2014. Face Anchoring, Static method of working and dimensioning. *Swiss Tunnel Congress.*

Tunnels and Underground Cities: Engineering and Innovation meet Archaeology,
Architecture and Art, Volume 11: Urban
Tunnels - Part 1 – Peila, Viggiani & Celestino (Eds)
© 2020 Taylor & Francis Group, London, ISBN 978-0-367-46899-6

Caltanissetta Tunnel – project implementation according to safety issues

A. Focaracci
Prometeoengineering.it Srl Rome, Italy

A. Antonelli
Empedocle 2 S.c.p.a, Caltanissetta, Italy

ABSTRACT: The twin-tube Caltanissetta tunnel (~4km), located on the SS 640 of Porto Empedocle, has been realized by the TBM "Barbara", one of the largest and most powerful in Europe (diameter 15.08 m).

The paper will highlight how the whole implementation process of the work has been characterized by a continuous confrontation with safety issues in order to identify the optimal solutions. Before the excavation, the methodology of Technical Risk Analysis (TRA) was used to estimate the uncertainty of different solutions in terms of time and costs. During the excavation phase, following the occurrence of cracks on the prefabricated segments of the tunnel lining caused by the thrusts of the jacks for TBM advance, studies and analyses were carried out on the possible optimization of the construction methods. In addition, due to the complex geological, geotechnical and hydrogeological conditions of the area, a redistribution of the emergency exits has been necessary and compensated by the adoption of safety measures defined by the risk analysis of the tunnel, carried out according to the IRAM methodology.

1 INTRODUCTION

The Caltanissetta tunnel is part of the extension work of the S.S.640 "Porto Empedocle", in Sicily, consisting in the upgrading of the existing S.S.640 from a single carriageway road, two-way traffic to a double carriageway road. The adaptation project provides, in correspondence with the city of Caltanissetta, an off-site variant of the S.S. 640 which develops between the junctions of Caltanissetta Sud and Caltanissetta Nord and constitutes, in essence, a by-pass of the Nissena urban area. The dual carriageway variant includes the Caltanissetta tunnel, built using a mechanized excavation method with double-shield TBM machine.

The twin-tube tunnel has a length of about 4000 m, and one-way traffic with 2 lanes in each direction. The excavation of the work was performed using the "Barbara" TBM, of the H.P. EPBM (High Performance Earth Pressure Balance Machine) type, one of the largest in Europe with a diameter of 15.08 m for an excavation area of about 177 square meters.

The entire construction phase of the work was developed through a continuous comparison with the safety issues: a technical risk analysis regarding the transversal section of the tunnel was conducted to define the tunnel diameter excavation and consequently the TBM characteristics; the presence of cracks on the segments due to the considerable thrusts generated by TBM Barbara required a detailed design and specific studies that determined the optimization of the reinforcement within the segments; eventually, because of the geological, geotechnical and hydrogeological nature of the areas affected by the work, and due to the registration of the infrastructure to the TERN network (Trans European World Network) at a later date respect to the completion of the project, the safety features of the tunnel had to be reconsidered.

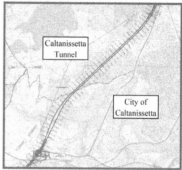

Figure 1. Intervention Area – Caltanissetta (Sicily).

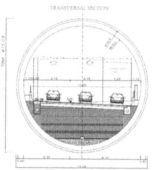

Figure 2. TBM "Barbara" and Tunnel cross section.

2 TECNICAL RISK ANALYSIS

The realization of a work in the underground is always strongly influenced by the geo-logical-environmental context and the occurrence of a "geological unforeseen", with the consequent changes to the project, can cause significant increases in time and costs of con-struction. The "TRA" method (Technical Risk Analysis) aims to identify the uncertainties of the project, quantifying them in terms of execution times, so as to establish, already in the pre-contractual stage, the limits of mutual responsibility between the Contracting Authority and the Company. With this methodology, carried out through a probabilistic analysis, it is possible to examine the uncertainties that most affect the determination of the times and costs of the work, and to define a "cloud" of time values- costs corresponding to the com-binations of possible project scenarios, from the most pessimistic to the most optimistic ones. By means of the Technical Risk Analysis it was thus possible to compare the solution of the Final Project with a different proposed solution in terms of variability of time and costs of implementation.

2.1 Analysis methodology

The technical risk analysis provides a tool for the choice between several design hypotheses through the evaluation of the main risk factors. The results of the analysis make it possible to identify the geological-geomechanical or technological parameters to be specified or deepened and to include the possible project variation in terms of time and costs within acceptable ranges, both for the contracting authority and for the executor.

The technical risk analysis develops in the following phases:

- Acquisition and evaluation of geological and geomechanical data deriving from on-site and laboratory geotechnical surveys conducted, supplemented by literature data, to characterize the rock masses to be crossed and identify all situations that cannot be excluded in reality;
- Identification and evaluation of the risk factors that can be found along the route of the work and subdivision of the route in areas with homogeneous geological-geomechanical and hydrogeological characteristics;
- Statistical analysis of the single most significant geotechnical parameters for each individual area, with the relative variability defined by the specific probability distribution functions;
- Attribution, for each homogeneous route, of the most probable stress-strain behavior category (defined by the mean value of the probability distribution) as a function of the possible variability of the characteristic quantities; application of the "characteristic curve method", inserting in the analytical calculations the input geotechnical parameters as the random variables each described by a probability distribution function;
- Choice of the most probable typological sections with the relative variability: for each of these the fields of costs and realization times are calculated;
- Graphical representation of the results (times and costs necessary to realize the individual sections) in a time-cost scattergram, referring to all the possible projects obtained by combining the non-excludable situations found in the analyses with reference to the previous points.

2.2 Project hypothesis analysed

The final project of the Caltanissetta tunnel planned the realization of the work by mechanized excavation using a TBM with a diameter of 13.40 m. The road section of the executive project included a 9.75 m wide platform consisting of two lanes of 3.75 m each, a shoulder on the left of 0.50 m and a shoulder on the right 1.75 m wide (Figure 3 on the left). Every 600 m the lay-bys were planned, to be built by widening the tunnel. Instead, the proposed variant solution provided for the use of a TBM with a diameter of 15.08 m, so as to have a platform with a total width of 11.25 m divided into 2 lanes 3.75 m wide, flanked by a shoulder on the left by 0.50 m and an emergency lane on the right 3.25 m wide, continues along the entire length of the tunnel; consequently the lay-bys were eliminated (Figure 3 on the right).

2.3 Analysis of the deformation behavior of the rock mass in the two design hypotheses

To evaluate the most probable response to the excavation, considering the variability of the geomechanical parameters, a bundle of characteristic curves was constructed. This bundle of

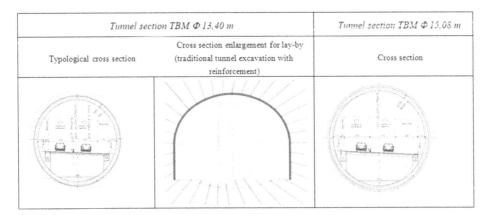

Figure 3. Comparison between cross sections of the executive projects (on the left) and the proposed solution (on the right).

curves, both near the front and at a certain distance from it, was made by using the Monte Carlo technique and introducing, in the analytical formulation of the curves, the probability functions that best approximate the distributions of the parameters of the rock.

Through the probabilistic study it is possible to identify the combinations of resistance and deformability parameters that determine the most favourable or unfavourable conditions in terms of tensile-deformative response of the rock mass. The different deformation responses are associated with different construction speeds and the consequent costs.

2.4 *Analysis of risk variability*

For both the design hypothesis described before, a risk model has been implemented by varying the input parameters in a probabilistic way. In particular, an in-depth study of the geomechanical study was conducted to identify the variability of the deformation response of the rock mass to the excavation, deriving from the variability of the geomechanical parameters deduced from the conducted geognostic surveys.

Moreover, in the case of mechanized excavation, the TBM is designed to cope with the various situations that can be foreseen in the design phase, but obviously the different deformation response translates into a variability of the possible advancement speeds of the TBM, ie a variability of the construction costs work.

The uncertainties described above have been included in a risk model, already used in similar contexts which, through the Monte Carlo method simulating 1000 possible situations that cannot be excluded a priori, allows to evaluate the effect of the aforementioned uncertainties in terms of costs and timing execution of the work (Figure 4).

The statistical analysis provided a set of pairs of time/cost values corresponding to a set of possible alternative scenarios to the most probable one. The results of the simulations carried out were represented on a time/cost diagram (Figures 5–6), which highlights the average project and the two extreme situations: the "very bad" project (the one that requires more time and costs) and the "optimal" project (the one that requires less time and costs). The method also makes it possible to verify the effect on the size of the cloud of the reduction of some risk factors, the consideration of which determines the non-exclusivity of scenarios furthest from the average value.

In particular, the percentage changes in terms of production time are shown in the ordinates and the percentage variations in terms of cost in relation to the average project are shown in the abscissa. For an easier comparison, the results of the two hypotheses were reported on a scattergramm cost times in absolute value and a scattergamm reporting the percentage changes in cost times compared to the average project (Figure 5).

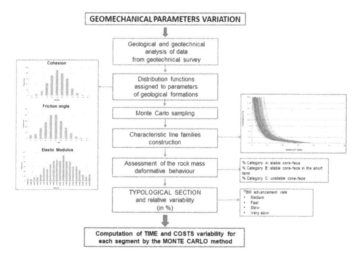

Figure 4.   Flow chart of the method.

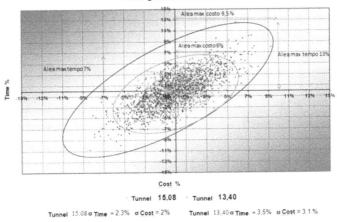

Figure 5. Comparison of projects in terms of percentage time-cost variation compared to the average project.

Therefore, from the analyses carried out, it is clear that the final project design hypothesis has greater uncertainty regarding execution times and costs. This uncertainty can be linked to traditional consolidation and excavation operations, which are carried out on an already partially disturbed rock area and which can be significantly influenced by the local conditions of the soil.

In accordance with these analyzes, the tunnel was then constructed in accordance with the proposed design, which reduced the uncertainties associated with the excavations and at the same time maximized the safety conditions for the users during the operation.

3 OPTIMIZATION OF REBARS IN PRECAST SEGMENTS

The construction of tunnels using increasingly advanced and technological TBM (Tunnel Boring Machine) machines allowed continuous passive monitoring to be carried out, and consequently improved the management of the project and its adaptation to the actual soil conditions. By means of a TBM cutter it is therefore possible to fully mechanize the excavation operations and the construction of the final lining. However, especially for significant excavation diameters such as that of the Caltanissetta tunnel, the thrusts applied by the jacks on the prefabricated segments for the advancement of the TBM are very high and can induce cracking phenomena in the segments.

The final covering of the "Caltanissetta" tunnel consists of a 60 cm thick ring, made up of 8 + 1 prefabricated reinforced concrete segments, including the keystone. They are trapezoidal in shape so that they can be positioned in different positions, to be able to better adapt to the shape of the tunnel profile. For the advancement of the "Barbara" TBM, the maximum thrust was equal to 236 kN, and the exceptional thrust amounted to 269 kN.

To identify the probable causes that caused cracking in the segments, studies were carried out on their structural behavior.

In particular, through 3D modelling of the prefabricated segments (Figure 7), the geostatic and hydrostatic thrust condition were analysed, as well as the thrust condition of the TBM machine during assembly, also taking into account possible assembly defects. The FEM analysis (Finite Element Method) allowed to study of the stress and deformation state in the different load conditions to which the segments are subject during their nominal life, which derive both from transitory situations, such as the transport phases, the process of laying and

Figure 6. Cracking pattern in the precast segments in tunnels excavated by TBM.

pushing the jacks, either by permanent actions, such as the pressure of the ground under operating conditions or the fire load generated as a result of an incident.

It is evident that one of the main reasons for the cracking conditions in the segments of tunnels realized by a TBM derives from the presence of assembly defects. For example, this happens when the support on the previous ring (especially at the keystone) does not occur correctly, leaving a prominence (the keystone remains slightly higher than the rest of the ring segments), or a depression (more likely, when the keystone penetrates too deeply into the ring). Moreover, due to the planimetric layout of the tunnel and the directional adjustments that the shield must make to maintain the axis of the path, it is frequent the need for directional corrections, which are obtained by varying the values of hydraulic pressure within the thrust jacks divided into different groups, causing non-symmetrical pressure on the rings.

A 3D model was then developed to simulate the stresses induced by the thrust applied by the jacks on the segments both in the condition of perfect contact between the segments and in the case of imperfect alignment of the keystone (Figure 8).

The results of the analysis carried out for the case of optimal thrust conditions show how stresses in the direction of the tunnel axis are all of compression, while in the orthogonal direction traction zones concentrated in the side on which the thrust is applied (Figure 9 on the left) arise. In the absence of assembly defects, the compressive stress values fall within the limits given by the class of concrete used and the tractions, always lower than those prescribed by the regulations, can be easily supported by the steel reinforcement.

In the case of presence of assembly and/or abnormal thrusts (Figure 9 on the right), the values of the stresses, compression and traction at the edge in contrast with the previous ring,

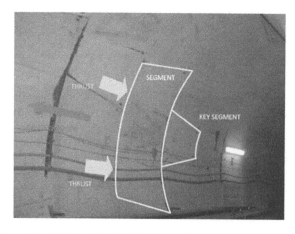
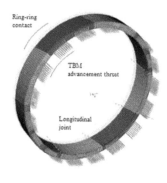

Figure 7. FEM model for TBM thrust simulation.

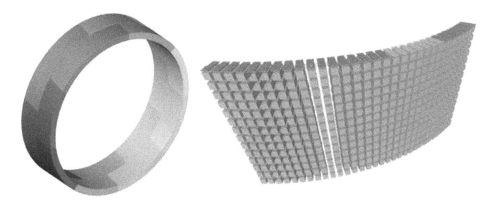

Figure 8.   3D FEM TBM ring model.

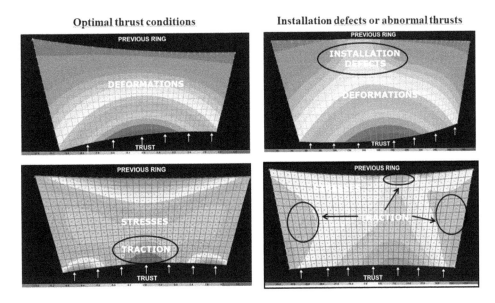

Figure 9.   Stresses within a precast segment subject to TBM thrust.

rise to adapt the stress regime to the lack of a part of the support and to collect the excesses of thrust. The tension peaks that are generated at the edges of the area where the support is lacking (the area opposite to the thrust zone) may cause cracks to originate and propagate if the traditional reinforcement, arranged in a diffused manner inside the precast segment, do not optimally contrast the state of tension that is generated.

The considerations, following the results of the analysis, lead to say that to solve the problems of cracking, found at the edges of the prefabricated segments, it is necessary to resort to the following solutions:

- pay more attention to the assembly of the rings, paying attention that the face of the ring on which the thrust will be applied is as free as possible from steps and projections;
- reinforce the edge of the segments on the side opposite to the thrust, positioning a contrast reinforcement directed to the maximum thrust of the jacks;
- place a friction reinforcement in correspondence with the connector pockets;
- introduction of an electro-welded mesh with non-structural functions for greater protection of the corners of the prefabricated segments;

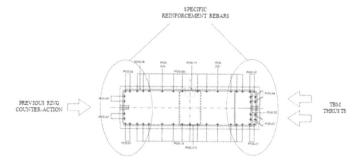

Figure 10. Specific rebars to contrast jack thrusts.

- increase the strength class of the concrete to improve the mechanical performance of the entire prefabricated concrete.

4 EMERGENCY EXIT REDISTRIBUTION

According to the Ministerial Decree 5/11/2001 "Geometric and functional rules for road construction", which establishes, as indicated in paragraph § 4.1.2, that "For twin-tube tunnels, pedestrian connections must be provided every 300 m and connections for the passage of rescue vehicles or service vehicles every 900 m", the infrastructural measures dedicated to the exodus of users in the event of an emergency within the Caltanissetta tunnel consisted of 9 pedestrian passages and 4 pedestrian / driveway passes.

With reference to t, the final project of the in line with the regulations, However, following the registration of the route on which the Caltanissetta tunnel is located within the TERN network, with a deed following the date of approval of the project, the tunnel has entered the field of application of Legislative Decree 264/06, the legislation reference in the field of road tunnel safety. This made it possible to present the proposal to bring the maximum distance between the emergency exits of the Caltanissetta tunnel at 500 m for the pedestrian emergency exits and 1500 m for the driveway passes, in accordance with the requirements of the standard.

Moreover, during the excavations for the construction of the two tubes of the tunnel, due to the complex geological, geotechnical and hydrogeological conditions of the areas and the lands involved in the excavation, ground subsidence manifested which imposed careful reflections on the location of the by-pass of connection between the tubes.

Consequently, a new distribution of pedestrian and driveway by-passes has been proposed which, in compliance with the requirements of Legislative Decree n. 264/06, does not involve the critical areas previously highlighted during the tunnel excavations.

In consideration of the greater distance between the emergency exits than the one provided for in the tunnel's executive project, and in order to guarantee in any case a level of safety equivalent to that of the approved project, it was necessary to adopt additional system safety measures.

The level of safety of the tunnel in the different configurations analysed was therefore assessed by means of a quantitative risk analysis, carried out according to the methodology established by Legislative Decree 264/2006 art. 13 and paragraph 3 of Annex 3, and better detailed by the "ANAS Guidelines for the design of road tunnel safety", known as IRAM (Italian Risk Analysis Method).

In particular, in order to guarantee an equivalent level of safety in the tunnel with respect to the executive project, the implementation of the innovative SCADRA system will be realized: the SCADRA is new layout of the SCADA, able to perform a Dynamic Risk Analysis based on the IRAM method.

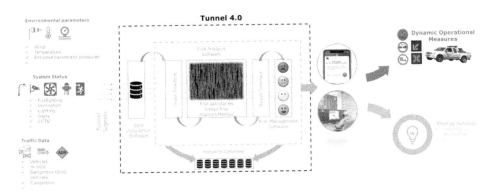

Figure 11. SCADRA operating scheme and principles.

Through the processing and analysis of the main parameters that can determine changes in the level of safety in the tunnel, continuous monitoring is carried out for both the status of the systems and the flows and traffic composition in order to implement preventive and corrective maintenance by decreasing operating costs, and reducing response times.

The change in traffic conditions and/or environmental conditions, or the malfunctioning of systems, can in fact determine a higher risk than the risk assessment carried out during the design phase. If the level of risk increase towards unacceptable limits, the SCADRA shall activate and/or report all the security measures necessary to achieve the required level of safety, such as, by way of non-exhaustive example:

- Traffic management and contingency measures (speed limit reduction, overtaking prohibition, increase in safety distance, etc.);
- Interruption of the energy saving mode of the plants;
- Communications to users;
- Fire brigade warning (if any);
- Pre-alert Institutions involved such as the Prefecture and the Fire Brigade.

Moreover, the SCADRA allows, in the normal operation of the tunnel, to implement plant management strategies aimed at saving energy, without compromising user safety.

5 CONCLUSIONS

All the construction process of the twin-tube Caltanissetta tunnel (4km), located on the SS 640 of Porto Empedocle, has been characterized by a continuous confrontation with safety issues in order to identify the optimal design.

The tunnel, realized by the TBM "Barbara", one of the largest and most powerful in Europe, crossed soil with complex geological, geotechnical and hydrogeological conditions that required particular attention in the design choices in order to limit surface disturbances and guarantees the best possible optimization between costs and time.

For this purpose a technical risk analysis has been firstly conducted and the project with a larger TBM was considered an improvement solution with respect to the approved final project in which the excavation of lay-bys fallowed the excavation of the tunnel.

Secondly, due to the large thrusts applied by the jacks for the TBM advancement a specific study was conducted to optimize the rebars configuration inside the precast segments to avoid cracking phenomena.

Eventually, a new layout of the SCADA, named SCADRA, designed to develop dynamic risk analysis according to the IRAM methodology established by Legislative Decree 264/2006 will be implemented in the tunnel to compensate for the longer distance between the emergency exits in the final configuration of the tunnel because of the presence of critical areas highlighted during the excavations of the two tubes.

REFERENCES

Lunardi, P. & Bindi, R. & Focaracci, A. 1989. Nouvelles orientations pour le Project et la construction des tunnels dans des terrainsmeubles. Etudes et expériences sur le preconfinement de la cavité et la pre-consolidation dunoyau au front, *Colloque International "Tunnels et micro-tunnels en terrain meuble"*; *Parigi 7–10 February 1989*

Lunardi, P. & Focaracci, A. 1998. Quality Assurance in the Design and Construction of Underground Works, *International Conference "Underground Construction in Modern Infrastructure"*; *Stockholm 7–9 June 1998*

Lunardi, P. & Focaracci, A. 2000. Modern Tunnelling in Italy for High Speed Railway Line, *Bauma 2001*; *Munich, April 2nd to 8 th*

Focaracci, A. 2005. La Gestione progettuale e costruttiva delle Gallerie, *Gallerie e Grandi Opere sotterra-nee, n 76*; *Bologne, August 2005*

Focaracci, A. & Mattei, M. 2007. Le gallerie oggi e nella storia, *Le Strade, n.6*; *Milan, June 2007*

Focaracci, A. 2008. La galleria Frena, *Le Strade n.12/2008*; *Milan*

Focaracci, A. 2013. Managenment of Tunnel advance by TBM. *Arabian Tunnelling Conference & Exhibition*; *Dubai United Arab Emirates 10–11 December 2013*

Focaracci, A. & Farro, G. 2014. Construction Risk Analysis Method (CRAM) for underground infra-structures, *City Expo*; *Moscow 15–16 October 2014*

Focaracci, A. & Salcuni, M. Design of the final lining of a great diameter tunnel excavated by TBM - *International Conference "The great infrastructures and the strategic function of Alpine Tunnels, Rome 22–23 October 2015*

Tunnels and Underground Cities: Engineering and Innovation meet Archaeology,
Architecture and Art, Volume 11: Urban
Tunnels - Part 1 – Peila, Viggiani & Celestino (Eds)
© 2020 Taylor & Francis Group, London, ISBN 978-0-367-46899-6

The use of the "Pipe Arch" technique for under-crossing in urban areas

A. Focaracci & C. Mattozzi
Prometeoengineering.it Srl, Rome, Italy

ABSTRACT: The project of the Campalto tunnel is characterized by the crossing of Gobbi street ensuring the opening to vehicular traffic during the work. The tunnel is located in the Venetian lagoon, within inconsistent soil with the presence of a significant hydraulic surcharge. To ensure the minimization of subsidence allowing the presence of traffic the highly innovative solution of the 'Pipe Arch' has been used consisting in a double arch of concrete micropiles reinforced with fiberglass and sub-horizontal metal insertions passing through the whole rock mass, from side to side. To protect the excavations and the near existing buildings plastic diaphragms in bifluid jet grouting $\phi 800$ mm and jet grouted $\phi 1500$ mm columns made by trifluid jet-grouting intervention were realized to counteract the thrust from the water. The article will highlight how the jet grouting treatments, calibrated through instrumented test fields, and the Pipe Arch technique have allowed the success of the work.

1 INTRODUCTION

The metropolitan area of Venice represents the fundamental hub of the North-East road and motorway network, acting as a hinge for traffic to and from the south of the country as well as a direct link between Italy and Eastern Europe. Within this road network, the state road S. S.14 "of Venezia Giulia" is an important link between the provinces of Venice, Udine, Gorizia, Trieste and the Slovenian territory. The Veneto stretch of the highway, called "SS 14 – Triestina", is configured as a connecting road between the most important centres of the province of Venice, characterized by important traffic volumes, both long-distance and local, in relation to the current characteristics of the track and the roadway are no longer sufficient to guarantee safe traffic transit. Functional discontinuities deriving from the urban connotation that the axis assumes at the crossings of the inhabited centres of Campalto and Tessera often occur, characterized by a situation of perennial congestion due to the overlap of internal traffic and crossing traffics. Therefore, it was necessary to study a variant of population centres, a beltway that would contribute to the rationalization and redistribution of vehicular traffic, discouraging medium-long distance traffic within inhabited centres and proposing diversified routes.

The Campalto variant is geographically identifiable as an East-West loop to the Campalto town centre, with the development of approximately 2 km, intended to intercept the traffic of origin and destination crossing out of the urban appurtenances, avoiding that it weighs on local roads. The overall variant to the SS 14 of Campalto and Tessera, in fact, would allow to remove the traffic from local roads, and to connect the centre of Venice with the airport area. The variant develops in the northern area with respect to the Campalto municipality, crossing scarcely anthropized areas, at the edge of currently agricultural areas. Approximately at half of its development, the route meets Gobbi street whose interference was solved through the construction of a 25.00 m long natural tunnel built using the Pipe Arch method. Moreover

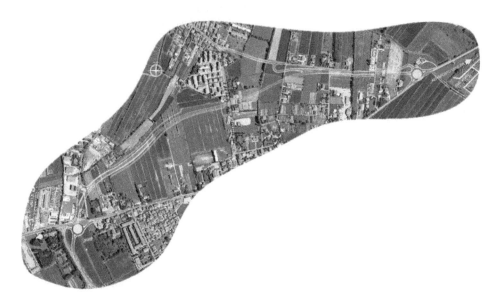

Figure 1. Urban intervention in Campalto municipality (Venice).

Figure 2. Sky view of the two elliptical wells to access the natural tunnel.

two artificial sections with a total length of 49 m and two access ramps, with a length of 430 m and 410 m, respectively, have been built. This work represents both from a technical and an economic point of view, the most demanding of the whole intervention.

2 GEOLOGY

The subsoil of the sector of the Veneto plain within which the present project is realized is made up of a powerful thickness, estimated in the order of hundreds of meters, of earthenes

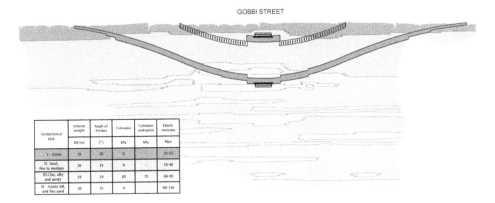

Geotechnical Unit	Volume weight	Angle of friction	Cohesion	Cohesion undrained	Elastic modulus
	KN/m3	(°)	kPa	kPa	Mpa
I - Cover	25	30	0		20-25
II- Sand, fine to medium	20	34	0	-	35-40
III-Clay, silty and sandy	19	24	10	75	60-90
IV -Sandy Silt, and fine sand	20	13	0	-	80-140

Figure 3. Geological – geotechnical profile.

belonging to the Veneto Pleistocene-Holocene sedimentary basin, constituted by soils poorly compacted and with scarce consistency. They are part of the fine-grained land group, comprising terms ranging from clays, with varying incidence of silty and sandy fraction, to silts, to fine sands, to organic soils.

The geometric arrangement of the various sedimentary bodies appears to be completely random, in virtue of the relative variations undergone by the sea level during the Quaternary period and, with it, by the sedimentation environment, regulated by the succession of sea ingressions and regressions, as well as by variations hydrographic courses of rivers. It is characterized by lateral eteropies and interdigitations between deposits with frequently lentiform geometry and by the presence of paleo alvei, often identifiable as elongated bands, morphologically detected with respect to the surrounding areas.

The route of the planned variant crosses fine and very fine-grained soils; in particular, in the initial stretch, from the grafting roundabout on the Sabbadino and Orlanda streets to the crossing in the natural tunnel of Gobbi street, terrenovities are reported outcropping, ranging from clayey silts to silty clays, of color from hazelnut to gray, for a thickness of about 3 m, deposited in the depressed interfluvial areas. Below these and, on the surface, in the intermediate and end section of the track, starting from the crossing of Gobbi street, the prevalence of silty sandy, superordinately silty-sandy, gray-to-hazel sandy terms is reported. The granulometry of the sand is very fine to medium-fine. The natural tunnel crosses soils with a silty sandy granulometry and the site is part of an aquifer whose piezometric surface is close to the countryside level.

The investigations carried out during the executive design phase have favored the deepening of the investigations in the areas that are more critical for design purposes, that is, in the section in the gallery. In light of these investigations, the following geological-co-geotechnical model has been defined, divided into 4 geological geotechnical units (U.G.):

– U.G. I: present below the ground level there is a layer of varying thickness from 1 to 3 m represented by covering and silty soils;
– U.G.II: to follow we note the presence of a layer with a thickness of 7 ÷ 10 m of mainly incoherent nature soils consisting of sand with a granulometry variable from fine to medium with increasing depth;
– U.G. III and U.G. IV: below the sand material bench and up to a depth of 33 ÷ 33.5 m from the ground level there are lands of a prevalently cohesive nature on average consisting of silty and sandy clays (U.G. III). Within this layer there are frequent intercalations of sandy silt and fine silty sand with a thickness varying from 0.5 m to about 2 m and a medium thickening degree (U.G.IV).

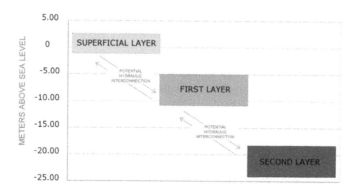

Figure 4. layers level.

3 HYDROGEOLOGY

Through the campaign of measures conducted in the design and construction phases, the presence in the area under examination of a stratum in the superficial levels at an altitude of $0.5 \div 1$ m from the countryside was confirmed. In relation to the lithological variations described above, the presence of a deeper multi-layer system is also known, with piezometric levels located at different altitudes: the general hydrogeological model elaborated considers a first surface layer in the land and in the first horizon sandy-loam depositional. These levels can be locally interconnected and connected to form a single aquifer. Subsequently, the first water table was found in the depositional level constituted by very fine to medium-fine sands, which is the first true incoherent horizon endowed with a certain spatial continuity. The second layer is located deeper in correspondence with the depositional levels E and E', that is alternations of horizons, fine sands and silts (E) and sandy and clayey silts (E') site of a confined aquifer with spatial thickness and continuity of a certain importance.

4 WORKS DESCRIPTION

4.1 *Access ramps to the natural tunnel*

Within the two ramps there are different types of intervention, whose different design solutions derive from the progressive deepening of the excavation quota. The design approach of the executive project was characterized by the crossing of Gobbi street and by the need to ensure the normal flow of vehicular traffic during the works. In this regard, a design solution was evaluated and chosen, through which the urban crossing of Gobbi street that has been

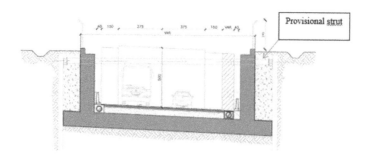

Figure 5. Ramp segment between U-wall and the provisional strut.

Figure 6. U-wall construction.

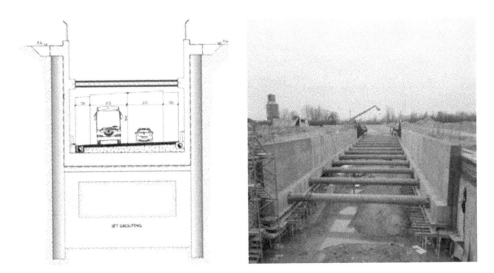

Figure 7. Ramp segment between definitive diaphragm.

carried out through the construction of a natural tunnel of about 25m in length, with the particular technique of the "Pipe Arch".

The solution involves the deepening of the excavations for the construction of the two approach ramps from which derive different types of works of art that are described below.

4.1.1 *U-wall segments*

These are the terminal segments of the underpass for which it is sufficient to create side walls that are cantilevered from the foundation to support the lateral excavation walls. They have a specific geometry in correspondence with the base soleplate in order to counteract the under thrust from the water. Depending on the excavation height the excavations was open or supported by sheet piles contracted by steel struts.

In the areas where the trench is deeper, therefore near the underpass, the support of the excavations has been realized with diaphragm bulkheads of 120 cm thickness and depths of 19 m and 20 m. The contrast to the thrust of the soil and the action of water is assisted by definitive struts placed on the beam connecting the diaphragms. In order to achieve an effective waterproofing between the concrete panels of 250 cm width, water stop joints have been

Figure 8. 3D view of the work.

placed and, in order to position the PVC sheet, the diaphragms have a r.c. coating made up of a 130cm height base slab and a 60cm side deck.

4.1.2 *Elliptical wells to access the natural tunnel*
The search for the most efficient solution, from both a logistical, operational and economical point of view, which would allow the processing of access ramps to be independent with respect to natural tunnel, suggested to adopt a solution that provides for the construction of elliptical wells within which to position the drilling and consolidation machines necessary for the construction of the natural tunnel.

As anticipated, the structures just outside the natural stretch consist of two r.c. wells, one on the Venice side and one on the Tessera side. Both wells are made by a crown of piles arranged in plan according to an ellipse with major and minor axis respectively equal to 28.2 m and 24.0 m oriented with the minor axis aligned with the longitudinal development of the underpass. The vertical facing is achieved by using CFA piles Φ 800 mm; in some particularly sensitive areas from the point of view of possible failures and the effects of consolidation, a further row of CFA piles has been planned to further protect existing ones. Starting from the ground level, all the volume of soil behind the piles, has undergone a sub-vertical fluid-fluid jet grouting treatment in order to improve the mechanical characteristics of the pushing soil and guarantee the hydraulic seal. Similar treatment with jet grouting Φ 1500mm mesh 1200 x 1200mm has been performed to ensure the stability of the excavation bottom with a 5m thick bottom layer between -26m and -21m from ground level. The wells are excavated for samples of 1.5 m and, at each advance, a covering wall is laid down for under-construction by reinforced concrete rings anchored to the existing piles. The downward operations of the excavation plan inside the wells have been accompanied by the progressive execution of the interventions at the front (micropiles, insertions and injections with mortars) for the consolidation of the volume of land within which the natural tunnel is located.

4.1.3 *Geotechnical analyses*
The problem in the various construction phases of the elliptical wells was analyzed in the hypothesis of the deformation plane state, through a finite element model with a three-node triangular mesh for each element; the reference section is that along the mine axis of the wells

and that intersects perpendicularly Gobbi street. The structural parts were modeled through Beam elements according to the Timoshenko formulation. The Mohr – Coulomb breaking criterion was used. The thrust exerted by the ground on the external walls depends on the deformation state of the ground and, consequently, is the result of the deformation of the mesh in the model. The numerical elaborations have been carried out considering the presence of groundwater, whose limit has been positioned at -1m from the countryside level in the upstream part of the bulkheads of the wells and at excavation level in the downstream section.

Given the elliptical geometry, the well is self-supporting and the efforts in the internal lining and in the connecting beam of the external piles are pure compression. The rings of 28.2 m in diameter and height 1.5 m made at each fall up to -14.5 m from the countryside, have been simulated through springs of equivalent rigidity, their operation simulates the presence of a "strut". The main phases considered in the calculation are:

– generation of the initial stress state;
– activation of the bulkhead and of the "strut" that simulates the head beam;
– partial excavation up to the depth of installation of the first ring in c.a.;
– activation of the "strut" that simulates the presence of the 1st ring;
– construction of the consolidation from the Mestre side to the Tessera side in the presence of traffic in the area of interference with the natural tunnel and then within the excavation volume of the future natural tunnel.
– deepening of the excavation up to the implementation of the ribs;
– activation of the "strut" that simulates the presence of the next rib ring;
– construction at the height of the excavation of the consolidations from the well on the Mestre side to that on the Tessera side
– repetition of the previous phases until reaching the final quota.

4.1.3.1 SUBSIDENCE ANALYSIS AND EFFECTS ON BUILDINGS

The execution of excavations always induces a variation of the stress state and consequently a field of deformations that also manifests itself on the surface. Surface displacements depend on many factors:

– excavation methods;
– mechanical characteristics, and in particular the deformability, of the land involved;
– variation of the stress state;
– changes in hydraulic conditions if the excavation is positioned under the water table and the coating is permeable; in these conditions the drainage of the water causes an increase of the effective tensions and therefore a lowering of the ground.

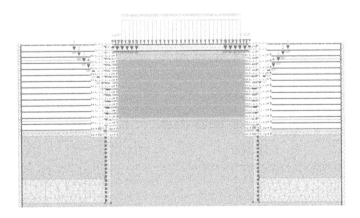

Figure 9.   Calculation model.

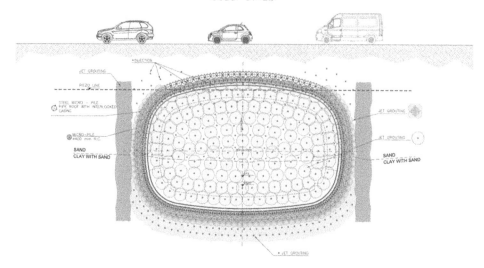

Figure 10.   Geometry and face reinforcement for the excavation.

The problem of subsidence prediction is of particular relevance for excavations in metropolitan areas, as the differential subsidence on the surface can cause damage to the overlying structures.

In the specific case the presence of buildings in the worst case at 2 m from the excavations for the realization of the wells required an analysis of the possible consequences of the waterproof treatment for each building.

In the implemented FEM model, the interfered buildings were schematized as simple weightless beams with height H and length L equal to those of the section of the building analyzed (Burland, 1997).

The calculation method used is based on the following assumptions:

– the displacements and deformations induced on buildings are assumed equal to those corresponding to the "green field" condition;
– the rigidity of the buildings analyzed is considered null, whereas in reality the foundation works interact with the ground, reducing the deflection ratio and the horizontal deformations obtained in the "green field" condition;
– no distinction between foundations on plinths, on beams and on stalls that clearly present different behaviors especially with regard to horizontal displacements;
– all buildings are considered as being arranged perpendicular to the axis of the excavated galley (a configuration that determines the worst conditions for the same building in terms of deformation);
– for buildings in c.a. the category deriving from the approach that provides the most difficult condition is assumed as a category of damage.

Once the movements have been determined and through the definition of damage categories known in the literature, threshold values have been defined that have been carefully recorded and interpreted in real time during the processing.

4.1.4 Excavation of the natural tunnel

In terms of economic, logistic and planning importance, as anticipated, the intervention is characterized by the crossing of Gobbi street and the need to ensure vehicular traffic during the work. In this regard, the best design solution was chosen with respect to the definitive project, through which the crossing in the urban area is underway through the excavation of a natural tunnel of about 25m in length, with the particular "Pipe Arch" technique

Figure 11.   Realization of the tube arches in a well.

This technique makes it possible to solve the problem of creating a cavity in difficult terrains consisting of very fine sand, immersed in stratum and in conditions of small coverages in relation to the size of the tunnel, relying on a solid, resistant structure even before the excavation is carried out. It should be considered, in fact, not a pre-consolidation but a support consisting of a double crown of micropiles, one of which consists of steel tubes, able to ensure the channeling of the tensions to the boundary of the cable and then artificially create the arch effect that the ground is not able to create naturally and the second of micropiles in reinforced concrete interpenetrated, such as to allow a transverse distribution of the excavation forces. Both the inner and outer crowns are coadiuvate closed transversal elements in a.r. made up of HEB220 ribs 0.5m pass to form a resistant framework and able to counteract the effects of surface subsidence.

In the design and construction of a cavity it must be kept in mind how the channeling of the flow of stresses deviated from its opening ("arch effect") can be controlled through the choice of the excavation phases and stabilization interventions and, consequently, how the deviation of the flow can be driven with the same instruments (excavation phases and stabilization instruments) in the different geostructural and geomechanical situations of the land in question. In addition, the "Pipe Arch" technique, neglecting the static contribution of the front (even if in the specific case of the Gobbi street underpass it will be congealed by jet grouting and cement and/or chemical injections) and operating a continuous confinement of the ground around the cable, allows to minimize and often annul the deformations that are induced in the soil before the arrival of the front itself, significantly reducing the thrusts generated on the lining, and allows, in short, to preserve almost unaltered the balances pre-existing to the excavation of the cavity.

4.1.5 Waterproofing and stability of the excavation bottom interventions

The urban context of the work and the constant presence of groundwater table at ground level have suggested the creation of waterproofing works to protect the excavations and consequently the existing structures, since avoiding to lower the level of the water table avoids building consolidation phenomena. In particular, on the backside of the piles plastic diaphragms made of jet grouting Φ 800 mm have been realized. In addition, starting from the frontal bulkheads of the Mestre and Tessera side, for a distance dependent on the depth of the excavation, a bottom intervention made of trifluid Jet grouting Φ 1500mm mesh 1200 x 1200mm of thickness between 3 and 5m has been realized to counteract the under-presses from the water. In the implementation phase the jet grouting treatments were calibrated through appropriate test fields instrumented to verify their suitability in terms of the geometrical size of the single column, mesh, definition of the operating parameters of drilling and injection. In particular, in correspondence with the underpass of Gobbi street in the natural tunnel, the waterproofing intervention was designed to obtain a waterproof parallelepiped containing the volume of the tunnel. In general, the jet grouting treatment has been used to create volumes of solidified soil with greater dimensions and resistance, while cement and/or chemical injections were used to contrast possible cracks and filtrations within the structure of consolidated terrain. In the latter case, different mixtures were used according to the medium to be treated (natural soil or consolidated soil) capable of blocking the filtration phenomena in the shortest possible time.

5 CONCLUSIONS

The execution of underground excavations in urban areas often presents many complex problems related to the presence of possible land interactions with the pre-existing in terms of allowable subsidence. In the undercrossing of Gobbi street, the Client's need to avoid interrupting the vehicular traffic and the particular geological formation of the area with very fine sand soils under the water table made the work a real engineering challenge.

These boundary conditions have suggested the choice of an innovative and experimental design solution with regard to the natural tunnel, the "Pipe Arch". This together with an important set of consolidations and waterproofing, has allowed to contrast the filtration processes and the effects on adjacent structures.

REFERENCES

Lunardi, P. & Bindi, R. & Focaracci, A. 1989. Nouvelles orientations pour le Project et la construction des tunnels dans des terrainsmeubles. Etudes et expériences sur le preconfinement de la cavité et la pre-consolidation dunoyau au front, *Colloque International "Tunnels et micro-tunnels en terrain meuble"*; *Parigi 7–10 February 1989*

Lunardi, P. & Focaracci, A. 1998. Quality Assurance in the Design and Construction of Underground Works, *International Conference "Underground Construction in Modern Infrastructure"*; *Stockholm 7–9 June 1998*

Lunardi, P. & Focaracci, A. 2000. Modern Tunnelling in Italy for High Speed Railway Line, *Bauma 2001*; *Munich, April 2nd to 8 th*

Focaracci, A. 2005. La Gestione progettuale e costruttiva delle Gallerie, *Gallerie e Grandi Opere sotterranee, n 76*; *Bologne, August 2005*

Focaracci, A. 2008. La galleria Frena, *Le Strade n.12/2008*; *Milan*

Tunnels and Underground Cities: Engineering and Innovation meet Archaeology,
Architecture and Art, Volume 11: Urban
Tunnels - Part 1 – Peila, Viggiani & Celestino (Eds)
© 2020 Taylor & Francis Group, London, ISBN 978-0-367-46899-6

Correlation approach between surface settlement and TBM pressure parameters

V.H. Franco
Universidade de Brasília, Brasília, Brazil

H.C. Rocha
Companhia do Metropolitano de São Paulo, São Paulo, Brazil

A.A.N. Dantas, T.A. Mendes & A.P. Assis
Universidade de Brasília, Brasília, Brazil

ABSTRACT: TBMs have become an excellent tool for a fastest development of the underground space in urban areas. An important aspect of its operation, either by EPB or slurry, is the control of the tunnel face stability aiming to minimize the effects on the surrounding ground both during boring and standstill. Moreover, a continuous monitoring campaign is mandatory to check if the assumed risks stand inside the limits of integrity on the nearby buildings. Hereupon, the need for a relation between the TBM face support pressure and the immediate surface settlement has become a matter of interest. Therefore, the following paper is presented with the goal to provide a correlation approach to relate these two important features of tunneling. To achieve that, data from the extension works of São Paulo metro Line 5 will be used to frame the basis of the correlation approach here proposed and proved to fit well.

1 INTRODUCTION

Undoubtedly, nowadays, the use of tunnel boring machines (TBMs) has played an important role in the increase of underground infrastructures. Long distance covering, fastest construction cycle per tunnel section, control of construction (e.g. excavated material, face stability, concrete lining quantity), safe work environment and the relative low impact on inducing of ground movements have been the reasons behind its success over conventional tunneling methods in the urban areas.

The use of either earth pressure balance (EPB) or slurry shield (SS) machines will depend particularly on the type of soil to be excavated and the presence of groundwater level above tunnel if any. Even though, Maidl et al. (2012) indicated that the range of application of EPB machines can be extended far into the area of application of SS machines by the addition of some chemical agents, to the muck in the excavation chamber.

Regarding tunneling effects, it is well known that the construction of tunnels, in an urban environment, inevitably induces ground movements, which according to Leca et al. (2007) is a consequence of the rupture of the ground mass stability that begins at the tunnel face and propagates to the surface in the form of settlements. For the case of TBMs, the stability of the tunnel face is achieved by the continuous application of support pressure from the excavation chamber. A detailed explanation on the mechanism functioning for applying support pressure, either on EPB or SS machines, as well as its theoretical estimation is provided by Guglielmetti et al. (2008) and Maidl et al. (2012).

On the other hand, the firsts to represent the settlements at surface were Peck (1969) and later Attewell and Woodman (1982) which considered the shape of surface deformation as a settlement trough. These authors used an empirical approach to study this. However, there exist others approaches that can be employed like analytical solution (Loganathan and Poulos, 1998), numerical approaches (Rowe et al., 1983; Lee and Rowe, 1990) and modeling test (Atkinson and Potts, 1977; Mair and Taylor, 1997).

In general, at the design stage and during tunnel construction, it is of common practice to analyze the surface settlement and the tunnel face stability separately. Therefore, arises the interest by the authors to inquire about this topic. In this contest, some references about correlation of these two important variables can be mentioned as: Macklin (1999) which, based on the concept of a Load Factor (*LF*) proposed by Kimura and Mair (1981) from results of geotechnical centrifuge tests, proposed a relationship to estimate the volume loss; Repetto et al. (2006), where through back analyses of the EPB performance used for the construction of a 7 km railway tunnel below the city of Bologna – Italy, presented a diagram to correlate the face support pressure with volume loss and, lastly, Fargnoli et al. (2013) which, by studying the case of the new Milan underground line 5, attempted to relate the surface settlement with the face pressure.

Thus, by considering the statements indicated above as well as the reviewed attempts, the present paper is intended to provide a correlation approach for estimation, during TBM tunneling, of surface settlement in relation to support face pressure. The aim of this approach is to provide an additional tool to evaluate ground movement due to mechanized tunneling in urban dense areas, where its estimation is paramount.

Considering the importance of framing the proposed approach to a case study, information regarding surface settlement monitoring campaign and support pressure data of an EPB machine for the extension works of Line 5 of the São Paulo metro system will be used.

2 PROJECT BACKGROUND

The tunnel project regards 11.5 km of extension works of São Paulo metro Line 5 in the highly dense south region of the city of São Paulo – Brazil (Figure 1). The extension works of the line involve the construction of eleven stations, thirteen ventilations and emergency exit shafts, one parking station and one depot "Pátio Guido Calói".

The construction of the 11.5 km extension line was divided in: *i*) 0.63 km excavated by the sequential method; *ii*) 5.13 km excavated by Ø 6.9 m diameter EPB machine (two single track tunnels) and *iii*) 5.74 km excavated by Ø 10.6 m diameter EPB machine (double track tunnel).

In a specific way, the following work deals with information regarding the excavation of the double track tunnel stretch between the ventilation and emergency exit shaft of Bandeirantes and Dionísio Da Costa (points A and B respectively from Figure 1). The intense instrumentation campaign carried out along the stretch line, the relative simplicity of using one EPB machine for the same stretch, the almost regular disposition of soil layers and the fact that was a reuse TBM machine with the same working crew were the reasons for selecting this part of the line to implement the propose approach.

2.1 *Geological and geotechnical features*

Briefly speaking, the ground around the tunnel stretch is formed by the São Paulo formation which consists in sediments of red porous silty clays, silty-sandy clays and clayey sands, and the Resende formation formed by sediments of gray and yellow silty clays and yellow and orange silty-clayey sands. The major part of the tunnel excavation goes through the transition zone between these two formations. Figure 2 shows the geological profile between Hospital São Paulo and Santa Cruz stations. Besides, Table 1 summarizes the main geotechnical parameters of the soils presented along the tunnel stretch.

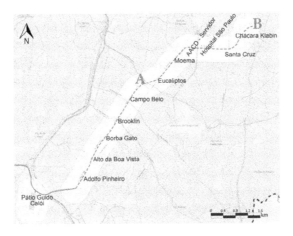

Figure 1. Layout of extension works of São Paulo metro Line 5. Case study area between Hospital São Paulo and Santa Cruz station (top right corner).

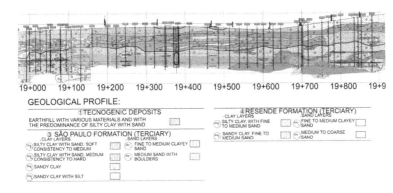

Figure 2. Geological profile between Hospital São Paulo and Santa Cruz stations (HSP – SCR).

Table 1. Geotechnical parameters of groundmass presented along tunnel stretch.

Geotechnical Parameters		$3Agp_1$[*]	$3Ag_{1,2}$[*]	$3Ar_{1,2}$[*]	$4Ag_{1,2}$[*]	$4Ar_{1,2}$[*]
γ	kN/m³	16.6	18.5	19.5	20.2	19.6
k_0	-	0.67	0.88	0.77	0.71	0.83
c	kPa	18	40	7	80	4
φ	°	24	24	32	26	32
E_0	MPa	20	120	185	230	162
v	-	0.26	0.30	0.31	0.28	0.28

* Geological units are indicated in Figure 2.

2.2 *TBM face pressure and monitoring data*

The TBM used for the extension of Line 5 is characterized to have 335 machine parameters (electronic sensors) that allow TBM operator and technicians to perform tunnel excavation in terms of guidance, excavation process, hydraulic and lubrication of mechanical parts as well as health and safety warnings.

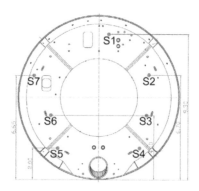

Figure 3.   Allocation of TBM face support pressure sensors.

Table 2.   Summary of best fitted PDF for face support pressure along the tunnel stretch.

TBM Sensors	PDF	Parameters Max. Likelihood	Mean (kPa)	Std Dev (kPa)	CV (%)
EPB - S1	Lognormal	$\mu = 5.22225$ $\sigma = 0.195442$	188.294	37.212	20
EPB - S2	Weibull	$\alpha = 7.15073$ $\beta = 231.885$	217.153	35.771	16
EPB - S3	Normal	$\mu = 250.92$ $\sigma = 30.8159$	250.92	30.816	12
EPB - S4	Lognormal	$\mu = 5.63268$ $\sigma = 0.154504$	282.766	43.951	16
EPB - S5	Gamma	$a\ (Shape) = 38.0527$ $b\ (Scale) = 7.2786$	276.973	44.899	16
EPB - S6	Normal	$\mu = 248.152$ $\sigma = 31.4733$	248.152	31.473	13
EPB - S7	Weibull	$\alpha = 7.22592$ $\beta = 230.101$	215.598	35.172	16

From all the TBM sensors, the face support pressure and grout injection pressure constitute the main TBM components for the achievement of ground stability during tunneling (Guglielmetti et al., 2008; Maidl et al., 2012 and Mollon et al., 2012). In this study the focus of implementation of TBM support pressure regards mainly the face support pressure. Figure 3 shows the allocation of the seven earth pressure balance sensors.

All along the tunnel stretch, a number of 2977 rings were installed. Thus, information about face support pressure during excavation for installation of these rings were available. Table 2 shows a summary of best fitted probability distribution function (PDF) for every pressure sensor of the TBM registered during tunneling. The values of coefficient of variation (CV) in this table help to illustrate the effects of variability. With CV < 20 % implies major control on applying face support pressure.

Regarding monitoring data, a total of 92 surface monitoring sections were analyzed. The transverse settlement trough from the measurements were used for the analysis. In this manner, Table 3 shows a summary of the maximum surface settlements. The length of the tunnel stretch (5.74 km) was splitted in seven zones. These zones were delimited between launching and arriving TBM structures (i.e.: Shaft and Stations), in order to have areas with homogenized ground behavior.

From the information shown in Table 3, it is possible to conclude that the zone between Hospital São Paulo (HSP) and Santa Cruz (SCR) stations provided a better result in terms of CV, which could be interpreted, from the statistical point of view, as a zone with a better homogenized behavior respect to the other tunnel zones.

Table 3. Summary of estimation of settlement curve parameters according to Peck (1969).

Zone	Sections	Smax (mm)		
		μ	σ	CV (%)
BAN – EUC	4	5.33	1.873	35
EUC – MOE	16	2.40	1.219	51
MOE – SER	24	3.00	2.654	88
SER – HSP	12	2.01	1.057	53
HSP – SCR	19	2.69	1.246	46
SCR – CKB	10	3.90	4.226	108
CKB – DDC	7	5.55	9.293	168

3 CORRELATION APPROACH

3.1 Proposed equation

The model presented by Atkinson (2007) shows that, during TBM tunneling, as the applied support pressure decreases the volume loss increases. Furthermore, when the support pressure is the same of the vertical stress, at the axis level, the volume loss will be negligible, and as well as the support pressure reaches the ground ultimate limit state the settlements will become very large up to its collapse.

In order to be able to represent mathematically this consistent behavior of ground movement due to TBM support pressure, the hyperbolic equation implemented by Duncan and Chang (1970), to describe nonlinear analysis of stress – strain behavior of soils, was taken as reference. So, by adapting this formulation to the model presented by Atkinson (2007), the following equation is proposed:

$$S_{\max} = \frac{(P_0 - P)}{a - bP} \tag{1}$$

where S_{max} = maximum surface settlement (mm), P = applied TBM support pressure (kPa), P_0 = estimated initial TBM support pressure for face stability, and a and b are curve-fitting parameters (the physical meaning of these variables are not subject of discussion in the present work).

Figure 4 shows the representation of the proposed mathematical formulation (from Eq. 1) to express the nonlinearity behavior of ground due to the applied TBM support pressure. This equation represents a good mathematical approach to reproduce model presented by Atkinson (2007). If the applied TBM support pressure equals the initial stress of the soil (P_0), thus the settlement at surface will be negligible.

By reducing the applied support pressure (P) is observed that the soil deformation follows an elastic behavior. After that the limit is reached and then the development of soil plastic behavior begins to be noticed, meaning in large surface settlement which won't be more recoverable. Large settlement and,consequently, tunnel face collapse is achieved when the applied TBM support pressure reaches a minimum value (P_{min}).

The slope of the proposed curve is obtained through the following mathematical derivation:

$$\frac{dS_{\max}}{dP}\Big|_{P \to P_0} = -\frac{1}{a - bP_0} \tag{2}$$

The value of minimum applied pressure (P_{min}) to have the maximum surface settlement (S_{max}) is mathematically demonstrated by first transforming P (from Equation 1) as a function of the maximum surface settlement (S_{max}), which is:

$$P = \frac{P_0 - aS_{\max}}{1 - bS_{\max}} \qquad (3)$$

Therefore, as S_{max} approaches infinity the value of P_{min} in which large settlement and/or tunnel face collapse will be achieved by the following expression:

$$\lim_{S_{\max} \to \infty} \frac{P_0 - aS_{\max}}{1 - bS_{\max}} = P_{\min} = \frac{a}{b} \qquad (4)$$

3.2 *Parametric analysis of the proposed equation*

In order to analyze the impact of the fitting-curve parameters, presented in the proposed equation, a series of parametric analyses are made to describe this influence. As shown in equation 1, the fixed input variable is the initial TBM face support pressure (P_0), which value depends on the case study under analysis.

In this regard, by considering the monitoring data analysis shown in Table 3, the tunnel stretch between Hospital São Paulo (HSP) and Santa Cruz (SCR) stations was chosen as reference to extract the information required to perform this analysis. Thus, the analytical method proposed by Anagnostou and Kovari (1996), based on limit equilibrium method, was implemented for the estimation of face support pressure.

Figure 5 shows the model geometry for estimation of the face support pressure (P_0), by considering the geological profile between HSP – SCR stations as shown in Figure 2.

Therefore, by implementing the method of Anagnostou and Kovari, the estimated value of face support pressure was 310 kPa. Figures 6 and 7 shown the effects of variation the fitting-curve parameters. By analogy with the Duncan and Chang formulation's, a constitutes the slope of the curve indicating a linear behavior and b correspond to the asymptotic tendency of the curve is achieved when the applied face support pressure is minimum (P_{min}), thus the surface settlement tends to be large or infinite (tunnel face collapse).

3.3 *Fitting the proposed equation to case study monitoring data*

Measured surface settlements monitoring data, along the tunnel stretch between HSP – SCR stations, were selected to show the fitting suitability of the proposed equation. As it well known, within the mathematical literature, the coefficient of determination of R-squared is an inadequate measure for the goodness of fit to apply directly in nonlinear models. Nevertheless, it is still a frequently tool use for the analysis and interpretation of nonlinear fitting to data.

An example, to the mentioned above, is given by Spiess and Neumeyer (2010) which performed thousands of simulations for their study on pharmacology and got to the conclusion that the use of R-squared to evaluate the fit of nonlinear models leads to an incorrect interpretation.

Under this context and by taking it to the present scenario, firstly a mathematical artifice was applied to linearized the propose equation and on this linear equation, the goodness of fit analysis was applied by using the minimization solver available in MS *Excel 2016*.

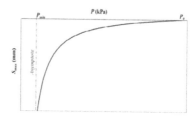

Figure 4.   Representation of the proposed Immediate Surface Settlement Curve (ISSC).

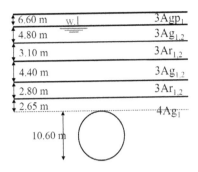

Figure 5. Tunnel model geometry between HSP – SCR stations.

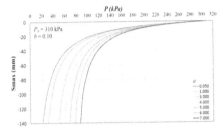

Figure 6. Effect of changing a on ISSC.

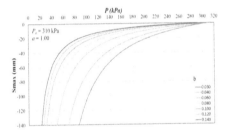

Figure 7. Effect of changing b on ISSC.

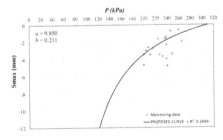

Figure 8. Best-fit curve to the monitoring data along the tunnel stretch between HSP – SCR stations.

Figure 8 presents the best-fit curve to the observed data, the value of R-squared presented correspond to that applied to the equivalent linear equation model which by relation can be considered the same for the proposed model. The value of fitting parameters a and b obtained were 9.850 and 0.211, respectively.

3.4 *Quantification of variability to the proposed equation*

The following analysis is based on the work of Zhai and Rahardjo (2013), which procedure applied for the understanding of the Soil Water Characteristic Curve (SWCC), that can be perfectly taken to apply to this case.

The idea of this procedure will be to represent the upper and lower bounds of the proposed equation (Equation 1). According Kool and Parker (1988), The bounds of a model is directly correlated to the confidence limits of the parameters involved in the equation. This can be done by estimating the variance of each parameter by approximately using t-statistic as follows:

$$a \sim [a_{\min}, a_{\max}] = \left[a - t_{\alpha/2} \sqrt{\mathrm{var}(a)}, a + t_{\alpha/2} \sqrt{\mathrm{var}(a)} \right]$$
$$b \sim [b_{\min}, b_{\max}] = \left[b - t_{\alpha/2} \sqrt{\mathrm{var}(b)}, b + t_{\alpha/2} \sqrt{\mathrm{var}(b)} \right] \tag{5}$$

where α is the confidence level (95%). Thus, by the combination of the maximum and minimum values of the fitting parameters on the proposed equation, the upper and lower confidence limits can be obtained.

Figure 9 shows the correlation between the maximum surface settlement with fitting parameters a and b, respectively.

Therefore, the combination of a_{max} and b_{min} will give the upper bound while the combination of a_{min} and b_{max} will give the lower confidence limit. The confidence limits are expressed as follow:

$$Upper : S_{\max} = \frac{P_0 - P}{a_{\max} - b_{\min} P}$$
$$Lower : S_{\max} = \frac{P_0 - P}{a_{\min} - b_{\max} P} \tag{6}$$

Finally, Figure 10 shows the variability of the proposed equation through computation of the upper and lower confidence limits. In order to be able to compute this procedure it was assumed a coefficient of variation (*CV*) of 20% for both the fitting parameters, which will allow to estimate the variance, and a $t_{\alpha/2} = 2.093$ ($\alpha = 95\%$ with n = 19 observed data assumed, from Table 3, as the degree of freedom).

4 CONCLUSIONS

The present paper was intended to show a proposed equation for relating the applied TBM face support pressure with the immediate surface settlement, which is credited to be of helpful for the personnel involved during tunnel construction with TBMs in urban areas. This formulation is based in the hyperbolic equation that provided an excellent result to the nonlinear constitutive model proposed by Duncan and Chang (1970) and proved to be an excellent tool to relate these variables.

A rationale procedure for the estimation of the fitting parameters was presented on the basis of having information regarding TBM support pressure and field monitoring data. To this case, it was taken data from the extension works of São Paulo metro Line 5. Thus, the values of fitting parameters a and b regarded only to the tunnel stretch that was studied. The analysis, with the proposed equations and upper and lower confidence limits, have shown that excavation proceeded inside the expected range of correlation. The approach here implemented to this case study can be applied to other tunneling projects.

The explanation of the physical meaning of the fitting parameters was not the subject of discussion in the present work.

For other TBM tunneling projects, the proposed equation could be used by considering the local characteristics where tunnel will be constructed, which is an important aspect because allow to estimate, for example, the initial support pressure (P_0) which intrinsically considers variables as tunnel geometry, TBM tunneling method (EPB or Slurry), soil stratigraphy, water

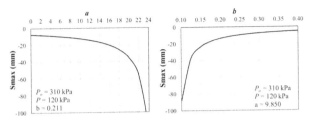

Figure 9. Relationships between maximum surface settlement (S_{max}) and fitting parameters a and b, respectively.

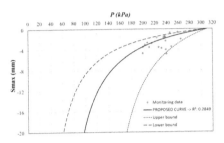

Figure 10. Illustration of the best fitted, upper and lower bound limits of maximum surface settlement between HSP – SCR stations.

level and geotechnical parameters. Values of the fitting parameters could be taken from the first monitoring measurements and while excavation is in progress, these fitting parameters can be updated allowing a better correlation.

Thus, the proposed equation can be implemented generically in every TBM tunneling project keeping in mind the ground conditions of the area, as well as that the fitting parameters a and b need to be calibrated for that specific area of study, which can be done by progressively updating the data from monitoring field measures.

ACKNOWLEDGEMENT

The authors are grateful to the Companhia do Metropolitano de São Paulo (METRÔ) for providing and authorizing the use of the data on this study to validate the correlation approach here proposed. The authors would also like to acknowledge the financial support of the Brazilian Research Agency – CNPq, as well as the Foundation for Support of Research DF (FAP-DF).

REFERENCES

Anagnostou, G. and Kovári, K. 1996. Face Stability Conditions with Earth-Pressure-Balanced Shields. Tunn. Undergr. Sp. Tech. 11(2),pp.165–173.

Atkinson, J.H. & Potts, D.M. 1977. Subsidence above shallow tunnels in soft ground. Proc. ASCE Geotech. Engng Div. 103 (GT 4), pp. 307–325.

Atkinson, J. 2007. The Mechanics of Soils and Foundations. London: Taylor & Francis, 2nd ed, 475p.

Attewell, P.B. & Woodman, J.P. 1982. Predicting the dynamics of ground settlement and its derivatives caused by tunnelling in soil. Ground Eng. 15(8),pp. 13–22 and 36.

Duncan, J.M. and Chang, C.Y., 1970. Nonlinear analysis of stress and strain in soils. Journal of Soil Mechanics & Foundations Division, 96(5), pp 1629–1653.

Fargnoli, V., Boldini, D. & Amorosi, A. 2013. TBM tunneling-induced settlements in coarse-grained soils: The case of the new Milan underground line 5. Tunnelling and Underground Space Technology, Elsevier, 38: 336–347.

Guglielmetti, V., Grasso, P., Mahtab, A., & Xu, S. 2008. Mechanized Tunnelling in Urban Areas: design methodology and construction control. London: Taylor & Francis, 528p.

Leca, E., & New, B. 2007. Settlements induced by tunneling in Soft Ground. Tunnelling and Underground Space Technology, Elsevier, 22: 119–149.

Lee, K.M. and Rowe, R.K. 1990. Finite element modelling of the three-dimensional ground deformations due to tunnelling in soft cohesive soils: Part I—Method of analysis. Computers and Geotechnics, 10(2),pp.87–109.

Loganathan, N. & Poulos, H.G. 1998. Analytical prediction for tunneling-induced ground movement in clays. Journal of Geotechnical and Environmental Engineering, 124(9),846–856.

Kimura, T. and Mair, R. J. 1981. Centrifugal testing of model tunnels in soft clay. Proc. 10th International Conference on Soil Mechanics and Foundation Engineering, Stockholm, Vol.l, pp. 319–322.

Kool, J.B., Parker, J.C., 1988. Analysis of the inverse problem for transient flow. Water Resources Research 24, 817–830.

Macklin, S.R. 1999. The prediction of volume loss due to tunnelling in overconsolidated clay based on heading geometry and stability number. *Ground engineering*, *32*(4), pp.30–33.

Maidl, B., Herrenknecht, M., Maidl, U. & Wehrmeyer, G. 2012. Mechanized Shield Tunneling. Berlin: Ernst & Sohn, 2nd Edition, 470p.

Mair, R.J. & Taylor, R.N. 1997. Theme lecture: Bored tunnelling in the urban environment. Proc 14th Int. Conf. Soil Mech. & Fdn Engng, Hamburg, 4, pp. 2353–2385.

Mollon, G.; Dias, D. & Soubra, A. H. 2012. Probabilistic analyses of tunneling-induced ground movements. Acta Geotecnica, Springer, 8:181. DOI:10.1007/s11440-012-0182-7.

Peck, R.B. 1969. Deep excavations and tunneling in soft ground, State of the Art Report. Proceedings of the 7th International Conference on SMFE. Mexico City. State of the Art Volume, pp. 225–290.

Repetto, L., Tuninetti, V., Guglielmetti, V. & Russo, G. (2006). Shield tunneling in sensitive areas: A new design approach for the optimization of the construction-phase management. Proc.: World Tunnel Congress and 32nd ITA Assembly. Seoul, Korea.

Rowe, R.K., Lo, K.Y. and Kack, G.J. 1983. A method of estimating surface settlement above tunnels constructed in soft ground. Canadian Geotechnical Journal, 20(1),pp.11–22.

Spiess, A-N. and Neumeyer, N. 2010. An evaluation of R^2 as an inadequate measure for nonlinear models in pharmacological and biochemical research: a Monte Carlo approach. *BMC Pharmacology*, 10:6.

Zhai, Q., Rahardjo, H., 2013. Quantification of uncertainty in soil-water characteristic curve associated with fitting parameters. Eng. Geol. 163, 144–152.

Tunnels and Underground Cities: Engineering and Innovation meet Archaeology,
Architecture and Art, Volume 11: Urban
Tunnels - Part 1 – Peila, Viggiani & Celestino (Eds)
© 2020 Taylor & Francis Group, London, ISBN 978-0-367-46899-6

Research and exploration on building large urban underground space by small TBMs-multi-section undercutting and prefabricated component lining method

Y. Gao, P. Cheng, Y. Li & C. Feng
Underground Space Architectural Design and Research Institute of China Railway Engineering Equipment Group, Co., Ltd., Zhengzhou, CHINA

ABSTRACT: With the growth of cities, the demands for building large underground space become increasingly intense. However, the traditional open-cut and undercut methods have certain limitations and are hard to fully satisfy the complicated boundary conditions of urban areas. For the purpose of a safer, more efficient, economical, and environmentally friendly development mode, by drawing on the merits of traditional methods and consider the development trend of smart construction and prefabricated structural application, we present a brand-new concept of building large urban underground space with small TBMs for multi-section undercutting and prefabricated component for lining, namely the Cut and Convert Method (CC Method for short). The main principle of the method is to divide a large underground space into small sub-units, and complete these units via subsection undercutting with small TBMs and prefabricated components; and then implement space and structure conversion to finally form a large integrated underground space. This paper introduces the overall idea and core principle of this method, illustrates the method in details by combining with the first application case, and discusses the application prospects and challenges. We hope to provide a new solution for developing large urban underground space.

1 INTRODUCTION

With the rapid urban expansion, the urban land is becoming increasingly tense, and urban space resources are increasingly scarcer. Meanwhile, as urban infrastructures age and people are demanding better living conditions and environmental protection, the development and utilization of underground space has become an inevitable trend for modern urban development.

As we know, Chinese and foreign scholars have done a lot of researches on the construction methods and achieved fruitful results. Japanese scholars Takuya Satou, Man Shimizu, et al[1–2] developed the HEP&JES method, which has been widely applied in undercrossing the existing traffic lines. Chinese scholars Li Zhaoping, Chen Jiuheng, et al[3–4] drew on the experience of foreign underground engineering prefabrication technologies, solved such technical difficulties as assembly hoisting, positioning and rectification, crack control, anti-seismic, and component connection, and proposed a subway station construction technology with prefabricated component lining, which has been successfully applied to many subway station projects, including Changchun Metro Line 2. With the joint efforts of Chinese scholars Wang Mengshu, Gong Xiaonan, et al[5–7], the Shallow Cover Underground Excavation (SCUE) method is quite mature and is developing towards the super shallow overburden and complex geological adaptability. Peng Limin, Gao Yi, et al[8–9] demonstrated the application of this technology in underground space development while studying the modern rectangular pipe jacking technology, and discussed the prospects of this method.

However, for the development and utilization of underground space in built-up urban areas, extremely unique and complicated boundary conditions are often encountered, and it is difficult to use land resources to maximize the comprehensive development of underground space with existing technologies. Furthermore, in the current development stage, the regularly used open-cut method still has huge adverse impacts on surrounding buildings, traffic, landscape, and pipelines, and greatly interferes on the construction process and the daily life of urban residents. However, the existing undercut method is mainly applied to tunnel projects with simple sections and limited spans, such as subway tunnels and underground passages. For large underground spaces featuring large spans, multiple spans, and complex sections, such as underground parking lots, subway stations, and underground malls, the existing undercut method cannot fully meet the needs for large underground space development due to its limitations, high risks, high cost, and low efficiency.

So, how to maximize the value of urban land resources? How to minimize the impact on the order of normal city life during city development process? This is a very important topic for the future urban construction.

In this paper, we present a brand new method of building large urban underground space with small TBMs for multi-section undercutting and prefabricated component for lining, namely the Cut and Convert Method, which has been successfully applied to practical projects. By means of this method, the large underground space with prefabricated component lining is better developed, thus avoiding the high cost and low reliability of large TBMs, and achieving the safe, efficient, and economical construction goals.

2 EXPLORATION

2.1 Lessons Drawn from Traditional Methods

As shown in Figure 1, a non-excavation method is required to reduce the impacts on the ground environment. For multi-span or large underground space projects with complex sections, CRD and PBA methods are usually used. These methods divide a large structural section into several small sections, which then are completed one by one, and finally connected together to form a large section. Compared with the one-time formed large structure, this effectively reduces the construction difficulties. As we all know, traditional methods usually incorporate small machineries and manual operation, featuring high labor intensity, high safety risks, harsh working environment, and low efficiency. Howbeit, their concept of splitting and benching operation is an effective solution for realizing large undercut sections.

2.2 Combination of Mechanized Construction

Compared with traditional CRD and PBA methods, the packaged mechanized construction (such as shielding and pipe jacking methods) possesses many advantages, including safety, high efficiency, favorable working environment, and low labor intensity. However, the section size formed by TBM and pipe jacking machine is limited, and it is hard to form a super-large section by one time. Meantime, it lacks the flexibility of traditional benching method since it often cuts circular sections, which are difficult to be integrated into a large sectional structure.

(a) Schematic of CRD method

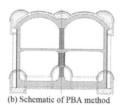

(b) Schematic of PBA method

Figure 1. Building underground space with traditional undercut methods.

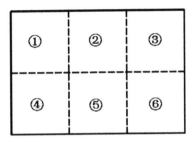

Figure 2. Preliminary conception of CC method.

Then, can we utilize the safe and reliable mechanized construction method, learn from the gradually excavation adopted by traditional benching methods, and find a construction method that combines the advantages of both? This is the main direction that the research team will follow. Nowadays, with the advancement of the rectangular section TBM technology, it is possible to combine the mechanized construction with multi-section excavation concept adopted in traditional benching method.

2.3 *Conception*

As shown in Figure 2, the rectangular section of a large underground space is divided into several small rectangular sections, which are built one by one by small TBMs (adaptive to the rectangular section) and finally connected together to form a large sectional structure. This is the preliminary conception. The Harmonica method from Japan is also based on this idea. Namely, a grid-like working space is formed via multi-section excavation with small TBMs and lining with profile steel segment. Then, the cast-in-place construction of the reinforced concrete main structure is completed in this space. In this method, the unique profile steel segment are mainly used as a temporary support structure, which features such weaknesses as low recycling rate, great waste, and inferior economic benefits. So, the main problem that the method needs to study and solve is to improve the economy.

3 PRINCIPLE

To directly utilize the main structure for multi-section excavation, maintain the flexibility of section division and the feasibility of mechanized construction, we need to re-define the underground engineering structure and propose a new structure-splitting and converting method. As shown in Figure 3, we illustrate the CC Method with the case of a single-layer dual-column three-span underground structure.

3.1 *Redefinition of Structure*

The outermost roofs, floors, and side walls of the underground project's main structure directly contact the surrounding rocks and bear the external load from the surrounding rocks. This load is then transferred to the internal beam columns or wallboards, and converted into the internal structural force. Then, the structural components directly contacting the

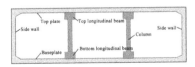

Figure 3. Single-layer dual-column three-span underground structure.

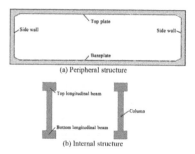

(a) Peripheral structure

(b) Internal structure

Figure 4.   Redefinition of structures.

surrounding rocks and bearing the external load from the surrounding rocks are defined as the peripheral structure. And the internal beams, columns, and wallboards are defined as the internal structure, as shown in Figure 4.

3.2   Splitting of Structure

As shown in Figure 5, a unit length of main structure is taken. the peripheral structure is split from the internal structure, so that we can analyze the peripheral structure separately. Then, the peripheral structure is split again, at a uniform size if possible, to reduce the types of sections and construction equipment. The split peripheral structure is prefabricated and integrated with the temporary structure into a closed frame, to form the sub-unit components as shown in Figure 6, and to meet the needs for mechanized construction with small TBMs.

3.3   Conversion of Structures

The sub-units are completed at a very close spacing with small TBMs to form an intermediate structural system which composed of the peripheral and temporary structures as shown in Figure 7, where the peripheral structure bears the external load of the surrounding rocks and transfers the load to the internal temporary structure. The load form and bending moment of the roof are shown in Figure 8.

The internal structure is built within the space of the intermediate structural system, as shown in Figure 9.

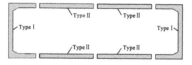

Figure 5.   Secondary splitting of peripheral structure.

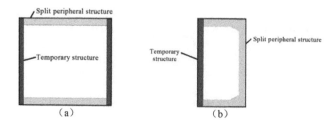

Figure 6.   Formation of sub-unit components.

Figure 7.　Intermediate structural system.

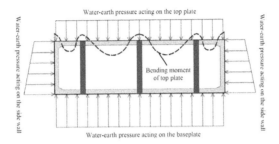

Figure 8.　Stressing mode of the intermediate structural system.

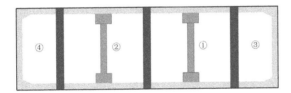

Figure 9.　Construction of the internal structure.

Figure 10.　Schematic of the final structure.

Finally, the temporary structures are removed in proper order, to convert the load of surrounding rock transferred from the peripheral structure to the internal structure.. Meantime, the peripheral structure is subject to joint treatment, to complete the conversion of the structural system and form the final main structure, as shown in Figure 10. After that, its structural stressing mode and bending moment of roof change accordingly, as shown in Figure 11.

Through structural splitting and conversion, the mechanized undercutting and prefabricated construction can be applied to the large rectangular sections of underground projects.

4 PRACTICE

After putting forward this method, the Underground Space Architectural Design and Research Institute of China Railway Engineering Equipment Group, Co., Ltd (CREG) has first applied it to an actual project.

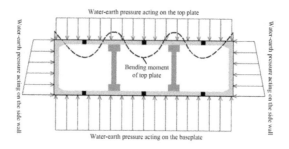

Figure 11.   Stress mode of the final structure.

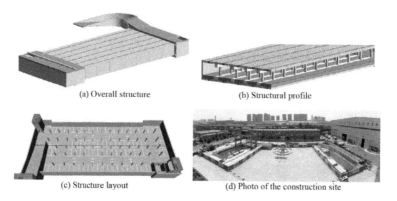

(a) Overall structure   (b) Structural profile

(c) Structure layout   (d) Photo of the construction site

Figure 12.   Schematics of the project.

Figrue 13.   The TBMs.

4.1   *Project Background*

The test project is an underground single-storey 5-column and 6-span parking lot, as shown in Figure 12.

The project is about 90m long and 35m wide, and has a total construction area of about 3,500 m^2 and 3m thick overburden. The main structure of the garage is located in the silt stratum, and the construction site lies on the CREG Plaza. To minimize the damages to the existing square landscape and the impacts on the office environment, the project team utilizes the working shafts at both ends and uses small TBMs for subsection construction.

As shown in Figure 13, this practice uses small rectangular TBMs; its section size is 5.00x5.70m, which can be freely split into two TBM with a section size of 5.00x2.85m.

4.2   *Splitting of Structure*

As per the cross section of the main structure as shown in Figure 14, a unit length of main structure of the underground parking lot is taken out for splitting.

The roof, floor, and side walls constitute the peripheral structure, and the internal top longitudinal beams, columns, and bottom longitudinal beams constitute the internal structure. The peripheral structure is split into 7 sub-units, including 5 standard sections and 2 small sections, which are integrated with the temporary structure made of profile steel to form the sub-unit components where tunneling machines can be applied, as shown in Figures 15, 16, and 17.

4.3 Conversion of Structures

The sub-units are undercut one by one with TBMs to form the intermediate structural system. As shown in Figure 18, the roof, floor, and side walls of the garage's main structure are basically completed. In this state, the roof and floor directly bears the load of the surrounding rocks and transfers it to the temporary structure. Figures 19 and 20 show the real pictures of the project.

In the internal space of the intermediate structural system that has been completed, the internal structure (namely the top longitudinal beams, columns, and bottom longitudinal beams) is built, as shown in Figure 21. For this parking lot, the top longitudinal beams and columns are made of profile steel, and the bottom longitudinal beams are made of cast-in-situ reinforced concrete.

Next, the middle temporary structure of profile steel is disassembled in proper order, and the peripheral structure (roof and floor) is subject to joint treatment. The split roof and floor are connected into a continuous slab structure. In this way, the load of the surrounding rock, that the roof and floor directly bear and transfer, is converted from the profile steel temporary structure to beams and columns of the internal structure, to complete the conversion of the structural stressing system. Furthermore, the small space of sub-units is also connected to

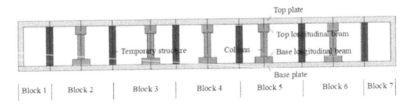

Figure 14.   Cross section of the main structure.

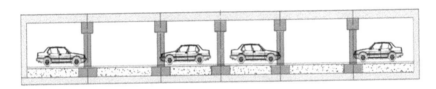

Figure 15.   Splitting of the main structure.

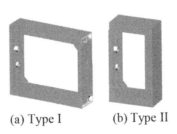

(a) Type I   (b) Type II

Figure 16.   Schematic of sub-unit components.

Figure 17. Photo of Type-I sub-unit.

Figure 18. Intermediate structural system of the underground parking lot.

Figure 19. External view of sub-unit construction.

Figure 20. Internal view of sub-unit construction.

form a large space needed by the garage. Finally, the 5-column and 6-span framed main structure as planned in the original design scheme is realized, as shown in Figures 22 and 23.

During the whole construction process, only two small working shafts are occupied for construction, and the intermediate main structure is subject to the fully mechanized SCUE construction, which is safe and environmentally friendly, without damaging the ground environment. The prefabricated structure features reliable quality and no leakage. During the whole construction process, the precision and the structural deformation completely meet the

Figure 21.   Photo of the internal structure.

Figure 22.   Schematic of the final main structure.

Figure 23.   Internal view after removing the temporary structure.

design requirements. Finally, the same structural form and spatial functions as the conventional open-cut method can be achieved. The construction of the entire main structure only takes 8 months, which is much shorter than the construction period required by the conventional undercutting method. The reason is that this method does not need the main enclosure structure and groundwater lowering, and its cost is as low as the open-cut method, yet far lower than the traditional undercut method. Through the successful application of this method, we can achieve the safe, environmental, and cost-effective construction of large underground space.

5 PROSPECT

The successful application of the large single-storey multi-span underground space project verifies the feasibility of the CC Method. Its unique undercut mode is particularly suitable for the development of underground space in built-up urban areas. Moreover, this theory can be further applied to large multi-storey multi-span underground space projects, which are illustrated in Figures 24 and 25 using a single-column dual-span subway station.

First, the peripheral structure of the subway station is split, to form the sub-unit components with the profile steel temporary structure. Next, six sub-units are built with rectangular TBMs to form the intermediate structural system. In the space of the intermediate structural system, the top longitudinal beams, middle plates, middle longitudinal beams, bottom longitudinal beams, and columns are flexibly constructed. Then the temporary steel supports are removed in proper order, to complete the conversion of the structural stressing system. At the

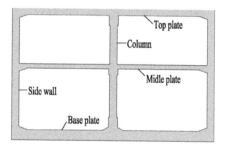

Figure 24. Section of the single-column dual-span subway station.

same time, joints of the peripheral structure are treated, then the single-column dual-span main structure is formed.

This method still faces many difficulties and challenges if it wants to meet the needs of diversified space and various boundary conditions for underground space development projects, such as underground parking lots, subway stations, underground expressways, and underground warehouse logistics:

1. It is necessary to further carry out adaptability and extension researches on underground projects featuring long-distance, curved and complex sections, and special geological conditions.
2. The underground engineering structure needs to be further studied, to make prefabricated components form a safe and reliable global stress system to hold the multi-storey underground engineering structure, and meet the anti-floating and anti-seismic requirements.

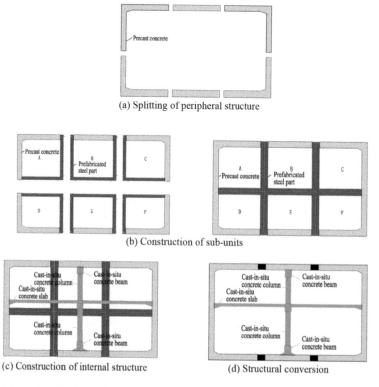

Figure 25. Schematic of building the single-column dual-span subway station.

3. More lightweight engineering equipment needs to be developed to better meet the narrow boundary conditions of construction sites in built-up urban areas.
4. New building materials and components need to be developed to further save costs and facilitate construction.
5. This method is required to be applied to more engineering projects, to continuously improve related technologies and processes, and forge them into mature, reliable, and widely applied construction technologies.

6 CONCLUSION

In this paper, by drawing on the merits of traditional construction methods, we present a brand new method of building large urban underground space with small TBMs for multi-section undercutting and prefabricated component for lining, which has been successfully applied to practical engineering projects. Therefore, we draw a conclusion as that:

1. This method can divide a large underground cross-section into multi-section suitable for undercutting with small TBMs, and convert the completed small sub-units into an integrated large underground space.
2. This method can realize large urban underground space construction with non-excavation method and lining with prefabricated structure for .
3. Through the successful application of this method to a practical project, we have verified its feasibility and obvious advantages against traditional methods.
4. Although this method can be further applied to complicated multi-storey long-distance underground space projects, it still faces many limitations and challenges that need to be further studied.
5. This method provides a brand new solution for the development of large urban underground space.

REFERENCES

Sato, T. & Minakawa, Y. & Furukawa, N. 2000. HEP&JES construction method. SED: Structural engineering data 11(15): 96–99.
Shimizu, M. & Watanabe, A. & Chiba, K. 2003. Strategy of mechanical undercrossing transection construction method. Construction Machinery & Equipment 39: 6–11.
Li Zhaoping, Wang Chen, Su Huifeng, et al. An Experiment Study on the Evolution Law of Concrete Structure Crack and Joint Seam Deformation for Tenon Groove Joints in the Prefabricated Metro Station [J]. CHINA CIVIL ENGINEERING JOURNAL, 2015(s1):409–413(in Chinese)
Chen Jiuheng. Construction Technique of Prefabricated Subway Station [J]. RAILWAY CONSTRUCTION TECHNOLOGY, 2015(11), 62–69(in Chinese)
Wang Mengshu. The Boring Excavation and Construction in Beijing Metro[J]. Chinese Journal Rock Mechanics and Engineering, 1989, 8(01):52–62(in Chinese)
Wang Mengshu. An Investigation on Design and Construction of Shallow Cover Underground Excavation [J]. UNDERGROUND APACE, 1992(2):289–294 (in Chinese)
Wang Zhida, Gong Xiaonan. Size and Effect of Step Length in Pedestrian Underpass[J]. BULLETIN OF SCIENCE AND TECHNOLOGY, 2009, 25(6):820–825 (in Chinese)
Peng Limin, Wang Zhe, Ye Yichao, Yang Weichao. Technological Development and Research Status of Rectangular Pipe Jacking Method [J]. Tunnel Construction, 2015, 35(1):1–8 (in Chinese)
Gao Yi, Feng Chaoyuan, Cheng Peng, et al. Study on the "Overall-Carrying-Soil Effect" of Shallow Buried Rectangular Pipe Jacking [J]. Chinese Journal of Geotechnical Engineering, 2018:1–6 (in Chinese)

Tunnels and Underground Cities: Engineering and Innovation meet Archaeology,
Architecture and Art, Volume 11: Urban
Tunnels - Part 1 – Peila, Viggiani & Celestino (Eds)
© 2020 Taylor & Francis Group, London, ISBN 978-0-367-46899-6

The new underground line of Sarmiento Railway Project in the city of Buenos Aires

G. Giacomin & M. Maffucci
Ghella S.p.A., Roma, Italy

M. Cenciarini & S. Vittor
Ghella S.p.A.-Sacde JV, Buenos Aires, Argentina

ABSTRACT: The paper describes the complex logistics and challenges to excavate the new railway of the Sarmiento line below the existing line without any disruption to the railway in operation. The Sarmiento railway with more than 300 daily trips, transporting 110 million people each year, lies entirely on surface and provides the main transport system for people moving between the western suburbs and the city center of Buenos Aires. The Earth Pressure Balance Tunnel Boring Machine (EPB TBM) chosen for the specific excavation has a diameter of 11,46m and is expected to excavate a total of 16km for the first Lot. The overall length of the three Lots adds up to 36.9Km. The geology is formed by silty clays and soft clay, silty sand and puelche (a monogranular sand with a water pressure of 2.5 bar). Along the line there are 15 huge stations excavated through conventional methodology in cavern and through top-down methodology. The main challenge of the project is to optimize production reducing the disruption for the community and stakeholders throughout the highly populated city of Buenos Aires.

1 INTRODUCTION

The enormous metropolitan area of the city of Buenos Aires and the growth of its population in the last 10 to15 years has required a new transport system. In the North - West of the city, between the Ciudad Autonoma of Buenos Aires and the municipality of CABA, the new underground line of the Sarmiento Railway passes through 6 different municipalities covering an area of approximately 6,5 million citizens. The new infrastructure is expected to thus improve the quality of life of 15.5% of the overall population of all of Argentina.

The new line will allow an important increase in the quantity of passengers as well as in train frequency, reducing travel times from 8 to 3 minutes. The great advantage of the new line is its lowering from surface to underground improving road traffic by avoiding 52 existing crossings. This specific improvement also allows a huge reduction of accident likelihood and at the same time gives the opportunity to use the freed surface area for additional road, public space, cycling lines, green area, etc.

It is estimated that the new line will generate 10.000 new jobs.

The project is designed to move 150 million passengers starting from 2022, with a frequency of one train every 3 minutes. The contract has been awarded in 2008 and after different delays due to the Country's political conditions, the new Joint Venture (JV) formed by Ghella S.p.A. and Sacde is executing the works from 2016.

1.1 Geological Formation

The geological conditions of the city of Buenos Aires are quite homogenous along all the alignment. In the geological profile, the formations are mainly clay (medium to fine), consolidated silt (known as Tosca), Sandy Clay and compacted sand.

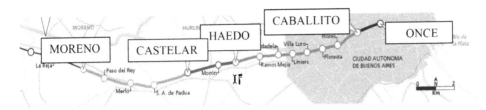

Figure 1. Plan View of the overall Sarmiento Project.

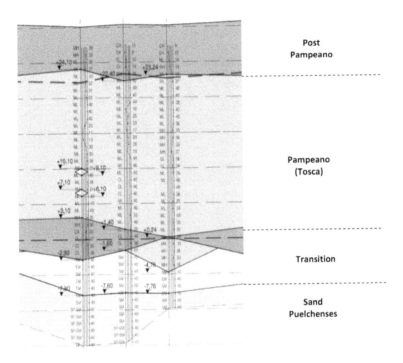

Figure 2. Typical Geological Profile.

- Post-Pampeana Formation: characterized by lacustrine fluvial deposits and clay, fine sandy, gray, greenish-gray sediments, with a thickness of up to 5-7m, and also in correspondence with the accumulations of the riverbed in the hydrographic basin;
- Pampeana Formation: characterizing mainly the surface formations of Buenos Aires, it is constituted by sandy, clayey silts, of medium high consistency, with a variable thickness of approx. 20-25m;

Puelches Formation: characterized on the upper part by a layer of consistent silty clays and on the lower part by deposits of sand. The waterproof clay layer has generally variable thickness between 3-4m and 6-8m. The successive layer of sand extends bibliographically to the upper part of the underlying Parana formation, with a thickness of approx. 20-30m.

The water level, as confirmed by the data from the surveys, is found in the upper part of the Pampeano and Post Pampeana sediments, at an average depth of 5m approx. The water table is generally flat with local depressions corresponding to the channels.

The works in the Sarmiento project affect only the first two hydrogeological levels described, developed mainly within the Pampeana formation; special attention is to be given to the "acquitard" permeable level that separates the two main aquifers (Epipuelche aquifer and Puelche aquifer) which would negatively impact on the water pressure conditions if entirely excavated.

Table 1. Description of characteristics of geological formation.

Description	g	E(Tn/cm²)	K₀	Cu/C'(Kg/cm²)	fᵤ/f'
Clay	1.9	1 – 1.5	0.5	0.8/0.0	5°/28°
Consolidated Silt (called "Tosca")	1.9	3 – 3.5	0.5	1.8/0.2	15°/30°
Clayed Sand	1.9	1.5 – 2.0	0.5	1.0/0.1	5°/28°
Consolidated Sand (called «Puelche»)	2.0	3.0 – 3.5	0.4	-/0.0	-/35°

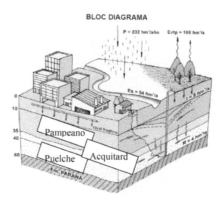

Figure 3. Typical Subsoil Formation.

2 STATIONS

The Sarmiento project includes 9 new stations: Moron, Haedo, Ramos Mejia, Ciudadela, Liniers, Villa Luro, Floresta, Flores and Caballito. With the same concept of the tunnel, also the stations are currently on surface and will be moved underground except for the last one, Caballito: 8 will be excavated in cavern with mining methodology and the surface one will be done in open air in trench.

The Stations are designed to be big enough to allow the stopover for the entire length of the train. The typical dimensions are 240 m in length and 21 m in width.

2.1 Mined Station

The design of the mined station is very complex and divided in several phases. Before the actual excavation of the cavern, construction activity begins with the installation of a complex network of drainage pipes below the bottom of the station, which will allow the excavation on a dry working platform.

The next phase consists in building the access ramp from the surface down to the access point of the station. The excavation works for the ramp are protected by retaining structures made of secant piles; once the ramp is 10m below the surface level, the access to the bottom of the station is done through mined tunneling.

Other two access points for the excavation are allowed through two vertical shafts at the two ends of the station. The shafts give access to the crown of the station to commence the excavation in partial sections, as shown in figure 4.

The next phase consists in the actual excavation of the cavern from the top, center, sides and bottom bench, as shown in figure 5.

Once the station has been excavated, it is fitted out with a permanent lining of cast-in-situ items consisting of a bottom arch, the column sides and the cradle, see figure 6.

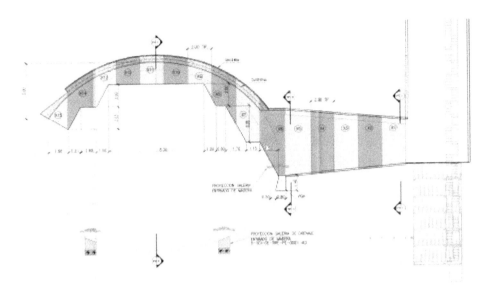

Figure 4. Typical Access in Mined Stations.

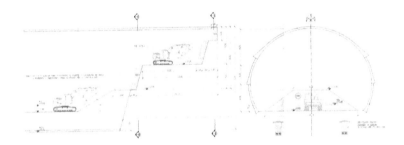

Figure 5. Typical Longitudinal and Cross Sections.

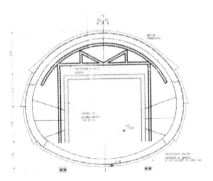

Figure 6. Typical Permanent lining.

2.2 *Surface Station*

The Caballito station will be built in trench between the transition of the underground works and the surface existing line. The construction sequence of this station is completely

Figure 7.   Typical Mined Station.

Figure 8.   Typical Surface Station.

different to the previous. Works commence from the foundation working their way up to the top level with formworks and concrete works.

3   TBM TUNNEL

3.1   *TBM EPB*

The TBM selected for the project is the 11.46m diameter S-728 EPB supplied by Herrenknecht and named Argentina. The segments for the lining are 1.8m in length and 36cm thick; the inner diameter is of 10.4m and the outer diameter is of 11.12m. The ring has a segment configuration of 6+1 key. The machine is equipped with 19 couples of thrust cylinders allowing a nominal thrust advance of more than 120.000kN. The main bearing is 6.6m in diameter and the nominal torque is more than 44.000kN.

The TBM has been refurbished in Herrenknecht's German factory, after delivering a previous tunnel project in Spain. During the refurbishment, different items were modified following a meticulous redesign between Ghella and the manufacturer, as listed below:

- Number of motors: reduction from 18 to 14 to achieve a new installed power of 4.900kW;
- Cutterhead dressing:
    - 13 gauge cutters of 17";
    - Replacing 23 double cutters with the same number of rippers;

- Removal of the transition conveyor and use of the conveyor in the back-up so to discharge in the muck car;
- Replacement of piston pumps for the pea-gravel injection with an injection pump for the B-Component grout.

During 2011 the Factory Acceptance Test was completed and the machine with all its ancillary equipment was shipped to Argentina.

3.2 *Assembly and Launching*

The TBM has been assembled in the Haedo launching shaft. The shaft's rectangular shape (105m length and 18m width) was specifically designed to accommodate the TBM and back up in its entire length, to speed up excavation and reduce the learning curve duration.

The assembly sequence was decided together with the supplier; two parallel teams worked concurrently in the shield.

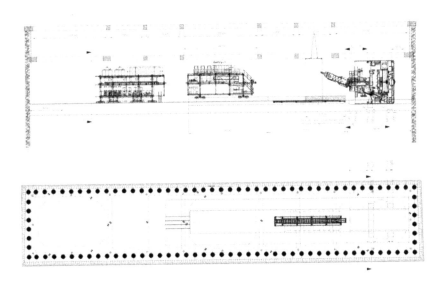

Figure 9. Launching Shaft: Side view and Plan view.

Figure 10. Site installation Haedo.

Figure 11.   Picture on the Assembly phase.

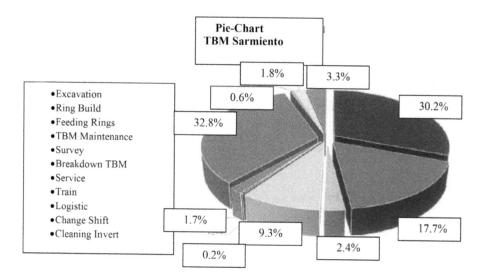

Figure 12.   Pie Chart TBM "Argentina".

3.3   *TBM Excavation*

The excavation of the tunnel started at the end of 2016 with the setup of the hydraulic, mechanic, electric and electronic features. Once also the ancillary equipment was tested and commissioned, such as the locomotive, B-Component plant, gantry crane, ventilation, etc, the TBM started boring at its cruise speed. Production shifts work 5days per week and proceed on the sixth day with ordinary maintenance. Each week, before restarting excavation, the crew practice operations in hyperbaric conditions monitoring the tools and verifying their wear by getting in the cutterhead. The aim is to reduce cutterhead interventions, improving TBM production and avoiding unplanned hyperbaric interventions.

During the excavation the surface settlement is continuously monitored. According to the design, the pressure in the face must be in the range indicated in the PAT (Plan for Advance of Tunnel). The same TBM excavation monitoring data is used to feed the PAT. To date, the TBM has been excavating with a pressure slightly above the range indicated in the PAT to avoid any kind of water inflow.

Figure 13. Tunnel View.

Within the alignment, the Contractor has installed more than 100 monitoring sections, most of them on surface and 10% underground; other monitoring stations are located on sensible buildings and each 2.000m of excavation through the instrumented rings inside the tunnel. The values monitored to date haven't detected any settlement.

In the last 20 months of excavation, Argentina has performed according to expectations. The pie-chart shown on the side indicates that the utilization rate of the TBM is around 60% (Excavation, Ring Build, Maintenance, Feeding Ring) above the average for a standard EPB with diameter greater than 11m.

3.4 *Backfilling Grout*

The fill is a two-component grout based on the proven advantages achieved in previous experiences by the Contractor such as the Legacy Way Project in Brisbane, Australia, the Follo Line Project in Oslo, Norway as well as a previous project in Argentina, the Maldonado Tunnel, in Buenos Aires. The backfilling of the annular gap provides the bedding for the segments and seals the annular gap from groundwater flows. It is important to inject the grout as soon as possible to optimize its performance. The injection occurs through grout lines, which are integrated into the machine's tailskin allowing the activity to happen while the TBM advances.

4 SEGMENT PRODUCTION

The main features of the segmental lining design are:

- Type: universal ring;
- configuration: 6+1 key;
- inner diameter: 10.40 m;
- outer diameter: 11.12 m;
- thickness: 36 cm;
- length: 1.80 m;

The segments are produced at the Haedo site area in a factory with a dedicated carrousel.

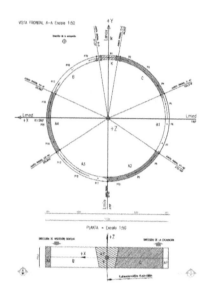

Figure 14. Cross-section of TBM Tunnel with the Segment configuration.

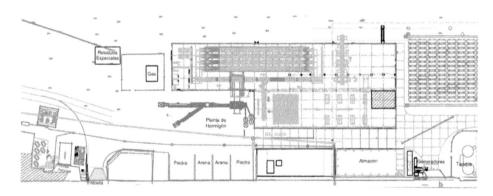

Figure 15. Segment factory.

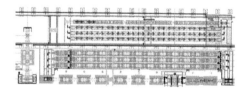

Figure 16. Typical carrousel in each factory.

4.1 *Segment Factory*

The decision to base the segment factory in the Haedo area allows to transport the segments directly to the launching shaft without any double handling of the elements. The factory area also includes a storage area, optimizing the logistics.

Contractor has provided adequate, qualified and experienced staff to manage, supervise and perform the entire production process. Several local laborers work in the production line, with the goal to train them to new skills and capacities.

Figure 17.   Segments storage.

The Haedo area also includes a batching plant, dedicated to the production of concrete for the casting of the segments.

The segmental lining is produced strictly in accordance with the design drawings and specifications. Contractor has optimized the area in order to have a minimum storage of 1.000 units of fully cured rings at all time.

The daily production rate is of approximately 14-15 rings per day, equivalent to 350m^3 of concrete per day. Each carrousel station has a duration of 12-13minutes. Given the Argentina weather, to obtain the best quality and curing of the segments, the oven temperature has been optimized with an ingress temperature of 35°C and an egress temperature of 50°C with a 95% humidity level. These values are extremely important to obtain, once the 7 to 8 hours of production cycle is complete, the minimum design level required to demold the segment at 15MpA. From a quality control point of view, the load of the sample test is between the range of 18-20MpA.

5   CONCLUSION

To achieve a good result and performance in the Sarmiento Project the design and planning of all the activities prove to be of main importance, starting from the TBM design, up to the logistical preparation which follows TBM excavation.

The optimization of the segment design and the correct installation of the precast factory result in better quality and reduced production cycle. The production of the TBM as of this publication is following an average above forecast; if excavation is to be continued with this production rate, the Contractor will complete the excavation before the Construction Baseline Schedule estimated date.

Tunnels and Underground Cities: Engineering and Innovation meet Archaeology, Architecture and Art, Volume 11: Urban Tunnels - Part 1 – Peila, Viggiani & Celestino (Eds)
© *2020 Taylor & Francis Group, London, ISBN 978-0-367-46899-6*

The complex network of sewage tunnel in the city of Buenos Aires, The Riachuelo Project

G. Giacomin & M. Maffucci
Ghella S.p.A., Roma, Italy

G. D'Ascoli, S. Sucri & F. Cichello
CMI JV, Buenos Aires, Argentina

ABSTRACT: The The paper describes the complex logistics and extreme underground activities of two TBMs (one EPB and one Mixshield) expected to bore a total length of 15km, and 2 Pipejacking ma-chine for a total length of 13km of link sewage tunnel in the middle of the city of Buenos Aires. The geology is formed by silty clays and soft clay, silty sand and puelche (a monogranu-lar sand with a water pressure of 2.5 bar). A significant stretch of the tunnels is excavated by a Mixshield TBM under the river at a depth of 30m. The remaining part of the excavation is per-formed in densely populated urban environment. Two Shafts built with diaphragm walls and peanut shape are foreseen for the launching of the both TBMs and 69 mined Shafts are re-quired for microtunneling section. The challenge of the project consists in optimizing the pro-duction reducing the disruption to the community and stake-holder in the huge and densely populated city of Buenos Aires.

1 INTRODUCTION

The Riachuelo Project is one of the main infrastructure that is part of a broader plan for miti-gating the serious environmental problems of the city of Buenos Aires, due to recurrent flood-ing of large urban areas and lack of an adequate sewer network in entire districts. The area of Matanza-Riachuelo, crossed by the Riachuelo river, is one of the most polluted in Buenos Aires where the health and the quality of life of the local population is seriously affected.

The Riachuelo Project is a large urban infrastructure, promoted by the Municipality of Buenos Aires via the local water authority AYSA (Agua y Saneamientos Argentino). It is expected to provide a definitive solution to the current limitations in the transport capacity of the sewage in a large part of the area, improve the quality of the service and relieve the Ria-chuelo river from the contamination due to the sewage tributaries. The Project is divided in three lots: Lot 1 is the sewer network covering the entire area of Matanza-Riachuelo; Lot 2 is the main treatment plant; Lot 3 is the water transport tunnel under the La Plata River for dissipating the treated water in the river, 12 km from the shore. The paper describes the main activities of the "Riachuelo Project – Lot 1: Left River Bank sewer collector" that will be built by CMI, a Joint Venture led by Ghella S.p.A. of Italy. The work consists in the realization of a complex sewerage network that develops on the left bank of the Riachuelo river in an intense-ly populated area, located south-east of the autonomous city of Buenos Aires, where there is a population of over 300,000 inhabitants.

The hydraulic system, which covers an area of about 200kmq, involves the construction of a series of tunnels excavated mainly with mechanized excavation techniques, adopting three different solutions according to the diameter of the collectors and the geology crossed. The distribution network, is generally constituted by pipes with an inner diameter of 800mm and 1100mm, executed by Pipe Jacking/Microtunnelling; the main collectors of inner diameter of

Table 1. Main Work Matanza-Riachuelo Project.

Item	Length (m)	Inner Diameter (mm)	Excavation Methodology
CMI-1	1651	800	Pipe Jacking
CMI-2	9.503	3.200	TBM EPB
DCBC	5.161	4.500	TBM Mix Shield
OC03	3.098	1.100	Pipe Jacking
OC03	366	800	Pipe Jacking
OC04	1.712	1.100	Pipe Jacking
OC04	1.409	800	Pipe Jacking
OC05	562	800	Conventional Method
OC06	1.415	800	Pipe Jacking
OC07	964	800	Pipe Jacking
OC08	14	800	Pipe Jacking
OC09	3x177	1.100	Pipe Jacking
OC09	18	3.5x3.5	Conventional Method
OC10	437	800	Pipe Jacking

3200mm (Colector Margen Izquierda -2 CMI2) and 4500mm (Desvio Colector Baja Costanera - DCBC) and total length of approximately 15km, collecting the fluids from the distribution network and transporting them to the main treatment plant, are excavated by TBMs of EPB type in the shallower stretch (CMI-2), under the more urbanized area, and by Mixshield TBM in the deeper stretch (DCBC), excavated for a significant length under the river bed. The en-tire sewage collection system on the left margin of the Riachuelo river, is summarized in Table 1 here below.

2 MICROTUNNELING

The distribution network is composed by micro tunnels with inner diameters of 800mm and 1100mm excavated with the methodology of pipe jacking.

With this method the tunnel is excavated by a micro shielded TBM. The thrust for advancing the shield in provided through the prefabricated concrete pipes that are jacked from a thrust frame installed in the launching shaft. Support of the face and transport of the excavated material is performed by circulation of bentonite slurry. The excavated material is finally separated from the slurry in a dedicated separation plant found on surface (Figure 2).

The advantage of using this method, is its compatibility with operating in the urban environment of a large and congested city like Buenos Aires, as a result of the extremely limited area required for site installation. Furthermore, most of the work is executed underground with minimal impact on the surface activities and on the existing utilities. The main activities taking place on surface are the excavation of the launching and receiving shafts and the related job-sites. As shown in Figure 1 a typical pipe jacking job-site, adjacent to the launching shaft, consists essentially of containerized installation including control container with control cabin, power pack, pumps and all the plant needed for the shield operation, separation plant including slurry production plant and pumps for the slurry circulation, electrical generator for power supply. Furthermore, a crane is required for the site assembly and for pipes and material supply to the shaft. Finally, areas for pipes and material storage and for muck disposal complete the installation. The receiving shaft jobsite is even more compact, since only a mobile crane is required for the extraction of the shield and some storage areas.

In the Riachuelo project the shafts have a depth of between 9 and 15m and a diameter of 4 to 5.50m while the pipeline is installed at a depth between 12 and 20m. The shafts are excavated in the most superficial layers generally constituted by a silty-clayey sequence of medium con-sistency followed by a layer of hard and "tosca" (silty clay), which is the layer crossed by the pipe jacking. Such vertical works are excavated top-down technique by conventional

Figure 1. Pipe-jacking Job Site.

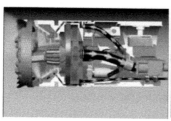

Figure 2. Slurry Circuit microtunneling.

method with spray concrete and mesh. A retaining structure of sheet piles or micropiles is installed as needed.

Distance between launching shaft and recovery shaft is of approximately 300-350mt. Once the shaft is built, it will be set up for the launching of the tunneling machine. This involves the installation of a raised mesh floor, seals, cradles, thrust frame, and other ancillary tunneling equipment such as service lines, lubrication control station and electrical control boxes. When the shaft is ready and equipped with all the plant the shield can be lowered. Using the jacking frame thrust rams, the cutter head will be pushed up as close to the launch seal as possible. This will be through the pipe break and emergency seal to the point where there is sufficient space behind the first can so that the second section can be installed. Once the power pack is installed, the TBM can begin cutting into the wall of the shaft. The micro tunneling will be connected to its services and commissioned. The pipejacking frame will be used to push the cutter head of the TBM through the launch seal and into the shaft wall to commence boring. The rams will be retracted to the point where the next pipe can be lowered onto the jacking cradle. RC pipes will be used to provide horizontal thrust to the TBM allowing it to advance. All machine operations are controlled remotely by the operator from the control cabin at surface. Guidance of the machine is provided by a laser and target system. The laser system is complemented by a water level system for maximum reliability of the vertical alignment. Correction of the shield alignment is performed by an articulation of the shield itself controlled by three cylinders.

Currently the pipe jacking machines are both in operation and have reached 81% and 89% re-spectively for the PJ800 and PJ1100 machines. Below is a bulleted summary of the tasks required:

1. Site installation;
2. Shaft excavation;
3. Thrust structure assembly;
4. Launch of pipe jacking machine;
5. First excavation with length of the shield;
6. Launch of the first tube (length 3m);

7. Second excavation length 3m.

Phases 6 and 7 are repeated until reaching the retrieval shaft. The excavation of the sections composing the complementary works takes place with shifts that guarantee a continuous production of 6 days a week, maintaining an average for both machines 35m/day, with peaks of 50m/day. In March 2017, 63m were excavated in a single day, corresponding to 21 pipes, establishing a world record recognized by Herrenknecht.

3 EPB TBM TUNNEL DN3200

The main collector of the sewage system consists of two different sections: a first section, named DCBC-1, that extends from the three-cell shaft, located in the Barracas district, to the 2.1 shaft, located in the district of Lugano at the crossing between Avenida General Fernando de la Cruz and Avenida Larrazabal. A second section from a three-celled shaft by the main treatment plant at Dock Sud to the three celled shaft at Barracas.

The first section (Barracas-Lugano), called CMI-2 is within the Capital District and develops along the left side of the Riachuelo river (Figure 3). The total length of this tunnel is 9.540m and is excavated by an EPB TBM Φ3900mm supplied by Herrenknecht. The excavation of the CMI-2 section started on 11/06/2018.

The section is found in the geological formation defined as polygenetic, derived from a geological process essentially characterized by the following phases:

1. Alluvial period (Olocene and Olocene medium):
2. Riachuelo formation: sequences of sandy layers and clayey interlacing's and silt-clay layers with sand lenses (Cretaceous).

The tunnel in the CMI-2 section has an average slope of 0,4% with an overbound ranging be-tween 21.5m at the launching shaft DCBC-1 and 19.5m at the retrieval shaft located in the Lugano district of the capital. The minimum radius along the alignment is 250m. The geological units crossed are mostly made up of fine granular mono-granulated sands called Puelchenses with a variable density from medium to strong, while some sections provide a uniquely clay geology with a variable consistency from medium to strong. The tunnel is being excavated by an EPB TBM baptized with the name Elisa, like the first woman engineer of South America Elisa Beatriz Bachofen. Lining is formed by prefabricated segmented rings,

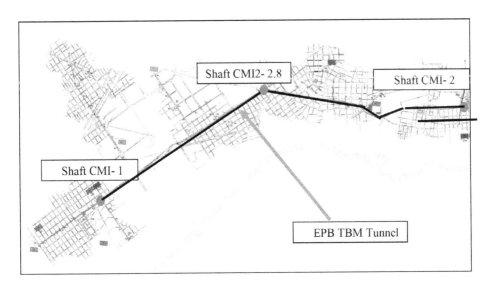

Figure 3.   Plan View EPB TBM Riachuelo Project.

with inner diameter of 3200mm, thickness of 200mm and with a length of 1200mm for a total extension of 9.450m. The TBM is an articulated shield type with a back-up consisting of 8 gantries for a total length of 96m.The thrust system consists of 10 thrust cylinders with maximum force of 13.305kN and 4 pairs of pistons of the articulation with a maximum thrust equal to 10.179kN. The annular gap is backfilled with a two-component grout pumped through 4 lines. Along the tunnel 2 Californian switches will be assembled every 3000m for train crossing, 4in addition to the first Californian switch located in the first part of the tunnel adjacent to the launching shaft.

4 MIX SHIELD TBM TUNNEL

The second section of the main collector (DCBC) is driven from Dock Sud to Barracas for a total length of 5.161. The DCBC tunnel can be divided into two sections: a first section (lot A in Figure 4) that extends from the Riachuelo treatment plant at Dock Sud until after the crossing of the Dock Sud Access Channel, and a second section (Lot B in Figure 4) that extends from there to tribular shaft of Barracas. The geology along the section A t is mainly composed of dense silty sand and very dense puelchenses (monogranular fine sand). The tunnel alignment of this stretch is driven for the most part under the river in the access tunnel to the port and across the Dock Sud access tunnel at an average depth of 26m under the water table. The long stretch to be excavated under water in monogranular sand, drove the JV to select a Slurry Mixshield TBM that in these contexts was considered preferable with respect to a TBM EPB.

The stretch B will present a similar situation as the stretch A for the initial part, up to the crossing of the Riachuelo river, 11m under the river bed, then the tunnel will be excavated inland up to the shaft of Barracas across a heavily urbanized urban area, with an average coverage of 20m. Figure 4 presents the general plan view of the extension of the part A and of the part B, with length of approximately 2.390m and 3.026m respectively.

The lining the A and B stretches will be formed with prefabricated rings with same geometry and different reinforcement, in order to meet the different soil conditions in different sectors of the tunnel. The segmental rings have internal diameter of 4500mm, thickness of 250mm and length of 1400mm. The TBM is mono shield type with a back-up consisting of 11 gantries for a total length of 154m, the thrust system consists of 10 pairs of tail pistons with force of 28.890kN at 380bar and 3 articulation pistons (bearing displacement system) with 6.465kN. Boring diameter is 5300mm. For this operation, the type of TBM needs to be suitable for all grounds to be encountered; face support pressure needs

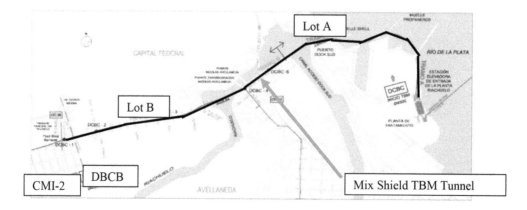

Figure 4.   Plan View Mixshield TBM Riachuelo Project.

to be accurately controlled at all times and is to be uniformly distributed in mixed ground conditions or when encountering highly permeable ground, also face support needs to be controlled and maintained continuously even during downtime. The backfilling grout is a two-component (grout + silicate) mix pumped across 4 lines. Two California switches will be installed in the tunnel to allow a more efficient traffic of train along the tunnel: the first one will be installed in the tunnel adjacent to the launching shaft and the second one will be approximately at the middle of the tunnel.

5 SHAFTS

The alignment of the main collector is crossing 7 intermediate shaft, in addition to the launching and retrieval shafts. In the project are foreseen the crossing of 6 intermediate shafts, some already excavated, others under construction (Table 2), plus the launching and retrieval shaft. The complete list is summarized below in Table 2.

In particular here below the sequence of construction for the box-shapedCMI2 – 2.8 launching and receiving shaft at San Lorenzo, for the DCBC launching three cells shaft at Dock Sud and for the DCBC-1 three cell shaft at Barracas are described. At the progressive pk 4 + 999 the tunnel intercepts the shaft CMI2 – 2.8 located in the San Lorenzo district called the shaft CMI-2.8, where the EPB TBM jobsite will be relocate. Therefore, the EPB TBM will enter in the shaft coning from Barracas and will be re-launched toward Lugano. For the realization of this rectangular shaft 33mx10m and a depth of about 30m the top-down excavation methodology will be adopted where the shaft of the parallelepiped structure will be executed by hydromill Bauer BC 35 mounted on a crane Bauer MC 96. The rig is capable to execute diaphragm walls up to 60 m depth. For the construction of the shaft, the following steps are foreseen:

– Excavation of primary panels thickness 1,20m depth 60m;
– installation of the reinforcement cages of the primary panels;
– in-situ Concrete pouring;
– Excavation of secondary panels, 1.20 m thick;
– reinforcement of the secondary panels;
– in-situ concrete pouring;

The above steps are repeated until the shaft box is closed

– drilling of a pumping well located in the center of the shaft to verify the correct execution of the panels and the waterproofing of the structure by lowering the water table inside the retaining structure;
– Excavation of the shaft (up to the level of the first level of struts depth 6m);
– Installation of support beams (struts);

Table 2. Shaft Data Matanza-Riachuelo Project.

Shaft	Diameter (m)	Base Slab	Excavation Methodology	Type
DCBC	Three Cells	bottom plug	Hydromill	Launching
DCBC-4	7,9	bottom plug	Hydromill	
DCBC-1	Three Cells	bottom plug+piling	Diaphr wall kelly grab	Launching/Receiving
CMI2-2.16	7,9	bottom plug+jet grout	Diaphr wall kelly grab	
CMI2-2.10	7,9	bottom plug+jet grout	Diaph wall kelly grab	
CMI2-2.9	7,9	TBD	TBD	
CMI2-2.8	33 x 10	bottom plug	Hydromill	Launching/Receiving
CMI2-2.6	7,9	bottom plug+jet grout	Diaphr wall kelly grab	
CMI2-2.2	7,9	bottom plug+piling	Diaphr wall kelly grab	
CMI2-2.1	13,4	bottom plug+piling	Diaphr wall kelly grab	Receiving

- Excavation of the shaft (up to the level of the second level of struts, depth 15m);
- Implementation of support beams (struts);
- Excavation of the shaft up to a depth of 23m;
- Construction of reinforced concrete bottom plug 0.70m thick;
- Construction of the reinforced concrete cradle (thickness 1m) for the passage of the TBM;

The retaining structure is extended up to 60 m depth in order to intercept the deep layer of plastic clay that will constitute a perfect watertight bottom plug.

The excavation involves the crossing of shallow silt-clay layers in surface resting on a layer of puelchenses and then very substantial clays that prevent the water from rising during the ex-cavation phase. The hydromill performs the excavation for the vertical panels using the ben-tonite slurry as a fluid for the support of the excavation walls. The presence of fresh bentonite in the line is guaranteed by an inverse Bauer BE 425 circuit installed near the machine. The launching shaft of the Mixshield TBM at Dock Sud, known as the DCBC shaft (Figure 5), has a three-cell structure consisting of two 21m diameter shaft and one tangent with a diameter of 45m. the construction sequence of the three shaft is similar to the one de-scribed above. In-stallation of struts will not be needed thanks to the circular shape of the cells. The first two cells, belonging to lot 1, have panels from 60m and thickness of 1,20m that cross a geology made up of a layer of shallow clay, with a thickness of 19m, a layer of puelchenses (monogranular sand) with a thickness of 16.5m and finally the panels are embedded in very substantial Miocene clays for a depth equal to 5.50m. In San Lorenzo shafts reached the excavation depth equal to 40m, which would correspond to the first meters of the layer of pulchenses, a reinforced concrete cradle of 2.50m thick will be realized for the launching of the TBM. Only the two smaller shafts will be used for assembly and launching of the Mixshield TBM. The larger Shaft will be handed over to the Lot 2 contractor.

The launching shaft of EPB TBM at Barracas, named DCBC-1 (Figure 6), has a three-cells structure consisting of 3 shafts of which the first two have an internal diameter 10,60m and the third 16.40m. The geology crossed in the construction of this shaft consists of a layer of silty sand with a thickness of 8m, a layer of clay silt 6m thick, a clay lens with a thickness of about 2m and finally a bench of very dense silty sand up to the depth of the panels equal to 34m. the retaining structure is constituted by diaphragm walls executed with Kelly Grab machine. The excavation of the shafts is performed under water by of long-boom excavator, for maintaining the balance of the water pressure. Once the excavation of is completed, the 4 m thick concrete

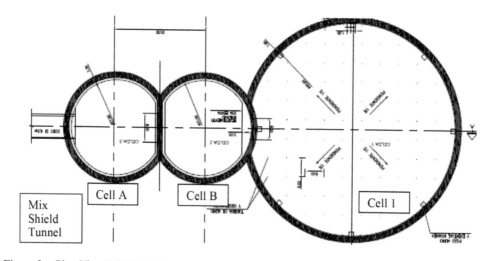

Figure 5. Plan View DCBC Shaft.

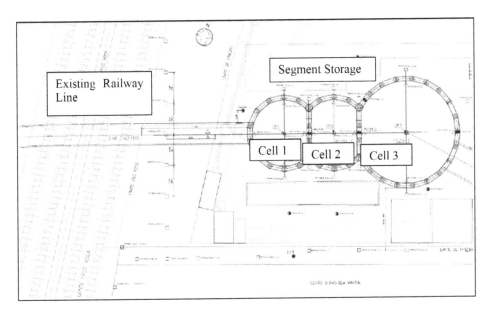

Figure 6.   Plan View DCBC-1 Shaft.

Figure 7.   Pipe Jacking Machine.

bottom plug is casted under water before pumping the water out from the shaft. Less the 50m form the launching shaft the EPB TBM excavated below the existing railway line, the area was consolidated with Jet Grouting and the pressure of the TBM was monitored continuously to avoid any kind of settlement. Finally, the internal concrete lining and concrete cradle

Figure 8. DCBC-1 Shaft.

for the TBM assembly and launching are realized. After the work was completed in March 2017, the TBM EPB was assembled in the following month of April

6 CONCLUSION

With the aim of mitigating the environmental problem, the World Bank has been supporting environmental development programs for several years, accompanied by the construction of large infrastructures that make it possible to reduce the contamination of the Riachuelo River. The most important program for this purpose is represented by the modernization of the Riachuelo sewage system, which consists of a series of underground works that serve an area of 200 square kilometers. The works, are included in the contract called "Lot 1: Left River Bank sewer collector" are currently under construction performed by of the CCMI Joint Ven-ture (Consorcio Colector Margen Izquierda). The total amount of works is approximately 500Mio's of US$.

The Joint Ventures know how and capability allows us to adopt within the same project numerous different technologies such as pipejacking, EPB TBM, Mixshield TBM and a hydromill. All these work is managed entirely and directly by the CMI JV. The project under construction consists of a series of underground work that complete the Riachuelo sewage system. In particular, a main collector with a total length of 15km of which 9.6 excavated with EPB TBM and internal diameter 3200mm and 5.45km excavated with Mixshield TBM with internal diameter of 4500mm through geology of saturated silty sand, fine sand and layers of clay under a constant hydraulic head equal to 26m (see Figure 8 the DCBC-1 launching shaft). To this main collector are connected directly, through 10 shafts, a series of secondary links for a total length of 13km by two pipe jacking with diameter 800mm (see Figure 7) and 1100mm which currently have already achieved more than 85% of the tracks generally crossing a geology consisting of hard-to-very hard-hardy tosca excavated with a variable overburden between 5 and 10m. The connection links between the main and secondary collectors on the main section are represented by structural shafts. Further, the sewer system consists of another 20 shafts with a diameter of 6m and depths of between 6 and 10m that correspond to those works necessary for pipe jacking launching and retrieval shaft.

REFERENCE

AATES, 8° Jornada de Tuneleria y Espacios Subterraneos, 4-5-6 Septiembre 2018, S. Sucri, G. d'Ascoli.

Tunnels and Underground Cities: Engineering and Innovation meet Archaeology,
Architecture and Art, Volume 11: Urban
Tunnels - Part 1 – Peila, Viggiani & Celestino (Eds)
© 2020 Taylor & Francis Group, London, ISBN 978-0-367-46899-6

The challenging yet successful excavation of the TBM tunnel for the Athens Metro Line 3, Extension Haidari-Piraeus

G. Giacomin
Ghella S.p.A., Roma, Italy

P. Pediconi & D. Goudelis
Ghella S.p.A., Athens, Greece

ABSTRACT: this paper describes the complexity of the excavation of the Athens Metro Line 3, Extension Haidari-Piraeus, including a 6.5km double track tunnel with a 9.48m diameter EPB TBM. The EPB TBM Ippodamo has completed its journey in approximately 4 years. The project includes six new stations. Maniatika Station is the only one built below surface in cavern while all the other ones through cut and cover. The mainline also sees the construction of eight ventilation shafts and two NATM sections, double and triple tracks, at the assembly and disassembly shafts for overall 340m. The Athens TBM has excavated through a densely populated urban area in one of the oldest cities in the world. Due to the geological, hydrogeological, and geotechnical conditions, an EPB TBM has been chosen based also on specific needs, such as avoiding disturbance and causing damage to buildings and structures located within the zone of influence. Through the limestone section, Karstic voids of large dimension have been encountered and from both surface and the tunnel face, these have been successfully treated and filled. Working below the water table, in extreme proximity to the highly congested Piraeus harbor, has been an incredible challenge completed and executed in a highly professional way. Furthermore, the progress of the works has been also affected by the difficult economic situation of the country during the period.

1 INTRODUCTION

The Haidari – Piraeus Contract forms the last portion of the Line 3 Extension, which extends from the south of Aghia Marina Station and terminates at Piraeus Port and further on to the east, up to Dimotiko Theatro Station (Figure 1). The alignment is approximately 7.55 km long and includes six stations. The beginning of the alignment is at Kilometer Position 1+418.55 (Aghia Varvara Shaft) and its end at Kilometer Position 8+968.95 (Deligianni Shaft).

The boring of the double track tunnel is constructed with a front face cutting machine EPB-TBM, operated by the Contractor J&P-Avax/Ghella, in charge of the Civil Works of the Project. The selection of this particular type of machine has been made according to the geological, hydrogeological and geotechnical conditions along the entire length of the line. Also the requirements that refer to the avoidance of disturbance and causing damage to buildings and structures located within the zone of influence, the minimization of sedimentation on the surface and the prevention of damage to utilities networks, while ensuring the safety of constructed tunnels have influenced the TBM selection.

The EPB-TBM used for the tunnel excavation, named after the ancient Greek civil engineer Ippodamos, is a Lovat Hydraulic machine, previously utilized for the construction of the 6 km tunnel for the High Speed Railway Line Bologna-Firenze and the 3 km Metro Line 5 in Milan.

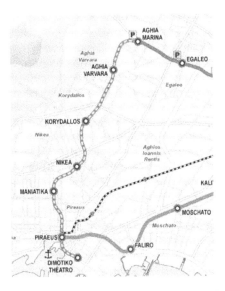

Figure 1. Plan View of the Extension Haidari-Piraeus of Athens Metro Line 3.

It has a 10.7 meter long shield, 9.48 meter diameter of excavation, maximum EPB pressure of 3.5 bar, total weight of 1230 tons and 300 meters minimum horizontal radius.

2 GEOLOGICAL AND GEOTECHNICAL CONDITIONS

One of the main challenges of the project was the heterogeneous geology within the entire alignment. Moreover, the geological formations appear with essential alterations even within few meters including fault zones, tectonic geological contacts, shear zones and unconformities. According to the findings of the geotechnical investigation, tunneling works had to be executed within five main geological formations, as presented below.

2.1 Athens Schist

The formation consists of a variety of metamorphosed formations. The upper level of the Athens schist comprises alterations of meta-sandstones and meta-siltstones while the lower level mainly consists of black clayey shale. Within these formations, there is a main Shear Zone which is 6–8m wide in the contact of upper and lower level and several smaller fault zones. The overburden within this formation is 11m which is the minimum height of overburden found in the entire project. Also, the average depth of the man-made deposits and scree is of about 6m but, locally, reaches a peak of 10m. The formation is classified as "Poor Rock" and locally "Soil".

2.2 Ultrabasic formations

Serpentinites, serpentinized peridotites and ophiolites are contained within this formation. The ultrabasic formations cover only a short length of the project, approx. 385m, but they have appeared in the contact zone between Athens Schist and Karavas Limestone, creating peculiar mixed face conditions. The formation extends from few meters below the surface up to the tunnel level and, generally, shows fairly low permeability. The scree has a significant extent which reaches, locally, 10m but its average height is less than 7m. Due to the mixed conditions and fractured overburden, this formation shows low rock mass quality and, therefore, is classified as 'Poor to Fair Rock'.

2.3 Karavas Limestones

The formation consists of the intact, thin–bedded to thickly–bedded and massive limestones. Wide, empty or filled with calcite, karstified zones are encountered during tunneling works, as described in chapter 5. Both excavation profiles and overburden are dominated by Karavas Limestone and cover an overall tunnel length of approx. 600m. The overburden thickness above the tunnel crown varies from 13 to 37.6 m, which is the highest found in the whole tunnel alignment. From the hydrogeological point of view, the formation is characterized as adverse due to the high permeability and the presence of karstic groundwater horizon. The average measured value of UCS within this formation is greater than 55 MPa, reaching a maximum of 160 MPa, which is by far the highest within the entire alignment. . Also, the results of the C.E. R.CHAR. abrasivity tests show 'high to very high' abrasiveness index. In Figure 2, this formation is located around Tampouria Station, the first name for Maniatika Station. Overall, Karavas Limestone shows very good rock mass quality and is therefore classified as 'Good Rock'.

2.4 Littoral Deposits

The recent littoral deposits are located in the wider area of the Piraeus port. From a lithological point of view, this formation consists of dense silty sand, stiff sandy silt to weak silty sandstone and medium – weak sandy siltstones. This formation includes primarily mixed

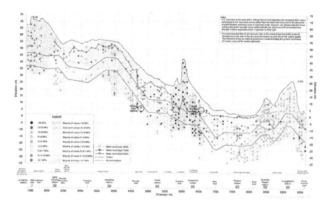

Figure 2.   UCS values distribution graph along the area of the alignment, based on laboratory tests results (Marinos et al.,2007)

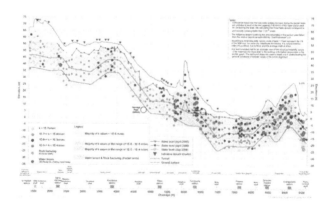

Figure 3.   Permeability distribution graph along the area of the alignment (Marinos et al., 2007).

Table 1. Geotechnical characteristics within the entire alignment of the project.

Geological Formation	Athens Schist	Ultrabasic Formations	Karavas Limestones	Littoral Deposits	Neogene Deposits
γ (kN/m³)	22.9	20.2	25	20	21.4
UCS (MPa)	0.2 - 5	2 - 15	35 - 163	0 - 5	5 - 25
GSI	15 - 34	38 - 47	64 - 73	N/A	33 - 52
k (m/s)	5x10-7	1x10-6	2x10-3	1x10-4	5x10-6
RQD (%)	0 - 25	25 - 50	75 - 90	N/A	50 - 75
E (GPa)	0.63	1.7	17.8	0.05 - 0.27	0.48

soil conditions with fine-grained soils with low to medium plasticity (W_L<50% και PI<17) while the geomaterials show a medium sticky behavior. It also shows high permeability (Figure 3) and the tunnel within this formation is always located more than 10m below the water table. Additionally, marsh deposits of very poor quality overlay the littoral deposits. Due to the proximity to the Piraeus port, the sea water infiltrates to the underground water table.

2.5 Neogene Deposits

The Neogene deposits include mainly marly deposits and marine marls (Piraeus Marl). This geologic unit shows a great variety of lithologies and consists of alternations of mainly marly limestones, marls, claystones, siltstones, sandstones and conglomerates. Due to the geological variety, the laboratory tests show a wide range of results. However, mainly competent Piraeus Marls prevail within this section and, therefore, the formation is classified as 'Fair Rock'. Moreover, the formation appears with low permeability, except for the fault zones and fractured marly Limestone locations. The formation shows medium and locally high, sticky behavior leading to a high clogging risk. The tunnel within this formation is always located below the water table.

In view of this heterogeneous geology along the alignment, the TBM has been prepared to excavate in a safe and productive manner from soft ground to hard rock. The cutter-head configuration, the treatment of the face and muck in the chamber, the operational parameters to be followed and monitored by the crew have had to be continuously changed and adapted to the wide range of different geotechnical conditions. Trying to summarize the entire geology of the project, the tunneling works have been completed within the following ranges, as presented on the Table 1 below.

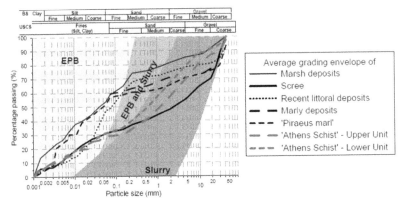

Figure 4. Average grading envelope of geomaterials and EPB or Slurry suitability limits (Marinos et al., 2007).

The average grading envelopes of the geological formations (with the exception of the Karavas Limestone) are shown in the Figure 4 diagram, together with the operational range of the closed shield TBMs, as proposed in the actual literature.

As estimated during the preliminary study for the project (Marinos et al., 2007), the utilization of the EPB-TBM was appropriate and successful. Nevertheless, the frequent change of material in the excavation face and tunnel surrounding has imposed a challenging monitoring and response plan, to adapt immediately the operational parameters to the required standards.

3 ORGANIZATION AND PROGRESS OF TBM WORKS

The TBM excavation works have started at the Launch Shaft, on the North side of the alignment, within a limited construction area where the TBM has been assembled and lowered 35 meters below surface. The length of the bottom slab of the Shaft along the longitudinal direction was 40 meters; the 145 meter-long TBM had to be launched through a consecutive process of partial assembly and staged excavation, using temporary services and de-mucking system, while fully applying EPB support to the unstable ground, the black clayey shale of the Lower Unit of Athens Schist.

Once the entire TBM was assembled, the final conveyor system has been installed, made of a 19° inclined section in the shaft and directly conveying the muck up to the spoil storage on surface. With such configuration, the TBM completed the first half of the tunnel drive mainly excavating through lower and upper Units of the Athens Schists, passing through the Aghia Varvara, the Korydallos and and the Nikeia Stations. Ippodamo has excavated 3100 meters in 1 year with an average production rate of 260 meters/month, reaching a best month production of 410 meters, on a 5 and ½ days per week operation.

At the Nikeia Station, works were interrupted to relocate all TBM supporting plants and free the first section of the tunnel for the fit-out and electromechanical works, while the TBM cutter-head was rearranged and reinforced to ensure adequate performances for the following excavation through the Karavas Limestone formation of Maniatika. During this phase the project was forced to suspension, due to the Greek Financial crisis.

Restarting from Nikeia Station, the TBM bored through the Karavas' Limestone Formation characterized by an extended karstic network until breakthrough into Maniatika Station, as described in chapter 5 of the present article. The final 2130 meter-long stretch of the tunnel, in the Piraeus Harbor area composed of Littoral Deposits and Piraeus Marl, has been completed in 9 months, with an average production rate of 235 meters/month and reaching a best month production of 435 meters.

On the overall assessment of the TBM operation progress, it is possible to identify five different phases characterized by different advance rates: the assembly and start-up with logistic constraints, the first section of the alignment in Athens Schists, the suspension of the project,

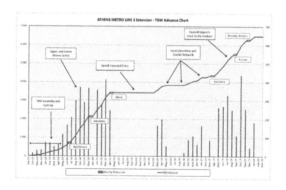

Figure 5. TBM Advance chart.

the Karstic Limestone and the final section in the Littoral Deposits and Marl of Piraeus. Putting aside the start-up and the suspension, the comparison between production rates in the Athens Schist, in the karstic Limestone and in Coastal Deposits gives a very clear image of the impact of the geology on the operation of the EPB-TBM (see Figure 5).

The tunnel crew has managed successfully the heterogenous formations, the transition zones and in general the wide range of geotechnical features encountered in the first and last sections of the alignment. However, the excavation of the very hard and karstic Limestone has had a relevant impact on the advance, despite the adequate preparation of the cutter-head which ensured a very successful cutting rate.

4 TECHNICAL CHALLENGES

4.1 Pre-existing underground structures

The neighborhoods interested by the underground works had been built and developed during the previous centuries, when many different types of utilities and structures were built. The tunnel excavation has successively passed below active railways, sewages and pipelines.

The most relevant situation was the crossing 3.2 meters below the Central Sewer Pipe of Athens which was built in the 1930s. As a result of preliminary inspections and numerical analysis to predict the interaction between the TBM tunneling and the sewage, operation works were performed without interruption and following very strict requirements on face and grout pressure limits (upper and lower) and volume loss, achieving maximum settlement at the surface of 1.3mm.

In addition to the existence of deep public utilities, for which the geometry and location were known and verifiable, many other abandoned wells and sewages, frequently privately built, either carelessly backfilled years earlier or left empty, were spread along the alignment and not well documented.

The potential interception of these old networks, occasionally connected to the surface, have been a risk for the proper functioning of the TBM in EPB mode and for the integrity of the pavements and foundations above. It has been necessary to carry out an extensive and continuous investigation, organized with the following steps:

- survey with the residents and visual inspection of the basements and courtyards of the buildings above the tunnel, to identify the localizations and potential interferences;
- detailed indirect investigation with geophysical method GEORADAR, with the use of low-frequency antenna for deeper penetration into the ground in order to determine the accurate position, geometry and have a preliminary indication of the density of the filling material;
- when the tunnel section was intercepting or passing very close to pre-existing abandoned utilities, additional measures were implemented to ensure the appropriate density and mechanical properties of the fill, like grouting or concreting, from surface when possible or else from the tunnel;
- intensive monitoring was also organized in the most sensitive situations.

4.2 Cutter-Head Maintenance

During the excavation works through and under the heterogeneous geological conditions and complex structures, an accurate program for inspection and maintenance of the cutter-head has been organized. From a safety and operational point of view, cutter-head maintenance is preferable to take place with atmospheric conditions or under pressurized air below 1 bar, considering the requirements of face stability and settlement limitations.

The entire tunnel profile has been evaluated taking into consideration all the boreholes and laboratory information, the structural conditions of the buildings, the permeability of the rock mass and the prediction on cutting tools consumption.

With the support of limit equilibrium analysis, following the Anagnostou-Kovari method for face stability conditions with earth pressured balance shield, the entire alignment was categorized in sections where different types of interventions could be applied, as follows:

– Atmospheric intervention full face;
– Atmospheric intervention only half empty chamber;
– Hyperbaric intervention, with a fraction of hydrostatic pressure, only half empty chamber;
– Hyperbaric intervention with full hydrostatic pressure.

In all circumstances, the geological assumptions used in the model have had to be verified with a preliminary hyperbaric inspection under full hydrostatic pressure and the behavior of the excavation face has been kept under constant observation by the geologists.

The effect of face depressurization has been additionally analyzed by coupled numerical seepage flow-stress analyses in three-dimensional steady-state conditions, for some representative sections, and the resulting values of groundwater draw down and surface settlements have been compared to continuous real-time measurements.

A big portion of the maintenance works could be carried out in atmospheric environment and the monitoring results have confirmed the adequacy of the models, always on the conservative side.

Hyperbaric interventions have been successfully performed in the black clayey shale of the Lower Unit of the Athens Schist at the beginning of the TBM drive, and in the Littoral Deposits of Piraeus area in proximity of the sea.

4.3 *Piraeus Tower*

The tower of Piraeus, built in the 1970s, is an 80 meter-tall building, abandoned soon after construction due to excessive and differential settlements. Its foundations are based on the Littoral Deposits of the Piraeus in front of the harbor, approximately 70 meters distant from the sea. The tunnel alignment axis is located half way between the wharves and the tower, 16 meters below the street level.

After a supplementary and detailed geotechnical investigation, the assessment of the TBM passage has considered safe to proceed the works without ground improvement, but the excavation has nevertheless been carried out following a very close monitoring on the settlements and on the operational EPB parameters.

4.4 *Geotechnical Monitoring*

The continuous and extensive monitoring of the tunnel, surface and buildings has been essential to ensure that the TBM operations had the minimum impact on both surface and underground.

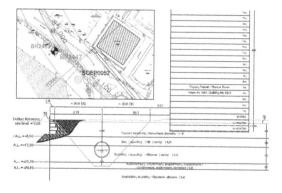

Figure 6. Cross section of the tunnel between the port and the tower of Piraeus.

A dense network, consisting of settlement points, 3D targets, extensometers, strain gauges and piezometers, has been setup giving measurements in accordance with TBM location. In the most critical locations, an automated system has been integrated providing more precise and frequent results in real time.

To minimize the risk of settlements or face collapse during the mining works, the Additional Face Support (AFS) system has been added to the TBM which was automatically activated in case of sudden EPB pressure drop lower than the alarm level. This system works by injecting bentonite into the excavation face, using a propelling pump with a capacity $10m^3/h$ at 10bar, and it has been able to maintain the EPB pressure within the required levels.

As a result of highly skilled and well-judged operation throughout the entire alignment:

a) All settlements have resulted below the 80% of the expected top limit;
b) The maximum surface settlement has been of -7mm;
c) The displacements on the buildings have been negligible;
d) The water level measurements have been constant (atmospheric interventions excepted);
e) EPB face pressure and volume of excavated ground have been continuously within the required design levels;
f) There hasn't been any lateral convergence inside the tunnel.

5   EXCAVATION THROUGH KARSTIC LIMESTONE

The geological formations of the 4th interstation, between Nikaia and Maniatika stations, in order of reach during the excavation, have been the Upper Unit of the Athens Schist (150 meters), the Ophiolites and Serpentinites of the Ultrabasic Formation (250 meters) and the Karavas Limestone (550 meters). In accordance with the extensive geotechnical investigations performed during the preparation of the project, the rock was expected to be very competent, massive and locally thin or tick bedded, with medium to high compressive strength and few karstic phenomena, sporadically encountered with the boreholes having a maximum extension of 0.5 meters.

Entering the Limestone Formation, the groundwater level dropped at the invert level, mostly due to the dewatering at the Maniatika Station excavation.

Just 40 meters away from the switch to Limestone, the TBM advance has been interrupted because of the interception of a very wide and unexpected karstic network, that has required adequate interventions to ensure stability of the surrounding ground, safety of all personnel involved and integrity of the machine during progress.

In addition to the measures required in the specific location, the remaining section of the alignment through the limestone had to be further investigated and reassessed in relation to new evidences on the size of the karstic phenomena, in order to mitigate the risk and minimize the impact on the time schedule.

5.1   *First Event*

At the kilometric position 5+757, during the excavation of ring no. 2538, the operator in charge of the TBM drive has noticed a sudden increase of the advance speed and a significative reduction of the quantity of material extracted through the screw conveyor. The excavation has been immediately interrupted and after the usual controls, it has been decided to inspect the face to have more information about the abnormal conditions observed.

A karstic cavity was discovered, with a sub-vertical development, extending for half the excavation face, 5 meters in longitudinal direction, and from above the tunnel crown down below the invert. The karstic void's walls and the rest of the face were very blocky and locally the limestone was weathered with partially open joints and calcite (Figure 7)

With a long sequence of drilling and filling, both from surface and from the tunnel, with concrete and bi-component grout, protecting the TBM with injection of polyurethane foam on the cutter-head and bentonite around the shield, the area around the location of stoppage was investigated and consolidated. The final total quantity of cementitious material pumped has been more than 1500 cubic meters, injected through 16 sub-vertical boreholes from the street and 18 sub-horizontal probe drills from the TBM (Figure 8).

The majority of the volume of material was injected below the building located on top of the tunnel alignment, approximately 10 meters ahead of the cutter-head location.

To complete the investigation of the ground and ensure that no major voids were still existing, in the locations where it was not possible to drill from the surface due to existing buildings, a geophysical survey was conducted using cross-hole seismic tomography.

The presence of voids in rock formation causes a considerable drop in the seismic P-wave velocity (Vp). Thus, a mapping of Vp in sections, produced by the implementation of seismic tomography, could detect and delineate the karstic zones.

Even though the indirect investigations are generally able to provide results that need to be evaluated together with other information and used with proper knowledge, the seismic method could usually allow to differentiate between zones of intact solid rock and "disturbed" zones with probable presence of voids or fractures or cavity filled with different materials.

After the described intervention, the TBM has been able to resume excavation without having any impact on the structures above and maintaining the design alignment both on the TBM drive and with the installed rings.

5.2 New Risk Assessment and Methodology

Considering the new evidence on the possible extension of karstification in the Karavas Limestone formation, it became necessary to reassess the tunneling operation and define a new design for the implementation of all the necessary measures, ensuring that all residual risks were still at an acceptable level.

As per the expected geology, the TBM has been set to operate in EPB closed mode, which was not feasible in the actual conditions, or in open mode with capability to perform investigative probe drilling ahead. The machine has also been equipped with a non-destructive system for continuous geo-electric investigation in front of the cutter-head; however, it has not been able to detect the big karstic cavity in the first event, due to the small difference in value for the electric resistivity between solid rock and void, when above the water table.

The possible maximum size of the karstic voids that could have been encountered anywhere around the tunnel profile or inside the section, was now presenting important risks, in relation to:

- loss of control on TBM operation;
- uncontrolled settlement of the TBM shield and fall into an empty cavity;
- impossibility to support an unstable excavation face;
- blocks of rock falling on the cutter-head and jamming the TBM;
- uncertain static adequacy of segmental lining, both in the short and long term.

In addition to the primary considerations addressing safety and settlement issues, it has also been necessary to identify the optimal solution in terms of budget and time schedule.

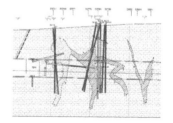

Figure 7. View of the karstic cavity from the cutter-head.

Figure 8. Longitudinal section of the intervention.

Following a deeper study on the specific geology and the alignment, also based on the experience with the first big cavity encountered, and after optimization of the methods used to solve that situation, the contractor has developed a cyclic methodology, combining indirect geo-physical investigations, vertical drillings from surface and horizontal drilling from the tunnel, that has provided a safe and successful approach to the challenging conditions.

5.2.1 Geo-physical non-destructive methods

Two methods have been used to narrow down the high-risk sections along the alignment and target the destructive investigations.

The geo-radar, through the use of low frequency antennas for deeper penetration into the ground, has provided interesting results in the shallowest layers, up to 15 meters deep. It has been used to identify the vertical karstic developments, that are often extending from the sea level (approximately at the tunnel invert level) up to the more permeable layers closer to the surface.

Furthermore, the vertical karstic features have sometimes been utilized as discharge point from the old residential developments, then abandoned and filled in more recent stages. Those kinds of situations could have potentially led to damages to buildings or internal pavements as consequence of the sudden emptying of the fill towards the tunnel.

The second method used, only in the zone identified with high karstic risk, due to the significative time and investment required, has been the geophysical survey using cross-hole seismic tomography. An extensive campaign, made of 24 vertical sections between 15 boreholes, has been performed across 200 meters of the tunnel alignment, starting from the back of TBM shield (including the first event).

The karstic networks are normally made of vertical conduits and horizontal channels; those channels usually extend within the limestone around the elevation of the running underground water. Having the sea level elevation just below the tunnel invert, was defining an important risk related to the potential presence of cavities below the TBM.

In addition, to investigate the area below the tunnel invert with destructive methods from the tunnel, it is required, for geometrical reasons, to frequently stop the tunnelling operations. Systematic drilling from the surface, up to 50 meters deep in hard limestone, has not been everywhere feasible due to the buildings, while the impact on the community would have been significant.

The diagrams of the seismic P-wave velocity, measured along the vertical sections, has provided useful information about the horizontal karstic network; of primary importance was the possibility to identify the sections of alignment were the probability of wide karstic voids was negligible.

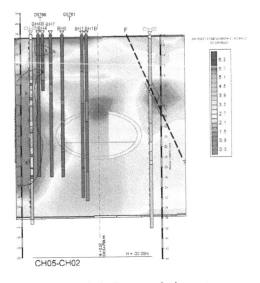

Figure 9. Seismic tomography with cross-hole: P-wave velocity contours.

5.2.2 Sub-vertical drillings from surface

The next step, having all the data coming from the indirect investigations, has been to reach all the underground locations where geophysical anomalies had been identified, accordingly with the existing buildings and traffic conditions, drilling sub-vertical destructive boreholes from the surface to 5 meters below the tunnel invert and filling with concrete or mortar when required.

A total of 68 boreholes have been drilled and 1800 cubic meters of concrete and mortar poured; 61 of them were located within the first 200 meters high risk section.

5.2.3 Sub-horizontal drillings from the TBM

Following all the steps described above, the TBM drive has been carried out alternating cycles of excavation, 10 to 20 meters long, with probe-drilling inspections and interventions. Depending on the risk related to any specific location and the interventions performed in advance from the surface, 1 to 3 probe drills have been made to verify the presence of karstic features and, when voids were detected, a scheme of additional holes, distributed around the void and the tunnel section, have been drilled and filled with bi-component grout injection.

The number of probe drillings have varied between 3 to 28 and the maximum injected grout has been 271 cubic meters in a single location and approximately 950 cubic meters in total for the 32 cycles.

Using the bi-component grout has given the advantages of utilizing the systems already operational in the TBM for back-filling of the annular void; ensuring easy supply of material since the component A was pumped through pipelines along the tunnel; having a cementitious material that could be pumped in the probe holes ahead of the TBM, but at the same time setting in a short time, filling only the voids affecting the surrounding of the tunnel and resisting the flushing effects of the running water.

5.3 TBM progress and results

The remaining 500 meters of tunnel in the karstic limestone have been completed in approximately 5 months, without any additional stoppages or risks along the way; this productivity appears low when compared to usual pressurized shield TBM operation, however satisfactory in this specific case, considering the cyclic alternation between advance and drilling, in addition to the unfavorable excavation of the hard rock mass with the EPB cutter-head and the environmental constraints.

Due to the high compressive strength and abrasivity of the Karavas Limestone, the number of cutter discs (15.5') has been increased to 75, with 90mm average spacing between the tracks, achieving an average cutting rate of 12 mm/min. The cutter head maintenance has been much higher compared to the rest of the tunnel: the average cutter discs replacement has been of approximately 1 cutter disc every 2.5 meters, equivalent to 1 disc change every 180 cubic meters.

Furthermore, the TBM operation, for a specific section with low cover and compact rock and during a period of 3 weeks, has had to be interrupted during nights, due to the excessive noise generated on surface by the excavation works.

Figure 10. Geological Plan view and investigation map: green square=ophiolite; blue bricks=limestone; pink lines=cross-hole sections; blue dots=vertical drillings; dotted black lines=faults and discontinuities; yellow area=excavated tunnel; light blue areas=limestone outcrops.

Figure 11. Picture of the excavation face showing the proper filling of the karstic network.

No settlement has been measured on the streets or buildings and, with the application of extensive secondary grouting, the back-filling behind the rings has been re-checked and completed, guaranteeing the integrity of the lining. From the frequent face mapping available along this section, and from the stable trend of the TBM excavation parameters, it has been possible to verify the successful implementation of the methodology.

6 CONCLUSIONS

Tunnelling in urbanized areas is always a challenging task, which requires detailed and continuous planning, attention and preparation. The case of Athens Metro Line 3 extension presented peculiar situations as a consequence of the extreme geological variability, excavating in soft soil to hard rock, below the water table and in close proximity to the sea. It is mandatory to formulate and implement a detailed risk management plan, from the early stage of the project and preliminary design, to the construction stage, transferring to each consecutive step all the available knowledge and considerations made in the definition of the construction methods.

Very successful results had been obtained in the delivery of this underground infrastructure, with no major health and safety incidents occurred, minimal impact on the residents and excellent quality of the final product.

The described project was also affected by external factors, of political and financial nature, and required very hard work and full collaboration for all the parties involved, to ensure the successful completion, representing an important experience for the tunnelling industry.

REFERENCES

Assessment of Ground Conditions with Respect to Mechanised Tunnelling for the Construction of the Extension of the Athens Metro to the City of Piraeus. - P.G. Marinos, M. Novack, M. Benissi, G. Stoumpos, D. Papouli, M. Panteliadou, V. Marinos, K. Boronkay, K. Korkaris, 2007
Face Stability Conditions with Earth-Pressure-Balanced Shields. - G. Anagnostou and K. Kovari, 1996
Athens Metro. Extension of Line 3 Section Haidari – Piraeus. Assessment of the feasibility of open mode or low pressure TBM operation (Various Reports). - G. Anagnostou, 2014–2015
Athens Metro. Extension of Line 3 Section Haidari – Piraeus. Geological, Geophysical and Geotechnical Investigations and Studies for the determination of Karstic Voids (Various Reports). - EDAFOMICHANIKI SA., 2016
Athens Metro. Extension of Line 3 Section Haidari – Piraeus. Design – Implementation Methodology for Investigation and Location of Karstic Voids and their tackling. – GEODATA SpA., 2016

Tunnels and Underground Cities: Engineering and Innovation meet Archaeology,
Architecture and Art, Volume 11: Urban
Tunnels - Part 1 – Peila, Viggiani & Celestino (Eds)
© 2020 Taylor & Francis Group, London, ISBN 978-0-367-46899-6

Effect of excavation stages and lining sequences on ground settlement in a twin tunnel

A. Golshani & M.G. Varnusfaderani
Faculty of Civil and Environmental Engineering, Tarbiat Modares University, Tehran, Iran

S.M. Poorhashemi
Executive Manager for Underground Projects of Tehran Municipality, Tehran, Iran

ABSTRACT: This research aims to declare a case study on a twin tunnel constructed by New Austrian Tunneling Method. The twin tunnel is sketched with connected initial linings and separate final linings, due to dimension and obstacles restrictions. Three scenarios of excavation were numerically modeled in order to reduce settlement. The first scenario comprises entire excavation and installation of the initial lining of each section in order. The second scenario simulates the effect of keeping Center Cross Diaphragm during excavation till the entire construction of initial lining. The third scenario evaluates the effect of final lining construction for a section before entire initial stabilization of another. Numerical results show that the third scenario reduces the surface settlement about 44% comparing to the first one, while the second scenario has no significant effect. Furthermore, the ratio of the surface settlement obtained from the third scenario reduces about 40% than the first one.

1 INTRODUCTION

Tunneling process with its all benefits to urban areas, has always its own difficulties either since engineers are required to consider strictly some more restrictions specified to these areas, in order to design and construct safely. These restrictions to underground structures, for instance, some stem from the existence of surface urban facilities among which are main roads, residential spaces, in particular, with high rise buildings. In better words, this proves the importance of interaction considerations between subsurface and surface structures which should certainly be accounted for numerical analysis and designing and then be checked during construction and operation. Thereby, the parameters of ground subsidence and induced surface settlement in consequence of the soil-structure interaction should be considered as criteria for checking the accuracy of design and proper performance of the structure, and then taking a proper action to redesign the soil structure conforming to the field recorded settlements, if it is necessary. The monitoring procedure is, therefore, an important long-term process evaluating the surface settlement and the convergence of the tunnel section during and after construction, through which, both the sufficiency and efficiency of the rehabilitation and new reinforcing methods would properly be checked relying on numerical sensitivity analysis. As far as the constant control of ground surface settlement is concerned, various projects and case studies have been conducted, especially in Iran, some which have presented valuable results in specific soil overburden conditions as well as in obtained field recorded data comparing to numerical results. Zolghadr et al. (2013) employed a numerical model for analyzing the behavior of the portal of a road tunnel and then controlled with monitoring data, based on the case study of Niayesh tunnel in Tehran, Iran. In this study also the pile system with NATM stage construction was proposed for portal stabilization. Based on their study the maximum settlement of a 6-story building after construction was about

Figure 1. Overall view of the project site (Satellite map).

44 mm and the final factor of safety was about 1.39 in which corresponding numerical result was larger than the monitoring data with the maximum settlement of about 27 mm just after construction completion (Zolghadr et al. 2013). Golshani et al. (2014) conducted a case study on the construction of a crowded junction with shallow soil depth, located under the Valiasr main roundabout in Tehran, Iran. By their studies, it was revealed that the sequential excavation method accompanied by the forepoling technique is efficient and shall be performed for safe construction of under passing with low soil overburden (Golshani et al. 2014).

This paper aims to report an especial twin tunnel located in the main street where certainly there were several hardships during design and construction procedure in order to reduce its influence on top and adjacent structures, and consequently, research was conducted simultaneously with the analysis of the relevant numerical models to select the most optimized excavation orders and efficient supporting systems, before constructing. In addition to numerical sensitivity analyses, adequate engineering experiences were also required to propose different supporting elements analogous to previous projects at the same location. This is the approach recommended to determine pre-selected design schemes of previous projects in similar geological condition (Zhu et al. 2017). Regarding the mentioned sensitivity analysis, different similar studies have also conducted, especially in the fields of twin tunnel interaction.

The article focuses on Arash- Esfandiar- Niayesh project in the city of Tehran, Iran, starting from Modarres highway and running to Niayesh highway with a total length of 1870m (see Figure 1). The particular twin tunnel crossing under the Valiasr Street was excavated based on New Austrian Tunneling Method (NATM) which was the main method applied for the construction of this project (Telford 2004, Hung 2009). Hereinafter, the intended tunnel is named the main tunnel. Chehade & Shahrour (2008), numerically studied the influence of relative position and construction procedure in terms of twin tunnel interaction in three different configurations. They actually found that the construction procedure affects the soil settlement and internal forces as well as revealed the critical position with highest and lowest v in settlement and bending moment (Chehade & Shahrour 2008).

Concerning underground obstacles, it should be noted that the main tunnel runs under an existing hydraulic canal (Velenjak canal) and crosses a sewer pipe, which is sensitive to induced settlement.

2 METHODOLOGY

2.1 Steps of analysis

In this research, two total stages were considered for the numerical analysis, as follow:

- First, developing initial condition; the in-situ geostatic stresses were modeled, assuming the coefficient for lateral soil pressure as much as $k_0 = 1 - sin\ \phi$ and the existing surface load (Jaky 1944). It should be noted that the Velenjak canal has already excavated, thus its effect was initially considered in this step.
- Second, modeling of the tunnel excavation and initial and final stabilization; the main twin tunnel is actually composed of south and north sections. Regarding NATM based

excavation, every two sections were divided into several top and bench galleries, which were excavated in different orders based on their efficiency to reduce settlement. In addition, the Center Cross Diaphragm (CCD) and final lining were meanwhile utilized with three different scenarios of excavation (more details about scenarios are presented in section 3-1).

Figure 2a, shows the south and north sections of the main tunnel with the galleries of excavations and the proposed CCD and final lining. The main adopted-approach of the excavation was first partial excavation and stabilization of south section and then beginning for the north one.

2.2 Numerical model property

Soil constant medium is usually well analyzed by Finite Element Method (FEM), and for the intended numerical analysis, a 2D finite element program (PLAXIS version 8) was therefore adopted to develop a numerical model (Brinkgreve et al. 2004). Typically, triangular 15-node elements are considered to provide an accurate calculation of stresses and failure loads. The model geometry and its configuration are schematically shown in Figure 2b. The main tunnel runs under a hydraulic tunnel (Velenjak canal) at a vertical distance of nearly 1.0m. According to the tunnel location shown in Figure 1, the maximum surcharge load is attributed to main street as high as 2ton/m^2 (i.e. equivalent traffic loading of the main street) as well as the maximum soil overburden approximately equals 12.5m, according to the longitudinal profile of the project route.

2.3 Soil Parameters

Based on how big and important the project is, various soil investigation procedures of any kinds of field and laboratory tests are required to be included in. Here, just like any other similar project with similar soil conditions, different in-situ tests were conducted such as Standard penetration, Pressuremeter, permeability, in-situ shear box tests as field investigations and other laboratory ones including mechanical and shear strength tests. The recommended plan for the machinery and manual boreholes are presented in Figure 3. As presented in this figure, four out of total seven identification boreholes were drilled manually (i.e. TP-A1 to A4) and the rest were machinery (BH-A1 to A3), and consequently, disturbed and undisturbed soil samples were provided regarding the type of drilled boreholes. Based on the performed site investigations, the soil parameters were approximately estimated, based on which a filling layer of about 2m was determined for the first layer, while the rest of soil layers were mainly of sandy gravel.

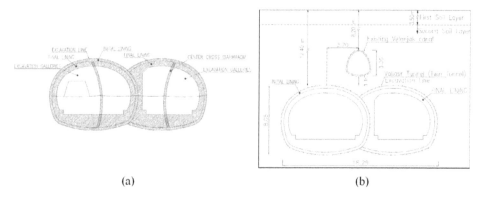

(a)	(b)

Figure 2. (a) Typical of tunnel section, initial and final lining and excavation galleries (b) Geometrical properties of the numerical model.

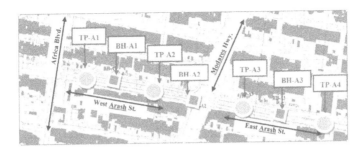

Figure 3.   Plan of boreholes (BH) and test pit (TP) locations (red marked points).

Table 1.   Parameters of the soil materials.

Symbol	Soil parameters	First soil layer	Second soil layer
ϕ	Internal friction angle (degree)	30	37
C	Cohesion (kg/cm^2)	0.1	0.25
γ_m	Natural density (gr/cm^3)	17	18
ν_{ur}	Poisson ratio of unloading/reloading	0.2	0.2
E_{50}	Secant deformation modulus (kg/cm^2)	400	700
E_{ur}	Unloading stiffness (kg/cm^2)	1200	2100
m	Power of stress level of stiffness	0	0.5
ψ	Dilatancy angle (degree)	0	7

The values of soil parameters are summarized in Table 1. It is noteworthy to say that different analogous projects with similar soil conditions have revealed that the hardening soil model is the best constitutive model well describing soil behavior in Tehran, and consequently, with regard to its positive points for the existing geological condition, the parameters are estimated based on HS constitutive model (Schanze 1999).

2.4   Tunnel model and reinforcing system

The main tunnel has a total length of about 50 m with 18.3 m wide and 8.1 m height. In terms of stabilization, it is initially supported by a 30m thick shotcrete lining and finally by reinforced concrete with a thickness of between 50 m to 70 m, and accordingly, all are numerically modeled by plate elements with linear elastic behavior. For the initial support, the used concrete shall be of class C25, and the used reinforcement shall be of type III, and accordingly for the final support, except for the concrete class that is of C35. The parameters of the tunnel lining are summarized in Table 2 (ACI Committee 1999).

Micropile as another reinforcing system is also applied in order to control displacement rate and consequently, various sensitivity analyzes were separately developed to decide on its best performance, but these are beyond the scope of this research and would subsequently be reported in another paper. Here, based on selected characteristic for micropile, just the

Table 2.   Parameters of the tunnel lining structure.

Tunnel Lining		Thickness	E_c (kg/cm^2)	EI (kN.m^2/m)	EA (kN/m)	W (kN/m/m)
Initial lining	wall	30cm	23875196.3	26860	7162559	7.2
	CCD	25cm		15248.4	5855392	
Final lining		50cm	282495	147133	14124757	12
		60cm	282495	254245.6	16949708	14.4
		68cm	282495	370106.3	19209669	16.3

Table 3. Parameters of the micropile structure.

Micropile	E_{eq} (kg/cm^2)	E_{eq} I (kN.m^2/m)	E_{eq} A (kN/m)	W (kN/m/m)
Steel Bar (ϕ32) @0.5m	764990.33	469.1	949200	0.47
Steel Bar (ϕ28) @0.5m	703291.94	431.26	872644.64	0.44

Remarks:

$E_{eq}= E_{casing} \times (A_{casing}/A_{total}) + E_{Grout} \times (A_{grout}/A_{total}) + E_{bar} \times (A_{bar}/A_{total})$
$A_{casing}= 13.2$ cm^2, $A_{total} = 62.04$ cm^2, $A_{grout}= A_{total} -A_{bar} -A_{casing}$

relevant mechanical and geometrical properties are briefly presented in Table 3 (AASHTO 2002, Sabatini 2005).

3 NUMERICAL ANALYSIS

3.1 Analysis procedure

Multiple line tunnels, in particular, twin tunnels, impact the adjacent infrastructures and superstructures by induction of huge ground loss and consequently, large surface settlement. As an early estimate of the settlement, the numerical methods are widely used, other than empirical ones which potentially give inaccurate results (Chapman 2007). Further, this study also focused on assessing the efficiency of each excavation and construction orders and selecting the most efficient method to reduce the induced field surface settlement effectively. Although the numerically obtained magnitudes of settlements do not accurately match with the actual filed ones (Chapman 2004), three different numerical models each with specific construction scenario were accordingly developed to calculate settlements, for which the specifications are described as follow:

The first Scenario (see Figure 4 Figure):

1. The sequences of excavation are proposed so to excavate and stabilize partially each south and north sections
2. The CCD is simultaneously eliminated with excavation stages
3. The final lining is installed after full excavation and entire initial stabilization of both sections.

The second scenario (see Figure 5 following the stage number VI in Figure):

1. The sequences of excavation are proposed so to excavate and stabilize partially each south and north sections
2. The CCD is eliminated after full excavation and initial stabilization for each south and north section.
3. The final lining is installed after completing the excavation and entire initial stabilization of both sections.

The third scenario (see Figure 6 following the stage number VIII in Figure 5):

1. The sequences of excavation are proposed so to excavate and stabilize partially each south and north sections
2. The CCD is eliminated after full excavation and initial stabilization for each south and north section.
3. The final lining is beginning to be installed upon completing the excavation and entire initial stabilization of each south and north section.

3.2 Numerical results

Based on three different scenarios numerically developed, the obtained results were analyzed in terms of the ratio of the surface settlement introduced by line graph in Figure 8. First, the

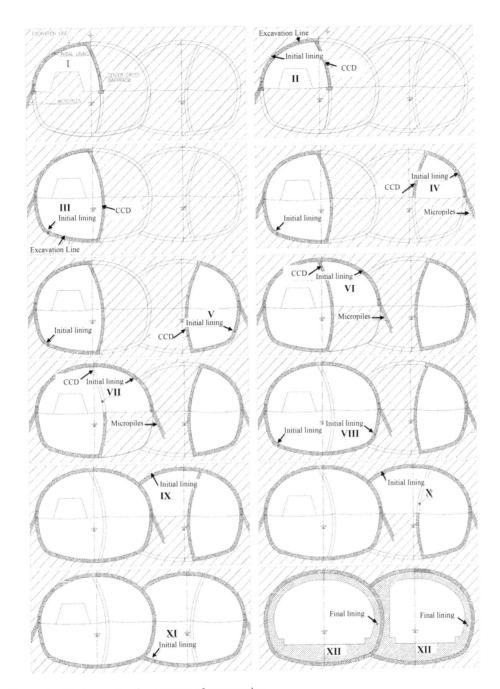

Figure 4. First scenario; the sequence of construction.

contours of vertical displacements are required to be checked in Figure 7. As shown in this figure, the maximum vertical displacements on top of the tunnel is about 6.4 and 6.0 cm respectively for the first two scenarios, and it is substantially reduced to about 3.4 cm. As far as the final ground surface settlements are concerned, the maximum values in Figure 8 are in order 4.8 and 4.5 cm for the first scenarios meaning that, the second considered scenario is not more efficient than the first one, and finally, the values for the third one is reduced to about 2.4 cm,

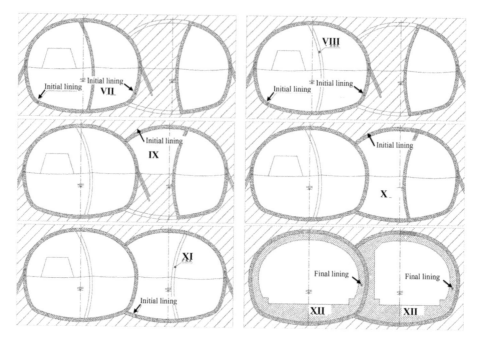

Figure 5. Second scenario; the sequence of construction.

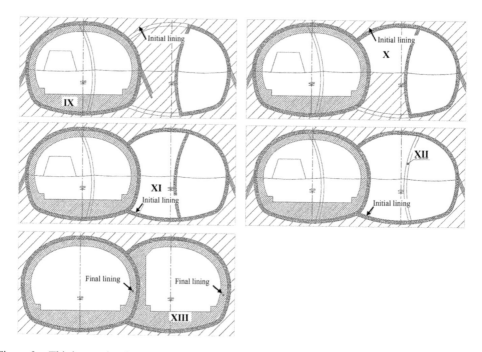

Figure 6. Third scenario; the sequence of construction.

indicating more its efficiency than the other scenarios. Also, in terms of distribution of surface settlement, by the first and second scenarios, displacements are severely localized in an area just above the tunnel axis, whereas, it is more dispersed around the tunnel axis in the third one.

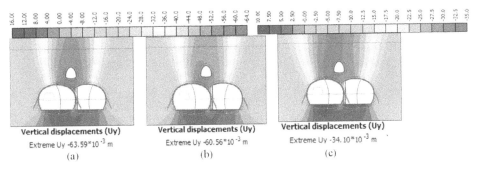

Vertical displacements (Uy)
Extreme Uy -63.59*10^{-3} m
(a)

Vertical displacements (Uy)
Extreme Uy -60.56*10^{-3} m
(b)

Vertical displacements (Uy)
Extreme Uy -34.10*10^{-3} m
(c)

Figure 7. The contour of vertical displacement; (a) First, (b) Second, and (c) Third scenarios.

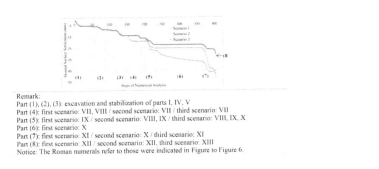

Remark:
Part (1), (2), (3): excavation and stabilization of parts I, IV, V
Part (4): first scenario: VII, VIII / second scenario: VII / third scenario: VII
Part (5): first scenario: IX / second scenario: VIII, IX / third scenario: VIII, IX, X
Part (6): first scenario: X
Part (7): first scenario: XI / second scenario: X / third scenario: XI
Part (8): first scenario: XII / second scenario: XII, third scenario: XIII
Notice: The Roman numerals refer to those were indicated in Figure to Figure 6.

Figure 8. Settlement history of the ground surface based on three scenarios.

When it comes to displacement ratio, there are significant differences between the three scenarios, particularly in final steps, as they are identically modeled and numerically analyzed at first steps (Figure 8). As shown in Figure 8, some steps containing abrupt changes, are marked from one to seven. For the first three marked steps, the ground behavior is the same. But, the disparity in area number four which occurred between the first scenario and the two others indicates the effectiveness of the CCD, as it is initially eliminated in the first scenario, and then the excavation part VIII is continued. The CCD, therefore, caused to reduce the surface settlement by 15%, while the corresponding ratio remained the same. On the other hand, the areas number five, seven and eight represent the effective role of the final lining installed meanwhile the initial stabilization in the third scenario. Considering the fifth region, the final lining of the south section was installed upon destroying the CCD in the third scenario, while for the second one, the excavation and initial stabilization were continued to proceed even after the destruction of CCD. Clearly, the induced surface displacement of the third scenario reduced by 20% than the first one and the corresponding ratio reached just little under 0.2 mm/step, as it was little over 0.3 mm/step based on first and second scenarios. Consequently, it rather proves the advantage of performing the final lining prior to the entire implementation of initial stabilization. As can be seen from this graph, the values of the first scenario in area number six gradually declined due to removing CCD in the north section and subsequently, other huge differences occurred in areas number seven and eight, except for the second and third scenarios in which the CCD is first destroyed before installing the final lining. Despite maintaining the CCD till the end of stabilization in second scenario, the corresponding values of settlement surged rapidly to 44 mm with a ratio of about 0.7 mm/step, while for the first one is about 0.4 mm/step, but concerning the combined effect of final lining and CCD, the line graph shows a settlement of about 27 mm with a lower ratio of about 0.2mm/step. The aforementioned trend continues till the end of final stabilization, meaning that the surface settlement reaches finally 30.3mm with a ratio of about 0.6 mm/step, whereas these values are about 49 and 51 mm and approximately with the same ratios. From the

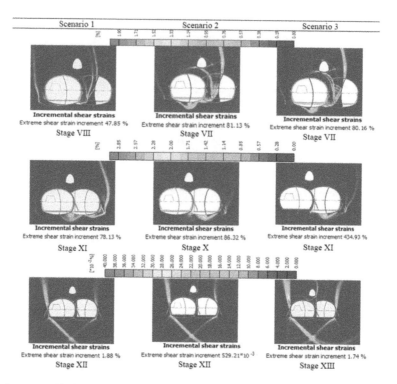

Figure 9. Contours of induced shear strains based on safety analysis.

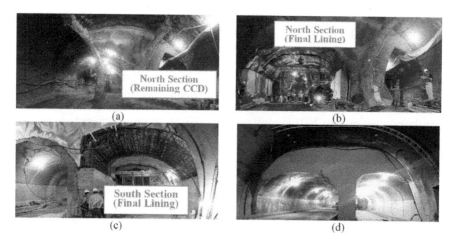

Figure 10. Tunnel excavation and construction (the 3rd scenario); (a) entire excavation and initial stabilization of north section with remaining CCD, (b) installation of the final lining of the south section meanwhile the initial stabilization, (c) entire implementation of final lining for south section, (d) a view of entire implementation of final lining.

geotechnical perspective, the shear strains induced in a state of failure were additionally examined in different stages of construction, so that the trend of displacement history were compared with. Figure 9, therefore, shows contours of induced shear strain corresponding to each abrupt drop in settlement history, and it is also calculated by safety analysis based on the method of phi-c reduction.

As shown in this figure, for the same phases of construction in three different scenarios, the zones of mobilized shear strength were typically thickened and increased in maximum value, as the extra technical provisions were considered. The maximum induced shear strains in failure state have, in fact, increased drastically in the third scenario than the two others, particularly in stage XI and this is approximately rational for the second scenario than the first one, obviously meaning that applying rigid supports such as installation of final lining and keeping CCD cause to activate more the potential shear capacity of the soil around in state of failure and consequently, more efficiently mobilize soil shear strength to minimize the settlement. Figure 10 finally shows some outstanding stages of excavation and construction based on the third scenario.

4 CONCLUSIONS

By a way of conclusion, due to difficulties with maintaining the CCD before the entire excavation and full stabilization based on the second scenario, it is not worthy effort, as the settlement values do not reduce adequately, and on average, the corresponding ratio does not change significantly. In contrast to the second scenario, the third one is actually effective, since it causes surface settlement reduction to 44 percent and the ratio of settlement declines averagely about 40 percent. Considering the points discussed in the second and third scenarios, the final stabilization is required to be performed upon the destruction of CCD, in order to have the most efficient method of excavating and stabilizing and to keep the safety in terms of allowable surface settlements.

REFERENCES

ACI Committee, 1999. Building code requirements for structural concrete: (ACI 318–99); and commentary (ACI 318R-99). American Concrete Institute.

Brinkgreve, R.B.J., Broere, W. & Waterman, D. 2004. *PLAXIS 2-D Professional Version 8.0–User's Manual*. PLAXIS bv, The Netherlands.

Chehade, F.H. & Shahrour, I. 2008. Numerical analysis of the interaction between twin-tunnels: Influence of the relative position and construction procedure. *Tunneling and Underground Space Technology*, 23(2), pp.210–214.

Chapman, D.N., Rogers, C.D.F. & Hunt, D.V.L. 2004. Predicting the settlements above twin tunnels constructed in soft ground. *Tunneling and Underground Space Technology*, 19(4/5), pp.378–380.

Chapman, D.N., Ahn, S.K. & Hunt, D.V. 2007. Investigating ground movements caused by the construction of multiple tunnels in soft ground using laboratory model tests. *Canadian Geotechnical Journal*, 44(6), pp.631–643.

Golshani, A., Hosseini, M. & Majidian, S. 2014. Construction of under passing a crowded junction with shallow soil depth (case study, Tehran). *Geotechnical Aspects of Underground Construction in Soft Ground*, p.479.

Hung, C.J., Monsees, J., Munfah, N. & Wisniewski, J. 2009. Technical manual for design and construction of road tunnels–civil elements. *US Department of Transportation, Federal Highway Administration*, National Highway Institute, New York.

Jaky, J., 1944. The coefficient of earth pressure at rest. *Journal of the Society of Hungarian Architects and Engineers*, pp.355–358.

Sabatini, P.J., Tanyu, B., Armour, T., Groneck, P. & Keeley, J. 2005. *Micropile design and construction*. US Department of Transportation, Federal Highway Administration, Washington, DC, Report No. FHWA-NHI-05-039.

Schanz, T., Vermeer, P.A. & Bonnier, P.G. 1999. The hardening soil model: formulation and verification. *Beyond 2000 in computational geotechnics*, pp.281–296.

Telford, T., 2004. *Tunnel lining design guide*. British Tunneling Society and the Institution of Civil Engineers.

Transportation Officials, 2002. *Standard specifications for highway bridges*. American Association of State Highway and Transport Officials (AASHTO).

Zhu, H., Chen, M., Zhao, Y. & Niu, F. 2016. *Stability Assessment for Underground Excavations and Key Construction Techniques*. Springer.

Zolghadr, E., Pasdarpour, M., Majidian, S. & Golshani, A. 2013. Numerical Modelling of NATM Urban Tunnels and Monitoring-Case Study of Niayesh Tunnel. *18th International Conference on Soil Mechanics and Geotechnical Engineering*, Paris.

Tunnels and Underground Cities: Engineering and Innovation meet Archaeology,
Architecture and Art, Volume 11: Urban
Tunnels - Part 1 – Peila, Viggiani & Celestino (Eds)
© 2020 Taylor & Francis Group, London, ISBN 978-0-367-46899-6

Study of existing tunnel effect on adjacent ramp stability during construction – a case study – Arash tunnel

A. Golshani & S. Majidian
Tarbiat Modares University, Tehran, Iran.

B. Alinejad
Science and Technology University, Tehran, Iran.

ABSTRACT: To investigate the effect of tunnel existence on the ramp stability, two models have been simulated. The case study is a part of Arash tunnel project constructed in Tehran, Iran. The first model contains construction of tunnel, conduit, and ramp, respectively, while in the second model (without tunnel) after conduit implementation, ramp is constructed. This article is aimed to discuss the displacements induced in ramp walls, and ground settlements due to existence of the adjacent tunnel, and supporting elements in two numerical cases. Numerical results demonstrate that the displacement of ramp walls, and ground settlements have increased in the tunnel existed model. Also the location of maximum displacement of ramp walls and ground surface are different in two cases. Eventually the settlement results of the actual constructed model (first model) has been compared to monitoring results.

1 INTRODUCTION

Nowadays, in developing countries along with cities, and population growth, the need for underground facilities has been increased. The main idea is to use the underground space for infrastructure usages, such as traffic tunnels, metro tunnels, transportation stations, utilities, storages, and etc. Traffic tunnels are one of the main development schemes of growing cities. Arash project is a traffic tunnel that has been designed to facilitate the connection and reduce the traffic volume between East Arash Street and Niayesh highway with a length of 1.7 km. This project is an east-west traffic tunnel, with different accessibilities in its path. The plan of Arash project has been shown in Figure 1.

There are two ways for investigating the tunnel effect on ground settlement known as analytical modeling and empirical relationships. According to (Peck,1969), it has been observed that the shape of the settlement surface in the initial stages of tunnel construction resembled an inverted normal Gaussian distribution curve. Rankin (1988) has been investigated the empirical relationships for tunnel settlement and compared them to analytical and case study results. According to W.J. Rankin during the initial stages of tunnel development, the settlement is accompanied by horizontal displacement. Also, W.J. Rankin concluded that the overall transverse settlement to the detectable limits of surface settlement is about three times the depth of the tunnel.

The objective of this article is to study the 2-lane tunnel's interaction on ramp wall and ground displacement for a specific area in Arash project. The survey is performed for West Arash Ramp which is exactly on top of the 2-Lane tunnel that underpasses the west Arash Street. The intended location has been displayed in Figure 1. The longitudinal view of the study area has been demonstrated in Figure 2. In this figure the ramp and tunnel crown junction is visible. As the top ramp and the bottom 2-Lane tunnel proceeds to the east of the project reaches to the two-story tunnel of modares.

Figure 1. Location of Arash Tunnel project with intended location specification.

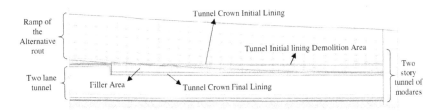

Figure 2. Longitudinal view of the intended area (the position of the ramp on top of the 2-Lane tunnel).

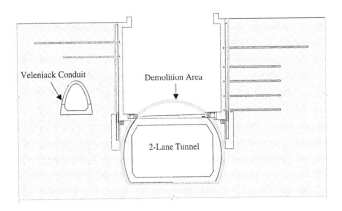

Figure 3. West Arash ramp section on top of 2-Lane Tunnel (first model).

For investigating the tunnel interaction on the ramp's wall displacement two numerical modeling has been prepared. The first model simulated with the existence of the tunnel, and the second model simulated without considering the tunnel.

Since Velenjack conduit located beside ramp's left wall (Figure 3), nailing as a supporting system for the left wall has been used to the allowable height above conduit. Thus, for supporting the rest of the left wall, soldier piles were utilized. The ramp and tunnel conjunction caused discontinuity of the ramp's walls with the slab. As shown in Figure 3, there is no ramp slab implemented, instead, the top of tunnel final lining is actually performed multifunctional as tunnel crown and also ramp slab. n fact, the ramp slab is the top part of the tunnel's final lining. For wall stabilization soldier piles has to be used on both sides of the ramp at every 2.5 meters. In the second model (without a tunnel) walls and slab of the ramp are joined together in the U shape. The U shape of the second model ramp also helps to reduce ramp wall displacement and its stability.

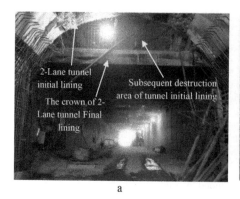
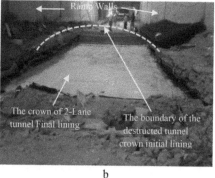

a

b

Figure 4. The tunnel crown initial lining destruction pictures, a) view from inside of the 2-lane tunnel, b) view from open trench ramp.

The destruction area of the tunnel crown initial lining has been demonstrated in Figure 4. The Figure 4a represented the subsequent destruction area of tunnel initial lining from the inside view of the tunnel, while in the Figure 4b the destructed tunnel crown of the initial lining has been shown schematically, also the tunnel crown of the final lining has been specified.

2 METHODOLOGY

The impossibility of West Arash Street occupation leads to the following construction sequences. As mentioned before, we tend to compare two simulations, the first one with considering a tunnel, the other one is without the tunnel.

The construction stages of the first model are respectively as following: 2-Lane tunnel, Velenjack conduit, and ramp construction. While in the second model without assuming tunnel, the construction stages are respectively, Velenjack conduit, and then ramp construction.

As ramp goes forward from west to east of the West Arash Street, the bottom slab of the ramp enters the top area of the 2-Lane tunnel initial support as demonstrated in Figure 2. For resolving this problem, the top part of the tunnel initial crown has to be demolished. As ramp proceeds to the east, bottom part of the ramp is getting more and more into the top of tunnel initial lining so the different amounts of destruction are required for the tunnel's top part with the length of about 65 meters, shown in Figure 2. Eventually, at the easternmost of the West Arash Street both of ramp and tunnel proceed to connect to the two-story tunnel of Modares. The tunnel final lining formwork for this length has been changed as represented in Figure 3. Due to the existence of weak soil area and Velenjak canal, some other obstacles, and also the impossibility of constructing a continues U shape ramp in the survey region, soldier piles has been implemented for supporting initial lining in both sides of the ramp.

The construction sequences of the West Arash ramp for the first simulation has been illustrated in Figure 5, and explained as follows:

1. 2-Lane tunnel initial lining construction
2. 2-Lane tunnel final lining construction
3. Velenjack conduit construction
4. Ramp initial lining construction, and top of tunnel initial lining destruction
5. Ramp final lining construction

The construction stages for the second model are explained below:

1. Velenjack conduit construction
2. Ramp initial lining construction, and top of tunnel initial lining destruction
3. Ramp final lining construction

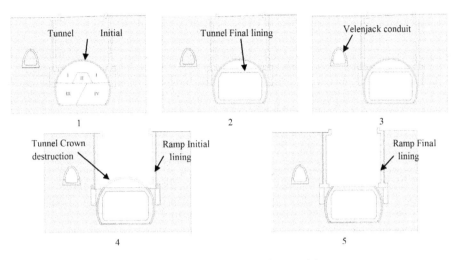

Figure 5.   West Arash ramp construction sequences for the first model.

3   NUMERICAL MODELING

The simulation performed with the finite element method. The two-dimensional finite element software Plaxis V8.5 is used for numerical analysis of the ramp (Brinkgreve et al.2002). The basic features of Plaxis program have been evaluated and verified in this project. Calculations are based on three FHWA publications, one is soil nail walls-reference manual (Lazarte et al. 2015), the other is technical manual for design and construction of road tunnels and civil elements (Hung et al. 2009), and finally for soldier piles designing the permanent ground anchors (Cheney 1988) has been used.

The excavation depth of the ramp is around 11 meters. For considering the boundary effect in numerical analysis, the soil area extends 5, 2.5 times in x, y directions, respectively.

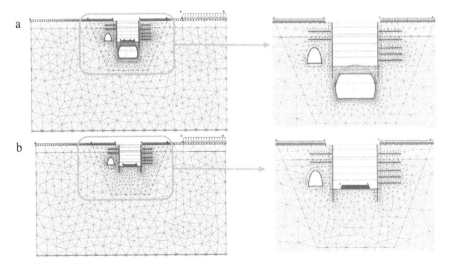

Figure 6.   West Arash ramp simulations, a) with considering a tunnel, b) without considering a tunnel.

4 MATERIAL PROPERTIES AND DESIGN PARAMETERS

For each simulation 15 node elements have been used. The soil constitutive model (material behavior) is considered as hardening soil (Schanz et al. 1999). Soil layers and properties result from the geotechnical survey. The position of boreholes are represented in Figure 7 and, Figure 8. The longitudinal outcrop for the whole project has shown in Figure 8. Soil layers parameters have been indicated in Table 1.

4.1 *Input parameters*

The initial support of the first model includes nails, soldier piles, concrete piles, initial lining with the thickness of 15cm with one layer of Φ8 bars. The soldier piles used in the ramp's wall is from IPB260 type at every 2.5 meters. Soldier pile details have been represented in Figure 9, which strengthened in web and flags, near ramp's bottom slab position, resisting against shear forces and bending moment based on AISC (2010), and FHWA (Cheney,1988). The right picture of Figure 9 is IPB260's position in the

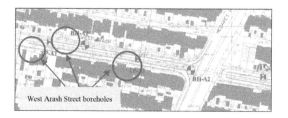

Figure 7.   Arash Project geotechnical boreholes position.

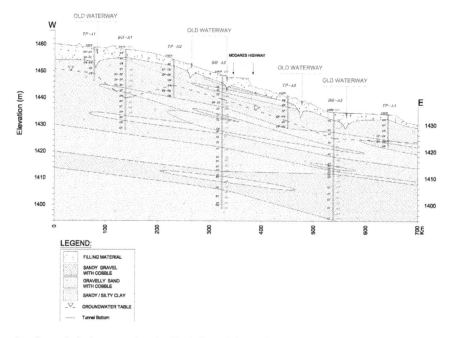

Figure 8.   Geotechnical outcrop longitudinal view of the project.

Table 1. Soil layers geotechnical parameters.

Upper layer		
ϕ = 30°	Internal friction angle	Thickness
C=0.10 kg/cm²	Cohesion	3.5 m
γ_m= 17 gr/cm³	Natural density	
v = 0.2	Poisson ratio of unloading/reloading	
400 kg/cm²	Secant deformation modulus	
1200 kg/cm²	Unloading stiffness	
ψ = 0°	Dilatancy angle	

Second layer		
ϕ = 37°	Internal friction angle	3.5 –
C=0.25 kg/cm²	Cohesion	down to bottom
γ_m= 18 gr/cm³	Natural density	
v = 0.2	Poisson ratio of unloading/reloading	
700 kg/cm²	Secant deformation modulus	
0.5	Power of stress level of stiffness	
2100 kg/cm²	Unloading stiffness	
ψ = 7°	Dilatancy angle	

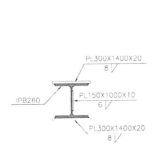 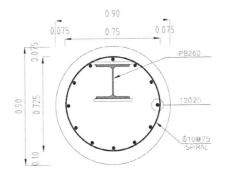

Figure 9. Soldier pile details for ramp walls.

concrete root pile below the 0.00 level of the ramp. The properties of steel piles presented in Table 2.

Tunnel initial and final lining and ramp's initial lining has been calculated due to ACI 318-02 (ACI Committee. 2002). Ramp walls Initial lining properties with the thickness of 20cm

Table 2. Steel piles properties.

Area of cross section A (cm²)	Moment of inertia I (cm⁴)	Axial stiffness EA/L (kN/m)	Bending stiffness EI/L (kN.m²/m)	Weight W/L (kN/m/m)
118	14920	9.44E+05	11936	0.372

Table 3. Ramp walls initial lining properties.

Specific strength f′$_c$ (kg/cm²)	Yield stress f$_y$ (kg/cm²)	Axial stiffness EA/L (kN/m)	Bending stiffness EI/L (kN.m²/m)	Weight W/L (kN/m/m)
250	4000	4.684 E+06	7807.189	4.708

Table 4. Tunnel initial & final lining properties.

Lining Type	Material properties		Lining thickness (cm)	Geometrical properties		
				EA/L (kN/m)	EI/L (kN.m^2/m)	W/L (kN/m/m)
Initial Lining	f_y (kg/cm^2)	4000	20	4.684 E+06	7807.189	4.708
	f_c (kg/cm^2)	250				
Final Lining	f_y (kg/cm^2)	4000	55	1.96×10^5	1.55×10^7	13.2
	f_c (kg/cm^2)	350	70	4.04×10^5	1.98×10^7	16.8
			87	7.75×10^5	2.457×10^7	20.88

Table 5. Nail properties.

Nail Parameters	Bar 32 @2m	Bar 32 @1.5m	Bar 32 @1.25m
EA/S_H =E_{eq}/S_h ×(($\pi D^2$$_{DH}$)/4)	1.016E5	1.355E5	1.626E5
EI/S_H =E_{eq}/S_h ×(($\pi D^4$$_{DH}$)/64)	46.98	62.653	75.183
W/S_H	0.073	0.097	0.116

has been mentioned in Table 3. 2-lane tunnel initial and final lining properties has been represented respectively in Table 4.

For assigning the nail properties the Plaxis reference (Issue 25/Spring 2009) has been considered for axial and flexural stiffness (Peck, 1969). Nail properties which have been used in simulation indicated in Table 5, for computing these parameters the below equations have been used.

4.2 *Loading*

Both simulations have been exposed to static loading, including soil surcharge load known as a dead load, traffic and adjacent structures load known as a live load. A seven, and a two-story buildings load respectively at right and left sides of the ramp, have been applied to the simulation as represented in Figure 6.

5 MONITORING PROCEDURE

The settlement pins were used for recording field data of ramps. In this article, the monitoring data for an easternmost chainage of the ramp which has been demonstrated as following has been used to compare with numerical data in the following sections.

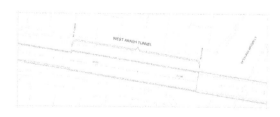

Figure 10. The position of monitoring tools on West Arash ramp.

6 RESULTS

The tunnel and ramp interaction is studied in this article. The tunnel existence effect on ramp wall displacement has been numerically investigated. Also, the displacement results of the first

Table 6. Total, horizontal & vertical displacement of the first model (with the tunnel).

Model total displacement	Wall horizontal displacement	
	Left	Right
Total displacements (Utot) Extreme Utot 42.33*10^{-3} m	Horizontal displacements (Ux) Extreme Ux 25.78*10^{-3} m	Horizontal displacements (Ux) Extreme Ux -29.00*10^{-3} m
Ground vertical displacement		
Left	Right	
Vertical displacements Uy Extreme Uy -33.67*10^{-3} m	Vertical displacements Uy Extreme Uy -28.38*10^{-3} m	

Table 7. Total, horizontal & vertical displacement of the second model (without a tunnel).

Model total displacement	Wall horizontal displacement	
	Left	Right
Total displacements (Utot) Extreme Utot 29.78*10^{-3} m	Horizontal displacements (Ux) Extreme Ux 14.40*10^{-3} m	Horizontal displacements (Ux) Extreme Ux -4.74*10^{-3} m
Ground vertical displacement		
Left	Right	
Vertical displacements Uy Extreme Uy -22.26*10^{-3} m	Vertical displacements Uy Extreme Uy -26.16*10^{-3} m	

Table 8. Displacement results of two models.

Model No.	Max wall horizontal displacement (mm)		Max ground settlement (mm)	
	left	right	left	Right
1. with tunnel	25.78	29	33.67	28.38
2. without tunnel	14.40	4.74	22.26	26.16
Increase percentage comparing to No.2	79%	512%	51.26%	8.5%

model (with considering tunnel) is compared to monitoring results. The displacement results have been illustrated in Table 6, Table 7, and summarized in Table 8.

From displacement results represented in Table 8, the following results are achieved; Tunnel's impact on the ramp's wall displacement and ground settlement are obviously observed

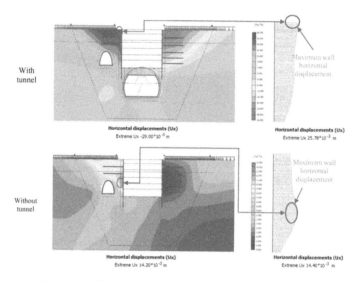

Figure 11. Maximum horizontal displacement occurrence point on the left wall.

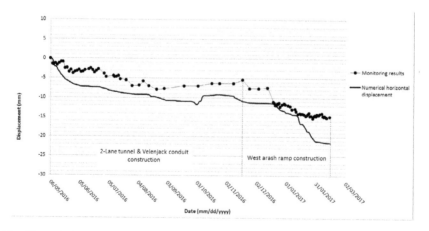

Figure 12. Numerical and monitoring analysis results comparison for the 2100 km chainage.

from analytical results obtained. Tunnel's existence has an enormous effect on the right wall horizontal displacement of the ramp. Whereas, on the left side because of Velenjack conduit presence, the effect of the tunnel on the ramp is less.

The existence of the Velenjack conduit on the left side of the wall has a direct effect on the left wall's displacement in compare to the right wall. As the tunnel axis is exactly under the ramp with no deviation from the center line of the ramp, we expect that the tunnel effect on both sides should be approximately the same, however, results indicate that the Velenjack conduit has a great influence on left wall's displacement. For the second model (without a tunnel) with merely Velenjack conduit existence, the displacement of the left wall is about three times the right wall. As mentioned before, because of the Velenjack conduit presence the proper nailing system at the left side of the ramp is impossible, so left wall's displacement is much more than the right wall. However, the tunnel's effect on the right side of the wall is more obvious.

The existence of the tunnel also influences the location of maximum horizontal displacement occurrence point, which presented in Figure 11. The maximum horizontal displacement for the tunnel existence case occurs in the top of the wall, while in the without tunnel case it happens in the bottom half of the wall beside Velenjack conduit. In the first model (with the tunnel) the existence of a tunnel beneath the ramp causes discontinuity between the ramp walls and slab, which makes them perform separately. The discontinuity of the ramp walls and slab causes the maximum horizontal displacement of the walls to occur on top of the walls. In the second case, as we predicted, the absence of nailing system beside Velenjack conduit causes the maximum displacement to occur on the position next to the conduit on the wall.

The numerical displacement results have been compared to monitoring data in Figure 12. The settlement results of monitoring data and numerical simulation of the first model (considering tunnel, Velenjack conduit, and ramp) have approximately the same pattern. In spite of the same pattern of displacement, Numerical modeling in the last stages of ramp construction resulted in larger displacements comparing to monitoring data, this might be due to better soil properties at the site compare to the parameters obtained from geotechnical experiments, which has been observed during excavation.

7   CONCLUSION

The tunnel interaction with the adjacent ramp has been investigated in this study. For this purpose, two numerical simulations accomplished, the first one is the ramp with considering tunnel effect, and the other one is the ramp without considering tunnel. Since the presence of Velenjack conduit in the left side of the ramp makes some difficulties in nailing procedure, and also the impossibility of ramp slab implementation above tunnel crown causes a discontinuity in the initial lining of the ramp. These difficulties lead us to design soldier piles on both sides of the ramp to provide stability. The impossibility of nail implementation besides Velenjack conduit position effects left ramp horizontal displacement. The most considerable impact of tunnel presence has been observed on the displacement of the right wall of the ramp. As a result, displacements induced to the ramp walls, and ground settlement for the first model (with the tunnel) is about 5 times comparing to the second model (without a tunnel).

REFERENCES

ACI Committee. 2002. Building code requirements for structural concrete (ACI 318-02) and commentary (ACI 318R-02). American Concrete Institute.
Aisc, L. 2010. Specification for structural steel buildings.
Brinkgreve, R.B.J., Broere, W. & Waterman, D. 2004. *PLAXIS 2-D Professional Version 8.0–User's Manual.* PLAXIS bv, The Netherlands.
Cheney, R. S. 1988. Permanent ground anchors. *Federal Highway Administration, Demonstration Projects Division.*

Hung, C. J., Monsees, J., Munfah, N., & Wisniewski, J. 2009. Technical manual for design and construction of road tunnels–civil elements. *US Department of Transportation, Federal Highway Administration*, National Highway Institute, New York.

Lazarte, C.A., Robinson, H., Gómez, J.E., Baxter, A., Cadden, A. and Berg, R. 2015. Soil nail walls-reference manual. *US Department of Transportation Publication No. FHWA-NHI-14-007, Federal Highway Administration*. Washington DC, USA.

Peck, R. B. 1969. Deep excavations and tunneling in soft ground. *Proc. 7th ICSMFE*, 225–290.

Rankin, W.J. 1988. Ground movements resulting from urban tunneling: predictions and effects. *Geological Society, London, Engineering Geology Special Publications*, 5(1),pp.79–92.

Schanz, T., Vermeer, P.A. & Bonnier, P.G. 1999. The hardening soil model: formulation and verification. *Beyond 2000 in computational geotechnics*, pp.281–296.

Tunnels and Underground Cities: Engineering and Innovation meet Archaeology,
Architecture and Art, Volume 11: Urban
Tunnels - Part 1 – Peila, Viggiani & Celestino (Eds)
© 2020 Taylor & Francis Group, London, ISBN 978-0-367-46899-6

Tunnel design and construction in swelling ground with low overburden and hydrothermal ground water conditions

K. Grossauer & E. Carrera
Amberg Engineering Ltd., Regensdorf, Switzerland

W. Schmid
Implenia Schweiz AG, Wallisellen, Switzerland

T. Zieger
Swiss Federal Railway Authorities, Lucerne, Switzerland

ABSTRACT: The new Boezberg tunnel is the largest and most important project of a large railway upgrade program of the Swiss railway authorities in order to further increase the transfer of transalpine traffic through Switzerland from rail to road. The tunnel was designed and built in an environmentally sensitive area with complex geological and hydrogeological conditions. One of the main challenges is swelling and complex ground conditions due to the presence of Opalinus clay, Anhydrite and Gypsum. Tunnels in comparable conditions have been exposed to long-term effects from swelling phenomena. In addition, the hydrogeological conditions with highly mineralised hydrothermal water in the catchment area of a close thermal spa imposed several measures in order to protect this zone from any impact of the tunnelling activities. In the first 180 m of tunnel, extensive exploratory drillings from the surface were performed parallel to the tunnel excavation in order to precisely determine the starting position of the TBM excavation. By means of systematic underground exploratory drillings a zone of swelling ground was identified below the tunnel with a limited extent. For this area the final lining was designed as a monolithic tube over a length of 37.5 m with continuous reinforcement in longitudinal direction in order to distribute and transfer any swelling pressure into zones with no swelling. The critical zone below the tunnel is permanently monitored by means of deformation meters.

1 INTRODUCTION

1.1 *Project description*

The Swiss Federal Railway Authorities - SBB - was commissioned by the Federal Office of Transport of Switzerland to develop the so called "4-meter-corridor" between Basel in the north western part of Switzerland and Chiasso next to the border to Italy (see Figure 1). This corridor allows transporting truck semi-trailers with a height of 4 m via rail along Switzerland and hence substantially contributes to a reduction of road traffic across the Alps.

The clearance profile of the existing Boezberg tunnel which was built in 1875 and accommodating two tracks doesn't fulfil the requirements for this corridor. In previous project phases the SBB assessed different options in order to integrate the existing tunnel into the corridor. The primary goal was to enlarge and refurbish the existing tunnel during operation. This option was discarded due to the long duration of the construction phase. Finally, a new twin-track tunnel parallel to the existing one was considered as best option including also an adaption of the old tunnel to a rescue tunnel which is also to be used for operational purposed for the new tunnel.

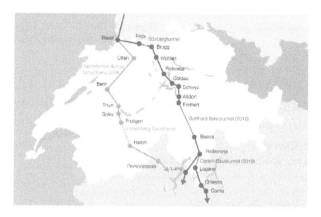

Figure 1. Overview 4-meter-corridor in Switzerland (red line, with main stations indicated with red dots).

The new tunnel with a total length of 2'693 m runs parallel to the existing one with an average distance of approximately 50 m (see Figure 2). Both tunnels are connected by 5 cross passages which allows to reach the rescue tunnel in case of emergency. The railway and safety equipment is also placed in the cross passages. This allows operating the new tunnel already after completion of construction before the existing tunnel is refurbished and converted to a rescue and maintenance tunnel. Two short cut-and-cover sections with a length of 36 m and 44 m are located on both portals. In order to reduce maintenance costs, a slab track system is installed as a low vibration track system over the entire length of the tunnel excluding the portal areas.

The SBB awarded the project to the contractor within a new contractual framework. The contractor is responsible for all construction works for the tunnel and structural works, and the installation of the railway, safety systems and tunnel equipment. Additionally, the contractor undertakes the detailed design as well as the site supervision. The project was awarded end of 2015 to the contractor Implenia Schweiz AG which subcontracted the design works to the design joint venture *IG BözbergPlus* (Amberg Engineering, Basler&Hofmann, Preisig AG, Heierli).

The main excavation works started in September 2016 and the tunnel excavation was completed by end of 2017. The commissioning of the new tunnel is scheduled for December 2020. The commissioning of the whole "4-meter-corridor" is also planned for December 2020. The existing tunnel will subsequently be converted to a rescue and maintenance tunnel and starts operation by end of 2022.

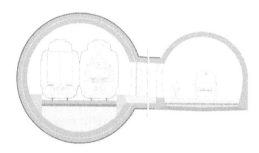

Figure 2. Cross section of the new tunnel and rescue and maintenance gallery (refurbished existing tunnel).

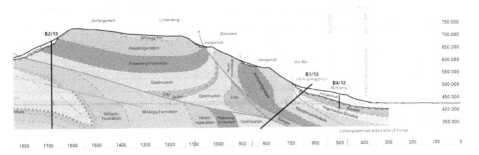

Figure 3.   Geological longitudinal section of the Faltenjura section, south portal area.

1.2   Geological conditions

The Boezberg tunnel is located in the Jura unit which extends all over the north-western part of Switzerland. The Jura is subdivided in two tectonic units - the Faltenjura and the Tafeljura [1]. The southern part of the tunnel crosses mainly limestone, dolomite, anhydrite and Opalinus clay corresponding to the Faltenjura. In the northern part sandstone, claystone, marl and limestone is encountered which corresponds to the Tafeljura. Both tectonic units are separated by a main thrust fault area called backfolding zone. In this area the orientation of the strata changes from horizontal to a sub vertical orientation – see Figure 3. In the first 175 m from the south portal quaternary deposits and strongly weathered rock are encountered. The overburden varies from a few meters in the portal area up to maximum 290 m in the central zone of the tunnel. For about 400 m the tunnel passes through the strata of the Hauptmuschelkalk (limestone) which belongs to the catchment of a close thermal spa. For this "spa protection zone" zone special measures are required in order to prevent drainage and any negative influence on the hydrothermal groundwater conditions as well as any long-term damage of concrete structures due to highly mineralised ground water.

1.3   Excavation method

The soft ground section starting from the south portal is excavated by means of conventional method with a total excavation area of 150 up to 180 m². The advance is done in top heading excavation with a temporary top heading invert under the protection of a pipe roof umbrella, bench and invert. The rock section following is excavated with a single-shield hard rock TBM with a bore diameter of 12.36 m. The support consists of segmental lining with 2 m ring length and without sealing. Each ring consists of five segments with a keystone in the invert. The

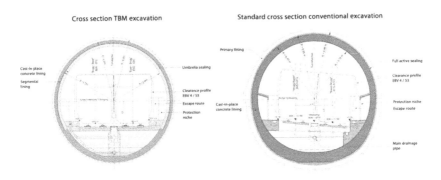

Figure 4.   Cross section of TBM tunnel (left, drained conditions) and conventional tunnel (right, undrained conditions).

segments have a constant thickness of 30 cm in the crown and the side walls. In the invert the thickness of the segments varies between 30 and 60 cm depending on the type of sealing against ground water (drained and un-drained type). The annular gap is filled with grout in the invert and pea gravel in the crown. To prevent water circulation in the thermal spa zone the pea gravel was subsequently grouted. A final cast in place concrete lining is installed over the whole tunnel length (see Figure 4). For about 400 m from the south portal the tunnel is completely sealed in order to prevent any water drainage in the spa protection zone. The remaining tunnel is drained with a standard plastic membrane and a dewatering system in the tunnel invert.

2 CHALLENGES IN THE SOFT GROUND SECTION

2.1 Extensive exploration during tunnel construction

During the excavation in the soft ground section at the south portal an extensive exploration program was performed in addition to the investigation program before starting the excavation with the objective to (a) identify any ground conditions prone to swelling in and/or below the tunnel, and (b) to determine a suitable starting position for the TBM. The TBM was already assembled in front of the south portal parallel to the conventional excavation and the TBM shall be moved from the portal through the excavated tunnel to the starting position after proper conditions are encountered allowing for a save start. Such proper conditions in this context mean (a) stable face conditions as the TBM used is a hard rock type machine without any face support, and (b) sufficient rock cover and overburden in order to omit any heaving of the tunnel due to swelling pressure in or below the tunnel invert.

As of the geological prediction and the experience made during excavation of the fist approximately 50 m of the soft ground section the transition from soft to competent rock was quite complex and heterogeneous. Typical for the geological conditions in the Faltenjura this section shows frequently changing conditions within even one excavation step with soft ground and strong rock due to tectonic activities in both transverse and longitudinal as well as deeply reaching weathering – see below Figure 5.

2.2 3D geological model and determination of TBM starting position

A spatial geological model was developed and continuously updated containing the relevant information from face mapping, previous exploration drillings as well as the drillings conducted during excavation, resulting in a continuously update of the geological prediction. Figure 6 show the spatial geological model including the position and orientation of the exploratory drillings. The model was combined with a 3D model of the tunnel structure containing both the temporary tunnel support measures as well as the final support. Frequently update and interpretation of the model allowed for accurately determine the required

Figure 5. Picture of the face during conventional excavation showing heterogeneous ground conditions with competent rock alternating with soft ground.

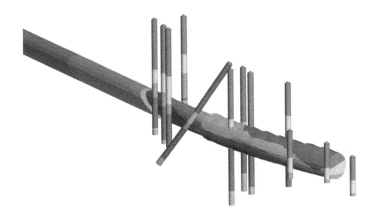

Figure 6.   Geological 3D model showing the additional exploration drillings.

continuation of the conventional excavation and the determination of the optimum starting position for the TBM with (a) competent rock cover of 4 m above the crown and (b) the tunnel invert being outside of the swelling prone Keuper formation in order to omit any swelling. Finally, the effectively encountered ground conditions differed only in a few meters from the predicted ones.

2.3   Measures regarding swelling in low-overburden conditions

According to the geological forecast the tunnel in the south portal area passes through Keuper formation which consists of anhydrite, gypsum and claystone. Tunnels in comparable conditions have been exposed to long-term effects due to swelling and associated high swelling potential and high swelling pressures of anhydrite (Anagnostou et al, 2010). When Anhydrite gets in contact with water a chemical reaction is triggered leading to re-crystallisation to gypsum. This chemical reaction can lead to serious volumetric expansion which in case of confined conditions (e.g. due to a stiff tunnel invert) yields high swelling pressure.

The extent and exact location of anhydrite within the Keuper formation was not detected in detail prior to the tunnel excavation. The exploration drillings during early design stages showed gypsum and claystone only. As a measure for reducing the risk of negative impact onto the final tunnel structure due to long term swelling exploration drillings parallel to the excavation of the soft ground section were performed in order to identify anhydrite lenses below the tunnel. In total 20 vertical drillings with a length from 15 m up to 60 m were performed from the top heading and from the surface.

Approximately 85 m from the south portal an anhydrite lens located approximately 6 m below the final tunnel invert was detected. It is known that anhydrite swelling can be a long-term process and estimating swelling pressures under laboratory conditions can last years or decades [Kovari, 2013]. Considering the construction schedule the swelling potential of the identified lens was determined based on the analysis of the mineralogical composition of samples taken as well as test from previous investigation and evaluation of experience with comparable conditions. Finally, the swelling pressure to be considered for the final design of the permanent tunnel inner was set to 800 kPa.

The soft ground cover above the tunnel in which the anhydrite lens was detected is 25 m, which does not provide sufficient resistance against the main hazard of lifting the whole tunnel due to the swelling pressure. Such a lifting of the tunnel would result in differential displacements between the blocks of the permanent lining and subsequently to differential displacements of the rail tracks with serious impact on the operation of the railway infrastructure. In order to prevent such a hazard the permanent concrete lining is monolithically

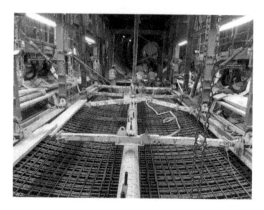

Figure 7.   Installation of the reinforcement in the monobloc.

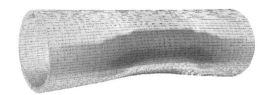

Figure 8.   Deformed shell model of the monobloc showing the deformation of the tunnel lining due to swelling pressure.

connected over 3 blocks, each block with a length of 12.5 m resulting in a total length of 37.5 m – the so called monobloc. With this measure a large weight of the ground above the tunnel can be mobilised allowing for sufficient resistance against the swelling pressure and the swelling stresses acting on a local area of the inner-lining are distributed over a longer section. A non-linear structural analysis of this monobloc was done utilising a 3D shell model utilising Sofistik software [3] and 2D FE-models (see Figure 8). Based to the analysis done the final vertical lifting of the entire tunnel section for the full swelling pressure is limited to less than 4 cm which is in accordance with the maximum value for the serviceability of the tracks.

2.4   Permanent monitoring of critical swelling zone below the tunnel

The behaviour of the anhydrite zone identified below the tunnel is permanently monitored. Three boreholes with a length up to 41m lowered from the tunnel level are equipped with high precision vibrating wire deformation meters and the cables are routed to the portal area. The monitoring results are evaluated frequently during tunnel construction and later on during tunnel operation in order to provide early indication of any heaving of this zone.

3   ASPECTS OF THE PERMANENT LINING

3.1   Waterproofing against hydrothermal groundwater

Over a length of approximately 400 m between the south portal and the non-permeable Opalinus clay formation the tunnel passes a groundwater protection area. The groundwater in this zone is connected to the hydrothermal aquifer of a close thermal spa. In order to prevent any drainage and any negative influence on the hydrothermal groundwater catchment high requirements are defined which include that groundwater drainage is not allowed during

Figure 9. Installation of full water proofing system with grouting pipes in the tunnel invert.

operation of the tunnel. In addition, the highly mineralised ground water (high content of sulphates) is demanding for the concrete of the segmental lining which is designed as a permanent structure over the whole service life of the tunnel.

A pressure resistant water proofing system around the tunnel circumference is installed between the segmental lining (shotcrete lining in the conventionally excavated section) and the concrete inner lining. It consists of two layers of plastic membrane (FPO) with a thickness of 3 mm and 2 mm respectively, and one protection layer on the exterior. The membranes are welded together in sectors with an area of approximately 75 m^2. Each sector can be tested with negative pressure during tunnel operation and in case of any leakage, be grouted with resin (see Figure 9). In order to prevent transmissivity in longitudinal direction grouted radial rings over the whole tunnel circumference are installed at the beginning and the end of the protection zone with a depth into the rock between 1 m and 3 m. Additionally, the pea-gravel in the annular gap of the segmental lining is grouted with mortar and waterstops are installed every 50 m between inner-lining and the sealing membrane.

3.2 *Fire protection of the final concrete tunnel lining*

According to the rescue and safety concept of the new Boezberg tunnel, high standards regarding fire protection are defined by the client. In the rock section the tunnel is designed to withstand the ISO-834 fire development curve for a period of 90 minutes with a maximum temperature rise up to 1'000 °C. In critical zones like the soft ground and the cut-and-cover section stricter design criteria are defined in order to prevent the tunnel collapsing in case of a fire event. In these sections the tunnel is designed considering the modified Hydrocarbon fire development curve with a maximum temperature rise of 1'300 °C for a duration of 120 minutes.

The temperature rise in the final concrete lining and the effect of a fire event is modelled considering the modified Hydrocarbon fire development curve. In a first step the temperature development and distribution for the concrete structure is determined by solving the thermal conduction equation by Fourier. Figure 10 shows the calculated temperature distribution inside the concrete structure for different time steps for a concrete thickness of 40 cm. Actually and at various positions measure temperatures at 120 minutes of fire loading resulting from fire test done on a set of concrete slabs are indicated with black rectangles. In a second step the structural analysis for the concrete lining is done by a non-linear analysis considering the temperature profiles and the temperature specific strength and stiffness properties of concrete and reinforcement according to the Swiss concrete code SIA. The temperature-induced deformations and constraints in the highly static indefinite system were also considered.

The fire analysis for the final concrete lining results in an increase of the reinforcement for tunnel sections with low overburden in soft ground due to poor lateral bedding of the tunnel structure.

3.3 *Measures and testing against concrete spalling*

In the case of a fire event and due to the high temperatures developing in the concrete structure, water evaporates resulting in high gas pressure that can lead to spalling of the concrete. In some cases spalling is propagating until the first layer of reinforcement. This spalling and the resulting reduction of the thickness of the concrete structure lead to a reduction of the bearing capacity and the stiffness of the concrete lining. For critical tunnel areas such as soft ground and cut-and-cover sections this could result even in a collapse of the tunnel structure. For sections in competent rock conditions fire exposure leads also to spalling. However, the probability of a tunnel collapse is comparable low. On the other hand, a fire event in such sections would result in high refurbishment costs due to the repair of the concrete surface. For those reasons PP fibres as protecting measure against spalling of the concrete is considered over the entire tunnel length. During a fire event the PP fibres melt and create new pores in the concrete structure. This allows developing gas pressure to escape and prevents against concrete spalling. The fibres used have a length of 6 mm, a diameter of about 34 μm and a melting temperature of about 150 °C. The dosage used is 2 kg/m^3 based on the experience in other projects.

The fire tests were conducted according to the Swiss standards SIA in order to verify the effectiveness of the plastic fibres and also to verify the numerically determined temperature distribution in the concrete structure over time. Concrete testing slabs (2.1 m x 1.9 m x 0.3 m) were casted on site and tested in the laboratory (see Figure 11). The samples are thermally loaded for 2 h using the modified Hydrocarbon and ISO-834 curve.

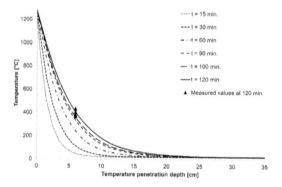

Figure 10.   Temperature distribution over the concrete thickness for various time steps.

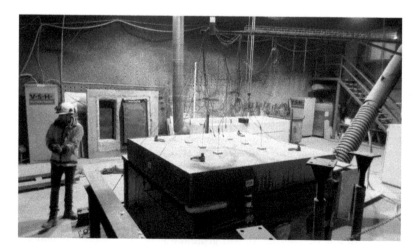

Figure 11.   Fire loading test of sample concrete slabs.

No spalling was detected after the test proving that the applied concrete mixture and type, quantity, and quality of PP fibres are appropriate. In total 6 temperature sensors were installed inside the concrete slab at a depth of 6 cm. The measurement results of these sensors verified the estimated temperature profiles during the detailed design – see Figure 10 above.

4 CONCLUSIONS

Extensive exploration of the complex geological conditions during the tunnel excavation and spatial geological modelling considering the relevant information provided valuable information for the determination of starting position of the TBM. This finally was quite valuable for the overall organisation of the excavation activities and allowed for an optimised schedule for erection and start of the TBM excavation.

In addition, the exploration yielded a critical geological zone with substantial swelling potential below the tunnel. By means of 3D modelling of the permanent concrete structure of the tunnel a monolithic load distribution structure was developed providing sufficient resistance against swelling pressure. No further measures such as tie backs are required. The effectiveness of this system and the behaviour of the critical zone is permanently observed with deformation meters which give indication of any heaving of the tunnel.

Focus is put on the bearing capacity and serviceability of the tunnel lining in case of a fire event. The tunnel structure in critical sections is designed to withstand fire loading and the effectiveness of PP fibres against concrete spalling is verified by means of fire test.

REFERENCES

Dr. von Moos A.G. 2014. Tender documents new Bözberg double track tunnel – geological report.
Kovari, K. 2013. Vorgehensweise zur Festlegung des Bemessungsdrucks auf das Sohlgewölbe in Anhydritführenden Gebirge.
Anagnostou, G., Pimenel, E., Serafeimidis, K. 2010. Swelling of sulphatic claystones – some fundamental questions and their pracitcal relevance. Geo-mechanics and Tunnelling, No. 5, p. 567–572.
Sofistik A.G. 2018.

Tunnels and Underground Cities: Engineering and Innovation meet Archaeology,
Architecture and Art, Volume 11: Urban
Tunnels - Part 1 – Peila, Viggiani & Celestino (Eds)
© 2020 Taylor & Francis Group, London, ISBN 978-0-367-46899-6

Drilling and blasting in close vicinity – monitoring of metro station Odenplan

B. Larsson Gruber
The Swedish Transport Administration, Solna, Sweden

ABSTRACT: Citybanan is a newly opened double track railway line in Stockholm, consisting of a 7 kilometers long double track tunnel, with two new railway stations underneath the urban city of Stockholm, with a very complicated design, such as; a vertical alignment varying from +10 meters above the sea level down to –40 meters below the sea level at the lowest point, and with several very near passes of existing infrastructure, consisting of various types of service tunnels and to that also the Stockholm Metro, with subway services running constantly from early morning until late night all week long.

This document will act both as a general view of the of the importance for the client of a project to establish a structural approach and create conditions for consultants and contractors to make it possible to execute drilling and blasting in close vicinity of an existing metro station Odenplan, with secure, solid, stable and traceable measurements of movements and settlements, that are reliable with respect to legally issues during the construction phase.

1 INTRODUCTION

1.1 *Historical*

The railway traffic from the southern to the north of Sweden and vice versa has since the beginning of the 1800's been running through Stockholm, the capital of Sweden, on a double track line above ground. From the beginning the capacity was sufficient enough, but during the years when the intensity of the traffic was increasing, there was a urgent need to extend the railway line through Stockholm with even more available railway lines. It is nearly always quite difficult to extend existing railway lines above ground through cities, so it was soon decided to find solutions to create a new railway line underneath the city.

1.2 *New layout*

The discussions about the new railway line through Stockholm started in the 1990's which ended up in three possible solutions, one above ground, and two undergrounds. With the necessary political discussions and financial issues to solve, all ended up after careful considerations in the beginning of the 2002 with a new layout underground for a railway line through Stockholm, with two new stations, that was decided and agreed between all necessary instances.

1.3 *Constructional layout*

As mentioned earlier above, this new railway line was to be constructed as a double track tunnel underneath the city of Stockholm, with two new railway stations, Station City as a four-track wide station and Station Odenplan as a two-track station, with the possibility in

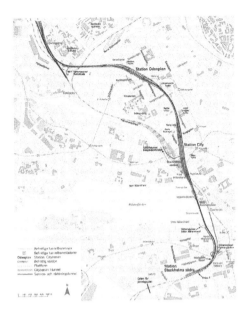

Figure 1. Layout of the new commuter line through Stockholm.

the future for the latter one to be widened up to four track. To make access for service and maintenance purpose and for the emergency services in case of an uneventful event, a service tunnel was to be constructed alongside the other two tunnels. The access to these new railway lines are created by dividing the commuter traffic from the regular railway line running through Stockholm above ground, meaning fact separating the faster and lighter commuter trains from regular passenger- and freight trains. This separation from the heavier traffic also meant that the vertical alignment of the tunnels could be more gradient, all to gain sufficient rock cover for the tunnels, but with the disadvantage with restrictions to use heavier train combinations in the tunnels.

1.4 Construction method

The geological situation for the tunnels was quite good, with a rock mass consisting of what is called Stockholm's granite, a solid structure with usually rock of good quality, but with presence of fracture zones, areas where the rock quality might be of poorer load capacity.

These two new tunnels, the train and the service tunnel, was designed to be constructed as for in Sweden the regularly way of construction method for rock tunnels, by pre-grouting the rock mass, thus followed by drilling and blasting, and a final reinforcement of the construction by bolting and shotcrete in combination.

1.5 Construction of the stations

These two new stations, Stockholm City and Odenplan, was to be located close to existing metro system, to gain an easy access to the overall commuting system for Stockholm, which in fact was the purpose with the construction of Citybanan. In general, the location for each station was placed approximately 5 to 15 meters below each existing metro station, with staircases, escalators and elevators designed to connect the railway station together with the metro station. By that design, inevitably there would also be some challenges to handle and solve during the separate construction phases.

2 METRO STATION ODENPLAN

2.1 *Background of the station*

The metro station Odenplan was constructed during the 1950's, as a station included in the so called "Green Line", stretching from south east to the north west of Stockholm, (482000 passengers per working day), with all traffic passing through the middle of Stockholm. Included to that are for example, stops at three major junctions with other traffic types; T-Centralen, Slussen and Fridhemsplan, three of the most frequent used metro stations in Stockholm, by the numbers per day of passengers varying from 183900 passengers to 53800 passengers per working days for the stations.

2.2 *Construction of existing metro station Odenplan*

Construction began in beginning of the late 1940's, with finalization and inauguration of the station in October 1952. The station is located quite near to the ground level, with the platform approximately 9 meters below ground level, built from above by access through an open shaft, and the connecting metro lines could be built in the surrounding rock. By that solution, the station is not located in rock, instead it is built as a concrete construction, and as drawings will show, the whole construction was designed to allow a further extension, with a parallel platform, (never built), thus leaving some crucial points laying on somewhat temporary foundations, which later on in the construction of Citybanan, proved to lead to preventive measures.

3 CITYBANAN – DESIGN AND FUNCTION

3.1 *Design and alignments*

The location of Citybanan through and underneath such a dense populated area as the city of Stockholm are quite a challenge from an engineer's point of view. In order to avoid existing infrastructures such as service tunnels, heating tunnels and motorway tunnels in combination with the projected and interpreted review of the assumed rock quality lead to a horizontal and vertical alignment that mainly was seeking a depth that was enough to go clear of obstacles, but not too deep, because of the need to have connections to each of the two metro stations that wasn't too far away for the publics availability to and from the commuter system. The main design of the railway line are by nature and regulations designed with large radius in both horizontal- and vertical alignment to allow a maximum speed of 80km per hour.

3.2 *Close connection to metro station Odenplan*

Layout of the new commuter station beneath the metro station was based to have access to the metro via escalators and elevators, with stair cases as an access in reserve and for use in case of an emergency. The railway station was designed with two accessible ways to the overlaying city area, one access via the metro will serve as one connection from the commuter station to the surrounding area above ground, the other connection is designed to gain access via an additional new entrance at Hälsingegatan, thus enable the commuter line to be of a more attractive way of transportation for new areas in the city.

4 CITYBANAN – CONSTRUCTION PHASE

4.1 *Production method*

As mentioned earlier in this document, the design for the production method for the tunnel was chosen to be of for conditions in Sweden normal drilling and blasting method, preceded

Figure 2. Production works beneath existing metro station.

by the pregrouting of the rock mass to get an almost dry tunnel, with water leakage within stipulated requirements set by the court of environment. Following the excavation was then the reinforcement of the tunnels by shotcrete and rock bolts combined to be part of the supportive system with a technical life span for the tunnel construction designed to be of a length of 120 years. A technical life span depending on forthcoming regularly inspections and maintenance during the running service phase.

4.2 Contract

In 2009, Bilfinger Berger, a German company, now merged up in the Swiss company Implenia, was awarded by the Swedish Transport Administration with the contract to build the northern part of Citybanan. A contract called B9509, Vasatunneln/Odenplan station. This contract consisted of approximately 2 kilometers of double track tunnels and 1.5 kilometers of service tunnel to be built via access from two access tunnels that also were to be constructed, included in the contract were also to design, construct and build the new commuter station Odenplan.

5 ORGANIZATION

5.1 The client's organization

As the design and construction of the new commuter station was very complex and interfering with several other actors and the public, it was decided early that the contract between The Swedish Transport Organization as client and the contractor Bilfinger Berger was to be of cooperation contract, which lead to an organization build up in a quite traditional way, involving supervisors of various technical functions such as rock, geological, survey etc. working close together with the contractors leading personnel.

5.2 The contractor's organization

Bilfinger Berger that was awarded with the contract had a recent experience from another railway project in the southern part of Sweden, namely the tunnel project City Tunneln, also governed by the Swedish Transport Administration, a twin tunnel constructed by TBM drilling, approximately 7 kilometers underneath the city of Malmö. Therefor the contractor's organization could transfer to Stockholm in a quite intact form, and from start perform tasks in a proper form.

5.3 Other actors

The owner of the metro station Odenplan was Stockholm County Council, abbreviated SLL, which acted in behalf both in the public interest, and following up that their requirements for their existing construction was taken care of both by the Swedish Transport Administration and the contractor Bilfinger Berger.

6 MONITORING OF EXISTING METRO STATION ODENPLAN

6.1 *Requirements in the contract*

6.1.1 *Monitoring by excavation of earthworks*
There was an originally set of requirements stipulated in the contract regarding monitoring of the existing structures, and the contractor started their work with that task as soon as excavation of earth works began early in 2010. Normal procedures for this existing structure such as monitoring of movements and settlements continued throughout 2010 and continued in to 2011 with no deformations to be detected.

6.1.2 *Monitoring during drill and blast of rock tunnels*
Soon as preparatory works for upcoming tunnel works beneath the existing metro station started, there was obvious signs that the existing construction was not of the standard that was to be expected. The originally set of requirements was only stipulated to take care of settlements, but further investigations started by the client ended up in a total review of the originally set of requirements that was included in the contract.

6.2 *New requirements in the contract*

6.2.1 *Monitoring during drill and blast of rock tunnels*
From originally only monitoring settlements inside the existing metro station, a total review ended up in an overall monitoring program of movements, settlements and inclinations of the whole station combined with a shutdown of the running traffic and platforms at a maximum length of 10 minutes. Since the owner of the metro station was quite keen to see that necessary precautions was taken care of, a series of meetings started between Stockholm County Council and the Swedish Transport Administration, involving a number of responsible supervisors from related disciplines, which all ended up in a strict document with procedures for upcoming works to follow.

6.2.2 *Methods of monitoring*
A various set of types of possibly movements were stipulated to be detectable; settlements – monitored using hydro static levelling sensors, movements – monitored by measuring prisms from total station, inclination – measured by angle sensors.

6.2.3 *Installation of equipment for monitoring*
Since all the installations for monitoring was to be installed inside the existing station, with running traffic during the whole day from 05.00 until 01.00, there was limited time available for installation, which demanded a fixed time schedule for all installation works to be executed.

The installation for monitoring movements consisted of a total of 15 total stations, a series of installed instruments all the way from the connecting metro tunnels through the station, all in all 14 installed instruments, and one additional instrument acting as a reserve in case of breakdowns or annual service of the instruments. For the points to be measured there was a total of approximately 60 prisms to be installed.

For measurement of settlements the installation consisted of two lines of hydro static levelling sensor lines, all in all each approximately 130 meters long.

To measure inclination, angle sensors were installed at a number of crucial points of the outer wall, in fact an installation made from outside and thus not interfering with the running traffic.

For the fixed installation of equipment for monitoring, a total of 7 nightly installations was needed, and carried out by both the ordinary personnel on site and by specialists from consultants that acted as subcontractors to Bilfinger Berger.

7 CONSTRUCTION OF THE TUNNELS

7.1 *Precautionary actions*

As the production of the tunnels came closer to the existing metro station preparations were made to ensure that all precautionary methods were performed and checking that all the systems were running in the way they were supposed to do. Besides all the monitoring to be done by surveying measurements, one should not forget all the measurements of vibrations that are included in the building contracts, a way of measurement all single blasts in the tunnel, regardless if it is blasting a full face salvo with 250 loaded boreholes, or just blasting one hole to take care of a small underbreak. These measurements of vibrations play a vital role in the regular follow-up of the contractors work, and also a very important measurement to ensure that the surrounding buildings will not be affected by the production underground. This type of measurement will not be described in detail in this document by the sheer amount of space that this would need.

7.2 *Blasting in close vicinity of the existing metro station*

When the production of tunnels had reached a distance from the existing metro station agreed well in advance between SLL and the Swedish Transport Administration, a distance that was interpreted in a specific chainage, a chain of actions took place in every single salvo of the tunnel faces, some actions well in advance, other after every salvo. This chain of actions was as follow;

7.2.1 *Planning of every salvo*
Well in advance, every salvo was time scheduled to fit in the allocated time frames when blasting was permitted during the allowed workday for underground works in Stockholm. These time frames, three in number, was designed to get a minimum influence on the running metro traffic. By reporting the scheduled blast some day in advance an information was sent out to the traffic control of the metro, all to planning the metro traffic.

7.2.2 *Action when blasting inside the production tunnel*
Inside the tunnel made by the contractor, the procedures for every salvo was quite normal according to normal routines for salvos. The main thing to observe is that a message that would confirm that the metro traffic and the platform itself was evacuated and this message was to be sent forward to the person responsible for the blasting.

7.2.3 *Actions in the metro tunnels and inside the metro platform before blasting*
Approximately 10 to 15 minutes before the scheduled time for blasting the running metro station was stopped by suitable holding points for the metro trains. For the platforms the actions involved special station hosts that had special tasks, such as searching the platform areas for unauthorized personnel (clear the area of people), closing the entrance gates for access by the public.

7.2.4 *Action to be taken after each independent salvo*
After each salvo the workflow was as follow; online checking of vibration measurements, trigger a complete round of measurements of all monitoring points and checking all measurements to be inside all the stipulated tolerances. Even if no alarming values are detected from the measurements of vibration or deformations, an optical inspection of the whole area is made by the responsible supervisor from the Swedish Transport Administration. Finally, when all systems say ok, traffic by the metro trains are set to go and the public are allowed to enter the platform areas and the metro traffic are running again in a normal way.

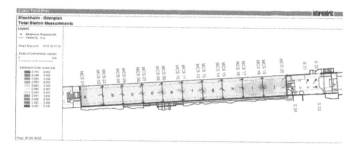

Figure 3. Part of report after blasting.

8 CONCLUSIONS

In order to have such a complex routine running during a longer period, it was very important to have a contractor which was both skilled in the technical matters and eager to find solutions that could work independently of what type of construction phase that occurred, and a client that was equal in competence so solutions could be weight in a production or economical way for the benefit of the economy of the project. Furthermore, all necessary action should be taken care of well in advance, especially when there is a risk for both the client and the 3rd party, the public interest.

REFERENCES

Banverket, Teknisk Beskrivning Berg och Mark, B9509 Odenplan och Vasatunneln
Banverket, BVH-GM 541, Handbok för geodetisk Mätning, 2009
Storstockholms lokaltrafik, SÄK-0409, Arbete på spårområde och inom säkerhetszon
Storstockholms lokaltrafik, SÄK-0139, Elsäkerhetsbestämmelser
Arbetsmiljöverket, AFS 2010:1, Berg- och gruvarbete

Tunnels and Underground Cities: Engineering and Innovation meet Archaeology,
Architecture and Art, Volume 11: Urban
Tunnels - Part 1 – Peila, Viggiani & Celestino (Eds)
© 2020 Taylor & Francis Group, London, ISBN 978-0-367-46899-6

Design and construction method of reinforcement around the lining concrete openings due to construction of additional evacuation tunnels in service

Y. Horioka
Hanshin Expressway Company Limited, Kobe, Japan

M. Ishihara
Hanshin Expressway Company Limited, Osaka, Japan

I. Otsuka
Taisei Corporation, Tokyo, Japan

ABSTRACT: At the Kobe Nagata Tunnel in Japan, four evacuation tunnels that will be connected to existing two main tunnels are planned to be excavated by the conventional tunneling method. The main tunnels were constructed about twenty years ago. After the Sasago tunnel accident where ceiling boards fell onto passing vehicles, ceiling boards in the Kobe Nagata Tunnel were removed and the ventilation system was changed accordingly. Therefore, additional evacuation tunnels must be constructed in operating tunnels. Two of the four tunnels are watertight and water cut-off grout injection will be done. Prior to excavation of the evacuation tunnels, the existing lining concrete and tunnel supports should be removed. Since the loads are acting on these existing structures, pre-reinforcement for the lining concrete around the opening is required against the loads. This paper shows an unprecedented design and construction method of opening up the lining concrete subjected to high pressure.

1 INTRODUCTION

Hanshin Expressway 31st, Yamate Line, is located in the Suma and the Nagata Wards in the western part of the Kobe City and forms a traffic network in the North-South direction. It starts from the Shirakawa JCT on Hanshin Expressway 7th, Kobe Line, and reaches the Komasakae entrance road connected with Hanshin Express 3rd, with a length of about 9.5 km.

In this route, the Kobe Nagata Tunnel (about 2.2 km long) consists of two tunnels on the North and South sides that have operated since August 2004. The location of tunnels is shown in Figure 1.

2 INTRODUCTION ON THE TUNNEL

2.1 *Existing main tunnels*

The existing Kobe Nagata Tunnel is a two-lanes road tunnel as shown in Figure 2. The excavation width and height are about 12 m, and 10.5 m respectively, while the excavation cross-section is about 110 m^2. The inbound and outbound lanes have almost the same tunnel cross sections. The tunnel specifications are shown in Table 1.

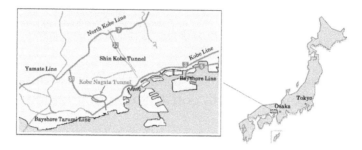

Figure 1. Location of the Kobe Nagata Tunnel.

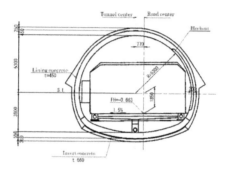

Figure 2. The existing tunnel cross section.

Table 1. Tunnel specifications.

	Inbound lane (South Line)	Outbound lane (North Line)
Service	August 1,2004	
Speed limit	60km/h	
Width	3.25m× 2+1.75m+0.75m	
Transverse gradient	3.0%	
Longitudinal slope	1.5%	
Extension	2,182m	2,118m
Traffic volume (Number of cars/Day)	11,700	13,300

2.2 *Evacuation tunnels*

The original concept of ventilation system was a cross flow system in which a ceiling plate was installed, and the polluted air passed through a space between the lining concrete and ceiling plate. The ceiling plate was removed, and the ventilation system had to be changed to a vertical flow system. Therefore, four additional evacuation tunnels (A to D) had to be constructed between the existing evacuation tunnels for the purpose of further improving safety inside the tunnel as shown in Figure 3. The general structure of the evacuation tunnels is shown in Figure 4.

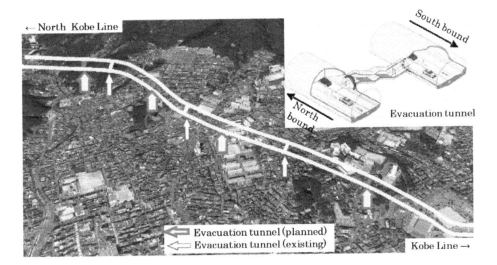

Figure 3.   Evacuation tunnels plan map.

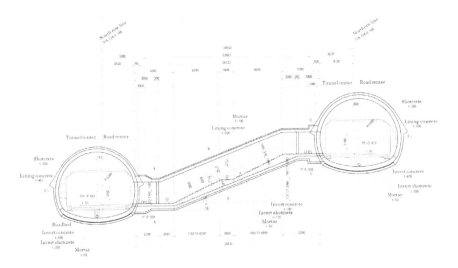

Figure 4.   Evacuation tunnels structure.

3   GEOLOGICAL CONDITION

3.1   *Evacuation tunnels A, B*

As shown in Figure 5, the tunnels will be excavated in the lower layer of the Osaka Group, alternated layers of sand, gravel, clay, and silt. The evacuation tunnel A was mainly in the gravel layer. On the other hand, the geology at the evacuation tunnel B consists of the strata of clay and sand layers. In addition, the level of the groundwater existing in the gravel and the sand layers is above the tunnel. The measures for seepage water will be necessary.

3.2   *Evacuation tunnels C, D*

The evacuation tunnels C and D will be excavated in the Rokko granite as shown in Figure 6. The Rokko granite in this area is distributed toward north from the Suma fault. Due to the

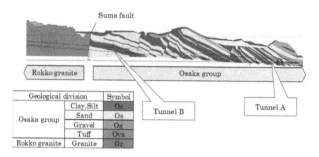

Figure 5. Geological section (evacuation tunnels A, B).

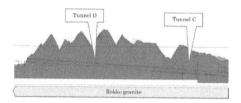

Figure 6. Geological section (evacuation tunnels C, D).

Table 2. Geological constants at the evacuation tunnels

	Tunnel A		Tunnel B		Tunnel C	
	Northbound	Southbound	Northbound	Southbound	Northbound	Southbound
Geology	Osaka group, Gravel , Clay		Osaka group, Gravel , Clay		Granite D2	Granaite C2
Unit weight γ (kN/m^3)	20	20	21	21	22	22
Young's modulus E(N/mm^2)	90	90	120	120	100	330
Cohesion c(kN/m2)	20	20	120	50	.	.
Friction angle ϕ (deg)	35	35	35	30	.	.

influence of the fault, a fracture zone with a considerable width consists of decomposed granite and clay-like rock. The fracture zone is so soft that it can be deformed or destroyed by finger pressure. In the fracture, pieces of hard rock remain locally.

In the construction of the existing tunnel, the southbound tunnel at the construction site of the evacuation tunnel C is in a class C1 – C2 rock, whereas the northbound tunnel is in a class D1 – D2 rock. On the other hand, at the construction site of the evacuation tunnel D, the southbound tunnel is in a class B hard rock, and the northbound tunnel is in a class C rock.

Table 2 shows the geological characteristics and physical properties of each evacuation tunnel.

4 REINFORCEMENT METHOD OF THE OPENING

Among the evacuation tunnels A to D, A and B are located in a sedimentary layer of the Osaka group. The existing tunnel is a watertight tunnel, and construction is required without lowering the groundwater level.

On the other hand, the evacuation tunnels C and D are located in the granite. The existing tunnel is a tunnel in a drained condition. As for D, it is in a hard granite (of classes B and C) and it is not necessary to consider loosening the ground when excavated. Below, we will describe the southbound evacuation tunnel A at the connected point with the existing tunnel at the deepest depth considered and in the severe load condition.

4.1 Selection of construction method

This work is to reinforce the opening created on one side of the existing two-lane tunnel. Normally, construction method is selected from the viewpoints of economy, construction feasibility, and construction period. However, when reinforcing the opening, it is inevitable to constantly regulate a single lane because of construction vehicles, arrangement of construction materials and provisional placement of excavation girders. Furthermore, depending on the works involved, it is necessary to stop traffic at night. A construction method with the smallest number of days of road closure necessary for the reinforcement of the opening portion is most desirable.

For the reinforcement method of the opening, we studied one with steel frame and another with concrete wall. These are the construction methods that can be done in the range of the passing lane. The former refers to the shielding opening reinforcing method[1].

In the method, beams are placed above and below the opening, and pillars for supporting the beams are arranged on the sides of the opening. The change in earth pressure accompanying the opening is transmitted from the existing lining concrete to the beam through the bracket on which the anchor is mounted and is supported by the pillars. Calculating and arranging the anchors necessary for reinforcement, they reach the traveling lane opposite to the opening. A road closure therefore becomes necessary to perform the opening reinforcement.

On the other hand, reinforcement with concrete walls resists against the change in earth pressure if placing anchors on the existing lining concrete and integrating them with the reinforced concrete walls. As shown in Figure 7 reinforcement with concrete walls enables construction within the range of the passing lane. After construction of evacuation tunnels, reinforced concrete walls are removed. Therefore, this method is adopted for the reinforcement of the opening.

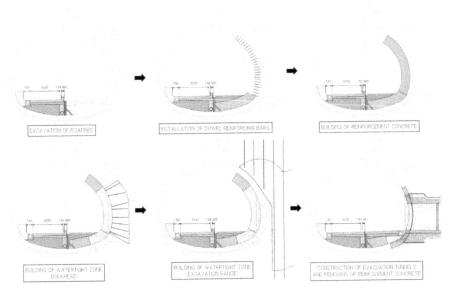

Figure 7.   Construction procedure.

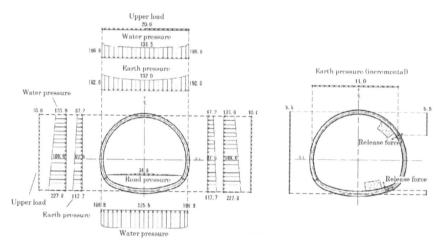

Step 1 : load condition (before creation of the opening)   Step 2 : change in load due to creation of the opening

Figure 8.   Loading condition.

4.2   Construction procedure

After constructing the reinforcement concrete, it is necessary to make a watertight structure for the evacuation tunnels A and B, so the water cut-off grout is injected. This forms a bulkhead from the inside of the existing main tunnels and creates a watertight zone from the ground throughout the excavation area. After completing the grout injection, the evacuation tunnels are excavated. Finally, reinforced concrete is removed. The construction procedure is shown in Figure 7.

4.3   Loading condition

The load acting on the opening part is considered using the same design method as the existing tunnel in service. However, since the opening is made in the existing tunnel following the excavation of the evacuation tunnel, the influence will be considered separately.

4.3.1   Earth pressure
The earth pressure acting on the existing tunnel is calculated from the Terzaghi's equation[2]. The initial load before the creation of the opening is first considered. Stress relief by the created opening and the change in load (during construction) by the evacuation tunnel excavation are then taken into consideration.

4.3.2   Water pressure
The design water level at the time of construction of the existing tunnel is higher, compared with that currently observed. Therefore, water pressure on the safe side is adopted.

4.4   Analysis method

The following items were taken into consideration for the selection of analysis method of the opening reinforcement.

1) To be able to grasp the stress state and behavior of the reinforced concrete of the opening and of the existing lining concrete.
2) To be able to analyze the influence on the entire lining concrete
3) To be able to consider the construction steps in the actual construction, and
4) To be able to consider the effect of construction joints.

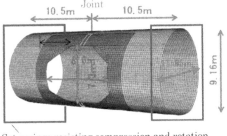 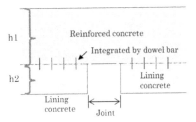

Set springs resisting compression and rotation.

Figure 9.   Analysis model.

Table 3.   Element stiffness.

	Lining concrete	Reinforced concrete	Joint
Sectional stiffness	Eh_2	$E(h_1+h_2)$	Eh_1
Moment of inertia of area	$Eh_2^3/12$	$E(h_1+h_2)^3/12$	$Eh_1^3/12$

Table 4.   Displacement of the opening.

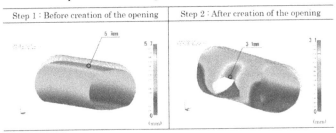

The three-dimensional FEM shell analysis method was adopted, since it satisfied all these conditions.

4.5   Analysis model

Since the joint is located in the opening reinforcement part, two tubes before and after the joint are to be analyzed (one tube 10.5 m). The reinforced concrete wall connected with dowel and the existing lining concrete are combined and modeled as an integrated wall by a shell element. Joints are modeled by the shell elements with a thickness of only the reinforced concrete walls. The analysis model is shown in Figure 9, while the cross-sectional stiffness of the shell element is shown in Table 3.

4.6   Analysis result

4.6.1 Deformation
In the initial state (Step 1) before opening the existing tunnel lining, the largest displacement was 5.7 mm at the tunnel top. After the wall concrete reinforcement was performed on the opening portion, (Step 2), the largest displacement was 3.1 mm at the upper portion of the opening.

Table 5. Analysis result (circumferential direction).

Bending moment	Axial force
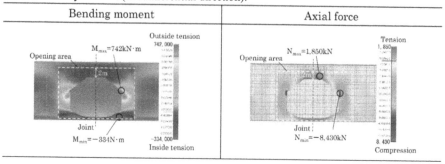	

Table 6. Results of stress (N/mm^2) examined.

	Concrete		Rebar	
	Max compression	Allowable stress	Max tension	Allowable stress
	σ_c(N/mm^2)	σ_{cR}(N/mm^2)	σ_s(N/mm^2)	σ_{sR}(N/mm^2)
Lining concrete	6.9	8.0	147.3	180.0
Reinforcement concrete	8.3	12.0	289.3	300.0

Although a large opening (8.6 m wide and 7.3 m long) was done, concrete lining deformation was very small (max. 3.1mm). The reason was the increase of stiffness. The existing tunnel lining was combined with the wall reinforced concrete, so the thickness of structure was increased from 0.55m to 1.55m.

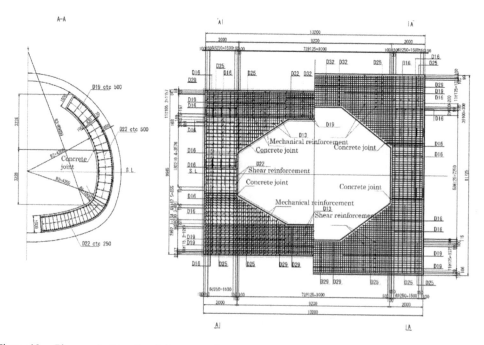

Figure 10. Placement of tensile reinforcement bars.

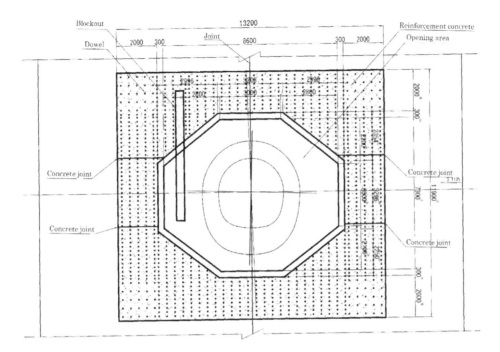

Figure 11. Placement of shear force reinforcement bars.

4.6.2 *Examination of bending moment*

Results of three-dimensional FEM shell analysis are shown in Tables 5 and 6. Since the reinforced concrete is removed after the opening is created, allowable stress of concrete and reinforcement bars was assumed 1.5 times the short-term stress. On the other hand, in the lining concrete, stress increment after creation of the opening was assumed to remain constant after construction of the opening. Therefore, the arrangement of the tensile reinforcement bars was determined so that the stresses induced would be less than or equal to the long-term allowable stress.

A large moment and axial force are generated at the corner of the existing lining concrete at the side of the opening both in the circumferential and axial directions. Integrating reinforced concrete with the dowel reinforcement to increase the rigidity of the section, the stress acting on the reinforcement bar of the existing lining concrete (circumferential direction: D 19 center to center 200 mm, axial direction: D 16 center to center 300 mm) was suppressed to: 147.3 N/mm^2 in the circumferential direction and 139.6 N/mm^2 in the axial direction respectively, both below the long-term allowable stress of 180 N/mm^2.

4.6.3 *Examination of shear force*

The shear force in the circumferential direction became large at the corner beside the opening. On the other hand, it was large at the tips of the upper and lower sides of the opening in the axial direction.

The hole for the dowel reinforcing bars of D 25 (drilling diameter: 34 mm) was drilled to a depth of 300 mm, about half the thickness of the existing lining concrete (550 mm).

4.7 *Reinforced concrete wall rebar pattern*

Arrangement of the reinforcement bars on the reinforced concrete of the opening is shown in Figures 10 and 11.

5 CONCLUSION

From the results of the study, the following summary and conclusions may be drawn:

1) Lining concrete (550 mm) and reinforced concrete (1000 mm) were integrated into the dowel reinforcement to increase the rigidity. As a result, the deformation after creation of the opening was only 3.1 mm.
2) By keeping the construction joints in the opening, they could be reinforced together. However, large tensile force acted the axial reinforcement bar on the joints, since it was not continuous in the joint. For this reason, the axial reinforcement bars with a diameter of 32 mm were arranged with an interval of 125 mm at the upper part of the reinforced concrete.
3) The number of dowel bars required for shear force resistance was calculated in the circumferential and the axial directions, respectively. They were arranged according to the area of the members.

This tunnel construction was very difficult, because the groundwater was high in the sedimentary layer of the Osaka group and the excavation tunnel was a watertight structure that had to be constructed while allowing vehicle traffic on one lane.

In addition, the dimension of the opening was 8.6 m wide and 7.3 m long, which was a very large structure for the existing tunnel with an inner diameter of 10.6 m. Because of the restricted condition as described above, reinforced concrete wall was built inside the existing tunnel lining to be integrated with it.

In the actual construction, there will be issues such as how to remove concrete constructed. To verify the effectiveness of the analysis, construction will be carried out while measuring the stress and displacements of the reinforcement bars of lining concrete and reinforced concrete.

We are planning to publish the results of a comparison analysis and construction measurement in future.

Since it is a large-scale reinforcement design/construction example under difficult conditions, it will be greatly appreciated if the proposed reinforcement method was used as a reference for other construction cases under similar conditions.

REFERENCES

Metropolitan Expressway Co., Ltd. 2018. Shield Design Guidelines
Japan Society of Civil Engineers 2016. Tunnel Standard Indication, Shield Construction Method. pp. 52–53
Railway Technical Research Institute 2005. Railway Structure Design Standards And Commentary

Tunnels and Underground Cities: Engineering and Innovation meet Archaeology,
Architecture and Art, Volume 11: Urban
Tunnels - Part 1 – Peila, Viggiani & Celestino (Eds)
© 2020 Taylor & Francis Group, London, ISBN 978-0-367-46899-6

Seismic performance of steel fiber reinforced concrete segmented lining tunnels

M. Jamshidi Avanaki
University of Tehran, Tehran, Iran

A.N. Dehghan
Islamic Azad University (Science and Research Branch), Tehran, Iran

ABSTRACT: In recent years, usage of fibers as reinforcement in replacement of conventional rebar in precast concrete tunnel segments has gained significant interest. In seismic regions, the structural performance of underground tunnels under earthquake events is of great concern for tunnel designers. This paper investigates the effects of incorporating Steel Fiber Reinforced Concrete (SFRC), as the tunnel's lining material, on its seismic vulnerability, compared to that of conventionally reinforced concrete. Results show that steel fibers enhance the seismic performance of concrete linings over conventional steel rebar, implying a technically preferable option for segmental lining tunnels in seismic zones.

1 INTRODUCTION

In developing countries, tunneling projects consume a considerable amount of national budgets, signifying the necessity for more research on cost reduction and productivity enhancement methods in this regard (de la Fuente et al., 2012). Introduction of fibers in the concrete mix as the main reinforcing configuration has shown to reduce costs and save times in tunneling projects (de la Fuente et al., 2012, Kasper et al., 2008, Burgers et al., 2007).

In the past few years, application of fibers as a replacement of traditional reinforcement for different structures under bending and shear forces, namely segmental linings of tunnel, has gained great interest (Buratti et al., 2013, Caratelli et al., 2011, Owen and Scholl, 1981, Yashiro et al., 2007). Fibers of different materials and geometries are used in Fiber Reinforced Concrete (FRC). In seismically active regions, the performance and vulnerability of infrastructure such as underground tunnels, under earthquake loads, is of great concern. Despite being less vulnerable than above ground structures, minor to extreme incidents of damage to underground tunnels has been reported in past earthquakes (Sharma, 1991, Owen and Scholl, 1981). Therefore, research on the structural performance of FRC tunnels is vital for a reliable design of such tunnels in seismically active regions (Jamshidi Avanaki et al., 2018b, Jamshidi Avanaki et al., 2018a).

The aim of this paper is to evaluate the seismic performance of TBM constructed tunnels with Steel FRC (SFRC) precast segmental linings. The focus of this paper is to investigate the effects of SFRC, as the tunnel's lining material, on its seismic vulnerability, compared to the conventionally Reinforced Concrete (RC) case. An experimental program is initially conducted to obtain the mechanical properties of the SFRC mix. The experimental results are then used in numerical simulations of a SFRC segmental tunnel under seismic actions and its performance is compared and analyzed against the RC case.

2 STEEL FRC CHARACTERIZATION

In this study, an experimental program is initially conducted to acquire the mechanical properties of the SFRC mix, primarily its post-crack tensile behavior. For this purpose, the basic engineering properties of the steel fiber-reinforced composite are determined using compressive and splitting tensile tests, and its flexural properties from 3-point bending tests. Properties of the steel fiber used in this research are presented in Table 1.

In Table 2, the materials and their proportions used in the concrete mix is given. The target 28-day mean compressive strength was set at 40MPa. A 0.5% volume content (equivalent to 40 kg/m³) of the steel fibers was used.

The concrete mix was sufficient for casting the following specimens:

- Two 150 mm cubes for compressive testing at 28-days.
- Two 150×300 mm cylinders for splitting tensile strength tests.
- Three150×150×550 mm beams for three-point bending tests over a span of 500 mm.

Cube and cylinder specimens of the SFRC and plain mixes were tested for compressive and tensile strength, respectively, according to the ASTM standards(ASTM C496, 1998, ASTM C39-96, 1998). The mean compressive and splitting tensile strength values of the FRC mixture at 28 days was measured 46.44 and 4.23 MPa, respectively. The 28-day mean compressive and splitting tensile strengths of plain concrete was about 41.1 and 3.40 MPa. The results indicated an approximate 13 and 24 % increase in compressive and splitting tensile strengths as a result of fiber presence in the concrete mix. The workability of fresh plain concrete, measured by the slump test, reduced significantly after addition of fibers (about 30%).

The flexural behavior of the SFRC composite was studied by 3-point bending tests according to RILEM TC 162-TDF (Rilem, 2003). In the 3-point bending test method, the tensile behavior of steel fiber reinforced concrete is determined by testing a simply supported notched

Table 1. Properties of fibers.

Fiber type	Geometry	Length (mm)	Diameter (mm)	Tensile strength (MPa)	Elasticity module (GPa)
Steel	Hooked	50	0.80	1169	210

Table 2. The materials used for the SFRC concrete mix.

Portland Cement type II (kg/m³)	Natural sand (kg/m³)	Gravel (kg/m³)	Water-Cement ratio	Steel fiber (%)
400	1172	632	0.375	0.5

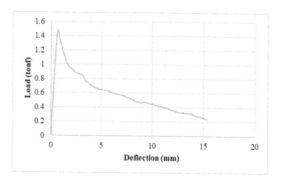

Figure 1. Load deflection curve for SFRC beam.

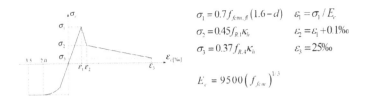

$$\sigma_1 = 0.7 f_{fctm,fl}(1.6-d) \quad \varepsilon_1 = \sigma_1 / E_c$$
$$\sigma_2 = 0.45 f_{R,1} \kappa_h \quad \varepsilon_2 = \varepsilon_1 + 0.1‰$$
$$\sigma_3 = 0.37 f_{R,4} \kappa_h \quad \varepsilon_3 = 25‰$$

$$E_c = 9500 \left(f_{fcm} \right)^{1/3}$$

Figure 2.   Stress-Strain relationship for fiber reinforced concrete (based on RILEM TC 162-TDF).

Table 3.   Stress-strain properties of the SFRC mix.

σ1	6.31 MPa
σ2	5.23 MPa
σ3	2.58 MPa
ε1	0.192 ‰
ε2	0.292 ‰
ε3	25 ‰

beam under three-point loading. Load and deflection at midspan was recorded continuously during the bending tests. The mean load–deflection curve of the SFRC beams is presented in Figure 1.

The RILEM procedure for developing stress-strain relationships for SFRC using load-deflection curves from 3-point bending test results, is employed for the SFRC mix. According to RILEM (Rilem, 2003), tension and compression stresses in steel fiber reinforced concrete are characterized based on the stress-strain diagram shown in Figure 2. The resulting stress-strain values of the SFRC mix is obtained and reported in Table 3.

3   THE SOIL-TUNNEL SYSTEM

A representative section from the segmental tunnel of the line 7 subway tunnel of Tehran (Iran) is selected for this study. A schematic description of the soil-tunnel section and rebar detailing of each segment is shown in Figure 3. The geometrical specifications of the segmental tunnel and geotechnical properties of the soil profile are gathered in Table 4 and Table 5, respectively.

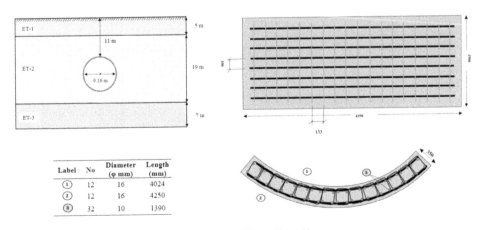

Label	No	Diameter (φ mm)	Length (mm)
①	12	16	4024
②	12	16	4250
③	32	10	1390

Figure 3.   The segmental tunnel lining and surrounding soil profile.

Table 4. Segmental tunnel properties.

Internal diameter	9.16 m
Segment thickness	35 cm
Segment width	150 cm
No. of segments (per ring)	6+1 key segment
Segmental joint configuration	Steel bolts (f_{yk} = 400 MPa)

Table 5. Geotechnical properties of the soil profile.

Soil Layer	Cohesion (KPa)	Friction Angle (°)	Elastic Modulus (MPa)	Poisson Ratio	Dry Specific Weight (g/cm³)
ET-1	31	28	35	0.35	1.7
ET-2	15	33	75	0.3	1.84
ET-3	30	33	50	0.32	1.9

4 NUMERICAL MODELING & SEISMIC ANALYSIS

A 2D numerical model of the soil-structure section is built using nonlinear continuum elements in a plain strain condition, considering the segmental joint behavior(Jamshidi Avanaki et al., 2018a) and capturing the SSI effects (Figure 4). The ABAQUS finite element software (ABAQUS Corp, 2012a) has been used for the numerical modeling and analysis phases. Both the tunnel lining and surrounding soil are modeled using solid continuum elements. The Mohr-Coulomb yield criterion was assigned to the soil elements. For the tunnel lining, properties of the considered SFRC concrete mix were incorporated via the obtained material model of section 2. As a basis for comparison, the conventionally reinforced concrete was also analyzed as the lining material. In the RC lining material, the mechanical behavior of the steel rebar was modeled using a bilinear diagram (elastic perfect-plastic behavior) with characteristic tensile strength, and elastic modulus values equal to 400 MPa and 210 GPa, respectively.

Between the three crack models available for modeling concrete, the concrete damaged plasticity model(ABAQUS corp, 2012b) was applied, due to its ability to characterize the complete inelastic behavior of concrete, both in tension and compression, including damage characteristics.

For the seismic analysis, seven ground motions were selected (Table 6), chosen to be compatible in soil conditions, fault rupture mechanism, seismic wave propagation and so on, with the tunnel site. Each earthquake record is scaled to four levels of Peak Ground Acceleration (PGA) to cover a range of seismic intensities: 0.2g, 0.5g, 0.7g, and 1.0g.

The peak soil displacement profiles, calculated from site response analyses, are applied on the lateral boundaries of the soil-structure model in a quasi-static manner to obtain the tunnels' ovaling response. A schematic representation of the ovaling response and an example

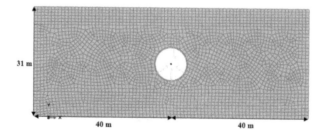

Figure 4. Numerical model of the soil-tunnel system.

Table 6. Specifications of the ground motions.

No	Date	Earthquake name	Record name	Magnitude (Ms)	PGA (g)	dt (sec)	T_{max} (sec)
1	01/17/94	Northridge	NRORR	6.8	0.51	0.01	39.97
2	06/28/92	Landers	LADSP	7.5	0.17	0.01	49.97
3	04/24/84	Morgan Hill	MHG	6.1	0.29	0.005	29.969
4	10/17/89	Loma Prieta	LPAND	7.1	0.24	0.005	39.595
5	10/17/89	Loma Prieta	LPGIL	7.1	0.36	0.005	39.945
6	10/17/89	Loma Prieta	LPLOB	7.1	0.44	0.005	39.94
7	10/17/89	Loma Prieta	LPSTG	7.1	0.50	0.005	39.945

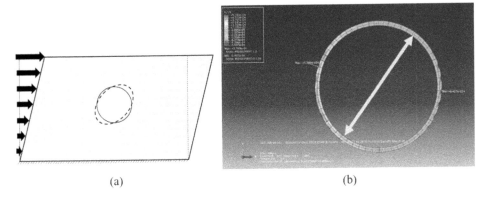

(a)   (b)

Figure 5. Typical 2D quasi-static seismic analysis: (a) ovaling response; (b) maximum diametric strain of the lining.

output of the tunnel lining are illustrated in Figure 5 (In figure (b), the surrounding soil elements have been hidden for better visualization). The analyses were repeated for variations of the tunnel lining material (SFRC and RC), earthquake record, and corresponding scale factor.

The transversal seismic response of circular tunnels is identified by the ovaling response of the lining. The induced seismic damage in circular underground tunnels is closely correlated to the predominant ovaling response (Hashash et al., 2001, Hashash et al., 2005). Therefore,

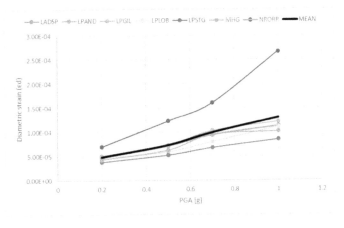

Figure 6. Diametric strain vs. PGA values for SFRC segmental tunnel.

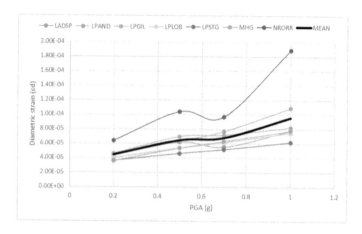

Figure 7. Diametric strain vs. PGA values for RC segmental tunnel.

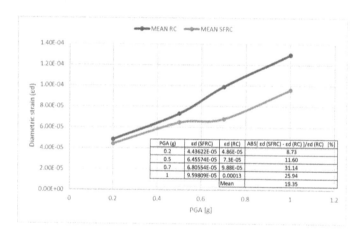

PGA (g)	εd (SFRC)	εd (RC)	ABS[εd (SFRC) - εd (RC)]/εd (RC) [%]
0.2	4.43622E-05	4.86E-05	8.73
0.5	6.45574E-05	7.3E-05	11.60
0.7	6.80554E-05	9.88E-05	31.14
1	9.59809E-05	0.00013	25.94
		Mean	19.35

Figure 8. Comparison of Diametric strain vs. PGA values for RC & SFRC segmental tunnels.

the diametric strain demand on the lining (ε_D), defined as the ratio of maximum diametric deformation to the lining diameter due to a certain seismic action, is considered as an appropriate indicator of the overall seismic performance of the tunnel.

Variations of ε_D due to the different earthquake records for the SFRC and RC segmental tunnel cases are presented in Figure 6 and Figure 7, respectively. For better visualization, the mean curves of the SFRC and RC segmental tunnels are plotted together in Figure 8.

5 CONCLUSIONS

The seismic performance of TBM constructed tunnels with segmental linings constructed using steel fiber reinforced concrete is investigated. Conventionally reinforced concrete lining is also analyzed for comparison. The seismic ovaling response is evaluated by measurement of the developed diametric strain in the tunnel ring, which is closely related to structural damage in the lining.

The results show that, SFRC lined tunnels generally show a lower potential of conceding damage than the conventional rebar case. This advantageous behavior is more evident as the seismic intensity increases. In particular, a 0.5% content of macro size steel fiber, reduces seismic diametric strain from 8 to 25 % at different levels of PGA. The obtained results in this research show that steel fiber reinforced concrete can be considered a technically preferable option over conventional rebar for the design of segmental lining tunnels in seismically active regions. Yet, further extensive research efforts are required on the various aspects of seismic structural performance to insure a structurally safe and viable design under earthquake loading conditions.

REFERENCES

Abaqus Corp 2012a. Abaqus v. 6.13 [Software]. Providence, RI: Dassault Systèmes Simulia Corp.

Abaqus Corp 2012b. Abaqus/Standard 6.13 User's Maunal.

Astm C39-96 1998. Standard test method for compressive strength of cylindrical concrete specimens.

Astm C496 1998. Standard Test Method for Splitting Tensile Strength of Cylindrical Concrete Specimens.

Buratti, N., Ferracuti, B. & Savoia, M. 2013. Concrete crack reduction in tunnel linings by steel fibre-reinforced concretes. *Construction and Building Materials* 44: 249-259.

Burgers, R., Walraven, J., Plizzari, G.A., and Tiberti, G. Structural behaviour of SFRC tunnel segments during TBM operations. World Tunnel Congress ITA-AITES, 2007. 1461-1467.

Caratelli, A., Meda, A., Rinaldi, Z. & Romualdi, P. 2011. Structural behaviour of precast tunnel segments in fiber reinforced concrete. *Tunnelling and Underground Space Technology* 26: 284-291.

De La Fuente, A., Pujadas, P., Blanco, A. & Aguado, A. 2012. Experiences in Barcelona with the use of fibres in segmental linings. *Tunnelling and Underground Space Technology* 27: 60-71.

Hashash, Y. M. A., Hook, J. J., Schmidt, B. & Yao, J. J. 2001. Seismic design and analysis of underground structures. *Tunnelling and Underground Space Technology* 16: 247–293.

Hashash, Y. M. A., Park, D. & Yao, J. I. C. 2005. Ovaling deformations of circular tunnels under seismic loading: an update on seismic design and analysis of underground structures. *Tunnelling and Underground Space Technology* 20: 435-441.

Jamshidi Avanaki, M., Hoseini, A., Vahdani, S. & De La Fuente, A. 2018a. Numerical-aided design of fiber reinforced concrete tunnel segment joints subjected to seismic loads. *Construction and Building Materials* 170: 40-54.

Jamshidi Avanaki, M., Hoseini, A., Vahdani, S., De La Fuente, A. & De Santos, C. 2018b. Seismic fragility curves for vulnerability assessment of steel fiber reinforced concrete segmental tunnel linings. *Tunnelling and Underground Space Technology* 78: 259-274.

Kasper, T., Edvardsen, C., Wittneben, G., and Neumann, D. 2008. Lining design for the district heating tunnel in Copenhagen with steel fibre reinforced concrete segments. *Tunnelling and Underground Space Technology* 23: 574-587.

Owen, G. N. & Scholl, R. E. 1981. Earthquake Engineering of Large Underground Structures. In: FHWA (ed.).

Rilem 2003. TC 162 - TDF. σ-ε design method-final recommendation. *Materials and structures* 36: 560-570.

Sharma, S., and Judd, W. R. 1991. Underground Opening Damage from Earthquakes. *Engineering Geology* 30.

Yashiro, K., Kojima, Y. & Shimizu, M. 2007. Historical earthquake damage to tunnels in Japan and case studies of railway tunnels in the 2004 Niigataken-Chuetsu earthquake. *Quarterly Report of Railway Technical Research Institute*

Tunnels and Underground Cities: Engineering and Innovation meet Archaeology, Architecture and Art, Volume 11: Urban Tunnels - Part 1 – Peila, Viggiani & Celestino (Eds)
© 2020 Taylor & Francis Group, London, ISBN 978-0-367-46899-6

3D numerical back-analysis on an experimental conventional tunnel in Paris Sanoisian "Green" Clay

J.P. Janin, A. Beaussier & H. Le Bissonnais
SETEC-TERRASOL, Paris, France

C. Gérardin & T. Charbonneau
RATP, Paris, France

ABSTRACT: In urban tunnel design and construction, one of the major issues is to minimize the surface settlements, which can generate serious problems on existing civil structures. As part of the extension of Paris metro Line 11, a shaft and a test tunnel were realized to improve the understanding of ground response of the Sanoisian "Green" Clay. In this paper, the authors present a 3D numerical back-analysis simulating the tunnel excavation. The study main objective was to best simulate the work carried out in order to calibrate the geotechnical model on the observed in situ measurements. The undrained effective stress approach coupled with consolidation analysis approach was the key to the success of the study which permits to fit the in situ measurements and to determine the geotechnical parameters, to be used in executive studies.

1 INTRODUCTION

The most important impact on the environment of a tunnel excavation is the creation of surface settlements. The prediction of ground movements caused by a tunnel excavation is a complex three-dimensional problem, depending on various factors such as the nature and characteristics of the soil, the geometry of the work and the excavation method adopted. This problem can be studied by empirical (Peck 1969, Rankin 1988), analytical (Panet 1995) and numerical approaches. Among these methods, three-dimensional numerical modelling is the only able to simulate the phenomenon in all its complexity (Mollon et al. 2011, Janin 2016). The real tunnel geometry, the initial stress state (even if anisotropic), the tunnelling method, the ground reinforcements, the phasing of the work can all be considered (Janin et al. 2015). However, in numerical methods, the correct choice of behaviour model, type of analysis (drained or undrained) and geotechnical parameters becomes crucial. The geomechanical model can be calibrated on the results of laboratory tests. Nevertheless, in the case of a particularly sensitive project and/or sensitive geo-materials, experimental works can be done to analyse the response of the massif on a large scale. This was the case of Paris metro Line 11, where a shaft and a tunnel have been realized to study the response of Sanoisian "Green" Clay.

The study main objective, presented in paper, is to analyse the in situ measurements carried out during the works and to recalibrate the geotechnical parameters by means of 3D numerical back-analysis.

2 PROJECT AND WORKS DESCRIPTION

As part of the extension of Paris metro line 11, a shaft and a tunnel were realized. These works, named "Calmette" had three functions: to study the ground response in a true scale, to be access structures during the excavation of line 11 extension and, finally, to serve as ventilation system of the metro line.

Figure 1.   Localisation of the project (on the left) and "Calmette" works photo (on the right).

The works main features as well as the construction chronology are:

- Shaft excavation, with a diameter of 10.2 m and 18.3 m of final depth (temporary support composed by HEB 180 ribs and 20 cm shotcrete). This phase took 1 month.
- Before tunnel excavation, installation of ground reinforcements composed by forepoling and fiberglass face bolting.
- Excavation of the tunnel upper part, section 10 m wide and 5 m high, over a length of 16.6 m, in conventional method with steps varying between 0.8 and 1 m. The temporary support consists of HEB 160 ribs and shotcrete. The tunnel cover is almost 10 m. This phase took 1 month too.

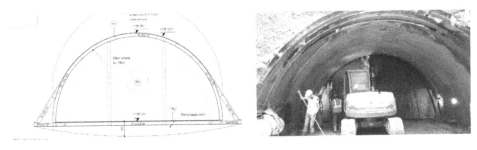

Figure 2.   Project tunnel section (on the left) and "Calmette" tunnel photo (on the right).

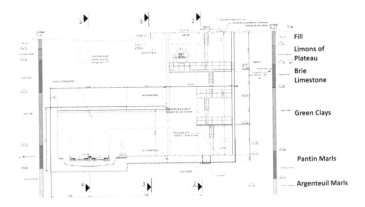

Figure 3.   Project longitudinal profile.

- After excavation of the upper section, the work was stopped for three month, in order to follow the evolution of the structures response and ground deformations.
- Installation of cast-concrete lining in the tunnel upper part and excavation of lower half section.

The 3D back-analysis, presented in this paper, concerns only the excavation of the tunnel upper part.

3 GEOLOGICAL AND GEOTECHNICAL CONTEXT

The project is located on the Romainville plateau. The shaft crossed the Limons of the Plateau, the "Brie" Limestone, the "Green" Clay and the "Pantin" Marls. The upper part of the tunnel was entirely excavated into Sanoisian "Green" Clays, whose behaviour and parameters were the subject of the back-analysis.

"Green" Clays are homogeneous and globally compact, with rare soft or sandy passages. Laboratory tests indicate that "Green" Clays are globally very Clayey (IP> 40), very plastic (WL> 50%) and quite well saturated (average Sr of 98%). The piezometers have highlighted two aquifers, separated by the impermeable wall of "Green" Clays. The superficial aquifer is located in Brie Limestone. The second aquifer is located in the Pantin Marls, under the tunnel invert.

Based on in-situ Menard pressuremeter tests, and laboratory mechanical tests (triaxial shearing and oedometric compression tests) on sample taken out of cored boreholes, the following geotechnical parameters were selected for the executive studies of the "Calmette" project.

In particularly, it can be observed that a slightly overconsolidated character was attributed to the "Green" Clay. In addition, undrained behaviour with total stress analysis was retained for execution design studies (Cu and short term modulus E).

4 IN SITU MEASUREMENTS

4.1 Monitoring system

An important monitoring system was set up to follow the massif and support responses during the "Calmette" works. It is composed by:

- surface and building X-Y- Z targets displacements measured by an automatic theodolites;
- inclinometers and vertical extensometers set up to a depth of 20 m compared to surface level;
- 4 convergence profiles in the upper half tunnel section and 2 ribs with strain gauges;
- 1 face tunnel extensometer, 24 m long, put in place before the excavation beginning.

4.2 Measurements analysis

The first part of the study consisted in the critical analysis of the measurements in situ. In the following paragraphs, the main monitoring measurements, used as the basis for the numerical back-analysis, are presented.

Table 1. Geotechnical parameters for executive study of "Calmette" project.

Layer	γ kN/m^3	c' kPa	φ' (°)	Cu kPa	E' MPa	E_{ST} MPa	K0 (-)	OCR (-)
Fill	19	0	25	-	5	-	0.5	1
Limons of Plateau	20	15	28	-	20	-	0.5	1
Brie Limestone	18	10	30	-	15	-	0.5	1
Green Clay	18	13	22	80	15	25	0.8	1.3
Pantin Marls	18	14	25	-	25	40	0.6	0.9
Argenteuil Marls	18	18	23	90	25	40	0.7	1.2

4.2.1 *Surface settlements*

Before the start of tunnel excavation, a resetting of settlements measurement was made. The realization of the ground reinforcements (forepoling and face bolting) caused indeed a generalized soil uplift of about 2 cm.

Surface settlements reach a maximum stabilized value, at the axis of the tunnel, of approximately 25 mm to 30 mm (Figure 4). Only the points in the last section (section 5) have been less settled since the excavation of the tunnel stopped before.

The graph also shows an important phenomenon: during the month after the end of the tunnel excavation, the settlement of the points, located mainly above the gallery, continue increasing. The settlement increment reached up to 5 mm. Then all the curves stabilized. This increase in settlements can be attributed to a consolidation phenomenon, given the very low permeability ($k \approx 10^{-8}$ m/s) of excavated soil ("Green" Clay).

4.2.2 *Face tunnel extrusion*

During the tunnel excavation phases, the evolution of the extrusion was measured using a 24 m long extensometer (with rings every meter). The extrusion was similar during tunnel excavation with a maximum value of about 8 mm, 1 m before facing (see Figure 9).

4.2.3 *Lateral ground displacements*

The ground displacements, measured with the inclinometers, in the tunnel section plane show a phenomenon at first sight quite unusual. At the lower part of the tunnel, displacements occur in the opposite direction to the excavation (see Figure 11 – Y transversal displacement). These displacements, in our opinion, are related to the shape of the gallery, which causes a concentration of the stresses in the ground at the level of the lateral supports, thus generating a kind of punching, given the weak characteristics of the "Green" Clay.

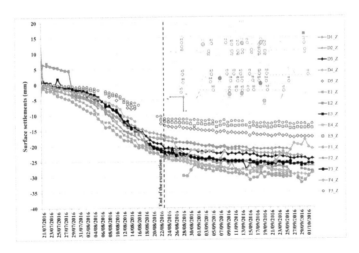

Figure 4.   Surface settlements evolution vs time.

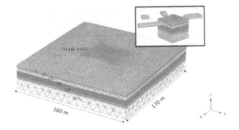

Figure 5.   3D Model (PLAXIS 3D).

4.2.4 *Support reaction*

The convergences and settlements of the ribs are very small but unfortunately this is linked to a "zero" measurement too late.

Thanks to strain gauge, a compression intrados stresses have been measured in rib C11 (put in place about half of tunnel length). Compression stress grew up approximately to 70 MPa.

5 3D NUMERICAL MODEL

5.1 *Model description*

To simulate the "Calmette" works a 3D model was create using the finite element code PLAXIS 3D (2016). The real geometry of the problem was considered, modelling the buildings, the shaft and the upper part of the tunnel with pre-reinforcements and temporary support. To avoid any boundary effects, the mesh extension is 170 m along the X-axis (tunnel excavation direction) and 160 m along Y-axis and 70 m along the Z-axis. In order to obtain more accurate results, a three-dimensional non-uniform mesh was used with smaller elements around the excavation. The model contained 330,000 tetrahedral elements and 440,000 nodes. All movements were fixed at the bottom of the model, and any horizontal displacements were blocked in the model's lateral faces.

5.2 *Ground constitutive model*

A non-linear elasto-plastic hardening soil constitutive model (HSM) implemented in PLAXIS code was chosen to represent the ground behaviour. This model, particularly during excavation, produces the ground deformation data that corresponds to the in situ measurements better than the results of the classical linear-elastic perfectly plastic model, which uses the Mohr–Coulomb failure criterion (Hejazi et al. 2008). The HSM constitutive model is often used to simulate the behaviour of both soft and stiff soils (Schanz et al. 1999). This model in particular distinguishes the stiffness modulus for the primary loading and for unloading/reloading.

As explained in the following paragraphs, to simulate the undrained behaviour of "Green" Clay two different analysis were used and compared: total stress or effective stress analysis.

5.3 *Simulation of excavation and support processes*

The shaft realization was simulated in a simplified way by means of macro-excavation phases, with stresses relaxation factor, followed by the activation of the temporary support, modelled by an equivalent plate.

Afterwards, the ground pre-reinforcements (forepoling and face bolting) were simulated explicitly by "embedded pile" structures to better simulate their effect. As shown, indeed, by several authors, homogenization approach often overestimates the effect on ground deformations (Dias 2011). An embedded pile consists of beam elements with integrated interface elements to describe the interaction between the soil and the surrounding pile. The beam elements are considered to be linearly elastic and the interaction between the piles and the surrounding soil is also described with linear-elastic perfectly plastic behaviour with a finite strength defined by the maximum traction allowed between the pile and the soil. This one was determined on the basis of pull-out tests performed in situ (T_{skin} = 26 kN/m). The other characteristics of the reinforcements (section, length, number and stiffness) are based on the elements actually set up on site.

The numerical simulation simulates the actual excavation steps and the ribs spacing, which varied between 0.4 m and 0.6 m (in the first part of the shaft opening) and between 0.8 m and 1 m in the tunnel. Each rib is modelled by "beam" element and shotcrete by "plate" element. The facing wall, put in place at the end of tunnel excavation, was simulated by a "plate" element too.

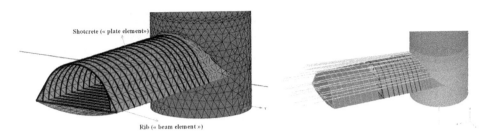

Figure 6.   3D Model – Pre-reinforcements and temporary support simulation.

Table 2.   Mechanical parameters of tunnel supports.

Support	Section m^2	Thickness m	E GPa	I m^4	Tskin kN/m
Tunnel rib	6.53 E-3	-	210	2.49E-5	-
Shotcrete	-	0.4	10	-	-
Forepoling	1.25 E-3	-	210	1.25E-7	26
Face bolting	7.20 E-4	-	40	0.0	26
Final wall	-	0.4	10	-	-

The Table 2 summarizes the main characteristics of each support.

6   COMPARISON BETWEEN THE NUMERICAL SIMULATIONS AND THE MEASUREMENTS

6.1   *Undrained total stress analysis*

In the case of excavation in a very low permeability soil, the study can be carried out with a total stress approach. It is a mechanical problem in which the reactions of a monophasic soil layer, having apparent characteristics in resistance and deformation, are studied. As the soil consists of a single phase, the excess pore pressure of water is not calculated.

Considering the low permeability of "Green" Clay and the low tunnelling speed, this approach can be adopted for our back-analysis (Panet 1995, Vermeer & Meier 1998). A first 3D simulation was thus carried using the geotechnical parameters selected for the execution studies of the "Calmette" project (see Table 1). In particular, the apparent resistance characteristic (Cu, $\varphi_u = 0$) and the short term deformation modulus E_{ST} were adopted for "Green" Clay. The HSM unloading modulus was considered to be equal to $E_{ur}{}^{ref} = 3\ E_{50}{}^{ref}$ (with $E_{50}{}^{ref} = E_{ST}$ and m=0).

The calculated settlements were compared with those recorded in situ, as a function of their distance from the tunnel face (Figure 7). The 3D calculation reproduces quite well the first part of settlements trends but then it underestimates final settlements (40% less compared to the settlement measured at the end of the excavation). In addition, with total stresses approach it is not possible to simulate the increase of settlements observed after the end of tunnel excavation and associated with a consolidation phenomenon (vertical portions of the curves). In order to overcome these problems, it was decided to adopt an effective stress analysis.

6.2   *Undrained effective stress analysis*

In this approach, the soil is considered as a biphasic medium: solid skeleton and water. It is a mechanical problem driven by the variations of the effective stress fields and interstitial pressures, these two phases being in interaction and also connected by a zero global volume deformation condition (due to the water incompressibility). The effective parameters of shear resistance (c' and φ') are attributed to the soil as well as a drained deformation modulus.

In Plaxis code, the "Undrained A" option permits to simulate undrained soil behaviour in effective stress analysis. The calculation code automatically takes into account the rigidity of the water and, by means of the determination of the negative and positive excess pore pressure, defines the maximum shear strength (τ_{max}) at any point of the soil. Contrary to the total stress approach, it is important to underline that the maximum shear strength is not an input data (Cu) but a consequence of the calculation.

The 3D simulation was thus redone. An undrained behaviour ("Undrained A") was attributed to the "Green" Clay layer, during the tunnel excavation steps. After the activation of the final wall, a consolidation phase was then introduced up to a maximum excess pressure value of 1 kPa.

The geotechnical parameters, presented above, were calibrated on in situ measurements. It is important to note that, unlike the choice made for "Calmette" project (see Table 1); "Green" Clay were considered to be normally consolidated. This choice was made after a new analysis of the triaxial CIU tests with pore pressure measurement results, showing a good alignment of points in Lambe diagram, without any sign of overconsolidation.

With the effective stress analysis, the evolution of the settlements fit better to the in situ measurements. In addition, this approach makes it possible to simulate an increase in settlement related to the consolidation phase. The settlement ratios S_{face}/S_{final}, measured in the zone next to the shaft (section 2) and towards the tunnel's end (section 3), are well reproduce by numerical simulation.

Table 3. Geotechnical parameters selected after numerical back-analysis with effective stress approach.

Layer	γ	c'	φ'	E'_{50}^{ref}	E'_{ur}^{ref}	K0	OCR	k
	kN/m³	kPa	(°)	MPa	MPa	(-)	(-)	m/s
Fill	19	0	25	5	15	0.5	1	1E-4
Limons of Plateau	20	15	28	20	60	0.5	1	1E-4
Brie Limestone	18	10	30	15	45	0.5	1	4E-5
Green Clay	18	13	22	5	30	0.63	1	1E-8
Pantin Marls	18	14	25	150	300	0.6	1	1E-6
Argenteuil Marls	18	18	23	65	130	0.7	1	1E-9

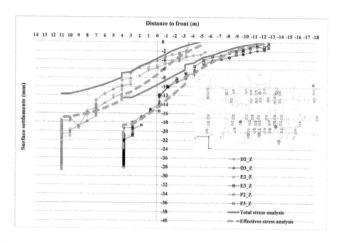

Figure 7. Surface settlements vs distance to front – Comparison between numerical simulation and in situ measurements.

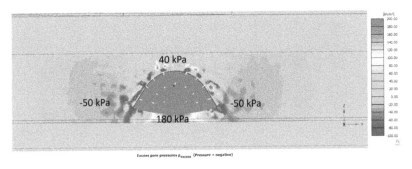

Figure 8. 3D simulation with effective stress analysis: Excess pore pressure before consolidation in tunnel section (compressive stress negative).

The analysis of excess pore pressure permits to explain the settlement increasing after the consolidation phase. Indeed, during the tunnel excavation, positive excess pore pressure (suction) is generated under the tunnel invert and, to a lesser extent, above the tunnel crown. This suction stiffens the ground, decreasing in the short term respectively the invert lifting and the crown settlement. However, at the same time, negative excess pore pressure (water compression) is generated, by arching effect, on the two sides of the tunnel. The settlement increase, obtained during the consolidation phase, is related in particular to the dissipation of this excess pore pressure.

As far as extrusion is concerned, it can be seen that the two approaches (total stress and effective stress analysis) simulate quite well the in situ measurements. This is explained again by the development of excess pore pressure around the tunnel. As shown in Figure 10, positive excess pore pressure (suction) is generated in front of the tunnel face. This suction stiffens the ground in front of the tunnel face and permits to obtain a short term extrusion similar to that calculated with total stress analysis (with an apparent short term modulus).

This positive effect of the suction in front of the tunnel face, on short term face displacements and stability, has also been studied by Schuerch et al. (2016).

The numerical simulation is able to well represent the displacements evolution measured by the inclinometer in the direction of tunnel excavation (X direction). In final phase, a "belly"

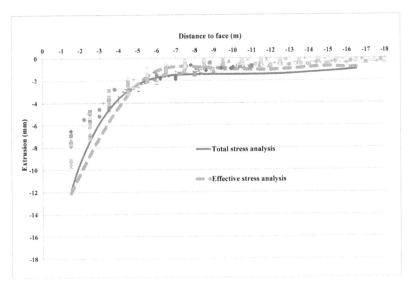

Figure 9. Extrusion vs distance to front – Comparison between numerical simulation and in situ measurements.

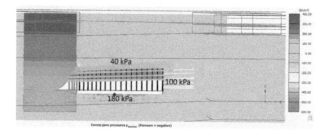

Figure 10. 3D simulation with effective stress analysis: Excess pore pressure before consolidation in longitudinal tunnel profile (compressive stress negative).

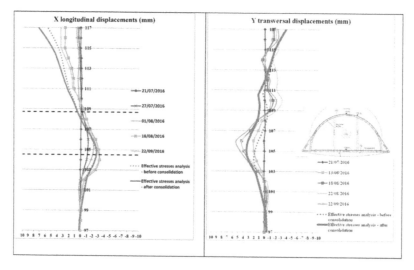

Figure 11. Inclinometer displacements along tunnel excavation (on the left) and inclinometer displacements in tunnel section (on the right) – Comparison between numerical simulation and in situ measurements.

remains in place at the tunnel level. This local displacement is an indicator of the volume loss occurring ahead of the face (extrusion), which cannot be recovered (Serratrice 1999, Janin 2015). In contrast, the top 8 m finally move in the direction of the tunnel face progression, once the face has passed the position of the inclinometer.

In the tunnel section plane (Y direction), the numerical simulation confirms the ground movements on the side opposite to the excavation, at the level of the tunnel lateral supports, and an increase of these movements in the same direction during the consolidation. The calculation validates also the hypothesis made previously for which this phenomenon is related to a ground punching related to the shape of the gallery, which causes a concentration of the stresses in the ground at the level of the lateral supports.

Finally, by means of the beam element simulating the rib C11, a normal compression force N = 300 kN and a bending moment of M = 5 kN.m are calculated, corresponding to an intrados compression stress of 63 MPa. This value is of the same order of magnitude as that measured in situ.

7 CONCLUSIONS

Due to a comprehensive monitoring section installed during the construction of "Calmette" works, the ground and tunnel support reactions during the excavation progress could be analysed. The study of the evolution of surface settlements, as a function of time and the distance to the face, shows in particular a consolidation phenomenon which took place in the "Green" Clays.

The real excavation work process and support installation were simulated in a 3D numerical FE model. The numerical simulation highlights the importance to simulate the undrained behaviour of "Green" Clays by an effective stress analysis. The development of excess pore pressure around the tunnel, during the excavation, explains the short and long term response of the ground. In particular, this approach permits to simulate the low extrusion observed in situ during the excavation progress and the final settlements measured after the consolidation process.

The calibration of the geotechnical parameters allowed to well reproduce different kind of in situ measurements (settlements, extrusion, inclinometer displacements, rib stresses). However, the drained modulus retained in the "Green" Clays seems to be weak compared to those obtained with the usual correlations linking the deformation modulus with pressuremeter modulus. This apparent underestimation can be explained principally by the following reasons:

- in the numerical simulation, an undrained behaviour was considered during the entire tunnel excavation and a consolidation phase was only performed after the installation of the final tunnel wall. In fact, since the tunnel has been realized in the traditional way, some of the excess pore pressure has already been dissipated during the excavation phases. Taking into account this consolidation after each excavation step would probably lead to retaining larger deformation modulus value;
- in the numerical simulation, it is difficult to simulate any problems that occurred during the works (e.g. sur-excavation, bad face bolting sealing, bad rib positioning).

REFERENCES

Dias, D. 2011. Convergence–confinement approach for designing tunnel face reinforcement by horizontal bolting. *Tunn Undergr Sp Tech*, 26: 517–523.

Hejazi, Y., Dias, D. & Kastner, R., 2008. Impact of constitutive models on the numerical analysis of underground constructions. *Acta Geotech*, 3 (4) : 251–258.

Janin, J.P., Dias, D., Emeriault, F., Kastner, R., Le Bissonnais, H. & Guilloux, A. 2015. Numerical back-analysis of the southern Toulon tunnel measurements: a comparison of 3D and 2D approaches. *Eng Geol*, 195: 42–52.

Janin, J.P. 2016. Apports de la simulation numérique tridimensionnelle dans les études de tunnels. *Rev. Fr. Geotech.*150, 3.

Mollon, G., Dias, D. & Soubra, A. 2013. Probabilistic analyses of tunneling induced ground movements. *Acta Geotech*, 8: 181–199.

Panet, M. 1995. *Le calcul des tunnels par la méthode convergence-confinement*. Presse de l'École nationale des Ponts et Chaussées.

Peck, R.B. 1969. Deep excavations and tunnelling in soft ground. In: *69 Proceeding of the 7th International Conference on Soil Mechanism Foundation Engineering, Mexico*, (3): 255–290.

Rankin, W.I. 1988. Ground movements resulting from urban tunnelling predictions and effects. In: *Proc Conf Enneering Geol Underground Movements, Nottingham (England)*, 79–92.

Serratrice, J.F. 1999. Suivi du plot Chalucet, application à la prevision des tassements de surface pendant le creusement du tunnel de Toulon. *Journée de Mècanique des Sols et des Roches d'Aix en Provence*, Aix en Provence (France).

Schanz, T., Vermeer, P.A. & Bonnier, P.G., 1999. The hardening-soil model: formulation and verification. In: Brinkgreve, R.B.J. *(Ed.)*, *Beyond 2000 in computational, geotechnics*: 281–290. Rotterdam: Balkema.

Schuerch, R. (2016). *On the delayed failure of geotechnical structures in low permeability ground*. PhD Thesis, ETH Zurich.

Vermeer, P.A. & Meier, C.P. 1998. Stability and deformations in deep excavations in cohesive soils. In: *Proceedings International Conference on Soil-Structure Interaction in Urban Civil Engineering, Darmstadt Geotechnics*, Vol.1, No.4.

*Tunnels and Underground Cities: Engineering and Innovation meet Archaeology,
Architecture and Art, Volume 11: Urban
Tunnels - Part 1 – Peila, Viggiani & Celestino (Eds)*
© 2020 Taylor & Francis Group, London, ISBN 978-0-367-46899-6

Time dependent modeling of salt-cavern natural gas storage

M. Javidi, H.R. Nejati, K. Goshtasbi & M. Ghasemi
Tarbiat Modares University, Tehran, Iran

ABSTRACT: Natural gas storage plays a vital role in maintaining the reliability of supply needed to meet the demands of consumers. Storage in rock salt caverns is one of the three underground storage methods for natural gas. In this study, mechanical behavior of salt storage cavern was numerically considered in a time dependent model using ABAQUS finite element software. A viscoplastic model was developed for simulation of a salt caver located in Nasrabad, Kashan, Iran and the stability of the cavern was evaluated using the strain rate criterion. Effect of maximum and minimum gas injection pressure, was investigated in the numerical simulation and found that the difference between the maximum and minimum gas pressure has a significant influence on the induced strain rate and stability of the cavern. Finally, a sensitivity analysis on the elastic and plastic parameters of the developed model was done and showed that the effect of elastic modulus on the induced strain rate is more than plastic parameters.

1 INTRODUCTION

Natural gas is one of the most important sources of energy in the world and its consumption and applications has rapidly grown up. Increasing in gas demand, especially in the cold months of the year, is one of the biggest challenges that natural gas producers counteract with it. In these months, increase demand for gas urge concerns about gas supplies for the consumption and industries. For example, in north-west Europe, nearly 67% of natural gas consumption occurs in cold months (October-March) (Hoffler and Kübler, 2007, US Energy Information Administration, 2002).

In order to this problem, there are three options. The first option is variable production, due to the technical conditions of the reservoirs and facilities and natural gas transmission lines, there is no possibility way for do this action. The second option is gas import, due to rising gas prices in cold months, low flexibility of imported contracts, the special design of gas pipelines and political crises, imports can also not be a good alternative for solve this problem. Third option is the storage of gas at the time of demand reduction and then extraction and consumption at the time of need for gas.

Given the disadvantages of first and second option, the need for gas storage is inevitable. In addition, for gas-exporting countries, gas storage becomes more important because it will firstly provide a reliable market for gas transportation to another country, which will lead to the commitment of macroeconomic contracts. Also, balancing in the gas transmission lines and optimally using its power are another advantages of gas storage. Currently, there are several ways in order to gas storage in the world, which is divided into two general surface and underground methods. The most common surface storage method for natural gas is the use of surface tanks. In the underground methods, gas can be stored in underground spaces such as empty oil and gas reservoirs, underground water reservoirs, salt caverns, abandoned mines and salt caverns (Veil, 1997). Several studies have been carried out on gas storage in salt caverns, which in of these studies behavior of salt rock had been modeled as a time-dependent material. In other words, the most important behavior of salt caverns used to store gas is their time-dependent behavior. In this study, the storage of natural gas in the salt cavern of Nasra-bad, Kashan, has been investigated and cavern stability under different scenarios of gas injection using time dependent behavioral of salt rock has been evaluated.

2 STORAGE OF GAS IN SALT CAVERNS

Since ancient times, salt has been extracted for commercial use using various methods. One of these methods is the salt dissolution of salt domes and salt beds. During this operation, a relatively large free space is created among the salt, which will later be used to store hydrocarbon materials. The salt's impenetrability is one of the benefits of this method. The disadvantages of this approach can be high cost at the beginning of the work (Veil, 1997).

In the dissolution method, by drilling a well from the surface of the ground to the desired depth and by setting up a circulating system, the pure water is sent to the depth of the well, which returns to the surface after salt dissolution. Natural gas can be sent and stored through the same well or other wells into the cavern. The quality of this operation depends on the ability of the salt to dissolve. The wash principle in the mining dissolve industry is such that every 7 to 8 cubic meters of fresh water pumped into the cavern will be able to dissolve 1 cubic meter of sodium salt. Usually the salt form is at a depth of more than 400 to 500 meters and sometimes more than 2000 meters. Mining dissolve industry is carried out in two ways: 1-direct circulation and 2- reverse circulation (Warren, 2006).

From other salt properties, time-dependent behavior is that more clear in the high depths. Creep over time reduces the volume of the cavern and can block the cavern in the long run time (Warren, 2006). The first use of salt caverns for storing materials was in Germany in 1916 and for oil storage. And so far, oil, light hydrocarbons, gas, compressed air and nuclear waste have been stored in salt caverns in the world (Xie and Tao, 2013).

3 SALT CAVE MODELING IN SOFTWARE

So far, gas storage caverns with different shapes and sizes have been carried out at depths of 300 to 2000 m. The volume of these caverns is between 5000 and 1000000 cubic meters (Berest et al., 2007). Its different shapes are mostly due to heterogeneous salt formation and its non-uniform dissolution. Caverns are usually capsular and they are drilled according of salt bedding and formation, horizontally or horizontally. In this study, the cavern designed vertically with a height of 140 meters and a diameter of 40 meters, so that the center of the cavern is located at a depth of 900 meters. This cavern is modeled and simulated in the ABAQUS finite element software. Extending the boundaries of the model from the sides of the model is 7 times the radius (140 m). Due to the symmetry in the shape of cavern and to reduce the volume of computations in the software, only one quarter of the simulator is simulated. In Figure 1, you will see an image of these cavern.

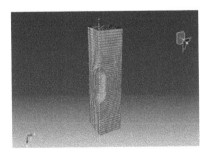

Figure 1. A view of the designed cavern in software.

The boundary conditions of the model are such that the movement of sides boundaries in two horizontal directions as well as the lower boundary movement of the model in the vertical direction is equal zero, and also imposed a 13.3 MPa pressure on the upper level of model that this pressure equal the overburden.

3.1 Behavior model

In this study, the cap plasticity model was used to study the mechanical behavior of the cavern. This model is one of the plastic behavior models for modeling frictional materials such as soils and rocks. Figure 2 shows the level of yield of the model in the cap plasticity model.

3.1.1 Creep in the extended Drucker-Pruger model

In the extended Drucker Pruger model, there are two different mechanisms for simulating creep in materials with long-term non-elastic deformation. One is the cohesion creep mechanism in the shear failure regions and the other is the mechanism of consolidation creep due to pressure load. Figure 3 shows the yield level of the extended Drucker Pruger model, with cohesion creep, consolidation creep and non-creep region (Documentation, 2017).

One of the most important characteristics of rock salt under the influence of stresses is their hardening behavior, which can be easily simulated in the extended Drucker Pruger model. The time-dependent strain rate in this model is also calculated from equation (1).

$$\overline{\varepsilon}^{cr} = A(\overline{\sigma}^{cr})^n t^m \tag{1}$$

Where A, n, and m are constants of rock salt parameters that can be determined by creep experiments.

3.2 Geo-mechanical parameters of salt rock

One of the most important and largest natural gas storage projects in salt domes in Iran is Nasrabad project, which is currently underway in its initial studies. In order to determine the geo-mechanical parameters of the salt rock, a series of uniaxial, three axis and creep tests are used to determine time-dependent and geo-mechanical parameters of salt rock. The results of laboratory tests on salt rock samples are presented in Tables 1 through 3.

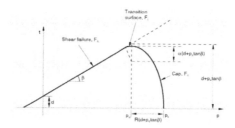

Figure 2.   yield surface of extended Drucker Pruger (cap plasticity) (Abaqus, 2017).

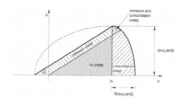

Figure 3.   Level of extended Drucker Pruger model yield with creep range (Abaqus, 2017).

Table 1. Geo-mechanical parameters of salt rock (Javidi, 2017).

Parameter	Quantity	symbol (unit)
Density	1990	$\rho\left(kg/m^3\right)$
Young modulus	1.2	E (GPa)
Poisson ratio	0.3	$v(-)$
coherence	1.7	C (MPa)
Friction angle	30	$\phi(°)$

Table 2. Hardness properties of salt rock (Javidi, 2017).

Yield stress (Pa)	Plastic strain
3,500,000	0
3,900,000	0.005
5,300,000	0.01
36,400,000	0.02
66,900,000	0.03

Table 3. Time dependent parameters of salt rock (Javidi, 2017).

A	m	n
9.5×10^{-36}	-0.3	2.2

3.3 Injection scenarios

Digging salt caverns is usually done using a dissolve extraction technique. Thus, after the drying step, the gas pressure in the cavern fluctuated between the minimum and maximum pressures. Two cycles of injection and withdrawal are considered for each year. In such a way that after the drying of the cavern, the injection is carried out within two months and the pressure is reached to the maximum pressure, and then for one month the cavern remains in the same state, after that inside pressure of cavern will reduce and reach the minimum pressure design, and then cavern remain in this state for one month again. This cycle of injection and withdraw has been called injection scenario.

Injection scenarios mean that each simulated caverns operates in a minimal and maximal pressure, which is technically one of the goals of the design of caverns for storage. In order to examine this issue, four scenarios of injecting are considered in this study: Scenario No. (1): In this scenario, the minimum and maximum injection pressures are 7 and 18 MPa respectively; scenario No. (2): In this scenario the minimum and maximum injection pressure of operation is 2 and 17 MPa respectively.; Scenario No. (3) In this scenario minimum and maximum injection pressure of 5 and 22Mpa respectively; and finally, the scenario No. (4) In this scenario minimum and maximum injection pressure of 5 and 18 MPa respectively.

After modeling the caverns, the above-mentioned injection scenarios have been imposed to the applied model of cavern and displacement values, plastic deformation and induction strain rates in the model has been obtained. Based on a suitable criterion, the optimal injecting efficiency scenario is determined. Figure 4 shows the time pressure graph of all types of proposed operation injection scenarios.

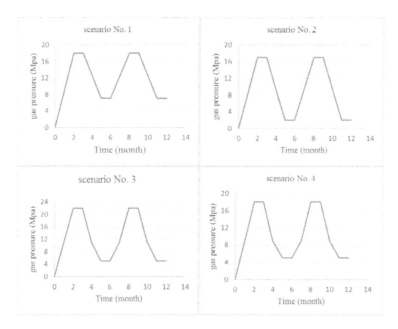

Figure 4.   Changes in gas pressure values in injection scenarios from 1 to 4 over a year.

3.4   *Presentation of numerical modeling results*

After modeling the geometry of the model and applying geo-mechanical parameters of rock salt as well as the, four scenarios of injection operation were applied as pressure on the cavern wall to the numerical model and the behavior of the environment around the cavern in the two-year time interval examined. Figure 5 shows the values of induced stresses in the cavern surroundings as a result of the application of different exploitation scenarios.

As mentioned, numerical analysis of the cavern in the two-year intervals was carried out using four gas injection scenario. Figure 6 shows the displacement curves over the time around the cavern through the application of four exploitation scenarios over a two-year period. Most of these displacements relate to elastic and plastic displacements that are created immediately after drilling in the cavern environment, but time-dependent displacements of the cavern environment is occur over time influence of the injection scenarios.

Another important parameter in the study of the stability of underground caverns, especially in time-dependent analyzes, is the strain rate created in the cavern environment. Figure 7 shows the induction strain rate in the cavern surroundings after applying the four exploitation scenarios.

One of the most important issues in the analysis of the stability of underground caverns, including storage caverns, is the use of a suitable criterion for analyzing the stability of the caverns. For analyzing the long-term stability of salt caverns, several criteria have been proposed, such as the criterion for reducing the cavern volume, and strain rate of the cavern. In this study of Strain rate criterion to study the long-term stability of the cavern under the influence of the four exploitation scenarios presented.

The strain rate criterion was presented by Berest and Brouard in 1998 by examining the amount of rock salt creep in the laboratory and measuring the convergence rate of the cavern. Based on this criterion: The strain rate of more than 10^{-3} per year represents a very strong creep. The strain rate is equal to 10^{-4} for the year indicating the creep of the rock salt standard and its amount is normal. A strain rate of less than 10^{-5} per year indicates a low creep value that will achieve long-term stability of the cavern (Berest and Brouard, 1998). The strain rates

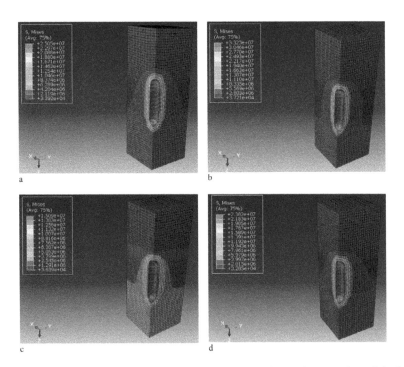

Figure 5. Induced stress contour in the surround of the cavern due to the operation of the inject scenarios A) Scenario No. 1 B) Scenario No. 2 C) Scenario No. 3 D) Scenario No. 4.

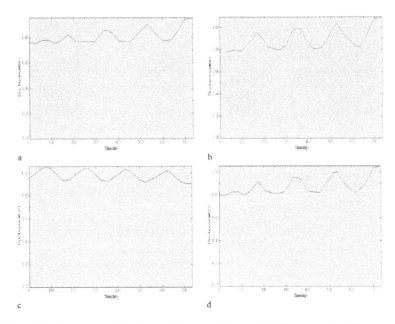

Figure 6. Time-displacement graphs after applying the four exploitation scenarios A) Scenario No. 1 B) Scenario No. 2 C) Scenario No. 3 D) Scenario No.4.

and classification above are valid for depth of 1000 meters. Since the cavern is located at a depth of 900 meters from the ground, this study uses this criterion to study the long-term stability of the cavern and determine the optimal exploitation injection scenario. According to Figure 7, the strain rate values of the cavern environment are determined after two years of injection and operation. The maximum strain rate values generated in each injection scenario are presented in Table 4.

Based on the data presented in Table 4, Scenario 1 has the lowest strain rate and Scenario 3 has the highest strain rate per year. Thus, scenario 1 is chosen as the most appropriate exploitation injection scenario, so that the stability of the cavern can be sustained over time.

In fact, the strain rate in the cavern surronding depends on the maximum and minimum pressure. For a more accurate investigation at this issue, the relationship between minimum and maximum injection pressures is plotted in Figures 8-A and B, respectively, and these

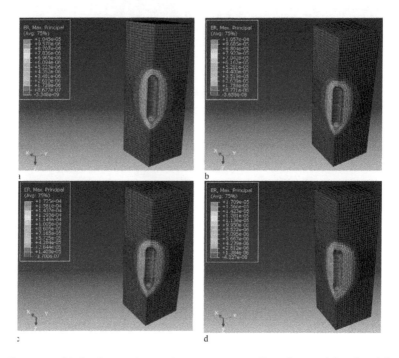

Figure 7. Contours of induction strain rate in cavern surrounding after applying four injection scenarios A) Scenario No. 1 b) Scenario No. 2 c) Scenario No. 3 d) Scenario No. 4.

Table 4. Maximum strain rate at each injection injection scenarios.

Scenario	Strain rate (1/year)
Scenario No. 1	0.52×10^{-5}
Scenario No. 2	0.53×10^{-4}
Scenario No. 3	0.86×10^{-4}
Scenario No. 4	0.85×10^{-5}

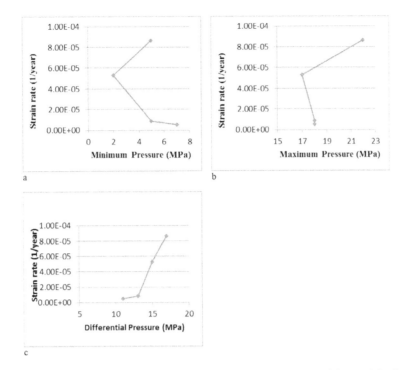

Figure 8. Strain rate variations around the cavern relative to (A) the minimum injection pressure (B) the maximum injection pressure (C) the minimum and maximum injection pressure difference.

figures show that there is a significant relationship between the minimum and maximum pressure at the injection with the induction strain rate. However, if the relationship between the pressure difference between the minimum and maximum pressure is plotted with the strain rate around the cavern, it is observed that there is a significant relationship between these two parameters, so if the difference between the injection pressure go up, the strain rate It will be increase (Figure 8-C).

4 CONCLUSION

Increasing the pressure of the maximum gas injection causes an increase in von Masse stress and an increase in the displacement of the cavern surrounding. By reducing the amount of minimum injection pressure in the reservoir, the displacement of the environment around the cavern increases as well. By increasing the difference between the minimum and maximum gas injection rates in the cavern, the amount of displacement, plastic strain, and strain rate increase. Based on the strain rate criterion, the most appropriate scenario of injecting is scenario (1) with the least difference between the minimum gas pressure and the maximum gas pressure.

REFRENCES

U.S. Energy Information Administration. 2002. *The basic of underground natural gas storage.*
Berest, P. & Brouard, B. 1998 A tentative classification of salts according to their creep properties. Proceeding of SMRI spring meeting, New Orleans, Louisiana, USA, 19-22.
Berest, P., Brouard, B., Karimi-Jafari, M., Van Sambeek, L. 2007. Transient behavior of salt caverns—interpretation of mechanical integrity tests. 44, 767-786.

ABAQUS Version. "6.17 Documentation. 2017. *Dassault systems.*

Hoffler, F. & Kubler, M. 2007. Demand for storage of natural gas in northwestern Europe: Trends 2005–30. 35, 5206-5219.

Javidi, M. 2017. An investigation on the failure of cap rock in gas storage cavern due to gas injection. Master thesis, Tarbiat Modares University.

Veil, J. A. 1997. Costs for off-site disposal of nonhazardous oil field wastes: Salt caverns versus other disposal methods. Argonne National Lab IL Environmental Assessment and Information Sciences DIV.

Warren, J. K. 2006. *Evaporites: sediments, resources and hydrocarbons,* Springer Science & Business Media.

Xie, J. & Tao, G. Modeling and analysis of salt creep deformations in drilling applications. Proceedings of the SIMULIA community conference, Vienna Google Scholar, 2013.

Tunnels and Underground Cities: Engineering and Innovation meet Archaeology,
Architecture and Art, Volume 11: Urban
Tunnels - Part 1 – Peila, Viggiani & Celestino (Eds)
© 2020 Taylor & Francis Group, London, ISBN 978-0-367-46899-6

Influence of pile row under loading on existing tunnel

P. Jongpradist
King Mongkut's University of Technology Thonburi, Bangkok, Thailand

N. Haema & P. Lueprasert
King Mongkut's Institute of Technology Ladkrabang, Bangkok, Thailand

ABSTRACT: This paper presents the study of the effect on existing water supply tunnels due to adjacent pile row under loading by 3D finite element method (3D FEM). The study covers the piles with diameter of 0.5 m and 1.5 m, spacing of 3 m and 4.5 m and tunnels having diameter of 2.0 m and 4.45 m. The pile tip elevations and clearances (*C*) are varied in the analyses. The Metropolitan Waterworks Authority (MWA) tunnel data and properties of Bangkok subsoil are considered in 3D finite element analyse. The analysis results are shown in terms of both safety factor and maximum changes of tunnel diameter. The tunnel influence zone due to nearby pile row under loading is suggested from the findings in this study.

1 INTRODUCTION

Due to the rapidly increasing population in the urban environment, the stable water supply system is necessary. To supply the water from the production plant to consumers in the highly populated city, water supply tunnels and conduits play an important role. In Bangkok, an efficient network system of tunnels and conduits has been established by the Metropolitan Waterworks Authority (MWA) for 30 years. Generally, the water tunnels were constructed beneath and are aligned below major roads to avoid being under the buildings. In recent years, new constructions continually increase to solve the transportation demands, e.g. flyovers, elevated trains, bridge and other infrastructures. In this soft ground condition, the pile foundations are required to support the new structures. The existing tunnels are inevitably close to new piles under loading as illustrated in

Figure 1. Thus, the new loaded piles would induce the effect on the MWA tunnels. An assessment of the effect of loaded piles on the stability and integrity of the tunnels is essential. The assessment is commonly performed in two phases. The first phase is to estimate the possible effect of the piles for modification of the pile position in the design. At present, this is conducted by describing the tunnel influence zone together with some criteria. A tunnel influence zone is defined by an area of ground (surrounding the tunnel) which future construction activities to be taken place could cause the stability problem on the tunnel structure or distress an operational system. If the potential impact cannot be disregarded, the second phase of assessment is required. The design must be changed or a comprehensive assessment must be performed.

Although a few patterns of tunnel influence zone are suggested (e.g., LTA 2004) based on shear plane, the zone which is applied to all construction types is proved to be unsuitable for piles. The tunnel influence zone subject to pile under loading can be smaller than that assumed by the shear failure plane. A more specific tunnel influence zone impacted by adjacent piles is thus essential for an effective issuance of construction permits of pile-supported structures in an area near the existing tunnel. Several researches have been conducted to understand the effect of adjacent loaded pile on existing tunnel and tunnel-soil-pile interactions. The finite

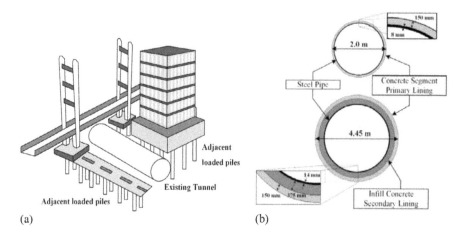

Figure 1. (a) New construction adjacent to existing tunnel in urban environment. (b) The cross section of MWA tunnel.

element method (FEM) is commonly used to analyze the effect of loaded pile on existing tunnel (Benton & Phillips 1991, Higgins et al. 1999, Schroeder et al. 2004, Yan et al. 2006). The effect on existing tunnel due to bored pile under loading located on one side of the tunnel was analyzed by 3D FEM (Lueprasert et al. 2015, Chai et al. 2014, Lueprasert et al. 2017, Heama et al, 2017) and better understanding on this interaction problem is continually obtained.

This study analyzes the effect of adjacent loaded pile row on existing tunnel by 3D finite element method. The pile row is located on one side of the MWA tunnel aiming to suggest the tunnel influence zone.

2 METHODOLOGY

Series of three-dimensional numerical analyses of the interaction between the adjacent loaded pile row and existing tunnel are carried out by finite element software PLAXIS (Brinkgreve et al. 2013). The tunnel response due to the effect of adjacent loaded piles is performed in terms of both safety factor of tunnel lining and maximum changes in tunnel diameter described below.

Two MWA tunnels having the outer diameters of 2.0 m and 4.45 m are chosen as the representatives in this study. The actual components of the water supply tunnels including the primary lining, infill concrete as secondary lining and innermost steel lining are modelled and analyzed. Figure 1(b) shows the details of cross section of MWA tunnels. The thickness of primary lining (concrete segment), secondary lining (infill concrete), steel pipe and tunnel depth (ground surface to center of tunnel) are listed in Table 1. The tunnels with diameter of 2.0 m and 4.45 m are constructed at a depth of 15 m and 20 m, respectively, below the ground surface in typical Bangkok subsoil as shown in Figure 2(a). The 2.0 m tunnel is located in soft clay while the 4.5 m tunnel is situated in between soft clay and stiff clay layers. A hydrostatic pore water pressure is considered in the analysis. Two types of pile are considered in this study. The 0.5 m pile with the length of 24 m represents the typical pile foundation of light to medium structures whereas the 1.5 m pile with the length of 40 m is for heavy structures in this subsoil condition.

Two categories of parametric study are considered in this study. In the first category, the effect of clearances, which is the distance between the closer edge of bored pile to edge of lining, on tunnel response are investigated. Two pile conditions defined by the pile diameter of

Table 1. The MWA tunnel parameters.

Outer diameter (mm)	Steel pipe (mm)	Concrete Segment Primary Lining (mm)	Infill Concrete Secondary Lining (mm)	Depth (Surface to center) (mm)
2000	8	150	-	15
4450	14	150	375	20

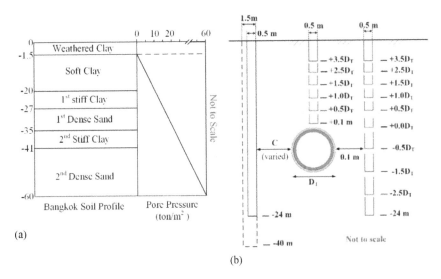

Figure 2. (a) Bangkok soil profile and pore pressure. (b) Position of pile tip and geometrical parameters.

0.5 m with pile length of 24 m and the pile diameter of 1.5 m with pile length of 40 m are of interest with varying the clearances (C) of 0.1 m, 0.5 m, $0.25D_T$, $0.5D_T$, $1.0D_T$, $2.0D_T$ and $3.0D_T$, when D_T is outer tunnel diameter.

For another category, the attention is paid to the effect of pile tip position on tunnel response. Only the pile diameter of 0.5 m is of concern. The piles whose tip are located above the tunnel crown and piles with clearance of 0.1 m are considered. A single pile row on one side of tunnel with various pile tip levels is considered as illustrated in Figure 2(b). In the row, 13 bored piles which can be sufficiently represented the behavior of pile row in 3D (Heama et al. 2018) are modelled. The working load as calculated by the α- method (Skempton 1959) is applied to the pile top.

Because the effect of adjacent loaded pile located in single side on existing tunnel can induce the shape of tunnel into ellipse or non-symmetric shape, the assessment methods proposed by Lueprasert et al. (2017) are adopted in this study to capture the tunnel deformation in terms of the maximum change in tunnel diameters as shown in Figure 3. The figure shows the maximum extension changes ($\Delta\phi_E$) and the maximum contraction changes ($\Delta\phi_C$) in tunnel diameter as can be calculated by Equations 1–2, respectively;

$$\Delta\phi_E = \Delta\phi_{E-MAX2} - \Delta\phi_{E-MAX1} \qquad (1)$$

$$\Delta\phi_C = \Delta\phi_{C-MAX2} - \Delta\phi_{C-MAX1} \qquad (2)$$

where $\Delta\phi_E$ = maximum extension changes of tunnel dimeter; $\Delta\phi_C$ = maximum contraction changes of tunnel dimeter; ϕ_1 = tunnel diameters before pile loading; and ϕ_2 = tunnel diameters after pile loading.

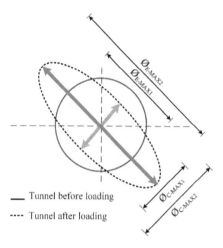

Tunnel before loading

---- Tunnel after loading

Figure 3. The maximum changes of tunnel diameter (modified from Lueprasert et al. 2017).

3   FINITE ELEMENT ANALYSIS

3.1   Finite element mesh

Figure 4 shows an example of 3D-finite element mesh used in this study. The dimension of FE model is 60 m in all directions (vertical, longitudinal and transverse directions). The boundary sizes of FE model are enough to fully model the tunneling effect (Mroueh & Shahrour 2008). The 10-node tetrahedral elements or the volume elements were used to simulate the soil layers, primary and secondary tunnel linings. The steel pipe was simulated by discretizing into 6-node triangular plate elements. The adjacent pile row was modelled by embedded beam elements (Tschuchnigg & Schweiger 2015). The monitoring section is at the center of longitudinal direction.

3.2   Analysis condition

For the displacement boundary condition, the sides of the model including front and rear sides were allowed for only vertical movement. The bottom of the model was restrained against vertical and horizontal movements. The top of the model is free for horizontal and vertical movements.

The initial stress condition is obtained through the overburden stress from the soil unit weight. The coefficients of earth pressure at rest, K_o of each layer are used to calculate the

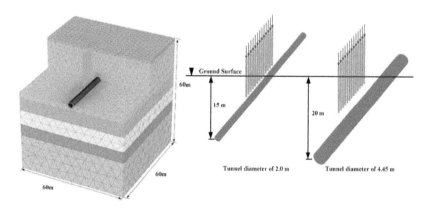

Figure 4.   FE mesh and model dimensions.

Table 2. Soil parameters for modeling (Jongpradist et al. 2013).

Soil layer	Wea. crust	Soft clay	1st Stiff clay	1st Sand	2nd Stiff clay	2nd Sand
Material model	MC	HS	HS	MC	HS	MC
E_{oed}^{ref} (kPa)	6000	5,000	60,000	80,000	60,000	80,000
E_{50}^{ref} (kPa)	-	5,000	60,000	-	60,000	-
E_{ur}^{ref} (kPa)	-	15,000	180,000	-	180,000	-
γ_{sat} (kN/m^3)	17	16	18	20	18	20
υ' (-)	0.32	0.33	0.33	0.3	0.33	0.3
ϕ' (°)	22	22	22	36	22	36
c (kPa)	8	5	18	0	18	0
m(-)	-	1	1	-	1	-
p^{ref} (kPa)	-	100	65	-	95	-

Table 3. Material properties of pile and tunnel lining.

Structure Type	Concrete Segment Primary Lining		Infill Concrete Seconded Lining	Steel Pipe	
	Diameter 2.0 m	Diameter 4.45 m	All Diameter	All Diameter	Bored pile
Material Model	LE	LE	LE	LE	LE
γ (kN/m^3)	24	24	24	78.6	24
E_{eqv} (kPa)	1.984×10^7	1.86×10^7	3.1×10^7	200×10^6	3.1×10^7
v	0.2	0.2	0.2	0.30	0.2

horizontal stresses. Undrained analysis was considered in this study. These conditions were used throughout of the analysis.

3.3 Material properties

The hardening soil model (HS) is assumed for the soft and stiff clays. The weathered clay and sand are modelled by Mohr-Coulomb (MC) assumption. The properties of soil and soil parameters are referred from previous study on tunnelling in Bangkok subsoil (Jongpradist et al. 2013; Prust et al. 2005) as shown in Table 2.

The material properties of MWA tunnels considered in this analysis are shown in

Table 3. The primary lining, secondary lining, steel pipe and bored pile were assumed to be liner elastic (LE) material model. The primary tunnel lining was simulated as continuous lining with reduced stiffness to take into account the segmental effect as suggested by Wood (1975). The interface friction (R_{inter}) between the structural elements and surrounding soil was chosen to be 0.9 (Brinkgreve et al. 2013).

4 ANALYSIS RESULTS

The safety factor of tunnel lining (SF), the maximum extension changes $(\Delta\phi_E)$ and the maximum contraction changes $(\Delta\phi_C)$ in tunnel diameter due to piles under loading are shown and discussed in this section. The results are presented in terms of influence of clearance and pile tip positions on the deformation of existing tunnel. Finally, the tunnel influence zone is suggested.

4.1 Investigation on tunnel stability

Figure 5 illustrates the safety factor (SF) of tunnel lining due to adjacent loaded piles row with a) various pile length in case of pile above the tunnel crown b) various pile length in case that the clearance of pile and tunnel is 0.1 m and c) various clearances to the tunnel. It is seen that the *SF* decreases as the pile is longer in case a) and the clearance is smaller for case c). Note that the initial *SF* (before subjecting to the loaded pile) of bigger tunnel is smaller than that of small tunnel. However, the decrease of *SF* of small tunnel becomes more pronounced. From the figures, it can be realized that the *SF* values are still large enough indicating that the tunnels are stable. However, the material properties of tunnel used in the analysis are ones after the construction, not the deteriorate condition.

4.2 *Influence of clearance between pile row and tunnel on tunnel deformation*

Figure 6 depicts the maximum tunnel deformation due to adjacent loaded pile row for various clearances. The clearances are normalized by the tunnel diameter (C/D_T) and plotted against the maximum tunnel deformation (mm) in y-axis. The positive and negative signs in y-axis denote the $\Delta\phi_E$ and $\Delta\phi_C$ respectively. The distribution patterns of maximum changes of tunnel diameter in both extension ($\Delta\phi_E$) and contraction ($\Delta\phi_C$) are very similar for all tunnel diameters. The impact on $\Delta\phi_E$ and $\Delta\phi_C$ of both tunnels due to the longer pile with larger diameter is larger than that due to the smaller and shorter pile. This is attributed to the larger load carried by larger and longer pile. The displacements of surrounding soil of larger and longer pile are thus more pronounced (Lueprasert et al. 2017). It is also seen that drastic increases of the maximum tunnel deformations ($\Delta\phi_E$ and $\Delta\phi_C$) exhibit when the clearance of pile and tunnel is less than $0.4D_T$. The rate of decrease of maximum tunnel deformations seems to be linear when clearance of pile and tunnel is greater than $1.0D_T$. Therefore, the clearances of $0.4D_T$ and $1.0D_T$ can be suggested as the boundary of the influence zone.

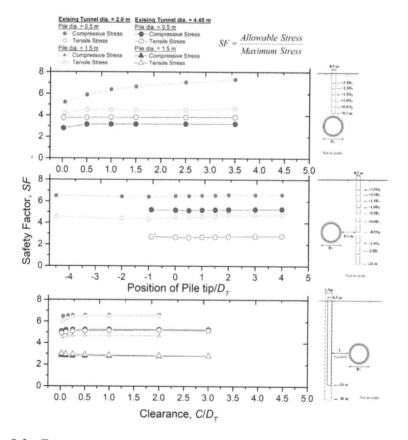

Figure 5. Safety Factor.

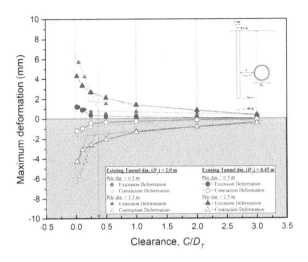

Figure 6. Change of tunnel diameter with various clearances.

4.3 *Influence of pile tip*

The maximum changes of tunnel diameter due to adjacent loaded pile row with all pile tip positions are shown in

Figure 7. The pile diameter of 0.5 m was considered in this section. The pile tips are normalized by the tunnel diameter (*Position of pile tip/D_T*) and plotted against the maximum tunnel deformation (mm) in x-axis. The positive sign is the $\Delta\phi_E$ and negative sign is the $\Delta\phi_C$ in x-axis. The $\Delta\phi_E$ and $\Delta\phi_C$ gradually increase with length of the pile. The $\Delta\phi_C$ are larger than $\Delta\phi_E$ with both sizes of tunnel. It can be seen that the tunnel deformation seems to linearly increase with depth up to the pile tip approaches $1.5D_T$. The rapid increase can be noticeably seen when the pile tip extends $1.0D_T$. Similar to the above section, the vertical clearances of $1.0D_T$ and $1.5D_T$ can be suggested as the boundary of the influence zone.

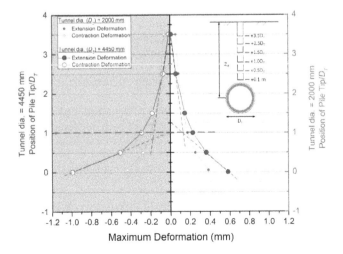

Figure 7. Change of tunnel diameters with various pile tip positions.

5 CONCLUSION

This paper analyzes the effect on existing tunnels due to adjacent pile row under loading by 3D finite element method. The tunnel diameter of 2.0 m and 4.45 m and the adjacent 0.5 m and 1.5 m bored piles subjected to working load were considered. The pile tip levels and clearances are varied. The tunnel responses are presented in terms of both safety factor and the maximum change of tunnel diameters. The main results of the analysis are as follows:

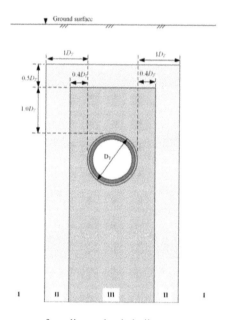

Figure 8.   The tunnel influence zones for adjacent loaded piles.

1. With three layers of tunnel linings and consideration of tunnel lining as newly constructed material properties, the tunnels are stable as indicating from sufficiently large *SF* even though the loaded piles are very close to tunnel.

2. For piles whose tip is situated above the tunnel crown, drastic increases of the maximum tunnel deformations ($\Delta\phi_E$ and $\Delta\phi_C$) exhibit when the distance of pile tip and tunnel crown is less than $1.0D_T$. The rate of decrease of maximum tunnel deformations seems to be linear when the distance of pile tip and tunnel crown is greater than $1.5D_T$.

3. For piles whose tip level is deeper that the tunnel spring line, drastic increases of the maximum tunnel deformations ($\Delta\phi_E$ and $\Delta\phi_C$) exhibit when the clearance of pile and tunnel is less than $0.4D_T$. The rate of decrease of maximum tunnel deformations seems to be linear when clearance of pile and tunnel is greater than $1.0D_T$.

4. Considering that the magnitudes of tunnel deformation become greater as the lining becomes deteriorated and the observed tendencies remain unchanged, the tunnel influence zone may be suggested using the above observation as shown in Figure 8.

ACKNOWLEDGEMENT

This work was financially supported by Metropolitan Waterworks Authority of Thailand in the fiscal year 2017. The permission for publishing a part of work in this article is acknowledged.

REFERENCES

Benton, L.J. & Phillips, A. 1991. The behavior of two tunnel beneath a building on build foundation, *Deformation of soils and Displacements of Structures. XECMFE. Florence.* 2: 665-668.

Brinkgreve, R.B.J., Engin, E. & Swolfs, W. 2013. *PLAXIS 3D Version 2013 manual.*

Chai, J.-C. Shrestha, S., Hino, T., Ding, W.-Q, Kamo, Y. & Carter, J. 2014. 2D and 3D analyses of an embankment on clay improved by soil–cement columns, *Comput. Geotech.* 68: 28-37.

Heama, N., Jongpradist, P., Lueprasert, P., & Suwansawat, S., 2017. Invsetigation on tunnel responses due to adjacent loaded pile by 3D finite element analysis, *Int. J. Geomate* 12(31): 51–57.

Heama, N., Jongpradist, P., Lueprasert, P., & Suwansawat, S., 2018. Investigation on Pile-Soil-Tunnel Interaction Due to Adjacent Loaded Pile Row by 3D, 4th *ICEAST 2018, Thailand.* pp: 792–795.

Higgins, K.G., Chidleigh, I., St John, H.D. & Potts, D.M. 1999. An example of pile tunnel interaction problems, *Proc. Int. Symp. on Geotechnical Aspects of Underground Construction in Soft* Ground. *IS-Tokyo'99. Kusakabe et al (eds).* Balkema: 99-103.

Jongpradist, P., Kaewsri, T., Sawatparnich, A., Suwansawat, S., Youwai, S., Kongkitkul, W. & Sunitsakul, J. 2013. Development of tunneling influence zones for adjacent pile foundations by numerical analyses, *Tunn. Undergr. Space Technol.* 34: 96-109.

LTA. 2004. *Code of Practice for Railway Protection, Development and Building Control Department,* Land Transport Authority, Singapore.

Lueprasert, P., Jongpradist, P., Charoenpak, K., Chaipanna, P. & Suwansawat, S. 2015. Three dimensional finite element analysis for preliminary establishment of tunnel influence zone subject to pile loading, *Maejo Int. J. Sci. Technol.* 9: 209–223.

Lueprasert, P., Jongpradist, P., Jongpradist, P. & Suwansawat, S. 2017. Numerical investigation of tunnel deformation due to adjacent loaded pile and pile-soil-tunnel interaction, *Tunn. Undergr. Space Technol.* 26(1): 166–181.

Schroeder, F.C., Potts, D.M. & Addenbrooke, T.I. 2004. The influence of pile group loading on existing tunnels. *Géotechnique.* 54(6): 351–362.

Skempton, A.W. 1959. Cast in-situ bored piles in London Clay. *Géotechnique* 9:153–173.

Mroueh, H. & Shahrour, I. 2008. A simplified 3D model for tunnel construction using tunnel boring machines. *Tunn. Undergr. Space Technol.* 23: 38–45.

Prust, R.E., Davies, J. & Hu, S. 2005. Pressure meter investigation for mass rapid transit in Bangkok, Thailand, *J. of the transportation research board, Transportation research of the national academies, Washington, DC, 1928.* pp:207–217.

Tschuchnigg, F. & Schweiger, H.F. 2015. The embedded pile concept – Verification of an efficient tool for modelling complex deep foundations, *Comput. Geotech.* 63: 244-254.

Wood, M. 1975. The circular tunnel in elastic ground, *Géotechnique.* 25(1): 115-127.

Yan, J.-Y., Zhang, Z.-X., Huang, H.-W. & Wang, R.-L. 2006. Numerical simulation of interaction between pile foundation and adjacent tunnel, *Underground Construction and Ground Movement (GSP 155).* pp:240-247.

Tunnels and Underground Cities: Engineering and Innovation meet Archaeology,
Architecture and Art, Volume 11: Urban
Tunnels - Part 1 – Peila, Viggiani & Celestino (Eds)
© 2020 Taylor & Francis Group, London, ISBN 978-0-367-46899-6

Did the selection of TBMs for the excavation of the Follo Line project tunnels satisfy the expectations?

A.K. Kalager & B. Gammelsæter
Bane NOR, Oslo, Norway

ABSTRACT: At the Follo Line project, four hard-rock double-shield TBMs have been operating from one single access-point, excavating in total 36 km of tunnels. This paper will describe the requirements for the machines and the results of the excavation, including performance in "extreme hard-rock conditions", in combination with continuous mapping of the geology and the performance of grouting in areas with water leakages. Other main activities, such as production of lining and depositing of excavated material, are performed within the rig area. The deposit of the excavated material is performed in accordance with specific procedures, with the intention of reusing it as a foundation for a future residential area. The paper will also address the challenges with structural noise and disturbances to neighbors during the construction of the tunnels.

1 THE FOLLO LINE PROJECT

The Follo Line project represents a unique project. The project is divided into different sub-projects, as shown in Figure 1. The main part of the project consists of the 20 km long tunnel sections, which is still under construction. The northern section has been excavated by drill and blast in combination with drill and split. The rest of the tunnel, 18.5 km is excavated by four tunnel boring machines, all of them operating from one single access point at the rig area at Åsland.

2 GEOLOGICAL CONDITIONS

The rock mass within this project area consists predominantly of Precambrian gneisses with banding and lenses of amphibolite and pegmatite. (Kalager, 2017) In addition, several generations of intrusions occur. Sedimentary shale occurs in the northern part of the tunnel, close to Oslo Central station.

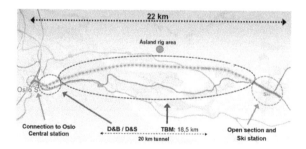

Figure 1. The Follo Line project is divided in four sub-projects. 18.5 km has been excavated by TBM.

Figure 2. Rig area, access- and logistic tunnels and the northern- and southern assembly caverns.

Generally, the rock mass is quite homogenous and competent, with moderate jointing. Laboratory tests show that the rock is abrasive and strong. The gneisses have a variation of the uniaxial strength from 100 to 250 MPa and the amphibolite in the area between 250 and 300 MPa.

Fracture zones have during several glacial periods been more exposed to erosion, which has resulted in deep valleys filled with marine sediments, mostly silt and clay.

Some of these fracture zones intersect the tunnel alignment, and in these zones, leakages were expected during the excavation of the tunnel. In some cases, these zones form a network of fracture zones with different orientation, both vertically and horizontally, and influence a large area or basin of marine clay. To avoid settlements and damages on buildings and infrastructure in these areas along the tunnel section, the requirements for leakage into the tunnel, both during the excavation and after completion, were very strict.

The overburden of the tunnel varies between 5 and 170 meters.

3 EXCAVATION METHODS

It was at an early stage of the project decided that due to time criticality and location close to other sensitive infrastructures, the excavation of the tunnel system in the northern part of the 20 km long tunnel section should be excavated by drill and blast in combination with drill and split.

For the remaining 18.5 km of the tunnel section, different excavation methods were considered. (Kalager, 2017).

Due to environmental impact, it was decided to perform the excavation by using four hard-rock double-shield TBMs operating from one single access-point, excavating in total 36 km of tunnels. The alternative to TBM-excavation was to perform the tunnels by drill and blast operating from six different locations along the tunnel section. Many of them would have been located within densely populated areas.

For the TBM-production, a network of access- and logistic tunnels, including two large assembly caverns, were excavated by drill and blast before the assembly and the start-up of the four TBMs. Se Figure 2.

Two machines were assembled in the northern cavern and performed the excavation in the northward direction and two machines were assembled in the southern cavern for the excavation in the southward direction. The TBMs should excavate approximately 9 km of tunnel each.

The four machines were ordered the 30th of March 2015. The first machine started up the 5th of September 2016 and by 30th November 2016 all four machines were in operation. Two years later, the 11th of September 2018, the two northbound machines had their break through.

4 DESIGN AND CAPACITY OF THE MACHINES

Based upon Norwegian experience from TBM-excavation of hydro power tunnels in the past, the four machines were designed for excavation in what many will define as "extreme" hard-rock conditions. For hard rock tunnel boring, a stiff cutterhead as well as a large diameter and high capacity main bearing capable to withstand extreme eccentric cutterhead loads is crucial for the tunnel boring operation. Some main design parameters for the TBMs is showed in Table 1 below.

Table 1. Some of the main design parameters for the four TBMs at the Follo Line project.

Item	Spec.	Item	Spec.
TBM cutting diameter	9,960 mm with new cutters	Weight of Cutterhead	265 metric ton equipped with cutters
Cutter size	19-inch wedge lock, back-loading	Cutterhead (CH) power	13 each VFD motors x 350 kW = 4 550 kW
Number of disc cutters	4 centers (x 2 discs) + 48 face + 15 gage = 71 cutting discs	Total power installed	Approx. 6900 kW
Load per cutter ring	315 kN	Nominal torque	11,115 kNm @ 3.67 rpm
Max. recommended CH load	71 x 315 = 22,365 kN	Max.overload torque	16,672 kNm @ 3.67 rpm
CH rotational speed	0 – 6.06 rpm	Water resistance	12 bar static
Main Bearing (MB)	3 axis roller bearing, 6,600 mm OD	Main bearing life time	> 20 000 hours according to ITA-tech guidelines
Total weight, TBM + BU	approx. 2,300 metric ton	Total length, TBM + Backup	Approx. 150 meters
Probe Drilling Equipment		Two drill rigs with rod adding system for drilling up to 35 m long holes for probing and pre-grouting through 38 ports in gripper shield with 11-degree angle to tunnel axis and or through 8 openings in the cutterhead.	

The experienced rates of penetration (ROP) and the advance rates, included time for probing and pre-grouting for the machines excavating in the northward and southward directions respectively are showed in Table 2 and 3 below.

Table 2. Rates of penetration by the end of August 2018.

	ROP [mm/min] North	ROP (mm/min) South
Average	31,05	31,21
Maximum	52,93	62,43
Minimum	14.94	14,95

Table 3. Advance rates including time for probing and pre-grouting by the end of August 2018.

	Average [m] North	Highest [m] North	Average (m) South	Highest (m) South
Day	14,1	31,0	12,7	32,1
Week	85	144	76	145
Month	364	568	330	542

5 HANDLING OF GEOLOGICAL CONDITIONS

5.1 *Continuous and systematic geological mapping*

As an important supplement to the geological information that had been collected before the start-up of the excavation, systematic geological mapping is daily being performed by experienced geologists.

5.1.1 *Optical tele-viewing*

Probe-drilling for detecting water ahead of the TBM and to prepare for systematic and continuous mapping by optical tele-viewing along the tunnel alignment, is performed every day during the maintenance-shift. (Fritsøe Lawton, Gammelsæter, Finnøy, Syversen, 2018). The detection of water is described below.

The results from the continuous fracture-mapping by optical tele-viewing (OTV) is used as input for the NTNU-model for compensation. This model is based on continuous fracture-mapping from an open TBM, where the rock-surface is available behind the cutterhead. (Bruland, 1998). For a shield-TBM, the rock-face is only visible through the cutterhead when the TBM is not excavating. To collect necessary and continuous information about the different fractures- and fissure systems along the entire tunnel section, 40 meters long probe-holes are drilled from behind the shield, forward ahead of the machines. An overlap of approximately 10 meters between the lengths are normally achieved.

The probe-holes are logged with Measure While Drilling (MWD), but this gives mainly information about weakness-zones and presence of water, and no precise information of the orientation or condition of the fractures. The MWD-data is therefore not suited for detailed fracture-mapping as required for the NTNU-model.

Instead, mapping of the probe-holes by an optical tele-viewer has given pictures with quite high-resolution scale, where fractures and their orientation can be mapped in detail. An example is shown in Figure 3 below.

The probe-holes prepared for OTV are bored upwards to achieve a drained hole. They are also flushed to avoid debris covering parts of the holes.

The OTV-logging provides a continuous geological data record along the entire tunnel section. The experience is that the data are detailed and of good enough quality to be used as input for the NTNU-model. The continuous OTV-logging provides a huge amount of information to be analyzed. To utilize all the data collected from the OTV, it is important that the analysis are performed by geologists with experiences from face-mapping and chip analysis as well.

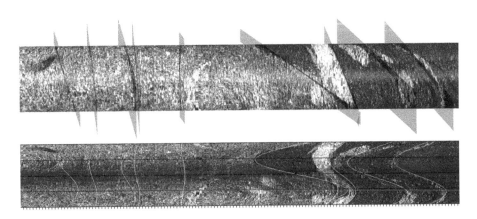

Figure 3.   Analyzed picture from OTV where fractures are mapped.

5.1.2 *Face mapping*

The purpose of performing face mapping is to gather general geological information and to get input to assess the fracture factor K_s, which is part of the compensation model. Mapping is performed every morning during the maintenance shift by geologists from both the Client and the Contractor. Both parties sign the agreed mapping form before leaving the TBM.

Depending on the excavation rate, there are 15–20 tunnel meters in average between each face mapping.

Access to observe the face is through the manhole and to a certain degree through muck openings as well. The cutter-head is then retracted from the face to make it possible to get an overview of a larger part of the face.

The geological mapping of the face gives information about presence of rock types and eventually of hard and abrasive minerals like quarts or garnet. Signs of weathering are often visible. Other important information is the number of fracture sets visible and the space between the fractures. In some cases, the mapping makes it possible to observe the roughness of the fracture planes and eventually infilling or aperture. It is also possible to verify if fractures or weakness planes contributes to fall-out or over-break. Water seepage from the face can also be identified. Figure 4 shows an example of how photographing of the face contributes to supplement the geological mapping.

Due to magnetism, it has not been possible to use a compass near the cutterhead. Only principal strike and dip orientation can be given.

5.1.3 *3D Photographing at face*

3D photographing of the face is performed regularly every day during the maintenance shift. Equipment and software from 3GSM are used. A camera is then mounted in the manhole of the cutterhead and a circular video is captured during one rotation of the cutterhead.

Advanced software generates scaled and oriented 3D images illustrating the various joint sets with their orientation from measurements taken. This is illustrated in Figure 5 below.

The width of the taken image is in the range of 0,5–1,5 meter depending on how far back the cutterhead has been retracted. The result is an illustration of the rock mass conditions. From the 3D images it is possible to identify over-breaks, perform geological mapping and to analyse fracture set orientation and to some degree fracture spacing as well.

5.1.4 *Chip analyses*

Chip analysis can be a valuable tool to obtain information on the rock breaking process and is therefore performed regularly. Normally 10–20 of the largest chips are collected from the TBM excavation and measured in three directions, x, y and z. The shape and size of the chip gives information or tendencies on fracturing factor, rock brittleness and hardness. The combination of chip shape and chip size can give tendencies on the efficiency of the boring process.

Figure 4. Fractures with different orientation identified at the face.

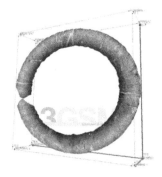

Figure 5. A 3GSM photo, a "doughnut" with fracture set identification.

5.1.5 Core drilling

Every 250 meter of the tunnel excavation, core drilling is performed at the front of the TBM's. Cores of four meters length are drilled perpendicular to the tunnel to get rock material for laboratory testing to determine Drilling Rate Index (DRI), Cutter Life Index (CLI) and mineral analysis. The DRI value is needed to be able to calculate the K_{ekv} as an input for the NTNU-model. For cutter life calculations, CLI and the mineral content is needed.

In addition, two meters long cores are bored for geological logging as a supplement to the geological information obtained from the daily tele-viewing, face mapping and chip analyzes

5.2 Monitoring of the pore-pressure, ground-water level and settlements

A network of registration wells for measuring the pore-pressure along the tunnel section and in sensitive areas connected to the tunnel by fracture-zones have been installed. The ground-water monitoring system consists of a combination of piezometers or stand-pipes in soil and deep rock wells. The first piezometers were installed in 2009. Early installation of monitoring is important to obtain a history of natural seasonal variations in the pore pressure. (Syversen, Lawton, Finnøy, Gammelsæter, 2018). The wells have been installed in the rock, with connection to fractures, and in the soil, that mostly consists of marine clay, as well.

The monitoring of the pore-pressure is a continuous and ongoing process throughout the project, and after finalization of the excavation until the water-balance is stabilized as well. All sensors are logged automatically every 10th minute, and the results are uploaded to a web-based GIS portal with a frequency of down to1 hour if deemed necessary.

In addition to the pore-pressure monitoring program, an extensive settlement program has been carried out. Nails have been mounted on the foundation of more than 2300 buildings. The nails are manually surveyed in due time prior to passing with the TBMs and after the TBMs had passed, and the readings are uploaded to a web portal. In addition to the manually measurement of settlements, a monitoring program utilizing satellite data (inSAR) from 2014 and up to date has been established to identify if settlements occurs on buildings along the tunnel section.

5.3 Probe-drilling and pre-grouting in areas with leakages

To fulfill the requirements regarding limited drop of pore-pressure and no damages to buildings or other infrastructure in the areas above or close to the tunnel, the tunnels are built as an undrained tunnel solution. Concrete segments with watertight gaskets are installed right behind the shield of the machine. When the lining is installed, the back-fill behind the lining is completed and the grout-ports are closed, the tunnel becomes water-tight and acts as an undrained tunnel.

Before the lining is installed, there is an open rock-face of approximately 15 – 20 meters between the tunnel-face and the last installed segmental ring. In a few areas where highly permeable fracture-zones intersects the tunnel, there has been some extensive leakages into the tunnel before the lining was installed. Such leakage usually follows the fracture-zones, and in the worst case a huge area within a distance of 1.5 km from the tunnel was seen to be affected.

The experience achieved in the beginning in areas with such leakages through fracture zones, was that the water ingress also resulted in outwash of the cement-based back-fill that was injected behind the lining. This outwash made it possible for continued water to flow behind the lining, which resulted in even more out-wash of material and in some cases a destabilizing of the lining as well. From an early stage of the excavation period, it was obvious that the contractor needed to improve their strategy and methods for handling the water as an integrated part of their construction.

On a daily basis, during the maintenance-shift, probe-drilling is performed in the rock ahead of the TBM to register the geological conditions, detect fractures with high permeability and identify if leakages can be expected. The number of probe holes depend on the sensitivity of the area above the tunnel and the expected geological conditions ahead of the TBM. The entire tunnel section is classified in different sensitivity-zones defined as small sensitive, moderate sensitive, sensitive and very sensitive. In general, one probe hole located on the top of the cutterhead or two probe holes in different locations related to the cutterhead are bored in areas defined as small sensitive. In moderate to sensitive areas, experience has showed that increasing the number of probe holes to four distributed in different positions around the cutterhead, gives quite reliable information about the geological conditions ahead of the TBM. In very sensitive areas the number of probe holes are set to six.

To reduce the amount of leakage before the lining is installed, pre-grouting is performed from the TBMs when identified as necessary, based on water ingress measurements from the systematic probe drillings. The trigger values of water leakage from the probe holes are based on the sensitivity class of the areas affected by the tunnel excavation. Based on experience achieved during the excavation, the trigger values for starting pre-grouting in the different sensitivity areas are set to 80 l/min from one probe hole in areas with small sensitivity, 40 l/min from minimum two probe holes in moderate to sensitive areas and 8 l/min in total from all the probe holes in very sensitive areas.

In some of the areas classified as high sensitive, mandatory pre-grouting is required.

Each TBM is equiped with two rock-drills for probing and for drilling the holes that should form the umbrella for pre-grouting. The double shield machines are designed with 38 holes around the shield where it is possible to perform holes for probing and grouting.

Every 500 meters, the two parallel tunnels are connected by cross-passages. The experience has showed that opening of this cross-passages often results in additional leakages. Even though pre-grouting is done as an umbrella from the tunnel around the portal of the cross-passage, leakages after opening-up the lining occurs. The water seemed to come through channels in the back-fill material between the lining and the rock. After considering different methods to stop the water in this portal-area for the cross-passages, contact-grouting, with low pressure, of the back-fill area around the opening is identified to give the best result.

In areas defined as very sensitive, it was decided to do systematic pre-grouting from the TBM in the areas around the future portals for the cross-passages as well as contact grouting.

This methodology for identifying water and limit the leakage has been developed and improved during the excavation phase, and the results appear to be positive. The drop of the pore-pressures stopped and were re-established when the performance of grouting was tailor-made to the geological conditions.

5.4 Infiltration wells

To compensate for the water leaking into the tunnel, and by that avoid a drop of the pore-pressure and development of settlements on buildings within the influence area of the tunnel, temporary infiltration wells have been installed at different locations. The wells are operated from the surface.

Many of the wells have been operated with good results, but not all of them. The key to success seems to depend on the quality of the installation and the match with the geological conditions. The infiltration wells are usually drilled 20–50 m into the rock. The intention is that they should cross identified permeable fracture zones that preferably are inclined under soil deposits.

Water with some overpressure is infiltrated from the rock well trough the fracture zone, up to the soil. Pressure and flow are carefully controlled to avoid piping effects in the soil. The infiltration of water is mostly activated in combination with pre-grouting to control the water balance in the area affected by the tunnel excavation.

It is a requirement that these infiltration wells shall only be used as a temporary mitigation to maintain the pore-pressure while the TBM passes by. After the lining is properly installed, there should be no need for them anymore.

5.5 Experience by the TBM-operation in the Norwegian hard rock and specific ground conditions

Most of the tunnel excavation in the Norwegian hard rock has traditionally been performed by drill and blast methodology. Therefore, a decision to use TBMs to excavate the main part of the 20 km long twin-tube tunnel at the Follo Line project caused a certain degree of skepticism.

Lessons learned from the excavation of this tunnel section by four double-shield TBMs is that there are some key-factors that must be present for achieving a successful result, namely establishing a good knowledge of the general geological conditions along the tunnel section as a fundament for the contract, systematic mapping of the conditions ahead of the TBMs during the excavation, systematic measurement of the pore-pressures and settlements, improved and tailor-made mitigations to limit the amount of leakages and timely and appropriate decision-making for activating the mitigations.

The machines and the equipment must be tailor-made for the specific ground conditions. Last, but not least, the experience, competence and skills of the personnel, on both the Contractor's and the Client's side, and the communication and co-operation between them, is also in many ways crucial for achieving a good result.

The experience from the excavation of the two Follo Line tunnels is that the Contractor improved their skills during the construction. Their procedures and performance for probe-drilling, detection of water and pre-grouting became more efficient after a while, and in total, the excavation must be defined as being a success.

6 SEGMENT PRODUCTION

The requirements regarding no leakage into the tunnel, no drop of the pore-pressure or development of settlements above the tunnel, demanded a watertight and undrained solution for the tunnel-lining.

The lining is designed as a ring consisting of seven segments. The thickness of the segments are 400 mm and the length of each segment is 1800 mm. The segments are reinforced by steel-bars in combination with steel-fibers in the concrete. (Gollegger, Pinillos Lorenzana, Cavalaro, Kanstad. 2019) The ring, and the gasket between each segment are designed to handle a water pressure of 16 bar.

The inner diameter of the lining is 8.750 meters. The gap between the lining and the rock is filled with a cement based two-component material, injected behind the segments a few rings after installation.

The experience is that the back-fill grout didn't work as intended in areas with high leakage. The result was that too much of the back-fill material was washed out. In some cases, this had influence on the stability of the rings as well.

By improving the procedures for detecting water and performing pre-grouting ahead of the face when necessary, the quality of the entire performance and delivery of the tunnel lining was improved. Ecco detecting equipment is used to check the quality of the back-fill, which makes it possible to identify whether the back-fill is homogenous with a good distribution or if there are spaces behind the lining that can act as drainage channels. If the quality is poor, additional contact grouting has to be performed.

Despite the lessons that needed to be learnt, the quality seems to be satisfying. One question to be raised is whether the problem with the fluid back-fill material could have been avoided if the material had consisted of more cement?

7 RE-USE OF THE EXCAVATED MATERIAL

Originally, it was a requirement that 10 – 15% of the excavated material should be crushed and used as aggregates for the concrete- and segment production. This re-use of the material was defined as an important environmental benefit for the project, but unfortunately this could not be achieved.

When analyzing the chemical composition of the material, 20% of the samples identified a higher amount of the unstable mineral Pyrrhotite that can be accepted in concrete when it occurs in combination with Sulfur. To achieve the required quality of the concrete, aggregates with no content of Pyrrhotite had to be procured from an external supplier.

Before the project started, there was some skepticism about the use of material excavated by TBMs. In close cooperation with geotechnical expertise, a procedure for alternating filling and compaction was developed and tested with good results. One of the experiences was that during heavy rain and with high water content in the material, it was difficult to achieve the expected result. Under such conditions, the excavated material is transported out for external deposition. Despite the limit regarding the water content, the major part, approximately eight million tons, of the excavated material is re-used within the rig-area as a basement for a future residential area. This limits the total transport volume during the construction phase and contributes to future gains connected to the development of the residential area and a saving in project costs.

8 ENVIRONMENTAL IMPACT AND COMMUNICATION

To achieve an efficient excavation of the tunnel and due to the schedule of the Follo Line project, it was required to have a 24/7 production for the TBM-drilling.

In some locations along the tunnel section, the two twin-tube tunnels are excavated under densely populated areas. In Norway there are restrictions regarding the level of the structural noise that the neighbors can be exposed to. The structural noise is measured in dB and defined as A-weighted sound pressure level, L_{pAeq}. The trigger-values are defined in Table 4 below.

Estimations for the expected values of structural noise depends on different parameters like the distances between the TBMs and the buildings, the geological conditions, the foundation of the buildings and on which level in the buildings people have their living rooms and bed rooms.

A prediction model has been built, and it is used as an important tool for the communication with the neighbours that are expected to be affected by structural noise from the tunnel excavation. This is a 3D-model including the tunnel alignment. The terrain above the tunnel section has been mapped and is combined with geographical data from official authorities like cadastre data and number of inhabitants at each address point. In addition, information of the geological ground conditions is also linked to the model. The TBM progress is updated every day and makes it possible to measure the real distance between the TBMs and the buildings along the tunnel section.

Based om the information from this model and the experience achieved about the migration of the structural noise through the different types of geological conditions, it is possible to calculate expected levels of structural noise that will affect the different buildings on the surface.

Before entering in to the different neighborhoods, it is quite clear which structural noise level the individual buildings will be exposed to. The general experience is that people living within a distance, horizontally and vertically, of approximately 70 meters from the TBM can expect to be exposed to structural noise exceeding 40 dB. People living between a distance of 70 and 120 meters from the TBMs can expect to be exposed to levels of approximately 35 dB and people living more than 200 meters from the tunnel and the TBMs will probably not be exposed to levels exceeding the trigger values. In addition to the distance to the TBM, the geological conditions, the foundation of the buildings and on which level in the building people are living, influence the calculated and the measured values of structural noise.

Table 4. Trigger-values for indoor structural noise.

Building type	Noise req. daytime (L_{pAeq} 12h, 07–19)	Noise req. evening (L_{pAeq} 4h, 19–23) or Sundays/public holidays (L_{pAeq} 16h, 07–23)	Noise req. night (L_{pAeq} 8h, 23 - 07)
Residential buildings	40	35	30
Work space requiring low noise levels	45		

As a successful strategy for the Follo Line project, the neighbors are informed in due time before the TBMs are expected to approach the different areas along the tunnel section. Those who will be affected to structural noise levels exceeding the trigger values, especially during night, are offered alternative accommodation, mainly in nearby hotels.

This strategy to excavate 24/7 and compensate the disturbances of the neighbors by offering them alternative accommodation has been accepted by the health authority in the affected municipalities. The argument is that it is better to pass the different residential areas as fast as possible, mainly within three weeks, instead of exposing the neighbors for structural noise over a longer period.

The experience so far is that in total a limited number of neighbors have accepted the offer to stay at a hotel while the TBMs passes the area where they live, but there are variations. In areas where the distance down to the TBMs are between 120 and 200 meters, less than 1% of the inhabitants have so far used the opportunity to sleep in more quiet environments. In some of the areas where the TBMs passed within a distance of 30 to 70 meters, between 50 and 60% wanted to stay at a hotel while the excavation took place under their neighborhood, but in other areas where the TBMs passed within this small distance, only 10 % accepted the offer. The general experience is that in spite of the expected noise levels, most of the neighbors wanted to stay at home as long as possible.

9 CONCLUSION

The experience by performing the main part of the 20 km long tunnel at the Follo Line project by four double-shield hard-rock TBMs, operating from one centrally located access point har demonstrated that this project was tailor-made for this excavation method.

An important key to the success is that the machines were designed for the specific geological conditions. They have so far been working as expected, but experience also shows that the machines seemed to be operated conservatively within their design parameters, and it can be concluded that performance could probably have been even better.

When it comes to handling of leakages, the contractor improved its skills and methodology and the result is satisfactory. The impact on the environment is probably equal to or better than what could have been expected if the tunnels had been excavated by drill and blast.

The re-use the excavated material within the rig area, with its environmental and cost benefits, was an exclusive opportunity related to the TBM-excavation from one single access point. This would not have been possible if the tunnels had been excavated by drill and blast from six different locations.

Communication with the neighbors has been an important tool to pave the way for acceptance of a 24/7 excavation. The process has so far been managed efficiently and successfully resulting in very few complains and no negative press coverage.

The conclusion is that the expectations for the drilling of the Follo Line tunnels by the four double-shield TBMs have been fulfilled.

REFERENCES

Bruland, A. 1998. Hard rock tunnel boring. Doctorial thesis 1998:81. Trondheim: NTNU.

Kalager, A.K. 2017. Projects with different parameters – Different excavation methods. Bergen: WTC 2017.

Fritzøe Lawton, M. & Gammelsæter, B. & Hoff Finnøy, A. & Syversen, F. 2018. Continuous mapping with OTV from hard rock double shield TBM as input for compensation. Dubai: WTC 2018.

Syversen, F. & Lawton, M. & Hoff Finnøy, A. & Gammelsæter, B. 2018. Water ingress and groundwater control in double shield TBMs at the Follo Line project – Norway. Dubai: WTC 2018.

Gollegger, J. & Pinillos Lorenzana, L.M. & Cavalaro, S. & Kanstad, T. 2019. Fiber reinforced concrete segmental lining – Evolution and technical viability assessment. Naples: WTC 2019.

Tunnels and Underground Cities: Engineering and Innovation meet Archaeology,
Architecture and Art, Volume 11: Urban
Tunnels - Part 1 – Peila, Viggiani & Celestino (Eds)
© 2020 Taylor & Francis Group, London, ISBN 978-0-367-46899-6

Construction of a mountain tunnel in the neighboring of residential areas

T. Kamikoshi, O. Takei, Y. Yoshikawa & A. Komatsu
Japan Railway Construction, Transport and Technology Agency, Nagasaki Construction Site Office

Y. Ougi
Kajima Umebayashi Nagasakiseibu Joint Venture

ABSTRACT: Shin-Nagasaki Tunnel is a mountain tunnel which has an inclined shaft for construction. Main tunnel between the start portal side and the intersection of inclined shaft was excavated right under residential areas from ranging between 85 and 135 m depth by blasting. Excavation from the intersection to the start portal side was blasted by high precision electronic detonator for environmental conservation to densely populated areas. The main tunnel between the end portal and the intersection of the inclined shaft was excavated directly under residential areas from ranging between 5 and 35 m depth by mechanical excavation and was employed the all ground fasten method for countermeasure against small overburden. This paper reports consideration of construction method, deformation monitoring and excavation situation of the mountain tunnel right under residential areas.

1 INTRODUCTION

Kyushu Shinkansen NishiKyushu route is the high-speed railway line connecting Takeo city, Saga prefecture and Nagasaki city, Nagasaki prefecture, and about 67 km between Takeo Onsen and Nagasaki is under construction for opening in 2022.

Shin-Nagasaki Tunnel is a double-track cross section tunnel of the Kyushu Shinkansen (NishiKyushu route) with an extension of length 7460 m located under in residential areas in Nagasaki city (Figure 1). Shin-Nagasaki Tunnel is divided into east and west work sites, and this report focuses on the west sites named Shin-Nagasaki Tunnel (West). The Shin-Nagasaki Tunnel (West) is 3590 m on the end portal side of the Shin-Nagasaki Tunnel. The geology of the tunnel in the work site is mainly consists of tuff breccia and andesite (Figure 2). The tunnel had the inclined shaft for construction. The intersection point of the shaft with the main tunnel was located 710m from the end portal. Excavation between the intersection

point and the start portal side ran neighboring residential areas of overburden 80m and directly under residential areas of overburden ranging between 85 and 135 m depth at a length about 680m. Excavation between the intersection point and the endportal ran under residential areas of overburden from ranging between 5 and 35 m in depth at a length about 390m.

In this paper, we report various countermeasures implemented to excavate right under the densely populated residential area while securing progress in a severe construction period.

2 OUTLINE OF EXCAVATION ON THE START PORTAL SIDE OF RESIDENTIAL AREAS

Excavation between the intersection of the shaft and the start portal side ran 2880m. This section is divided neighboring residential areas of overburden 80m (A area), and directly under

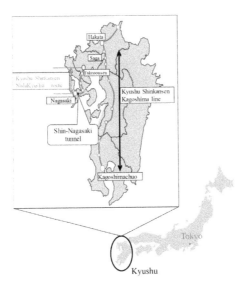

Figure 1.   Location map of Shin-Nagasaki Tunnel.

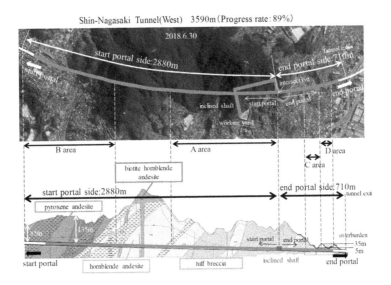

Figure 2.   Plane view and geological profile.

residential areas of overburden ranging between 85 and 135 m in depth (B area). Blasting was employed despite locating under residential areas, because excavating target ground was between middle hard rock and hard rock moreover this section was long and secured work progress. Therefore, this section was used controlled blasting to reduce vibration.

2.1   Consideration of controlled blasting

In this section, we considered controlled blasting with a high precision electronic detonator to reduce vibration caused by blasting to neighboring residents (Table1). A feature of the high precision electronic detonator is possible to set an interval by the 1ms (precision± 0.01 %) at a site. However the conventional electronic detonator used in standard control blasting has a fixed time interval of about 30ms. Therefore, it is promising to reduce the vibration value by

Table.1. Comparison of various detonators.

Item name	Electric detonator	Conventional electronic detonator	High precision electronic detonator
Second interval	250ms	30ms	1ms
Set maximum stages	20 stages	200 stages	500 stages
Blast duration	3second	30ms	7ms
Amount of charge per stage(W)	18kg	2kg	2kg
Characteristic	Vibration can be separated by stage	All hole separation possible. Fixed second time interval at factory shipment.	All hole separation possible. Time interval settable on site.
Magnitude of vibration	Bad	Good	Good
Bodily sensation vibration	Fair	Fair	Good
Cost	1	1.13	1.16
Evaluation	Bad	Fair	Good

suppressing the amplification of the vibration and suppress the bodily sensation vibration by shortening the blasting continuation time. In the application of high precision electronic detonator, the number of stages increases without changing the number of holes, therefore the explosive per stage is about 2 kg in high precision electronic detonator compare to about 18kg in conventional electric detonator (Figure 3).

2.2 Setting vibration control values

There are no legal regulation values of vibration on the blasting of tunnels in Japan. Therefore it was decided according to the recommendation values and similar construction cases of the Japan Explosives Industry Association (Zako,1984).

Where Limit values of vibration=Daytime 0.2(cm/sec), Nighttime 0.1(cm/sec).

2.3 Test blast

In order to verify the effectiveness of the high precision electronic detonator, a vibration meter was installed at the position shown in Figure 4 and conducted a test blasting to measure the vibration velocity. Moreover, according to previous research (Kitamura,2014), test blasting was carried out by setting the time intervals expected to reduce the vibration level to 7ms and 17ms.

Item name	Electric detonator	High precision electronic detonator
Number of stages	12	106
Number of holes	106	106
Charge	90~120kg	90~120kg
Amount of charge per stage(W)	18kg	2kg

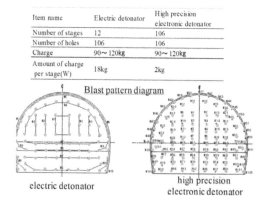

Figure 3. Blast pattern diagram.

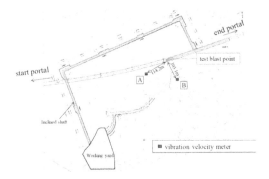

Figure 4. Measurement points at the time of test blast.

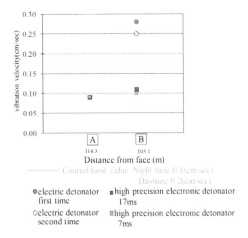

Figure 5. Vibration velocity result at test blast.

It revealed that the measured values at the high precision electronic detonator for the vibration velocity was significantly lower than the measured values at the electric detonator, and almost the same values as the nighttime limit values was obtained (Figure 5). Based on these result a high precision electronic detonator was an effective way to reduce vibration.

In addition, no significant difference was observed in measured values at the time intervals of 7ms and 17ms. However, in order to reduce the noise from the viewpoint of human sensory threshold, it is effective to shorten the duration of blasting to less than 1 second (Oishi, 2009). For this reason, the time interval was set to 7ms in order to reduce the duration of blasting to less than 1 second.

2.4 Consideration of applicable range

The distance D (m) from the blasting position exceeding the limit values was calculated back from the vibration prediction equation 1 based on the measurement result of the examination and the limit values. Then, the application range of controlled blasting with high precision electronic detonator was set, by using D as a separation distance from the target house. The range of application differs at day and night depending on the limit values of day and night time. High precision electronic detonators were used only after 22 o'clock for application range in the nighttime control blasting. Within the application range of the day and night time control blasting, high precision electronic detonators were used all time. The nighttime controlled blasting range was set as 163 m from the blasting point, and day and night controlled

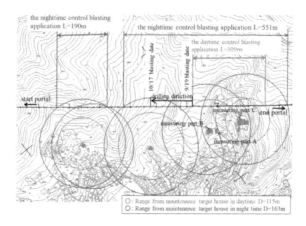

Figure 6. Control blast range in A area.

blasting range was set as 115 m from the blasting point. The applicable range of A area is shown in Figure 6.

$$V = K \times W^{0.75} \times D^{-2} \tag{1}$$

Where V= Vibration velocity (cm/sec); K= Coefficient which changes according to geology etc. (K=640); W= Amount of explosive per stage (kg) and D= Distance from blasting point (m).

2.5 Construction result

In A area, the measurement result of control blasting application was generally lower than the day and night limit values (Figure 7). Moreover, neighboring inhabitants did not claim about vibration of excavation at nighttime, excavation progress was in timeline.

In B area, the K value was revised based on the control blasting application result of this work sites, K = 880, daytime separation distance D = 139 m, nighttime separation distance D = 197 m (Figure 8). At the measurement point A, it was lower than the daytime and night-time limit values (Figure 9). At the measurement point B, the daytime and nighttime limit values were mostly lower but some were exceeded at daytime limit values (Figure 10). It is considered that vibrations were easy to be transmitted due to the area right under the residential area of the embankment.

On the ground surface, although the residents felt vibrations and noises, there were no complaints from them, therefore excavation could be made without losing progress. In addition to adopting controlled blasting, we were able to get the locals' understanding by having local

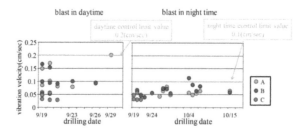

Figure 7. Control blasting result in A area.

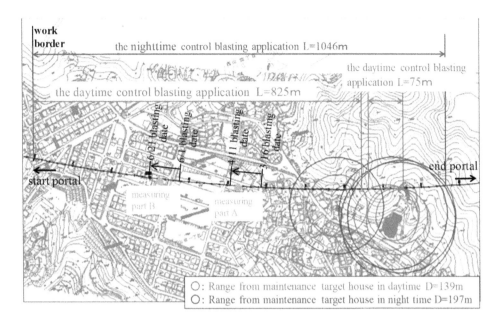

Figure 8.   Control blast range in B area.

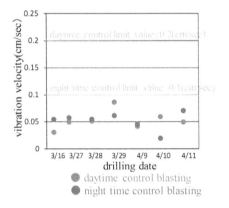

Figure 9.   Control blasting result in B area at the measuring part A.

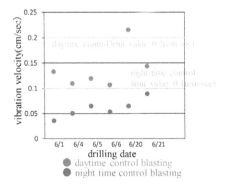

Figure 10.   Control blasting result in B area at the measuring part B.

briefing sessions and setting up display panels for raising awareness in advance of the blasting of the tunnel construction.

3 OUTLINE OF SMALL OVERBURDEN EXCAVATION UNDER RESIDENTIAL AREAS AT THE END PORTAL SIDE

Excavation from the intersection of the shaft with the main tunnel to the end portal side ran extension 710m. The end portal side from the intersection is overburden from ranging between 75m and 5m in depth. Moreover, there were slopes with undulations of 30 m or less, passing through it to the end portal (Figure 11). The slope is specified sediment related disaster special warning area around the residential area and there are an old wooden house built on embankment with the stone masonry wall (Figure 12, 13).

The geology of excavation target had many cracks and partly crushed in spite of the hard andesite observed from the surface layer. Considerations for vibration and noise are necessary because it is excavation along and directly below the slope residential high density area.

From the above, considering the excavation method, deformation monitoring and consideration to the residents' environment are necessary to excavation just under the dense residential area of the small overburden while ensuring the process.

3.1 Consideration of small overburden excavation

Vibration and noise countermeasures to small overburden under steep slopes and sloping residential area, mechanical method was employed one party at daytime and did not excavate at nighttime. Because the unconfined compressive strength of the excavation target ground is about 75N/mm² at maximum this time, the excavation machine model adopted a 350kw class road header (Figure 14).

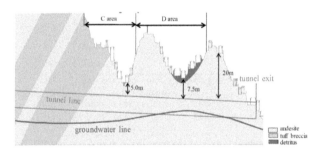

Figure 11. Longitudinal view of the end portal.

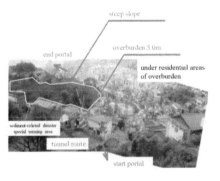

Figure 12. Tunnel ground condition on the end portal side.

Figure 13. Surface condition of 5 m overburden.

Figure 14. 350kw class road.

As a result of FEM analysis (nonlinear elastic model) on the condition that the geology was assumed to be heavily joined andesite,it was found that there is concern about the stability of the crown and face as the area of the overburden less than 10 m is loosened to the vicinity of the ground (Figure 15,16). In the section less than 10m in overburden, all ground fasten method that is steel pipe forepiling was adopted (φ 114.3 mm, t = 4.2 mm, L = 12.5 m @ 450 mm, silica resin) (Figure 16,17). Considering the influence range of collapse of 45° in the all ground fasten method range, margin section of 10m was provided at the start portal and the end portal sides of the section with less than 10m of overburden.

Measurement management was employed monitoring the convergence, ground surface settlement and deformation of slopes stretch measurement. Measurement management values were set up in consultation with each administrator (Table 2, Figure 18). In addition, the ground surface observer was placed for 24 hours, and the surface part was checked at all times for any abnormality.

3.2 Consideration of local residents

Before starting construction, we got the locals' understanding by having briefing sessions and installing display panels for raising awareness. In addition, temporary evacuation places were secured in advance as measures to be taken when residents go beyond the range of tolerance for vibration and noise.

	analysis section	section 1 <C area>	section 2 <D area>
conditions	overburden	5.0m	9m
	condition geology	heavily joined andesite	heavily joined andesite
analysis results	ground surface settlement	3.1mm	4.6mm
	crown settlement	3.7mm	6.0mm
	Destructive proximity	 topsoil	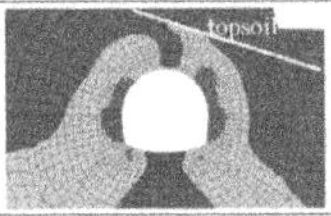 topsoil
Consideration		the overburden less than 10 m is loosened to the vicinity of the ground	

Figure 15. Result of FEM analysis.

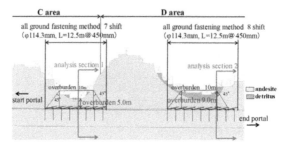

Figure 16.   The all ground fasten method application place.

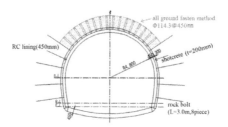

Figure 17.   Face of the all ground fasten method application place.

Table 2.   Management values of measurement of surface displacement.

| | ground surface settlement | | | | | slope stretching (strain) |
	municipal road	gas pipe	water supply pipe	sewer pipe	house foundation retaining wall	
control level 1	25mm(50%)	25mm(50%)	25mm(50%)	25mm(50%)	15mm(50%)	$1*10^{-7}$
control level 2	37.5mm(75%)	37.5mm(75%)	37.5mm(75%)	37.5mm(75%)	22.5mm(75%)	$1*10^{-6}$
control level 3	50mm(100%)	50mm(100%)	50mm(100%)	50mm(100%)	30mm(100%)	$1*10^{-5}$

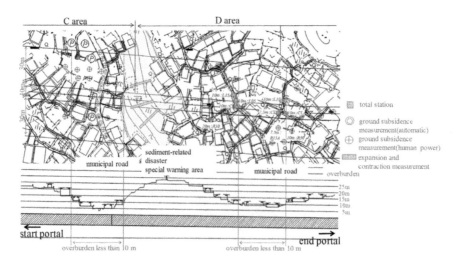

Figure 18.   Measurement points on the ground.

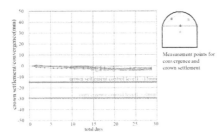

Figure 19. Result of crown settlement and convergence.

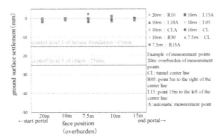

Figure 20. Result of ground surface settlement in D area.

3.3 *Consideration of local residents*

According to the preliminary geological survey, the geology was mainly andesite, however the facing geology during excavation in C and D areas were mainly weathered tuff breccia.

D area was added 1 shift of the all ground fasten method, because the cracks were interposed a lot of clay, and they were in a state of easy loosening.

Convergence and crown settlement were within control level 1 in both areas (Figure 19). The excavation in the C and D areas had a maximum value of -2 mm for ground subsidence and was within control level 1 (Figure 20). Furthermore, the amount of expansion and contraction of the slopes was within control level 1. During the excavation, monitoring was carried out by the ground surface and slope observers, but there was no particular abnormality.

4 CONCLUSION

While the construction schedule was tight, various hard measures were taken by examining the excavation method for the influence on the surrounding environment according to the local situation. In addition, we took communication by preliminary explanation to get the locals' understanding. As a result, we were able to excavate without losing progress.

We would like to continue construction while closely communicating with the local residents and make it in time for the completion of the Kyushu Shinkansen NishiKyushu route.

REFERENCES

Ken Zako, 1984. Happa shindo no shuhen e no eikyo to taisaku.
Yasutomo Oishi,Masataka Goto and Kazuya Takeda, 2009. Ongaku johoshori saizensen! 'Utagoe' to 'hanashigoe' wa do chigau ka? Ningen no 'koe' o rikai suru konpyuta no genjitsu o mezashite. Institute of Music Information Science: DTM Magazine
Yoshinobu Kitamura, Yasunari Tezuka, Keita Iwano, Yuuki Sano and Yoshikatu Maeda, 2014. A case study of evironmental impact on residental areawith highly acculate electronic dentinator: The 24th Tunnel Engineering Research Presentation

Tunnels and Underground Cities: Engineering and Innovation meet Archaeology, Architecture and Art, Volume 11: Urban Tunnels - Part 1 – Peila, Viggiani & Celestino (Eds)
© 2020 Taylor & Francis Group, London, ISBN 978-0-367-46899-6

Building responses due to deep excavation for Muzium Negara MRT Station, Kuala Lumpur

C.M. Khoo, A.C.S. Teoh & S.T. Poh
Mass Rapid Transit Corporation, Kuala Lumpur, Malaysia

ABSTRACT: Muzium Negara MRT station was constructed within 16m of the National Museum, a historically significant building in Kuala Lumpur. The 27m deep station excavation was constructed using diaphragm walls and top-down method. The museum building is founded on shallow footing in Kenny Hill Formation. Built in the 1960s, the building may have had previously undergone some degree of settlement apparent from a series of upgrading works over the years before the station box construction. It was therefore considered it would be more sensitive to ground movements making the challenge of controlling the impact of construction-induced settlement even greater. This paper presents the actual performance of the settlement sensitive heritage building responses to a deep excavation carried out in close proximity.

1 INTRODUCTION

Muzium Negara Station is one of the underground metro stations under the Klang Valley Mass Rapid Transit - Sungai Buloh-Kajang Line. This underground station located adjacent to the one of the most important of Kuala Lumpur's heritage buildings, the National Museum (*Muzium Negara*), has brought a new set of interesting insights in terms of the responses of the settlement sensitive heritage building to a deep excavation carried out in close proximity (see Figure 1). It is inevitable that any deep excavation would induce ground movements on the peripheral of the excavation. The consequence of this ground response due to the effects of the construction of the diaphragm wall and subsequent deep excavation could have a detrimental impact on the structural integrity of any structure within the influence zone of the excavation. To compound these problems further, the old heritage building is founded on shallow footing at the existing higher ground at one side of the station excavation. The building may have previously undergone some degree of settlement apparent from a series of upgrading works over the years before the station box construction. Hence, it is more sensitive to ground movements and makes the challenge of controlling and reducing the impact of construction-induced settlement even greater.

From the onset of the project, the potential problems faced by the building were identified and a carefully risk-managed approach for settlement control and minimizing construction impact was undertaken together with extensive instrumentation monitoring of the structure and the surrounding ground. The station construction scheme initially planned using the bottom-up method was however changed to a top-down method using diaphragm wall as both temporary and permanent earth retaining system. This paper presents the actual performance of the settlement sensitive heritage building responses to the deep excavation carried out in close proximity.

2 THE BUILDING AND ITS PRE-EXISTING CONDITIONS

The museum building is a 3-story reinforced concrete (RC) structure with one level basement. The width and length of the building are about 20m and 120m respectively. The maximum

Figure 1.   The National Museum before and during construction of the underground MRT station.

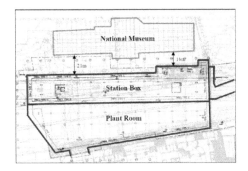

Figure 2.   Location plan.

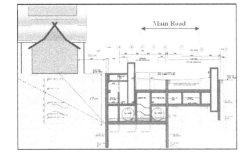

Figure 3.   Critical cross-section.

height of the building at the ridge is 25m from the ground level. The front of the building is closer to the edge of the station box and faces a busy road. With its length running parallel at a distance between the edges of the building to the proposed station box of about 16m to 21m, the museum building was within the influence zone of less than one time the depth of the station excavation. Figures 2 and 3 respectively show the location plan and critical cross-section of the proposed Muzium Negara Station side by side with the existing National Museum building.

Built in the 1960s, the structural framing of museum building was expected to be of conventional reinforced concrete (RC) beams/slabs supported on RC columns, while its roof structure is formed by sloping RC slabs. Based on trial pit excavation, the building is supported on isolated footings at about 2.5m below the ground. However, few of the columns are spaced at 2m c/c, it is practically assumed that the foundation at those locations to be on combined footing although there were neither foundation drawings nor records to confirm it.

Lately, numerous high-rise buildings with deep basements were built across the road at the lower ground platform opposite the National Museum site. These developments combined with its own series of upgrading works over the years, may suggest that the building has suffered settlement in the past. However, it is difficult to ascertain how much settlement the building has undergone. When the pre-construction condition of the museum building was assessed, it was found to be generally in good condition with a small number of minor cracks along the non-load bearing brick wall that had been repaired and plastered over. There was no visible distress exhibited by the RC structure.

3   GENERAL GEOLOGY AND SITE TOPOGRAPHY

The ground investigation conducted confirmed that Muzium Negara Station is located in the Kenny Hill Formation. The site mainly consists of a thin fill layer overlying 6m to 13m thick

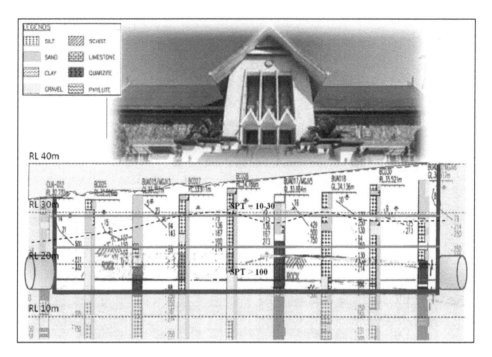

Figure 4. Interpreted geological profile.

gravelly SAND and sandy SILT deposit of SPT values ranging from 10 to 30. This is followed by very stiff sandy SILT layer generally of SPT > 100 with gravel material. Moderately to highly weathered sedimentary rocks (e.g. sandstone, shale) were occasionally encountered from the borehole explorations. Figure 4 shows the interpreted ground stratification in front of the museum building. As noted earlier the building is located at a platform on the higher ground to the north of the station box, the topographic elevation at the excavation line is sloping from the highest level of RL40m at the eastern wall towards the lowest RL33m at the western wall. It is worth noting there is a plateau at the north-eastern direction behind the museum building.

4 DESIGN MANAGEMENT APPROACH

The potential settlement problems associated with the construction of the underground station close to the building were realized very early within the project, an ideal solution was to try to manage these risks and minimize them prior to the onset of actual construction. Although some of the major decisions over the design approach had been taken before the contract award, further measures to reduce ground movements were incorporated during detailed design development to ensure that the requirement to safeguard the performance of the building could be satisfied. Much of the works in studying the building responses and the sources of movement was carried out to validate the feasibility as part of this overall project optimization. One of the key elements was the changing of the bottom-up construction method to the top-down method. This provided the assurance that a more robust and stiff system was being used with the permanent works put in place rapidly inhibiting potential ground movements. Mahalingam *et al.* (2015) presented the design management approach undertaken including the methodology of assessment of the potential damage on the museum building and the measures used to combat the impact of the excavation works vis-à-vis its various mechanisms involved in causing the ground movements.

As the design developed this risk-based approach strategy was further developed working closely with the contractor. A high level of instrumentation was committed which embraced real-time facility to the monitoring of the structure and ground including a large number of piezometers across the various stratigraphy. These instrumentation monitoring allowed for a high level of construction control over the works, thus enabling a rapid response to the project of any abnormal movement trend and a close check on the construction impact.

The construction was successfully carried out by firstly the installation of 1.2m thick diaphragm wall along the perimeter of station box with 2–3 rows of temporary ground anchors supporting the side of the higher ground. Bulk excavation was then commenced with top-down fashion by casting of roof slab/slabs before proceeded with subsequent downward excavation for the casting of concourse slab. The design required a layer of temporary strutting to be installed prior to excavation to the final formation level for the casting of the base slab. Upon backfill of the station box and road traffic was diverted on top, excavation for the plant room was carried out similarly with top-down construction method.

5 INSTRUMENTATION MONITORING

Instrumentation of the museum building is an important component in the overall strategy for managing the settlement risk to the building. Despite a number of measures had been designed to control settlement, but uncertainties may still exist particularly with respect to the building response to settlement and exactly how the settlements will manifest themselves once the construction actually commences. Consequently, an extensive and carefully designed instrumentation regime was integral to ensuring that the inevitable movement differences between design and construction could be reacted to and solutions sought swiftly to ensure that settlement of the building was kept within acceptable limits.

The usual suite of monitoring instruments was used to monitor the museum building, but each instrument was located at a specific place or depth to maximize its' potential benefits and feedback on useful information. The building was monitored with a number of discrete electro-level beams, tilt meters, building settlement markers, and optical prisms. An automated total station was set up for real-time surveying and linked to a computer display in the office. This served to alert if any rapid movement of the building occurred. Apart from monitoring of the building response, comprehensive ground instrumentation was put in place, which includes a series of settlement markers, inclinometers, extensometers and water standpipes for groundwater monitoring. Vibrating wire piezometers were also installed around the excavation to pick up, as early as possible, any under drainage and consolidation settlements. As the

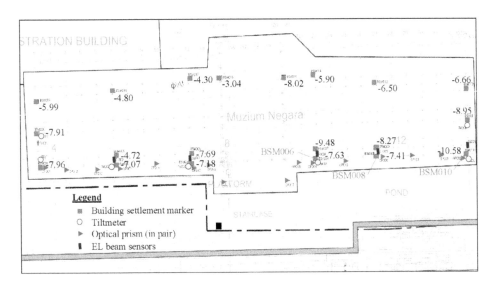

Figure 5. Layout of structure instrumentation.

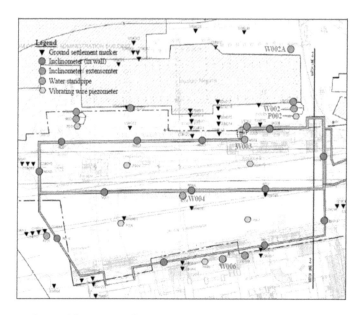

Figure 6.   Layout of ground instrumentation.

nature of building response to the construction-induced settlement would be relatively slow, the large number of settlement points installed allowed for full coverage of the building, covering any areas of concern and any potential differential settlements. Figures 5 and 6 respectively show the layouts of structure and ground instrumentation.

6   PERFORMANCE OF BUILDING RESPONSES

The recorded settlements of museum building throughout a duration of approximately 4½ years including 3 years post- critical excavation stage are presented in Figure 7. The building settlement profile clearly shows that significant movement occurred during the bulk excavation. It was observed that there were distinct phases of settlement which are clearly linked to the construction events. Building settlements were at first registered in tandem with the ground movements attributed to the installation of diaphragm wall coupled with temporary ground anchors and excavation to station roof level. Subsequently, phases of significant settlements were measured which corresponded with the excavation to concourse level and base slab level.

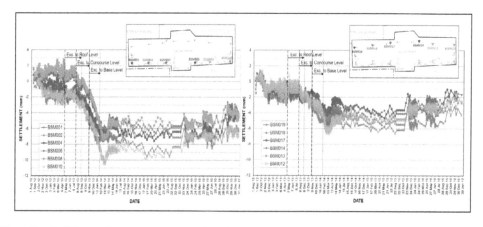

Figure 7.   Building settlements vs. construction time.

The largest settlements of 8mm to 10mm were observed for the columns that are on the front of the building which was next to the station excavation. At the same time as settlement of up to 7mm was observed at some points at the rear of the building, which was furthest from the excavation. The magnitude of the settlement was generally greatest at the east wing as compared to the west wing of the building, in which the former was closest to the excavation and deepest excavation depth, with the sign of extended settlement for some times after the bulk excavation for the station box had been completed. The highest absolute settlement of 10.58mm was registered by BSM010. As this section of the station has the thickest fill of loose deposit coupled with the deepest excavation and the more permeable nature of the ground, it was expected that this corner of the building would experience the most settlement. Nevertheless, the settlement was relatively uniform with a differential range of 3mm and this has resulted in not as much of differential settlement on the structure.

Figure 8 shows the monitoring data for a series of water standpipes located across the building and station box transversely. The periods of drawdown coincide with the significant settlement; this is somewhat establishing the link between piezometric drawdown and settlement where major excavation activity was being carried out. The most notable drawdown occurred when excavation to formation level took place and this occurrence was also extended some times after the bulk excavation for the station box had been completed. This exacerbated situation could be due to the presence of particular permeable strata allowing rapid water flows. Incidentally, the drop could also be due to prolonged dry spelt experienced in early of the year 2014. Remarkably even with the significant drop of piezometric level of 10m to 12m, there is hardly decreasing of pore water pressure registered by the VW piezometer P002 at 26m below ground (see Figure 9). This pointed to a general drawdown of the water table around the station probably due to under-drainage in the more permeable strata below the excavation. It was observed that the groundwater subsequently restored to the natural level rapidly. The groundwater table exhibited the second episode of slight drawdown, which corresponded with the excavation for plant room.

It is interesting to note that the northern diaphragm wall penetration has somehow dammed up the groundwater flowing from the upstream high ground, as illustrated in Figure 10. This supposition was also supported by the incidence of seepage backflow in the building basement during the initial stage of the diaphragm wall installation. Indeed, the trifling heave of the building as registered by BSM006 and BSM008 prior to the bulk excavation (see Figure 7) was exactly at the area where seepage was reported. Point should be noted that the localized steep drawdown measured at W003 prior to the bulk excavation (see Figure 8)

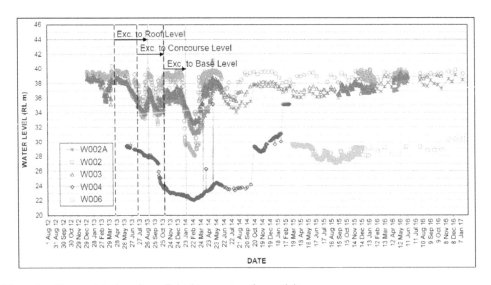

Figure 8.   Piezometric drawdown linked to construction activity.

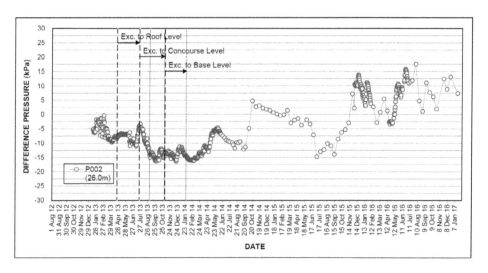

Figure 9.   Differences in pore water pressure.

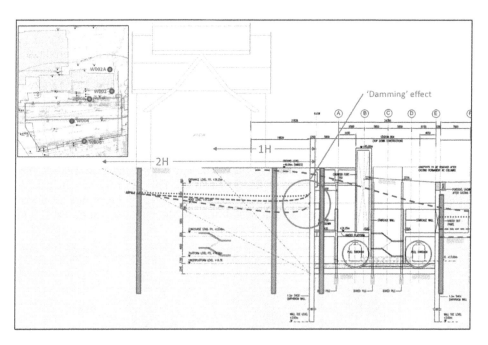

Figure 10.   'Damming' effect postulated from standpipe monitoring.

was due to the excessive seepage recorded in one of the ground anchor holes and the ground-water table was quickly recovered upon sealing off.

In summary, the settlements of the museum building were considerably smaller than those predicted and noticeably less than the surrounding ground. These additional settlements were essentially tolerated by the building which is founded on the shallow footing. The overall differential settlement of 3mm is considered insignificant in relation to the capacity of the structure. The building remained fully functional throughout the construction period and suffered no structural distress. It is also worth noting that about 30%-40% of the settlements suffered by the building have recovered post-excavation, almost three years later. The restoration of

Figure 11. Artist impression of the Muzium Negara MRT Station.

groundwater table could have led to the recovered building settlement over the time. There was never an instance that hinted the serious concern of consolidation settlement in Kenny Hill meta-sedimentary formation as a consequence of seepage flow due to piezometric draw-down caused by the station excavation.

7 CONCLUSION

This paper documented the actual performance of the National Museum building in respond to the deep excavation carried out in close proximity for the construction of Muzium Negara MRT station. The pattern of building settlements corresponded to the depth of excavation and distance from excavation albeit the overall settlement was relatively uniform caused not as much of the concern of differential settlement. The highest absolute building settlement of 10.58mm was registered by BSM010 located at the front edge of the east wing of the museum building. It was sensibly expected that this corner of the building would experience the most settlement in light of the deeper excavation and nearest distance to excavation coupled with the thickest fill of loose deposit and the more permeable nature of the ground.

The settlement profile clearly shows that notable movement occurred during the bulk excavation in conjunction with the significant piezometric drawdown registered by the instrumentation monitoring. Throughout this episode, however, the settlement readings of the building still continued to remain stable and slightly rebound together with the recovery of the piezometric level. Through the adoption of appropriate and measured responses and effective employment of the risk-based approach, the museum building was successfully protected against any detrimental effect caused by the deep excavation for the underground station. On completion of the works, it was demonstrated no structural damage to the museum building but identified a number of hairline cracks and small increases in existing cracks on the apron structure, all easily repairable. Figure 11 illustrates the view of the entrances to Muzium Negara MRT Station next to the National Museum building.

REFERENCE

Mahalingam, T., Khoo, C.M., Chockalingam, P. and Cheng, S.S. 2015. *Design Management and Case Study of Underground Construction adjacent to Settlement Sensitive Building.* Proceedings of International Conference and Exhibition on Tunnelling & Underground Space 2015, Kuala Lumpur: 280–284.

Tunnels and Underground Cities: Engineering and Innovation meet Archaeology,
Architecture and Art, Volume 11: Urban
Tunnels - Part 1 – Peila, Viggiani & Celestino (Eds)
© 2020 Taylor & Francis Group, London, ISBN 978-0-367-46899-6

Record-breaking results through teamwork on Istanbul's Dudullu-Bostancı Metro Line

E. Koç
Senbay Madencilik-Kolin-Kalyon JV, Istanbul, Turkey

B.A. Matheson
Terratec Ltd., Hobart, Australia

M. Bringiotti
Geotunnel S.r.l., Genoa, Italy

ABSTRACT: At 7am, on March 03, 2018, one of two TERRATEC 6.56 m diameter Earth Pressure Balance Machines (EPBMs) being used by the Şenbay Madencilik-Kolin-Kalyon JV on the new Dudullu-Bostancı Metro Line, in Istanbul, Turkey, completed an outstanding advance of 26.6 m (19 rings) in just 12-hours. The TBM and its crew worked non-stop – alternating between 20-minute mining and ring building cycles – throughout the night shift to accomplish a new production record for a TBM of this size and class in Turkey. The 14.2 km-long Dudullu-Bostancı Line runs north to south under the densely-populated Anatolian side of Istanbul and is located entirely underground at an average depth of about 30 m. Achieving consistently good TBM production rates in this dense urban environment – with limited space on a busy open-cut station site – is a testament to the importance of teamwork and well-planned logistics, and ultimately led to the early completion of tunnelling on an extremely challenging contract schedule.

1 PROJECT BACKGROUND

Istanbul was a pioneer of the urban underground railway. The city's first metro tunnel, between Karaköy and Galatasaray, is one of the oldest in the world – second only to the early London Underground lines – having gone into service in 1875. However, despite this, it wasn't until the late 1980s that construction began on a true 'modern' mass transit railway system for the city.

Since then, the population of Istanbul has almost doubled, from about 8.5 million to nearly 16 million. Traffic congestion has reached an all-time high and the city is now facing a major transportation crisis. Therefore, in recent years, work has been underway to meet the Istanbul Metropolitan Municipality's (IMM) goal of expanding the city's existing six-line, 145 km-long, metro service to an inter-connective network that covers more than 480 km by the end of 2019. Five new metro lines and three extensions of existing metro lines are currently under construction on both sides of the Bosphorus Strait.

One of these new lines is the Dudullu-Bostancı metro line (Figure 1), which runs north to south across the densely-populated Anatolian side of the city. The 14.3 km-long line, along with its 13 new stations, is located entirely underground at an average depth of about 30 m and will ultimately be fully-automated – with driverless trains, communications-based train control (CBTC) and platform screen doors at stations – providing numerous connections to other Istanbul transportation systems, such as the Bosphorus ferry (at Bostancı Harbour), the Marmaray railway, and the Kadıköy-Kartal and Üsküdar-Çekmeköy metro lines.

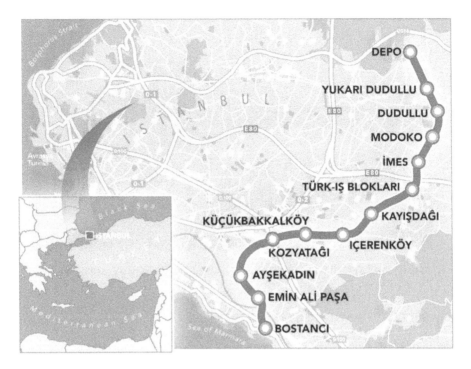

Figure 1. Map showing the location of the Dudullu-Bostancı metro line.

Figure 2. Aerial view of the Kayısdagi Station site.

The IMM awarded the construction contract for the line – which also includes E&M works, construction of Underground Transfer Centres (Car Parks), a Depo Area and a Management and Control Centre Building – to the Şenbay Madencilik, Kolin and Kalyon Joint Venture (SKK JV) on February 12, 2016. Two weeks later the JV mobilised on the future site of the centrally located Kayısdagi Station – which is surrounded by residential and business properties (including a school that abuts the open-cut station box) – where the project's TBM tunnelling operations would be based (Figure 2).

Figure 3. The TERRATEC S51 TBM cutterhead is lowered into the Kayısdagi Station box.

2 TBM SELECTION AND DESIGN

In order to meet the Dudullu-Bostancı metro line's fixed 2019 opening date, a challenging construction schedule was put in place, which required the project's four TBMs to commence tunnelling 10 months after contract award and complete excavation within 28 months. Members of the JV had been greatly impressed with the performance of the Terratec machine they had used on the new Mecidiyeköy-Mahmutbey Metro line – which is being built by the Gülermak, Kolin and Kalyon JV on the European side of the city (Bilgin et al., 2017) – and were keen to employ further Terratec machines on their second project for the IMM. However, with TBM delivery being on the critical path to project completion, the JV decided that the schedule risk should be mitigated by splitting the TBM order between two suppliers. Therefore, an order for two new 6.56 m diameter EPBMs was placed with Terratec and two refurbished 6.57 m diameter EPBMs were purchased from another manufacturer.

The Terratec mixed-face TBM cutterheads featured an opening ratio of about 35 percent, designed to best manage Istanbul's variable geology (Figure 3), and were fitted 17" back-loading disc cutters that were interchangeable with knife-edge bits for areas of soft ground. Other features included VFD electric cutterhead drives – with a total installed power of 960 kW and a high-speed cutterhead – high torque screw conveyors, and active articulation systems. Total thrust was 40,000 kN, with a nominal torque of 5,459 kNm, and max. torque of 7,097 kNm.

3 GEOLOGY & PREDICTION OF RISKS

The geology along the new line comprises Paleozoic aged Aydos, Gözdağ, Kurtköy, Tuzla, Trakya and Kartal Formations. The Aydos and Gözdağ Formations are in the upper level of the Kurtköy Formation and mostly consist of high strength and laminated shale layers and very abrasive rock consisting of quartz and arenite that is highly rich in feldspar. In the Tuzla Formation limestone can be found along with thinly laminated mudstone. The Trakya Formation consists of laminated interbedded siltstone and mudstone and is very abrasive and fractured in some places. The main geological formation to be excavated was the Kurtköy Formation. Arkose is the main unit of the Kurtköy Formation and generally consists of purple-coloured gravel, sandstone and mudstone.

With the expected ground conditions in mind SKK's TBM management team aimed to mitigate any problems that may arise during excavation. Grizzly bars were prepared for the

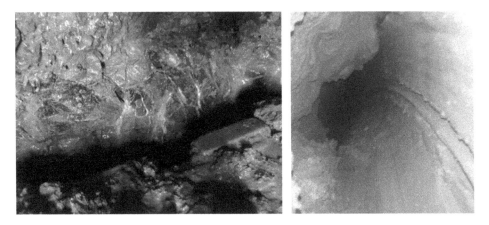

Figure 4. Rippers were prepared to replace the disc cutters for a known 500 m zone of soft clay.

TBM cutterheads, to prevent blockages in zones of fractured and laminated rock conditions, and were ultimately used for about 60 percent of the alignment. Ripper tools were also at hand, in order change out the 17" disc cutters for a known 500 m-long zone of very soft clay, reducing the possibility of clogging (Figure 4).

4 SITE SET-UP AND LOGISTICS

The precast tunnel lining rings that were installed by the TBMs consist of reinforced concrete trapezoidal segments (5+1), with an outer diameter of 6,300 mm, an inner diameter of 5,700 mm and a width of 1,400 mm. They were produced by SKK JV at its factory in Ferhafpasa, which is located about 10 km away from the project worksite.

While planning the logistics and TBM supply the TBM management team had to consider the traffic congestion in Istanbul. With the need for on-time segment delivery and muck removal from site, and the inevitability of severe traffic jams along trucking routes, the decision was made to conduct 70 percent of the operation during the night shift. Six trucks with trailers were used to transport the segments and these pushed hard during the night, as did the muck transports; driving 104 km round trips to a disposal site north of Istanbul.

The muck pit was allocated two clamshell excavators in order to manage the muck levels discharged by the conveyors, while the belt tower was limited to 5 m in height, in order to reduce noise levels within the residential area (Figure 5).

Although the Kayısdagi site encompasses a relatively large area, the 3,500 m^2 concrete batching plant, the 237 m long open-cut structure (which includes the station and switchbacks), and the 1,830 m^2 muck pit and conveyor ramp, take up most of the available surface

Figure 5. The muck pit and gantry cranes at the Kayısdagi site.

area. Added to this, was the additional challenge of large ground deformation loads on the station box that required the introduction of a safety exclusion zone alongside the TBM grout plant on the road side of the station structure. This enabled the JV to keep the number and size of the support struts used within the box to a sensible level, but it also eliminated much of the space that had originally been intended for segment storage, leaving room on site for just 50 sets of tunnel segments for four TBMs.

Twin gantry cranes were selected for the segment and material lowering operations. These were sited above the switchbacks and equipped with double remote-control units in order to reduce the risk of accidental damage while lowering the segments and placing them onto Multi-Service Vehicles (MSVs) as fast as possible while working in narrow spaces between the support struts. The first operator would control the crane while lifting with the hook on the surface. Then, during the lowering operation, a second operator took control once the hook had passed through the strut level and placed the segments or materials on the MSVs.

5 TBM DRIVES

The TBMs were delivered on time and assembly of the machines began at the end of 2016. By January 2017, the first two TBMs – the Terratec S50 TBM and one of the refurbished machines – had been launched on their 3,987 m journey south-west towards the coast. Shortly thereafter, the Terratec S51 and the other refurbished TBM began mining north-east on a 4,407 m trajectory towards Yukarı Dudullu station (with a planned section of NATM works due to complete the line to Depo).

TBM parameters and performance analysis were inspected by SKK's TBM management team via a program designed by the TBM Tunnels Chief Engineer, which collected all TBM data and, on monthly basis, data from all the shift engineers. This allowed the team to identify optimum operating parameters to reach maximum and continuous daily advances.

The Bostanci TBM drives progressed much as expected, achieving good advance rates and passing through stations on their way south (Figure 6), before terminating at a shaft located just north of Ayşekadın station in April 2018 (with the balance of the running tunnels to Bostancı due to be completed by NATM).

However, on the Dudullu side the abrasive nature of the ground proved challenging, especially between Kayısdagi and Modoko. On leaving Kayısdagi the geology encountered was

Figure 6.   The TBM sliding operation was performed over invert segments through the stations.

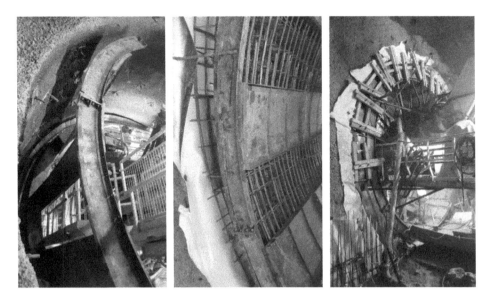

Figure 7.   The cast-in-situ TBM thrust ring used at the underground stations.

hard arkose sandstone and limestone. This is rich in feldspar and quartz grains that are highly abrasive and have a significantly negative impact on the operational life of the disc cutters (Seyedrrzaei et al., 2018). It also caused rapid wear to both the TBMs' cutterheads and screw conveyors, which had to be refurbished six times and five times, respectively (total for both machines). These works were carried out at the stations, with the team refining the refurbishment operations to a point where they could repair a screw conveyor in the matter of a week. Other remedial methods included the injection of anti-abrasive chemicals, and occasionally bentonite, during excavation to reduce friction. Once the TBMs passed Modoko however, the ground became gradually better and TBM utilisation rates improved.

5.1   Station Crossing & Relaunch

At all the underground stations, a novel cast-in-situ TBM thrust ring [designed by Enver Koç] was adopted. Using this method, a TBM relaunch was possible in two days, and without the need for enlarged NATM chambers or the installation of a heavy thrust frame. All that was required was the excavation of a 14 m-long A1 NATM tunnel (to house the TBM shield) and the radial installation of 60 x 32 mm diameter self-drilling rock bolts, with couplings attached to the heads, before the TBM reached the station. Once the TBM was in-place for relaunch, the rock bolts were extended (via the couplings) to reinforcement installed within a metal segment ring mould, which was manufactured in the site's workshop (Figure 7).

6   TEAMWORK FOR SUCCESS

Team spirit has been the key to success on the Dudullu-Bostancı metro, both in terms of the joint venture and subcontractor partnerships, the teamwork on site and the project's relationships with its suppliers, TBM manufacturers and their agents. At the outset of the project, SKK JV had appointed a project manager with extensive TBM experience and this decision provided swift and decisive action during procurement, employment and technical decisions relating to the management of the TBMs. The lines of communication between the mechanical and operational teams were also very strong and an ethos of 'being part of the solution not

Table 1. TBM Performance for Dudullu Bostancı EPB-TBMs

TBM Serial Number	Total Advance (m)	Total No. Rings	Maximum Daily Advance (m)	Ground Conditions
S50	3,987.87	2,850	37.2	Arkose & Limestone
EPBM 2	3,792.40	2,725	32.4	Arkose & Limestone
S51*	4,407.49	3,120	40.0	Arkose & Clay
EPBM 3	4,344.62	3,076	34.5	Arkose & Clay
Total	16,532.38	11,771		

* Shift record holding TBM

part of the problem' was adopted across the project. With this ideal in mind, decisions were made together as a team and focussed on results only.

TBM manufacturers need to be involved in the project as early as possible and (in the experience of the authors) should listen to the requirements of the contractor and adjust their mechanical solutions to suit the project's specific needs, from site access to ground conditions. Of course, a local presence, either in the form of a local office or a local representative, is key and ensures a 'can do' attitude to problem solving on site in real time (for example, sourcing consumables and spares locally, if necessary).

From the beginning, SKK's TBM management team aimed to gather together the most experienced workers and engineers possible for the project, but with 30 TBMs currently at work on various different projects in Istanbul, this proved a challenge. For the most part, newly graduated junior engineers – who had been identified following internships with the JV's partner companies – were selected to build the team. Following a month-long theoretical training period, during TBM assembly, the TBM management trained these young engineers as TBM operators and continued to mentor them throughout the project.

A daily bonus payment system was also applied to keep the teamwork level high. Engineers and workers were separated into three different groups of payment level and, on a daily basis,

Figure 8. Breakthrough of the S51 TBM at Yukarı Dudullu in June, 2018.

the JV paid bonuses to everyone for the 13th ring built, onwards, as long as there were no defects or offset on the built rings. These bonuses could increase a basic salary by approximately 10–15 percent on a monthly basis and really incentivised everybody to come together and fix problems. It provided excellent performance and was a major contributory factor in completing the TBM tunnelling two months earlier than scheduled.

In fact, in early March 2018, when progress rates had reached their peak, one of the shift engineers approached the TBM management and requested that the night shift be allowed to try and break the existing production record. This resulted in a record-breaking 19-ring shift that was completed by the Terratec S51 machine, which had mined about 75 percent of its alignment at this point. Equating to 26.6 m of excavation in just 12-hours, the TBM and its crew worked non-stop throughout the night – alternating between refined 20-minute mining and ring building cycles – to accomplish the record for a TBM of this size and class in Turkey. The excellent progress rates continued through to the completion of excavation, this June, when the S51 broke through into Yukarı Dudullu station (Figure 8).

ACKNOWLEDGMENTS

The authors are grateful for the support of the Şenbay Madencilik-Kolin-Kalyon Joint Venture (SKK JV) and, in particular: Project Manager, Mr. Mustafa Yurt; Tunnel Section Chiefs, Mr. Mehmet Ali Keleşoglu & Mr. Şahin Zerdeşt Arslan, Mr. Hüseyin Taşdemir; Shift Engineers Çağdaş Kaya, Serhat Palakcı, Erdal Gündoğan, Muhammed Arıoğlu, Satılmış Kalyon and Koray Yağlı; Terratec Site Operations Manager, Mr. Bill Brundan; and Mr. Erhan Ünlü and his team at Ertunnel.

REFERENCES

Bilgin, N., Acun, S., Ates, U., Murtaza, M. & Çelik, Y. 2017. *The factors affecting the performance of three different TBMs in a complex geology in Instanbul*. Proc. World Tunnel Congress 2017, Surface Challenges – Underground Solutions, Bergen, Norway, 9–15 June, 2017.

Seyedrrzaei, M., Koç, E. & Tumac, D. 2018. *Single Shield TBM Performance Analysis in Mixed ground Conditions: A Case Study in the Dudullu-Bostanci Metro Tunnel, Turkey*. Poster paper proc. ITA-AITES 2018 World Tunnel Congress, Dubai, 21–26 April, 2018.

Tunnels and Underground Cities: Engineering and Innovation meet Archaeology,
Architecture and Art, Volume 11: Urban
Tunnels - Part 1 – Peila, Viggiani & Celestino (Eds)
© 2020 Taylor & Francis Group, London, ISBN 978-0-367-46899-6

Ensuring the safety of the existing buildings during the construction of the underground in Moscow

D.S. Konukhov
LLC Mosinzhproekt, Moscow, Russia

A.G. Polyankin
National University of Science and Technology MISIS, Moscow, Russia

ABSTRACT: Plans for extension of Moscow Metro include construction of 154 km of new lines and 73 new stations in 2012–2020. About 10 stations have to be put into service each year. The main construction goes on in highly developed city areas. Working on 1 km of new lines demands taking into account 17...20 existing buildings. The complex plan of scientific support, elaborated in order to provide preservation and safe exploitation of buildings in construction area, goes along with all stages of design and construction works.

1   INTRODUCTION

Moscow is a city with almost a thousand-year history. Here, more than 12.5 million people live on an area of 2.5 thousand km^2. About 12 million of them live in the "old Moscow" area of about 0.9 thousand km^2.

Currently, Moscow is actively building new and extending existing metro lines (Figure 1). From 2012 to 2020, it is planned to build 154 km of new lines and 73 stations. This will reduce the load on the existing metro network, and also provide "walking accessibility" to stations for 93% of Moscow's population.

Currently, more than 50 km of lines and 30 stations have been built within the framework of this program. Construction work is carried out on more than 300 construction sites. At the same time, 29 tunnel boring machines (TBMs) with a diameter of 6 m are used for tunneling single-track tunnels and 2 TBMs with a diameter of 10 m for the construction of double-track tunnels. About 35 thousand specialists of various profiles take part in the construction of the underground.

2   MAIN ISSUE

Construction is conducted in a dense urban development. On the average 1 running km of line of subway under construction accounts for about 17 – 20 existing buildings. It is necessary not only to preserve them but also to ensure safe and comfortable staying in them of people during construction. For this purpose, a set of measures for scientific and technical support of construction (STSC) (Konyukhov D.S. 2017) is developed and being implemented, which includes:

   1. At the stage of design and survey works:
   1.1. Ensuring the completeness and sufficiency of engineering survey results.
   1.2. Forecasting of geotechnical risks taking into account all possible types of impacts.

Figure 1. The scheme of development of the Moscow Metro until 2020.

1.3. Consideration of modern structural, technical and technological solutions for the construction of metro facilities, the use of efficient and safe materials, construction machines and operating equipment.

1.4. Forecasting the impact of construction on the existing natural and man-made environment.

1.5. Providing a set of measures to minimize the impact of construction of underground facilities on the existing natural and man-made environment.

1.6. Formation of a set of special technical conditions, corporate standards and other normative and technical documents.

1.7. Certification of new structures and materials.

1.8. Expert-consultative analysis of project documentation in order to eliminate the risks of emergency situations, improve structural, space-planning, technological solutions for construction.

1.9. Preparation of a work program for STSC at the construction stage.

2. At the construction stage:

2.1. Analysis of the results of various types of monitoring and data on quality control of construction.

2.2. Instrumental support for monitoring and quality control of construction using geophysical and other non-destructive methods.

2.3. Assessment of the suitability for the operation of structures manufactured with deviations from the design.

2.4. Analysis of the causes and consequences (including long-term) of emergency situations.

2.5. Adoption of prompt decisions, development of recommendations and technical measures to eliminate the consequences of emergencies, as well as negative factors identified in the process of monitoring and quality control, as well as when deviating from design decisions.

2.6. Creation and updating of the information database based on the results of various types of monitoring and the consideration of these data in the subsequent design.

2.7. Execution of experimental research works.

3. Information support of construction.

As an example, let us consider the results of the tunneling of the Kozhukhovskaya Line (KZHL) tunnels of the Moscow Metro at a depth of about 3 m below the existing tunnels of the Tagansko-Krasnopresnenskaya Line (TKL) (Figure 2). The tunneling was carried out by the TBM "Herrenknecht" at a depth of about 14 – 19 m from the surface, with the installation of a high-precision prefabricated reinforced concrete waterproof lining with a diameter of 6 m.

The construction took place in the South-Eastern Administrative District of Moscow, the Zhulebino district in the territory between the Moscow Ring Road, Lermontov Avenue and the railway of the Kazan direction. Analysis of materials according to the historical use of the territory showed that in June 2013, approximately 500 m from the intersection site in question, during the construction of the intertunnel fence, the soil was taken to the bottomhole, which led to the deformation of the lining rings in an area about 160 m long. During restoration of lining and ground cementation, the processes of mechanical suffosion at the base of the left main line tunnel began to develop over a length of about 30 m, which led to deformation of the lining and the formation of "keys" up to 178 mm high. It was decided to excavate the pit, within which the deformed rings of the lining were dismantled and replaced with cast-iron

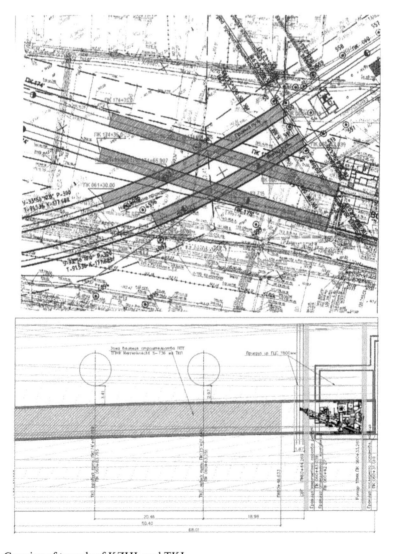

Figure 2.   Crossing of tunnels of KZHL and TKL.

tubing. In some parts of the preserved precast reinforced concrete lining, its reinforcement with fiber-reinforced shotcrete on the grid was performed.

Geomorphologically, the construction site is located within the third terrace above the floodplain of Moscow river which represents accumulative-erosive plain, composed of alluvial-fluvioglacial deposits. The natural terrain is technologically altered and planned by the existing buildings.

In the geological structure of the section up to a depth of 73.0 m water-saturated sandy and clayey Quaternary deposits are taken up, underlain by Upper and Middle Jurassic clays and limestones.

The passage of the KZHL tunnels was carried out under the existing tunnels of the TKL constructed in 2013. The lining is made of high-precision concrete blocks with a diameter of 5.9 m. Inspection of the technical condition of the tunnels revealed the following defects and damages in the structure of the lining and track concrete:

• areas of wetting in the joints of the rings and blocks, as well as in the junction between the lining and track concrete;
• lateral (in relation to the tunnel axis) cracks in track concrete with an opening width of up to 2 mm, with water filtration along individual cracks.

Geophysical survey of the state of contact "lining – ground" did not reveal zones of weakened contact of tunnels.

Estimation of the influence of construction on existing tunnels was carried out by the finite element method using the software complex Z_Soil 13.10 (Figure 3).

The structure of the prefabricated lining was modeled using area elements of the shell type with the specification of the actual geometric and physico-mechanical characteristics. The joints between the rings were modeled using non-linear joints according to the Janssen technique.

In Figure 4 and in Table 1 the calculated values of the additional displacements of the TKL tunnels after the completion of the construction are given.

Based on the results of calculation of internal forces in the lining of operating tunnels, the safety factors in strength were determined and are given in Table 2.

As can be seen from Tables 1 and 2 the tunneling of the KZHL tunnels at a depth of about 3 m under the operating tunnels of the TKL hardly affects the bearing capacity of the tunnel lining, but leads to deformation of tunnels that exceed the allowable values for operating conditions and can affect the safety of train traffic.

To minimize possible emergencies during the tunneling, a risk register was compiled using the methodology (Merkin V.E., Zertsalov M.G. & Konyukhov D.S. 2013).

As a result of the analysis, the following risks were classified as "high":

• Detection of plugged engineering-geological boreholes
• Detection of boulders and/or unaccounted communications/foundation elements that were not identified during survey.

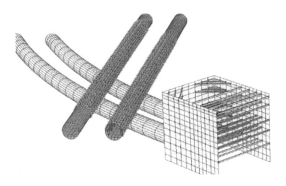

Figure 3.   Fragment of the design model at the time of completion of construction.

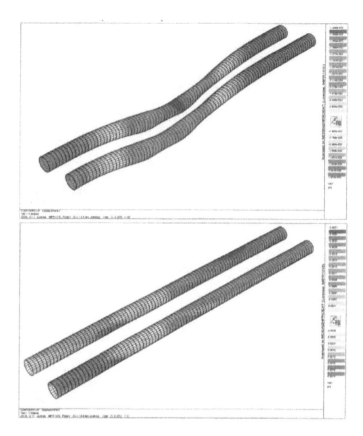

Figure 4. Vertical and horizontal displacements of tunnels of TKL after the completion of the construction of tunnels of KZHL.

Table 1. Maximum calculated values of additional displacements of the TKL tunnels at the time of completion of the tunneling of KZHL.

	Vertical, mm	Horizontal, mm
Left tunnel TKL	15,8	2,7
Right tunnel TKL	16,4	2,2

Table 2. Coefficient of safety for strength of lining of TKL tunnels.

	Before construction	After construction
Left tunnel TKL	1,7	1,68
Right tunnel TKL	1,88	1,85

- Detection of a lens of water-saturated sand, which was not detected during survey.
- Failure of the main elements of the TBM during tunneling under the TKL tunnels – engine, hydraulic system, wear of the incisors, etc. because of untimely performance of routine maintenance.

- Untimely/inadequate filling of the construction gap between the TBM and the ground with a grouting mortar

The following risks were classified as "average":

- Watering of an array in the breakdown of water-bearing communications
- Suffosion of the foundation of the TKL tunnels with the formation of voids under the TKL tunnels
- Poor ground conditioning
- The presence of an extended zone of disturbed soils in the bottomhole and the crown in front of the TBM formed after the tunneling of the TKL tunnels, as well as due to vibration effects, including TKL trains
- Infringement of the diagram of surface pressure, balance of extraction of soil and advancing of TBM

To minimize geotechnical risks, the following measures were proposed and implemented:

- railway traffic hault of the subway section of TKL from the station "Vykhino" to the station "Kotelniki" for the period of tunneling;
- additional geophysical studies of the soil array;
- strengthening of existing tunnels of TKL with metal frames;
- the development of a special technological regulation for tunneling under the tunnels of the subway;
- geotechnical monitoring during tunneling;
- monitoring of the observance of the requirements of the technological regulations.

Before the beginning of the tunneling unloading frames in the TKL tunnels were mounted (Figure 5) which allowed to reduce the vertical displacements of the left tunnel of the TKL by 6 times, the right one – 7 times (see Table 3), horizontal displacements were reduced to zero; frames also ensured safety and the stability of geometrical shape of the structure of operating tunnels.

Prior to the beginning of the tunneling, geophysical surveys were also carried out using the method of electromagnetic pulsed ultrawideband (EMP UWB) sounding. The EMP UWB sounding method combines modern achievements in the field of generation of nanosecond high voltage pulses, qualitative emission of an electromagnetic wave into subsurface structures and reception of broadband signals. The tasks of geophysical studies were to clarify the engineering and geological structure of the intersection part and to identify disturbed and watered geological formations composing the soil massif. The points of placement of the antennas of the measuring complex were located on the walls, track concrete, the upper and lower crowns of the existing tunnels. At each section, from 5 to 7 EMP UWB measurements were performed. In the sections of the EMP UWB sounding points the intervals of disturbed, watered and water saturated soils were identified, which were later displayed on cross sections (Figure 6).

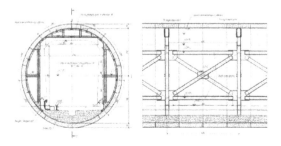

Figure 5.   Fragment of the design solution for the installation of unloading frames.

Right-track Left-track

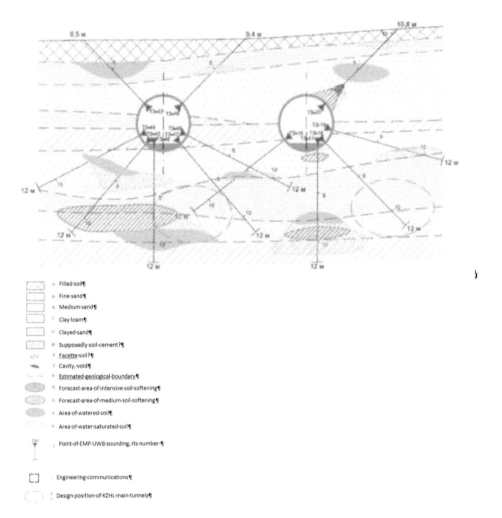

Figure 6. An example of engineering-geological section according to the EMP UWB sounding data before tunneling.

During the tunneling process, the state of the space behind lining was controlled by the method of seismic-acoustic sounding with the help of shock excitation of the lining. Observations were carried out daily during the tunneling process and then a control cycle of measurements was performed after the conditioned stabilization of the deformations.

The results of seismic-acoustic sounding in accordance to observation cycles are shown in Figures 7, 8. From the resulted graphic materials, a change in the state of the space behind lining in the process of tunneling is evidently visible.

Taking into account that the operating tunnels and one under construction are located mainly in water-saturated sandy soils, the weakened contact "lining-ground" can be interpreted as a combination of air bubbles exiting from the bottomhole space of the TBM and their ascent to the surface under vibration from the operation of the TBM mechanisms, with centripetal displacement of water-saturated ground mass during the operation of the TBM rotor. After completion of the tunnelng and stabilization of deformations, the state of the bottomhole space was practically restored.

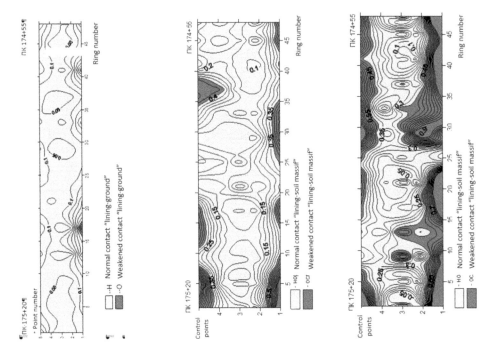

Figure 7. Results of the seismic-acoustic survey of track 1, cycle 0 (before tunneling), cycle 1, cycle 2.

Figure 8. Results of the seismic-acoustics survey of track 1, control cycle 3.

After the completion of the tunneling and the stabilization of the deformations, the soil massifs were again re-examined by the EMP UWB sounding method. As a result of the studies it was established (Figure 9) that:

- the intervals of disturbed, watered and water saturated soils revealed before the construction began remained largely unchanged;
- near the completed left main tunnel of KZHL, sections of unconsolidated and water saturated soils were formed, which is confirmed by seismic-acoustic sounding data;
- the number of sections of decompressed soils increased after tunneling;
- the moisture content of soils has decreased.
- To ensure the safety of construction, a technological regulation was developed, providing instructions, requirements and recommendations for shield tunneling, concerning:
- modes of tunneling, fulfillment of the basic operations of the technological cycle;
- diagrams of the pressure of the ground balance weight of the bottomhole on the tunnel route in the crown top of the tunnel and at the bottom the tunnel, as well as the limit values of the balance weight, ensuring the safety of work and the preservation of buildings

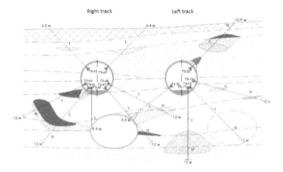

Figure 9. An example of an engineering-geological section according to the EMP UWB sounding data after tunneling.

and structures in the zone of influence of tunneling works. The calculation was carried out according to the technique of (STO NOSTROY 2.27.19-2011);
- composition of solutions for the balance weight of the buttomhole;
- composition of grouting mortars, their preparation and injection;
- tolerances for the conduct of the shield along the tunnel axis;
- measures to prevent an emergency situation during tunneling, including:
  - control of space behind lining with a grouting mortar with the use of a special device in the form of a "comb", which makes it possible to simultaneously inject through at least four holes in the blocks;
  - requirements for injecting bentonite based solutions through openings in the shell between the front and middle shields;
  - Compensation of soil overdig due to the gap between the rotor and the TBM body by injecting a bentonite solution into the space behind shield of the head part of the TBM through 4 injection ports;
  - Requirements for measures for the technological stopping of the TBM.

During the construction process, the technological parameters of the TBM's operation were constantly monitored, including compliance with the excavation cyclogram and the bottom-hole balance weight pressure diagram. Figure 10 shows a comparison of the calculated and actual bottomhole weight pressure from which it can be seen that when tunneling under the right TKL tunnel (track 1), the actual pressure of the balance weight was 0.1 to 0.2 bar less than calculated, and when tunneling under the left tunnel (track 2), the actual pressure was 0.2 bar higher than the calculated one, which, apparently, explains the difference in the settlement values of the operating tunnels. At the same time, judging by the balance weight diagram and the nature of the vertical displacements, when approaching the left tunnel, the pressure of the balance weight increased by 0.2 bar and the operating tunnel was lifted by 1.5–2 mm, and after the lining coming off from the tail part of the TBM shell the settlement of the tunnel increased by approximately 3.5–4.00 mm, which eventually led to the settlement of the left tunnel after the release of the TBM from under it by 2.3 mm.

For the entire construction period, a geotechnical monitoring system was organized that included the following:

- visual and instrumental monitoring of the technical state of tunnel structures with the record of defects and the time history (cracks, water inflow, etc.) with a frequency of 1 time per day;
- automated geodetic monitoring – automated geodetic observations of the horizontal and vertical displacements of the tunnels.

The results of the geodetic monitoring data are given in Table 3.

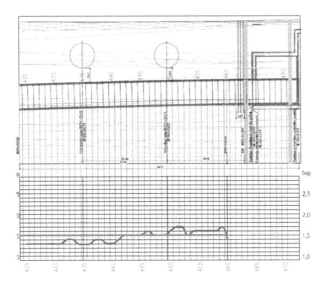

Figure 10.   Comparison of the calculated (red) and actual (blue) bottomhole balance weight pressure.

Table 3.   Additional displacements of TKL tunnels.

| | Vertical (settlement), mm | | |
| | Calculated | Measured | |
		At the moment of the tail part of TBM coming from under the existing tunnel	After stabilization of deformations
Left tunnel	2,6	2,3	2,9
Right tunnel	2,3	6,5	4,6

In addition to controlling the deformations of operating tunnels, observations of the settlement of the earth's surface were made during the tunneling. The forecast of the surface settlement is made according to the empirical formula:

$$s = 2,2643(L/h)^{-0.651} \qquad (1)$$

where s is the surface settlement, mm; L is the distance in the horizontal plane from the tunnel axis to the point on the surface along the normal to the tunnel axis, m; h is the depth of the tunnel axis, m.

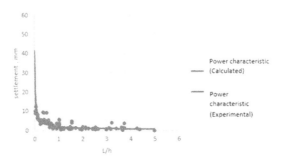

Figure 11.   Comparison of the forecast (according to (1)) and experimental data.

Figure 11 shows a comparison of the calculated and experimental curves $s = f(L/h)$ from which almost complete convergence of the predicted and experimental data is visible.

3 CONCLUSIONS

As a result of the implementation of measures for scientific and technical support of construction:

- the maximum stabilized settlement of the TKL tunnels were: 4.6 mm for the right tunnel, 2.9 mm for the left tunnel;
- the state of the soil massif and underground waters virtually did not change and stabilized after the completion of the tunneling.
- safe and having almost no settlement tunneling of the Kozhukhovskaya Line at a depth of about 3 m under the tunnels of the Tagansko-Krasnopresnenskaya line of the Moscow Metro.

REFERENCES

Konyukhov D.S. 2017. Scientific and technical support for the construction of metro facilities. The main types of work. *Metro and tunnels*, 3–4: 32–35.
Merkin V.E., Zertsalov M.G. & Konyukhov D.S. 2013. Management of geotechnical risks in underground construction. *Metro and tunnels*, 6: 36–39.
STO NOSTROY 2.27.19-2011. Development of underground space. Tunneling with tunnel boring machines (TBMs), using precast concrete segments (PC segments).

Tunnels and Underground Cities: Engineering and Innovation meet Archaeology,
Architecture and Art, Volume 11: Urban
Tunnels - Part 1 – Peila, Viggiani & Celestino (Eds)
© 2020 Taylor & Francis Group, London, ISBN 978-0-367-46899-6

Tunneling under existing Metro Line and Indian Railway Line for Mandi House – ITO Corridor of Phase-III, Delhi Metro Rail Corporation Limited

U. Kumar

Delhi Metro Rail Corporation Limited, New Delhi, India

ABSTRACT: Tunneling between Mandi House and ITO stations of DMRC involved tunneling underneath the existing Blue Line of Delhi Metro and the busiest corridor of the Indian Railway at Tilak Bridge. The absolute requirement for maintenance of operations, combined with the safe and economical construction of the tunnels, has mandated that the installation of tunnels be constantly evaluated. Two Mitsubishi Earth Pressure Balance Machines were chosen for this tunneling work and the precast lining of grade M50 concrete was used. The Blue Line Metro was located at 120m east of launching shaft, 30° skew with the tunnel alignment and Indian railway line was located around 500m east of launching shaft, 60° skew with the alignment. Trains run frequently on these lines, hence no track closures and reduction in the train speed was allowed through the project site. As a result, surface settlement and deformations had to be limited to negligible amounts during construction and over the lifetime of the structure. The tunnel was primarily excavated in Sandy Silt to Silty Sand. Groundwater depth was 4.7 m. PLAXIS was used for prediction of settlement analysis caused by tunnel boring. Soil volume loss was restricted to 0.3% to restrict settlements within permissible limits. Extensive Instrumentation and Monitoring was carried out and Alert Values were fixed.

1 INTRODUCTION

1.1 *General:*

Despite a distance of only 700m as shown in Figure 1, the tunnel from Mandi House to ITO on the heritage line of Delhi Metro rail Corporation was an engineering feat. Below ground, the tunnels crossed not only an existing underground Metro track but also a railway track at Tilak Bridge. What made the construction of the tunnels a challenge was the "no margin for settlement" goal of Delhi Metro. The Mandi House to ITO Underground Corridor included design and construction of tunnels between the associated stations. Two bored tunnels of 5.7m finished internal diameter were constructed.

The installation of tunnels under the metro tracks and Railway Track was one of the most challenging aspects of the DMRC Phase-III Project. The absolute requirement for maintenance of metro operations and railroad operations above, combined with the safe and economical construction of the tunnels, mandated that the installation of tunnels be constantly evaluated.

1.2 *Proposed Tunnels:*

Two Mitsubishi ϕ6410mm Earth Pressure Balance Machines (EPBM) were chosen for this tunnelling works. The EPBMs measuring 6.41m in outer diameter and 9.25m in length had a cutter head with an opening ratio of 0.7. The cutter head was driven by ten electrical motors.

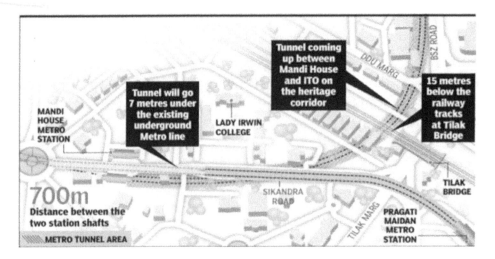

Figure 1. Pictorial Representation – Mandi House – ITO Tunnel Corridor.

Strategically placed pressure sensors in the pressure chamber gave indication of the earth pressure in the chamber. Sixteen shield jacks capable of exerting a combined thrust of 4800 tonne-force provided the machine thrust. The screw conveyors were able to handle up to 150mm sized boulders.

The lining consists of 6 reinforced precast segments: 3 ordinary segments, with two tapered and one tapered key segment. The segments were 1.5m in width and 280mm in thickness. The precast lining is of grade M50 concrete with epoxy coating on the extrados for the enhanced durability.

1.3 Existing Metro Line:

The existing metro line was located at about around 120m east of launching shaft, approximately 30° skew with the tunnel alignment a shown in the Figure 1. It was observed that trains run frequently on the metro line. Hence, no track closures and reduction in the train speed was allowed through the project site. Surface settlement and deformations must be limited to negligible amounts during construction and over the lifetime.

1.4 Existing Railway Line:

The existing railway line was located at 500m east of launching shaft, approximately 60° skew with the tunnel alignment. Tilak Bridge Railway Station is also located at project site. The two-lane railway is situated on an embankment as shown in the Figure 2, and adjacent to the dense residential area. It is observed that trains run frequently on the railway line. Hence, no track closures and reduction in the train speed was allowed through the project site. Surface settlement and deformations were limited to negligible amounts during construction and over the lifetime.

2 GEOLOGICAL CONDITION:

2.1 General Description:

According to the detailed soil investigations together with previous relevant geotechnical information, the tunnels were primarily to be excavated in SANDY SILT to SILTY SAND and Some traces of Clay.

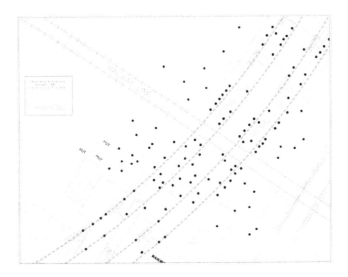

Figure 2. TBM Tunnel Crossing Metro Blue Line underground.

2.2 *Groundwater:*

Groundwater levels in the open boreholes were observed during the exploratory drilling and at the completion of borehole. The results indicated that at the time of investigation, Groundwater was at depth 3.5 to 4.7 m. It should also be pointed out that the groundwater is subject to seasonal fluctuations and fluctuations in response to major weather events.

2.3 *Borehole:*

According to detail geology investigation report, the nearest bore hole to the Metro ramp which was located approximately 30m from the actual work site, was selected in the analysis work for Blue Line Metro Crossing.

2.4 *Geotechnical Parameters:*

Metro Line - Soil parameters have been interpreted based on the results of laboratory tests. The available soil parameters given in Preliminary Geotechnical Interpretation Record (GIR) are summarized in Table 1. These were used to carry out the settlement analysis for evaluating impact of TBM for Metro Line Crossing.

Table 1. Soil parameters – Metro Line Location.

Soil Type	Consolidated undrained strength		Drained direct shear		E(MPa)
	C	φ	C '	φ '	
Filled up materials	2.2*	15*	0*	31*	10*
Clayey silt	0.35	30	0*	32*	10*
Silty sand	0*	33*	0	34	15*
Sandy silt	0.15	34	0	31	20*/15*
Boulder					20*

* Estimated Data for Metro Line Crossing -Modulus is estimated according to N value of SPT

Railway Line - Soil parameters have been interpreted based on the results of laboratory tests. The available soil parameters given in Preliminary GIR are summarized in Table 2. These were used to carry out the settlement analysis for evaluating impact of TBM for Railway Line Crossing.

Table 2. Soil parameters – Railway Line Location.

Soil Type	Consolidated undrained strength		Drained direct shear		E(MPa)
	C	φ	C '	φ '	
Filled up materials	2.2*	15*	0*	31*	10*
Clayey silt	0.35	30	0*	32*	10*
Silty sand	0*	33*	0	34	15*
Sandy silt	0.15	34	0	31	20*/15*
Boulder					20*

3   SETTLEMENT ANALYSIS FOR METRO LINE CROSSING:

3.1   Software:

PLAXIS was used for analysis of the settlement caused by tunnel boring.

3.2   Input data:

Metro Line Crossing:-
   When analyzing the building settlement caused by tunnel boring, the ramp structure can be simplified U type.

Table 3.   Input Data for Metro Line Crossing Prediction.

Structure height:	3m
Strength and modulus:	2900Mpa
Basic data of tunnel line:	
Modulus:	30000Mpa
Poisson's ratio:	0.15
Soil parameters -	Refer Table 1

3.3   Output data -

Metro settlement analysis: The results obtained from running the analysis on PLAXIS for the Metro ramp structure is depicted in the figure 3 and figure 4. The maximum displacement vector and Maximum Base Slab settlement limits were obtained after restricting the Volume Loss limits.
   Volume loss rate - 0.3%
   Tunnel depth at metro crossing position - 9.5m
   Deformation control standard:

Settlement(mm)	slope	Soil volume loss ration
5	0.3‰	0.3%

From the analysis results, we can know that if settlements can be in allowable deformation, TBM running can have no effect for Metro Line when TBM go through it.

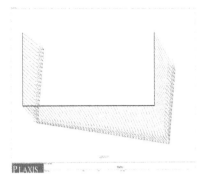

Figure 3. Max displacement vector - 5.5mm.

Figure 4. Base slab maximum settlement - 4.4mm, Tilt 0.3‰.

Accordingly Front face Pressure was pre calculated and communicated to operator for safe drive beneath the existing metro structure. The front face earth pressure was strictly adhered to so that the pressure in working chamber is maintained at the same level. The working hours were 24*7 whilst TBM was within the influence zone.

4   SETTLEMENT ANALYSIS FOR RAILWAY LINE CROSSING:

4.1   *Basic data of tunnel line:*

Refer Table 3.

4.2   *Soil parameters:*

Soil parameters listed in Table 2 was used.

4.3 *Output data:*

The results obtained from running the analysis on PLAXIS for the Railway Line Settlement Analysis is depicted in the figure 5. The structure has been considered linear and the maximum displacement vector and Maximum Base Slab settlement limits were obtained after restricting the Volume Loss limits.

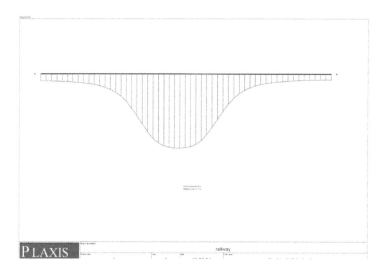

Figure 5. Maximum Settlement 6mm.

Settlement(mm)	Soil volume loss ration
6	0.4%

From the analysis results, we can know that if settlements can be in allowable deformation, TBM running can have no effect for Railway Line when TBM go through it.

5 INSTRUMENTATION AND MONITORING

5.1 *Metro Line Crossing:*

The Ramp Structure of existing Metro Line falls between Mandi House Station and ITO Station as shown in Figure 1. According to general arrangement of instrumentation and settlement monitoring plans as shown in the Figure 6, locations are predefined as per design requirements. The For Ramp Structure, series of 3D monitoring targets were fixed at construction joints of the ramp outside wall of metro line over the track alignment center of up line and down line up to required stretch width before tunneling below the structure. Additional to that ground settlement array were fixed as per approved Instrumentation and Monitoring Plan. All observations were downloaded and graphical presentations were made. Markings were done on the tracks inside the ramp. Crack meters were installed on the joints and other minor cracks to measure any variation. Measurements for the points installed inside the ramp and on the tracks were done in the night shift during the non-revenue hour of train operations.

The frequency of monitoring was fixed as shown in Table 4 however. In addition, following guidelines were used when deciding on the frequency of monitoring:

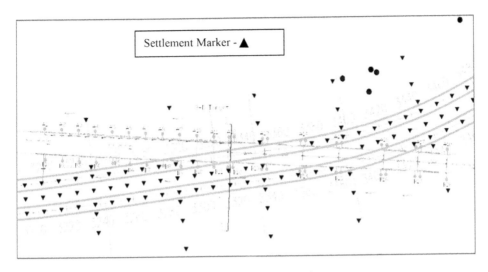

Figure 6.   – Instrument & Monitoring Plan for Metro Line Crossing.

Table 4.   Frequency of monitoring.

	Distance from Face	Frequency
Settlement Monitoring Points	-30 m to 0 m	4 Times Daily
	0 m to +30 m	4 Times Daily
	+30 m to +60 m	Once Daily
	> +60 m	Thrice Weekly

- Monitoring shall be repeated if readings were considered inconsistent.
- If reading shows show significant change, the frequency of monitoring shall be increased.
- If there is an undesirable occurrence such as excessive ground loss for bored tunnel or base heave or base below out is detected, the frequency of monitoring shall be increased.
- In no case shall the monitoring be stopped until the stretch width of the structure have been completed and further until 3 consecutive reading show no change

Alert Values:
The Monitoring Values for the drive were set based upon the results derived from settlement prediction using PLAXIS. The standard values set for Alert Values were as follows:

– Maximum settlement is ≤5mm
– Trigger value: 70% of maximum value i.e. 3.5mm
– Alarm value: 85% of maximum value I.e. 4.25mm

5.2   *Railway Line Crossing:-*

Railway crossing (Tilak Bridge) falls between Mandi House Station and ITO Station as shown in Figure 1. According to general arrangement of instrumentation and settlement monitoring plans, locations were predefined as per design requirements. For railway lines crossing monitoring, array of settlement markers series were fixed as per plan Figure 7. All observations were presented graphically. The additional guidelines for railway crossing were similar to as mentioned under Metro Line crossing.

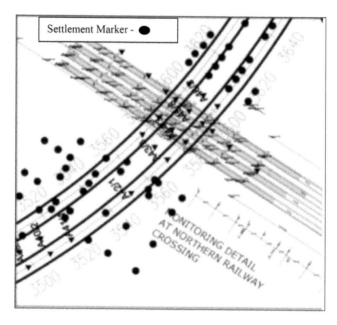

Figure 7. – Instrument & Monitoring Plan for Railway Line Crossing.

Table 5. Frequency of monitoring.

	Distance from Face	Frequency
Settlement Monitoring Points	-30 m to 0 m	4 Times Daily
	0 m to +30 m	4 Times Daily
	+30 m to +60 m	Once Daily
	> +60 m	Thrice Weekly

Alert Values:
The Monitoring Values for the drive were set based upon the results derived from settlement prediction using PLAXIS. The standard values set for Alert Values were as follows:

– Maximum settlement is ≤6mm
– Trigger value: 70% of maximum value i.e. ~4 mm
– Alarm value: 85% of maximum value I.e. ~5 mm

6 PRELIMINARY SETTLEMENT PREDICTION AT RAILWAY CROSSING & MITIGATION

Excavation of tunnels produces movements of the ground above and nearby. The movements are usually downwards (settlements) with a horizontal component of displacement towards the excavation. These movements have potential effects on the railway tracks above or near to tunnels only where the ground movements at one part of the building are different from those at another part. The potential impacts covered by this study are those due to settlement. One of the other impacts is that of heave occurring at the ground surface ahead and over the tunnel face. Due to granular nature of soils and generally low water table, the secondary potential settlement effect is considered as negligible.

6.1 Settlement/Ground Movement:

The empirical method proposed by O'Really and New (1982) was used to evaluate the predicted displacements. This approach fits a Gaussian (normal) probability density function to observed settlements, which is given by:

$$S_{\max} = \frac{V_s}{i\sqrt{2\pi}}$$

Surface settlement profile across tunnel alignment is given by:

$$S(x) = S_{\max} \exp\left(-\frac{x^2}{2i^2}\right)$$

Where V_s is the volume of the settlement trough, "i" the trough width parameter, is the horizontal distance to the point of inflexion of the settlement curve from the tunnel centerline, and x is the horizontal distance from the tunnel centerline. The maximum value of volume loss is taken as 0.3% and the ground movement due to tunneling has been calculated. This has been used to make the initial assessment of risk of damage.

6.2 Railway Protection Scheme:

Prior to the TBM passage, coordination between tunneling and railroad's schedules were done and limits were worked out defined by railway authorities. During the TBM passage, active work with the railway authorities was done on records for transportation capacity, operation of communication and coordination, to ensure the safety of railway and its smooth operation. In addition, extensive monitoring of settlement points on and adjacent to the tracks was done on a daily basis. The tunneling parameters corresponding to thrust, face pressure, grouting pressure and foam injection were optimized based on prior experience on drive. The instrumentation design included monitoring of movements for vertical change in diameter of tunnel, vertical settlement of tunnel lining impinging upon railway's structural gauge and twist of track. The TBM's worked 24*7 whilst within the influence zone of the Railway Line.

7 ACTION PLAN

For any event of calamity, the Action plan for Metro Line Crossing and Railway Line Crossing was prepared and kept ready. This briefly included the following measures:

- Analyze the cause of the alarm and the data's authenticity.
- Increase settlement points and frequency of monitoring for the risk location.
- Necessary Grouting – Primary, Secondary or from the ground level.
- Face pressure balance – Optimize chamber Pressure.
- Re-draw the drive scheme

8 RESULTS:

The tunneling works were completed successfully under the DMRC existing Blue Line metro and Indian Railway line at Tilak Bridge. The table showing maximum settlement against predicted settlement is as below:

Table 6. Settlement Record.

Location	Predicted maximum Settlement	Actual Settlement
Metro Railway Ramp	5.5mm	3mm
Indian Railway Line	6mm	6mm

9 CONCLUSIONS

The tunneling works under the DMRC existing Blue Line metro and Indian Railway line was a very challenging project and it gave several hands on experience with complex scenarios. The conclusion can be summed up as follows:

- Full scale instrumented test sections can be very beneficial to underground excavation projects by confirming economical and safe construction procedures, and potentially large cost saving can be realized.
- Verification of design assumptions and management of the construction in a safe and controlled manner help to safeguard existing adjacent and other facilities.
- Adjusting the TBM parameter of Face pressure, Thrust and Torque help in regulating the Mining speed and help in minimizing the ground deflection and settlement.
- Action Plan for any unforeseen situation shall be kept ready in an event of TBM mining under existing Services.
- A detailed instrumentation and monitoring scheme shall be finalized with appropriate equipment's to ensure working team is provided with correct and reliable data during execution. Also, instrument monitoring is a rigorous process which was religiously followed for identifying any movement of ground structure.
- For the purpose of monitoring the tunneling under the Metro Line and Railway track DMRC had defined a zone of influence. Method Statement with settlement analysis and mitigation plans was in place which ensured steady tunneling advance rate with safety in urban conditions.
- The exhaustive instrumentation monitoring as explained above and by taking protection measures, the tunnels were successfully constructed below the operational Blue Line Metro and Indian Railway Line. No damage has been observed on the structures coming in the zone of influence of tunnels and no change has been observed in the track level of existing metro and railway line.

Author Index